Game Theory and Machine Learning for Cyber Security

Game Theory and Machine Learning for Cyber Security

Edited by

Charles A. Kamhoua
Christopher D. Kiekintveld
Fei Fang
Quanyan Zhu

Library of Congress Cataloging-in-Publication Data applied for:
ISBN: 9781119723929

Cover Design: Wiley
Cover Image: © Dimitri Otis/Getty Images

Set in 9.5/12.5pt STIXTwoText by Straive, Chennai, India

10 9 8 7 6 5 4 3 2 1

Contents

Editor Biographies

Charles A. Kamhoua is a Senior Electronics Engineer at the Network Security Branch of the U.S. Army Research Laboratory (ARL) in Adelphi, MD, where he is responsible for conducting and directing basic research in the area of game theory applied to cyber security. Prior to joining the Army Research Laboratory, he was a researcher at the U.S. Air Force Research Laboratory (AFRL), Rome, New York, for 6 years and an educator in different academic institutions for more than 10 years. He has held visiting research positions at the University of Oxford and Harvard University. He has co-authored more than 200 peer-reviewed journal and conference papers that include 5 best paper awards. He is a co-inventor of three patents and six patent applications. He has been at the forefront of several new technologies, co-editing four books at Wiley-IEEE Press entitled *Game Theory and Machine Learning for Cyber Security, Modeling and Design of Secure Internet of Things, Blockchain for Distributed System Security, and Assured Cloud Computing.* He has presented over 70 invited keynote and distinguished speeches and has co-organized over 10 conferences and workshops. He has mentored more than 65 young scholars, including students, postdocs, and Summer Faculty Fellow. He has been recognized for his scholarship and leadership with numerous prestigious awards, including the 2020 Sigma Xi Young Investigator Award for outstanding leadership and contribution to game theory applied to cyber security, the 2019 US Army Civilian Service Commendation Medal, the 2019 Federal 100-FCW annual awards for individuals that have had an exceptional impact on federal IT, the 2019 IEEE ComSoc Technical Committee on Big Data (TCBD) Best Journal Paper Award, the 2018 Fulbright Senior Specialist Fellowship, two ARL Achievement Award (2019 and 2018), the 2017 AFRL Information Directorate Basic Research Award "For Outstanding Achievements in Basic Research," the 2017 Fred I. Diamond Award for the best paper published at AFRL's Information Directorate, 40 Air Force Notable Achievement Awards, the 2016 FIU Charles E. Perry Young Alumni Visionary Award, the 2015 Black Engineer of the Year Award (BEYA), the 2015 NSBE Golden Torch Award – Pioneer of the Year, and selection to the 2015 Heidelberg Laureate Forum, to name a few. He has been congratulated by the White House, the US Congress and the

Pentagon for those achievements. He received a BS in electronics from the University of Douala (ENSET), Cameroon, in 1999, an MS in Telecommunication and Networking from Florida International University (FIU) in 2008, and a PhD in Electrical Engineering from FIU in 2011. He is currently an advisor for the National Research Council postdoc program, a member of the FIU alumni association and Sigma Xi, a senior member of ACM and IEEE.

Christopher D. Kiekintveld is an Associate Professor of Computer Science at the University of Texas at El Paso, where he also serves as the Director of Graduate Programs for Computer Science. He received his PhD in Computer Science and Engineering from the University of Michigan in 2008. He also holds MS and BSE degrees from the same institution. Before joining UTEP in 2010, he was a postdoctoral research fellow for two years at the University of Southern California. His primary research interests are in artificial intelligence, especially in the areas of multi-agent systems, computational decision making and game theory, reasoning under uncertainty, and multi-agent learning. He also works on applications of artificial intelligence methods to security (including both physical and cyber), resource allocation, trading agents, smart grids, and other areas with the potential to benefit society. He currently directs the Intelligent Agents and Strategic Reasoning Laboratory (IASRL) at UTEP, and has mentored more than 30 graduate and undergraduate students. He has co-authored more than 80 articles in peer-reviewed conferences and journals (e.g. AAMAS, IJCAI, AAAI, JAIR, JAAMAS, ECRA), with an i10-index of more than 60. He has given many invited talks and keynote presentations at conferences, academic institutions, and research laboratories around the world, and has helped to organize numerous workshops and conferences including AAAI, AAMAS, and GameSec. He has also had a lead role in developing several deployed applications of game theory for security, including systems in use by the Federal Air Marshals Service and Transportation Security Administration. He has received recognition including multiple best paper awards, the UTEP College of Engineering Dean's Award for Research, the David Rist Prize from the Military Operations Research Society, and an NSF CAREER award.

Fei Fang is an Assistant Professor at the Institute for Software Research in the School of Computer Science at Carnegie Mellon University. Before joining CMU, she was a Postdoctoral Fellow at the Center for Research on Computation and Society (CRCS) at Harvard University. She received her PhD from the Department of Computer Science at the University of Southern California in June 2016.

Her research lies in the field of artificial intelligence and multi-agent systems, focusing on integrating machine learning with game theory. Her work has been motivated by and applied to security, sustainability, and mobility domains, contributing to the theme of AI for Social Good. Her work has won the Distinguished Paper at the 27th International Joint Conference on Artificial Intelligence and the 23rd European Conference on Artificial Intelligence (IJCAI-ECAI'18), Innovative Application Award at Innovative Applications of Artificial Intelligence (IAAI'16), the Outstanding Paper Award in Computational Sustainability Track at the International Joint Conferences on Artificial Intelligence (IJCAI'15). She was invited to give an IJCAI-19 Early Career Spotlight talk. Her dissertation is selected as the runner-up for IFAAMAS-16 Victor Lesser Distinguished Dissertation Award, and is selected to be the winner of the William F. Ballhaus, Jr. Prize for Excellence in Graduate Engineering Research as well as the Best Dissertation Award in Computer Science at the University of Southern California. Her work has been deployed by the US Coast Guard for protecting the Staten Island Ferry in New York City since April 2013. Her work has led to the deployment of PAWS (Protection Assistant for Wildlife Security) in multiple conservation areas around the world, which provides predictive and prescriptive analysis for anti-poaching effort.

Quanyan Zhu received BEng in Honors Electrical Engineering from McGill University in 2006, MA Sc from the University of Toronto in 2008, and PhD from the University of Illinois at Urbana-Champaign (UIUC) in 2013. After stints at Princeton University, he is currently an associate professor at the Department of Electrical and Computer Engineering, New York University (NYU). He is an affiliated faculty member of the Center for Urban Science and Progress (CUSP) at NYU. He is a recipient of many awards, including NSF CAREER Award, NYU Goddard Junior Faculty Fellowship, NSERC Postdoctoral Fellowship (PDF), NSERC Canada Graduate Scholarship (CGS), and Mavis Future Faculty Fellowships. He spearheaded and chaired INFOCOM Workshop on Communications and Control on Smart Energy Systems (CCSES), Midwest Workshop on Control and Game Theory (WCGT), and ICRA workshop on Security and Privacy of Robotics. His current research interests include game theory, machine learning, cyber deception, network optimization and control, smart cities, Internet of Things, and cyber-physical systems. He has served as the general chair or the TPC chair of the 7th and the 11th Conference on Decision and Game Theory for Security (GameSec) in 2016 and 2020, the 9th International Conference on NETwork Games, COntrol and OPtimisation (NETGCOOP) in 2018, the 5th International Conference on Artificial Intelligence and Security (ICAIS 2019) in 2019, and 2020 IEEE Workshop on Information Forensics and Security (WIFS). He has also spearheaded in 2020 the IEEE Control System Society (CSS) Technical Committee on Security, Privacy, and Resilience. He is a co-author of two recent books published by Springer: *Cyber-Security in Critical Infrastructures: A Game-Theoretic Approach* (with S. Rass, S. Schauer, and S. König) and *A Game- and Decision-Theoretic Approach to Resilient Interdependent Network Analysis and Design* (with J. Chen).

Contributors

Palvi Aggarwal
Department of Social and Decision Sciences
Carnegie Mellon University
Pittsburgh
PA
USA

Muhammad Aneeq uz Zaman
Coordinated Science Laboratory
University of Illinois at Urbana-Champaign
Urbana
IL
USA

Ahmed H. Anwar
Network Security Branch
Combat Capabilities Development Command
US Army Research Laboratory
Adelphi
MD
USA

Nurpeiis Baimukan
New York University Abu Dhabi
Abu Dhabi
United Arab Emirates

Tiffany Bao
School of Computing, Informatics, and
Decision Systems Engineering
Arizona State University
Tempe
AZ
USA

Anjon Basak
Network Security Branch
Combat Capabilities Development Command
US Army Research Laboratory
Adelphi
MD
USA

Tamer Başar
Coordinated Science Laboratory
University of Illinois at Urbana-Champaign
Urbana
IL
USA

Branislav Bošanský
Department of Computer Science
Czech Technical University in Prague
Prague
Czechia

Ing-Ray Chen
Department of Computer Science
Virginia Tech
Falls Church
VA
USA

Jin-Hee Cho
Department of Computer Science
Virginia Tech
Falls Church
VA
USA

Keywhan Chung
Coordinated Science Laboratory
University of Illinois at Urbana-Champaign
Urbana
IL
USA

George Cybenko
Thayer School of Engineering
Dartmouth College
Hanover
NH
USA

György Dán
Division of Network and Systems Engineering
KTH Royal Institute of Technology
Stockholm
Sweden

Gaurav Dixit
Department of Computer Science
Virginia Tech
Falls Church
VA
USA

Abdelrahman Eldosouky
Computer Science and Engineering
Department
University of Nevada, Reno
Reno
USA

Tugba Erpek
Intelligent Automation, Inc.
Rockville
MD
USA

and

Hume Center
Virginia Tech
Arlington
VA
USA

Fei Fang
School of Computer Science and Institute for
Software Research
Carnegie Mellon University
Pittsburgh
PA
USA

Jie Fu
Department of Electrical and Computer
Engineering
Robotics Engineering Program
Worcester Polytechnic Institute
Worcester
MA
USA

Cleotilde Gonzalez
Department of Social and Decision Sciences
Carnegie Mellon University
Pittsburgh
PA
USA

Marcus Gutierrez
Department of Computer Science
The University of Texas at El Paso
El Paso
TX
USA

Roger Hallman
Thayer School of Engineering
Dartmouth College
Hanover
NH
USA

and

Naval Information Warfare Center Pacific
San Diego
CA
USA

João P. Hespanha
Center for Control Dynamical-Systems, and
Computation
University of California
Santa Barbara
CA
USA

Yunhan Huang
Department of Electrical and Computer
Engineering
New York University
Brooklyn
New York
USA

Ravishankar K. Iyer
Coordinated Science Laboratory
University of Illinois at Urbana-Champaign
Urbana
IL
USA

Brian Jalaian
US Army Futures Command
US Army Research Laboratory
Adelphi
MD
USA

Susmit Jha
Computer Science Laboratory
SRI International
Menlo Park
CA
USA

Kun Jin
Department of Electrical Engineering and
Computer Science
University of Michigan
Ann Arbor
MI
USA

Zbigniew T. Kalbarczyk
Coordinated Science Laboratory
University of Illinois at Urbana-Champaign
Urbana
IL
USA

Charles A. Kamhoua
Network Security Branch
Combat Capabilities Development Command
US Army Research Laboratory
Adelphi
MD
USA

Murat Kantarcioglu
Computer Science Department
University of Texas at Dallas
Richardson
TX
USA

Thenkurussi Kesavadas
Healthcare Engineering Systems Center
University of Illinois at Urbana-Champaign
Urbana
IL
USA

Christopher D. Kiekintveld
Department of Computer Science
The University of Texas at El Paso
El Paso
TX
USA

Sandra König
Center for Digital Safety & Security
Austrian Institute of Technology
Vienna
Austria

Xenofon Koutsoukos
Electrical Engineering and Computer Science
Vanderbilt University
Nashville
TN
USA

Abhishek N. Kulkarni
Robotics Engineering Program
Worcester Polytechnic Institute
Worcester
MA
USA

Nandi Leslie
Network Security Branch
Combat Capabilities Development Command
US Army Research Laboratory
Adelphi
MD
USA

Yi Li
Electrical Engineering and Computer Science
Vanderbilt University
Nashville
TN
USA

Xiao Li
Healthcare Engineering Systems Center
University of Illinois at Urbana-Champaign
Urbana
IL
USA

Mingyan Liu
Department of Electrical Engineering and
Computer Science
University of Michigan
Ann Arbor
MI
USA

Shutian Liu
Department of Electrical and Computer
Engineering
NYU Tandon School of Engineering
New York University
Brooklyn
New York
USA

Patrick McDaniel
Pennsylvania State University
Computer Science and Engineering
Department
University Park
Pennsylvania
USA

Radha Poovendran
Department of Electrical and Computer
Engineering
University of Washington
Seattle
WA
USA

Stefan Rass
Institute for Artificial Intelligence and
Cybersecurity
Universitaet Klagenfurt
Klagenfurt
Austria

Anirban Roy
Computer Science Laboratory
SRI International
Menlo Park
CA
USA

Yalin E. Sagduyu
Intelligent Automation, Inc.
Rockville
MD
USA

Dinuka Sahabandu
Department of Electrical and Computer
Engineering
University of Washington
Seattle
WA
USA

Henrik Sandberg
Division of Decision and Control Systems
KTH Royal Institute of Technology
Stockholm
Sweden

Armin Sarabi
Department of Electrical Engineering and
Computer Science
University of Michigan
Ann Arbor
MI
USA

Serkan Sarıtaş
Division of Network and Systems Engineering
KTH Royal Institute of Technology
Stockholm
Sweden

Muhammed O. Sayin
Laboratory for Information and Decision
Systems
Massachusetts Institute of Technology
Cambridge
MA
USA

Stefan Schauer
Center for Digital Safety & Security
Austrian Institute of Technology
Vienna
Austria

Shamik Sengupta
Computer Science and Engineering
Department
University of Nevada, Reno
Reno
USA

Ryan Sheatsley
Pennsylvania State University
Computer Science and Engineering
Department
University Park
Pennsylvania
USA

Ezzeldin Shereen
Division of Network and Systems Engineering
KTH Royal Institute of Technology
Stockholm
Sweden

Yi Shi
Intelligent Automation, Inc.
Rockville
MD
USA

and

ECE Department
Virginia Tech
Blacksburg
VA
USA

Yan Shoshitaishvili
School of Computing, Informatics, and
Decision Systems Engineering
Arizona State University
Tempe
AZ
USA

Peicheng Tang
Department of Electrical and Computer
Engineering
Rose-Hulman Institute of Technology
Terra Haute
IN
USA

Sridhar Venkatesan
Perspecta Labs Inc.
Basking Ridge
NJ
USA

Gunjan Verma
US Army Futures Command
US Army Research Laboratory
Adelphi
MD
USA

Yevgeniy Vorobeychik
Computer Science and Engineering
Washington University in St. Louis
St. Louis
MO
USA

Mike Weisman
Combat Capabilities Development Command
US Army Research Laboratory
Adelphi
MD
USA

Bowei Xi
Department of Statistics
Purdue University
West Lafayette
IN
USA

Seunghyun Yoon
Department of Computer Science
Virginia Tech
Falls Church
VA
USA

and

School of Electrical Engineering and Computer
Science
Gwangju Institute of Science and Technology
Gwangju
Republic of Korea

Yan Zhou
Computer Science Department
University of Texas at Dallas
Richardson
TX
USA

Zeran Zhu
Coordinated Science Laboratory
University of Illinois at Urbana-Champaign
Urbana
IL
USA

Quanyan Zhu
Department of Electrical and Computer
Engineering
NYU Tandon School of Engineering
New York University
New York
New York
USA

Bolor-Erdene Zolbayar
Pennsylvania State University
Computer Science and Engineering
Department
University Park
Pennsylvania
USA

Foreword

While cyber security research has been carried out for the past three decades, the research area has been continually broadened when multidisciplinary approaches become the mainstream of research. This book captures the recent adoption of both game theory and machine learning techniques to address cyber security research challenges.

Game theory has been widely used to mathematically models strategic interaction among intelligent decision-makers. Cyber security research introduced a unique setting for formulating game theory models between attackers and defenders, and many times, these models may also need to consider regular users or insiders. The book provides comprehensive coverage on using game theory models to capture cyber security dynamics between attackers and defenders and to develop strategies for cyber defense, including using deception to defeat attacks.

Machine learning has interacted with cyber security in two separate research thrusts: (i) the study of robust, resilient, and trustworthy AI/ML, quite often referred to as adversarial machine learning; and (ii) using machine learning to solve cyber security challenges. The book covered both research thrusts comprehensively, highlighting critical issues in securing AI/ML systems and the use of AI/ML to enhance cyber defense.

Given its timely and comprehensive coverage of applying game theory and AI/ML to cyber security research, the book serves both a great reference for researchers in the field and a great introduction of the research topics to students or new researchers. It gives readers an excellent summary of the latest cyber security research front and presents results from multiple ongoing research problems.

Cliff Wang, PhD
IEEE Fellow
Division Chief, Network Sciences
Program Manager, Information Assurance
US Army Research Laboratory, Army Research Office

Preface

Cyber security is a serious concern for our economic prosperity and national security. Despite increased investment in cyber defense, cyber-attackers are becoming more creative and sophisticated. Moreover, the scope and repercussions of cyber-attack have increased over time. This exposes the need for a more rigorous approach to cyber security, including artificial intelligence methods, such as game theory and machine learning. Using a game-theoretic approach to cyber security is promising; however, one has to deal with many types of uncertainty such as incomplete information, imperfect information, imperfect monitoring, partial observation, limited rationality, and the computational complexity of finding equilibrium or optimum policies. On the other hand, artificial intelligence algorithms have typically been designed to optimize in a random stochastic environment. Recent advances in adversarial machine learning are promising to be more robust to deception and intelligent manipulation. However, they are still vulnerable to adversarial inputs, data poisoning, model stealing, and evasion attacks. These challenges and the high risk and consequence of cyber-attacks drive the need to accelerate basic research on cyber security.

This book contains 24 chapters that address open research questions in game theory and artificial intelligence applied to cyber security. We have included several introductory chapters that provide introductory material covering basic concepts in game theory and machine learning, as well as introductions to basic applications in cyber security suitable for use in educational settings, including as a textbook for graduate-level courses. This book addresses many emerging cyber security topics such as cyber deception, cyber resilience, cyber autonomy, moving-target defense (MTD), Blockchain, sensor manipulation, Internet of Battle Things (IoBT), software-defined networking (SDN), 5G Communications, and intrusion detection systems (IDS). The chapters present several cutting-edge results that use different modeling techniques such as hypergames, attack graphs, human experimentation, dynamic programming, adversarial machine learning (AML), generative adversarial network (GAN), reinforcement learning, deep reinforcement learning (DRL), and logistic regression. The results account for multiple practical considerations like attacker adaptation, scalability, and resource awareness.

The editors would like to acknowledge the contributions of the following individuals (in alphabetical order): Palvi Aggarwal, Ahmed H. Anwar, Nurpeiis Baimukan, Tiffany Bao, Anjon Basak, Tamer Başar, Branislav Bosansky, Ing-Ray Chen, Jin-Hee Cho, Keywhan Chung, George Cybenko, Gyorgy Dan, Gaurav Dixit, Abdelrahman Eldosouky, Tugba Erpek, Jie Fu, Cleotilde Gonzalez, Marcus Gutierrez, Roger A. Hallman, João P. Hespanha, Yunhan Huang, Ravishankar K. Iyer, Brian Jalaian, Kun Jin, Susmit Jha, Zbigniew T. Kalbarczyk, Murat Kantarcioglu, Thenkurussi Kesavadas, Sandra König, Xenofon Koutsoukos, Abhishek N. Kulkarni, Nandi Leslie, Xiao Li, Yi Li, Mingyan Liu, Patrick McDaniel, Radha Poovendran, Stefan Rass, Anirban Roy, Yalin Sagduyu, Dinuka Sahabandu, Henrik Sandberg, Armin Sarabi, Serkan Saritas, Muhammed O. Sayin, Stefan Schauer, Shamik Sengupta, Ryan Sheatsley, Ezzeldin Shereen, Yi Shi, Yan Shoshitaishvili, Peicheng Tang,

Sridhar Venkatesan, Gunjan Verma, Yevgeniy Vorobeychik, Mike Weisman, Bowei Xi, Seunghyun Yoon, Muhammad Aneeq uz Zaman, Yan Zhou, Zeran Zhu, and Bolor-Erdene Zolbayar.

We would like to thank Jaime Acosta, Michael DeLucia, Robert Erbacher, Stephen Raio, Robert Reschly, Sidney Smith, and Michael Weisman for technical review support. We would also like to extend thanks and acknowledgment to the US Army Research Laboratory's technical editors Sandra Barry, Sandra Fletcher, Carol Johnson, Lisa Lacey, Sandy Montoya, and Jessica Schultheis who helped edit and collect the text into its final form, and to Victoria Bradshaw, Mary Hatcher, Teresa Netzler, and Viniprammia Premkumar of Wiley for their kind assistance in guiding this book through the publication process.

Charles A. Kamhoua
Christopher D. Kiekintveld
Fei Fang
Quanyan Zhu

1

Introduction

Christopher D. Kiekintveld[1], Charles A. Kamhoua[2], Fei Fang[3], and Quanyan Zhu[4]

[1] *Department of Computer Science, University of Texas at El Paso, El Paso, TX, USA*
[2] *Network Security Branch, Combat Capabilities Development Command, US Army Research Laboratory, Adelphi, MD, USA*
[3] *School of Computer Science and Institute for Software Research, Carnegie Mellon University, Pittsburg, PA, USA*
[4] *Department of Electrical and Computer Engineering, NYU Tandon School of Engineering, New York University, New York, NY, USA*

Artificial intelligence (AI) technologies have significant promise for addressing many challenges in cyber security, and there are already many examples of AI used for both attack and defense. However, AI is still a rapidly developing technology and it has significant limitations as well that need to be considered carefully in a cyber security context. For example, AI systems that are deployed to do critical tasks such as malware detection or alert management will not be able to do these tasks perfectly, and they may even become new targets for attackers to exploit if they can be easily fooled or manipulated.

This book focuses on two primary areas of AI that are especially promising and relevant for cyber security: computational game theory and machine learning. These areas overlap to some extent, but game theory focuses more on decision-making in adversarial settings, while machine learning focuses more on using large data sets for prediction and adaptation. In both of these areas, we provide some introductory material, but focus most of the content on exploring recent research progress on models, algorithms, and applications of these techniques to cybersecurity. There is a balance of both theoretical contributions and analysis, experimental results and evaluation, case studies of applications, and discussions of ongoing challenges and future work. We also specifically discuss some of the connections between game theory and machine learning, as well as some of the unique issues involved in applying these techniques to cyber security. Our goal is that the reader should come away from this book with and improved understanding of these AI technologies, why they are important and useful for cyber security, what the state-of-the-art approaches are, what types of problems are currently being addressed in the literature, and what areas of future work are most promising and necessary for further advancement in the field.

1.1 Artificial Intelligence and Cybersecurity

Artificial intelligence (AI) refers to a set of computing technologies that are designed to accomplish tasks that require intelligence, which can be broadly defined as interpreting observations of the environment, using knowledge and problem solving to select actions to efficiently accomplish goals, and adapting to new information and situations as they arise. AI encompasses a large number of subdisciplines including knowledge representation, reasoning and decision-making, different types of learning, language processing and vision, robotics, and many others. While general-purpose "human level" AI has not yet been achieved and is not likely to emerge in the near future, AI technologies are rapidly improving in many areas and are being used in many new applications with increasing complexity and impact on our everyday lives. For example, AI agents have already met or exceeded the performance of humans in complex games of strategy such as Go (Singh et al. 2017) and Poker (Brown and Sandholm 2019), and they are also achieving success in many more general tasks such as medical diagnostics (Ker et al. 2017) and self-driving vehicles (Badue et al. 2019), to name just a few examples. For specific, well-defined domains and tasks, it is typical for AI systems to already be on par with human performance, and in many cases, AI systems excel at tasks in which humans perform very poorly. Given the rapid advances in AI, there is a great opportunity to deploy AI methods to make progress on the most challenging problems we face as a society.

Cybersecurity represents one of the great emerging challenges of the twenty-first century. This is driven fundamentally by the rapid development and adoption of computing and networking technologies throughout all areas of society. While providing dramatic benefits and fundamentally changing many areas of business, government, media, military, and personal life, the general paradigm has been to push out new technology and features as quickly as possible without great diligence given to security and privacy implications. The complexity of the interactions of the systems combined with the human users and administrators that interact with them has led to the current situation in which cyber attacks are a constant and very costly threat (Lewis 2018).

As we rely ever more on computing systems to perform critical functions (e.g., controlling the electric grid, driving a vehicle, or making medical diagnoses), the importance of addressing the cybersecurity problem becomes paramount. In addition, future conflicts among nation states as well as terrorist actors will involve attacks and defense in cyberspace as major elements. Those groups with the most robust and resilient computing infrastructure will have a significant security advantage in other areas as well.

AI is already used in many areas of cybersecurity, and new applications of AI will be a key part of the solution for improving cybersecurity in the future. Users, software developers, network administrators, and cyber response teams all have a limited ability to respond to cyber threats, and they need better automated tools for configuration, threat detection and evaluation, risk assessment, automated response, etc., to improve accuracy and reduce the costs of providing security. AI is especially important in cyberspace, because many of the events and responses in security situations take place either at speeds that are much too fast for an effective human response or at scales that are much too large for humans to process effectively. However, the many unique aspects of the cybersecurity domain present novel challenges for developing and deploying AI solutions. This book provides a broad view of recent progress in AI related to cybersecurity, including deep dives into many specific techniques and applications, as well as discussions of ongoing challenges and future work.

1.1.1 Game Theory for Cybersecurity

Many problems in cybersecurity are fundamentally decision-making problems under complex, uncertain, multi-agent conditions. Game theory provides a foundational set of mathematical tools for modeling these types of decision problems that can allow us to make better decisions (both as humans and automated systems), as well as understand the reasons for these decisions and what assumptions and data they depend on. While many problems in cybersecurity can be modeled as (approximately) zero-sum games between a defender and an attacker, there are also more complex games that are not zero-sum or that involve relationships among multiple attackers and defenders (Alpcan and Başar 2010; Manshaei et al. 2013; Pawlick and Zhu 2021). Game theory is not a single idea or approach, but a very diverse collection of modeling techniques and solution concepts and algorithms that can be applied to many different situations. Therefore, it is not a simple "off the shelf" technology that can be easily adapted to any problem. Rather, it is a powerful set of techniques that requires a clear understanding of the problem being modeled, as well as the strengths and limitations of different solution techniques to arrive at solutions that can be highly effective in practice.

In this book, we start with a general overview of basic game theory concepts and then cover a variety of specific modeling and solution techniques, as well as applications of these techniques in cybersecurity applications. These examples are intended to provide a good representation of common approaches in this very active research area, where new problems are being solved and new approaches are being developed at a rapid pace.

Now, we briefly discuss some of the main challenges that arise in applying game theory to cybersecurity, many of which are addressed in more detail in the first two parts of the book:

1. **Acquiring game models.** Any decision analysis technique including game theory first requires an accurate understanding of what the decision space is and what the consequences of different outcomes are. Specifying a specific cyber decision space can be difficult both due to the large and open nature of the potential actions (e.g., zero-day attacks by attackers), but also due to a limited understanding of the costs and consequences of different actions in a very complex environment. While this is not unique to game theory (any decision or risk analysis approach faces similar challenges), game theory often assumes a well-known model using common solution approaches. Understanding how to use machine learning, robust solution concepts, and other approaches to improve game models and be able to analyze models that may be imperfect in the game theory setting is necessary.

2. **Modeling uncertainty.** Game theory provides many tools for modeling and reasoning about uncertainty, including well-known approaches such as imperfect information games and Bayesian games. However, it can be hard to precisely characterize the uncertainty in a cybersecurity setting, and it is often the case that the players have a very different knowledge and understanding of the game they are playing. This disparity requires new concepts for handling different forms of uncertainty, as well as more scalable methods for computing solutions to games with uncertainty.

3. **Dynamic games.** Many cybersecurity games evolve over time, with many repeated interactions and opportunities in which to observe (partial) information about the opponent and adapt a strategic response to over time. These games become very large and complex due to the need to model learning behaviors and maintain belief states over time, similar to single-agent Partially Observable Markov Decision Processes (POMDPs). These games also begin to cross over into topics studied in machine learning, such as multi-agent reinforcement learning (MARL). More effective algorithms are needed for analyzing these games at scale, especially ones with theoretical guarantees and robustness to uncertainty.

4. **Bounded rationality.** Many core game theory solution concepts start by assuming that the players are perfectly knowledgeable and rational. This is clearly not the case when facing humans

or other imperfect opponents, though it can be useful as a worst-case assumption against an unrealistically powerful opponent. However, it is also desirable to be able to understand and exploit the weaknesses of limited opponents and act robustly against the potentially unconventional behaviors of these opponents. Therefore, we must be able to model and better predict the behaviors of various types of boundedly rational players and develop effective strategies for playing against them.

5. **Scalable algorithms.** Addressing many of the previous problems including uncertainty, dynamism, and bounded rationality is possible, but often results in a great increase in the complexity of the game model. In addition, cybersecurity domains are usually inherently very complex due to the very large number of possible states of systems and networks, and the very large number of both attack and defense actions that can be taken. Solving these kinds of cybersecurity problems at realistic scales is a key algorithmic challenge that requires understanding the structure of the problems and deploying state-of-the-art game theory solution techniques, as well as developing new techniques specific to these problems to enhance scalability.

6. **Evaluation.** A final challenge for any decision-making tool for cybersecurity is evaluation. While it is relatively easy to specify a theoretical model and evaluate the theoretical properties and test it in abstract simulations, to take the next step, we also need to evaluate these models in more practical settings with real systems, cyber attacks, and cyber defenses.

1.1.2 Machine Learning for Cybersecurity

Another fundamental problem in cybersecurity is using data to make predictions, identify patterns, perform classification, or adapt strategies over time. All of these tasks fall under the general domain of machine learning, which encompasses different paradigms including supervised, unsupervised, and reinforcement learning. Machine learning is a core discipline within AI that studies how agents can use historical data to adapt and improve their performance on future tasks. This area of AI has seen especially dramatic progress and success in certain tasks in recent years, especially with the advancement of "deep learning" approaches that focus on using the data to find sophisticated internal representations of features, rather than having a human specify features to the algorithm. Deep learning techniques have been very successful in many areas including specific cybersecurity tasks such as intrusion detection and malware analysis (Xin et al. 2018; Berman et al. 2019). However, deep learning does typically require large data sets and lots of computing resources to achieve good results, so it is not the best solutions for all problems.

In this book, we cover a range of topics in machine learning for cybersecurity, including examples of specific applications, theoretical and empirical evaluations of new techniques, the connections between machine learning and game theory in multi-agent settings, and analysis of some of the problems that can arise when applying machine learning in an adversarial context. Here, we briefly overview some of the main challenges for applying machine learning to cybersecurity problems that are addressed throughout this book:

1. **Problem definition and data sets.** Similar to the game theory approaches, machine learning approaches must also define exactly what the problem is and what data can be used to learn the desired model. First, one faces the problems of specifying exactly what you are trying to predict, how this will be useful to the decision-maker, and how you will characterize uncertainty about the predictions. Next, you need to be able to identify data that will actually be useful for making these predictions; usually this needs to be data in sufficient quantity and without too

much noise or data quality issues to be useful. In addition, it may be necessary to label the data based on human inputs, which can be expensive. In cybersecurity, there is the additional challenge that high-quality data sets are hard to acquire due to privacy and security concerns, and it is also hard to get direct data about attacker activities. All of this means that the first problem that must be solved before machine learning can be useful is to identify good data sources.

2. **Explainability.** Another key problem for machine learning methods is that the current approaches are opaque; they make predictions and classifications or recommend actions, but they cannot explain why. The models make predictions based on fine tuning many internal parameters (e.g., the weights in a neural network) for the data set, which is too complex for any human to interpret and clearly explain why these predictions are made. In critical environments where mistakes can lead to severe costs and accountability is necessary, this is a significant problem. Further, it is very difficult to convince humans to trust these systems when the results cannot be clearly explained and the reasons for mistakes cannot be addressed in a straightforward way.

3. **Adversarial samples.** Most machine learning methods have been developed for environments without adversarial opponents. For example, they mostly assume that the inputs they receive will resemble the historical training data. However, an adversary may exploit this by developing specialized adversarial examples to fool machine learning systems, for example, to evade an intrusion detection system. Developing an understanding of how the machine learning systems can be fooled and how we can mitigate these types of attacks is crucial when we rely on these systems as part of cyber defense.

4. **Data poisoning.** Another type of attack that can be conducted on machine learning algorithms is data poisoning, where attackers intentionally introduce misleading data into the information that is used to train the algorithms. This can reduce the performance of the algorithms, but could also be used to create specific types of errors that could later be exploited by an attacker.

1.2 Overview

This book is organized into six parts. The first two parts of the book focus primarily on game theory methods, while the next four focus on machine learning methods. However, as noted, there is considerable overlap among these general approaches in complex, interactive multi-agent settings. Game theory approaches consider opportunities for adaptation and learning, while machine learning considers the potential for adversarial attacks. The first part on game theory focuses on introductory material and uses a variety of different approaches for cyber deception to illustrate many of the common research challenges in applying game theory to cybersecurity. The next chapter looks at a broader range of applications of game theory, showing the variety of potential decision problems, beyond just deception, that can be addressed using this approach.

The remaining four parts of the book focus on a variety of topics in machine learning for cybersecurity. Part 3 considers adversarial attacks against machine learning, bridging game theory models and machine learning models. Part 4 also connects these two using generative adversarial networks to generate deceptive objects. This area synchronizes with the earlier work on game theory for deception since this method can be used to create effective decoys. The final two parts cover various applications of machine learning to specific cybersecurity problems, illustrating some of the breadth of different techniques, problems, and novel research that is covered within this topic.

Game Theory Fundamentals and Applications	PART I: Game Theory for Cyber Deception Chapters: 2,3,4,5,6	Part II: Game Theory for Cyber Security Chapters: 7,8,9,10,11,12
Adversarial Machine Learning and Connections to Game Theory	Part III: Adversarial Machine Learning for Cyber Security Chapters: 13,14,15,16	Part IV: Generative Models for Cyber Security Chapters: 17,18
Machine Learning Fundamentals and Applications	Part V: Reinforcement Learning for Cyber Security Chapters: 19,20	Part VI: Other Machine Learning Approaches to Cyber Security Chapters: 21,22,23

Part 1: Game Theory for Cyber Deception

The first part of the book focuses on using game-theoretic techniques to achieve security objectives using cyber deception tactics (e.g., deploying honeypots or other types of decoys in a network). We begin with a general overview of basic game theory concepts and solution techniques that will help the reader quickly acquaint themselves with the background necessary to understand the remaining chapters. Chapters 3–6 all develop new game models and algorithms for improving cyber deception capabilities; while they share many fundamental approaches, they all address unique use cases and/or challenges for using game theory in realistic cybersecurity settings. Chapter 3 focuses on using deception strategically to differentiate among different types of attackers, and addresses issues in both the modeling and scalability of solution algorithms for this type of deception. Chapter 4 focuses on the problem of modeling and solving games that are highly dynamic with many repeated interactions between defenders and attackers, and on scaling these solutions to large graphs. Chapter 5 brings in the aspect of human behavior, evaluating the game theory solutions against human opponents. Finally, Chapter 6 introduces a different framework for modeling the knowledge players have in security games and how they reason about uncertainty, as well as using formal methods to provide some guarantees on the performance of the system.

Chapter 2: Introduction to Game Theory, *Fei Fang, Shutian Liu, Anjon Basak, Quanyan Zhu, Christopher Kiekintveld, Charles A. Kamhoua*

Game theory mathematically models strategic interaction among intelligent decision-makers. It has wide applications in economics, sociology, psychology, political science, biology, and as we will introduce later in this book, cyber security. To facilitate the readers, we present the basics of game theory in this chapter. We will start by introducing two-player zero-sum normal-form games, the basic class of games that involve two decision-makers who will each make a single move at the same time, and get a payoff, with the two players' payoffs summing up to zero. We will then introduce the solution concepts in normal-form games, including the most well-known Nash equilibrium concept. We further introduce the extensive-form games, which are more complicated and more expressive than normal-form games. They explicitly represent the sequencing of players' moves and the information each player has about the other players' moves when they make a decision.

We will then introduce Stackelberg games and Stackelberg security games, a subclass of games with wide applications in security domains. We will also introduce repeated games where the players repeatedly play the same game. Finally, we will add Bayesian games that depict the uncertainty in players' types and payoff and stochastic games that capture the dynamic transition from one game to another. These models and concepts will be used frequently in later chapters.

Chapter 3: Scalable Algorithms for Identifying Stealthy Attackers in a Game-Theoretic Framework Using Deception, *Anjon Basak, Charles Kamhoua, Sridhar Venkatesan, Marcus Gutierrez, Ahmed H. Anwar, Christopher Kiekintveld*
Early detection and identification of a cyber attacker can help a defender to make better decisions to mitigate the attack. However, identifying the characteristics of an attacker is challenging as they may employ many defense-evasion techniques. We concentrate here on using a honeypot (HP) for detection and identification. Most current approaches focus on detection during an ongoing attack with more detailed identification done later during forensic analysis. Instead, we focus on formally modeling what types of actions the defender can take to support more detailed attacker identification early in the attack chain, allowing more information to target specific defensive responses. We construct our model to detect an attacker type early where an attacker type is defined by an Attack Graph (AG) specific to that attacker that describes their possible actions and goals in planning an attack campaign. Identification of different attack types in our model corresponds to identifying which of a set of possible attack graphs a particular attacker is using in a particular interaction. The defender chooses deception actions that try to force the attackers into choices that will reveal which type they using are early on, which can in turn inform later defenses. First, we propose a multi-stage game-theoretic model to strategically deploy HPs to force the attacker to reveal their type early. We present case studies to show how this model works for realistic scenarios and demonstrate how attackers can be identified using deception. Then, we present a general model, a basic solution algorithm, and simulation results that show that if the defender uses proactive deception, the attacker type can be detected early compared to only observing the activity of an attacker. Finally, we use some heuristics considering domain knowledge to reduce the action space to a reasonable amount to make the algorithm more scalable.

Chapter 4: Honeypot Allocation Game Over Attack Graphs for Cyber Deception, *Ahmed H. Anwar, Charles Kamhoua, Nandi Leslie, Christopher Kiekintveld*
This chapter proposes a novel cyber deception approach for network security. Cyber deception is an effective proactive network security and network resilience technique that misrepresents the true state of the network. Therefore, it limits the ability of any adversary to successfully gain useful information to credibly map out all the network topology features. The proposed cyber deception technique uses HPs to install fake and decoy nodes in the network. The network defender's goal is to find the optimal HP allocation policy, while reasoning about the adversary's attack policy. To this end, a two-person, zero-sum strategic game is formulated between the network defender and an attacker to solve for the optimal HP allocation policy.

The formulated game captures the network topological features and the difference between its nodes. The game also accounts for the cost associated with the defense action as well as the attack cost. On the attacker side, the game considers a practical threat model, where the attacker reasons about which node to attack given that the attacker is not naïve and expects that some nodes are HPs. Nash equilibrium defense strategies are analytically characterized and studied for the formulated game. This chapter investigates a one-shot game in which only one node to be defended or attacked given any current location within the network. Moreover, the game model is extended to model the case of defending and attacking a larger subset of nodes.

The computational complexity of the extended game is discussed, and a scalable algorithm is proposed to ensure that the defense approach applies to large-scale networks.

Finally, a dynamic game model where both players interact sequentially is introduced and the Q-minimax algorithm is described to learn the optimal strategies for both players. Numerical results are presented to validate our findings and illustrate the effectiveness of the proposed game-theoretic cyber deception defense approach.

Chapter 5: Evaluating Adaptive Deception Strategies for Cyber Defense with Human Experimentation, *Palvi Aggarwal, Marcus Gutierrez, Christopher Kiekintveld, Branislav Bosansky, Cleotilde Gonzalez*

This chapter evaluates adversarial decision-making against different defense algorithms that vary in three dimensions: determinism, adaptivity, and opponent customization. We utilize an interactive, abstract cybersecurity combinatorial Multi-Armed Bandit task, i.e., the HoneyGame, to conduct human experiments. The aim of the defense algorithms is to optimally allocate the defense resources so adversaries attack the fake nodes, i.e., HPs. This chapter measures the effectiveness of the defense algorithms by measuring how humans learn defense strategies and adapt their decisions. Different defense algorithms attempt to defer the attacker's learning. Results suggest that attackers learn the static algorithms quickly and struggle to learn the defenses of the more dynamic and adaptive algorithms.

The effectiveness of the adaptive defense algorithms relies on the accurate representation of adversaries and prediction of their actions. Currently, our adaptive defense algorithms rely on the Thomson Sampling algorithm to predict the adversaries' actions, but that could not capture human adversaries' actions perfectly. Based on new directions in the design of defense algorithms, we propose to use Instance-Based Learning Theory to develop models that emulate human behavior and inform the design of dynamic, adaptive, and personalized algorithms of defense.

Chapter 6: A Theory of Hypergames on Graphs for Synthesizing Dynamic Cyber Defense with Deception, *Jie Fu, Abhishek N. Kulkarni*

In this chapter, we investigate a formal methods approach to design effective defense strategies with active cyber deception, where the defender takes actions in reaction to the attacker's progressive attacks in real time. The use of formal methods ensures provably correct and secured system performance under any possible attack actions. To achieve this objective, we introduce an attack-defend game on a graph. This game extends an attack graph—a graphical security model—to incorporate the defender's countermeasures, enabled by software-defined networking and other dynamic defense mechanisms. In cyber deception, the defender introduces decoys in a network to create asymmetrical information between the defender and the attacker, as the attacker may not know if a host is a decoy without prior interaction or may even mistake a decoy to be the attack target. The key question is how could the defender leverage their private information for security advantages? To this end, we investigate the solution concepts of hypergames for attack-defend games on graphs. Informally, a hypergame is a game of perceptual games, each of which is perceived by one player given their information and higher-order information, i.e., what they know that the other players know. We examine the solution concepts of level-2 hypergames to design deceptive strategies. Assuming the attacker takes any rational strategy in their perceptual game, the defender solves a level-2 hypergame on a graph and obtains a deceptive strategy that exploits the attacker's misperception. This defense strategy not only ensures the security specification can be satisfied with probability one, but also ensures that the attacker's perception does not change until the game reaches a state when the attacker has no chance to carry out the attack successfully. We use illustrative examples to demonstrate the theoretical and technical approaches.

Part 2: Game Theory for Cybersecurity
The second part of the book represents a collection of different problems in cybersecurity that can be modeled and analyzed using the game-theoretic paradigm for thinking about attacker and defender strategies. The chapters represent a diverse selection of specific problems and techniques, though, of course, this cannot be considered exhaustive. However, they demonstrate the broad utility of the game theory approach to many different types of decisions and show a variety of specific examples of how this can be applied and how specific problems such as scalability can be overcome. Chapter 7 focuses on developing game models for detecting advanced threats that try to disguise themselves as benign processes and presents new algorithms for scalability in this context. Chapters 8 and 9 focus on the general problem of manipulating sensors and anomaly detection methods, which is relevant to both attackers and defenders. Chapter 10 considers game-theoretic methods to optimize moving target defense in the Internet of Things (IoT) setting. Chapter 11 applies game theory to the authentication domain, specifically considering how to make decisions in the continuous authentication paradigm. Finally, Chapter 12 explores game theory for modeling cyber autonomy and software security.

Chapter 7: Minimax Detection (MAD) for Computer Security: A Dynamic Program Characterization, *Muhammed O. Sayin, Dinuka Sahabandu, Muhammad Aneeq uz Zaman, Radha Poovendran, Tamer Başar*
This chapter introduces and analyzes a game-theoretical framework to detect advanced threats that can mimic benign system activities in a distributed and stochastic manner in order to evade detection. Accordingly, it introduces and analyzes minimax detectors that can assess system activities in terms of their riskiness (similar to signature-based detection) and likeliness (similar to anomaly detection) across modern computing systems in a cohesive and strategic way. The minimax detection (MAD) mechanism provides analytical guarantees for minimax performance by minimizing the system's detection cost under the worst actions of an adversary (maximizer), designated MAX, who seeks to intervene into the system activities. In accordance with this, the mathematical framework is naturally a zero-sum game, with detection cost as the objective function to be minimized by MAD and maximized by MAX. The challenge here, however, is that the game size is enormous, thus making computation of the equilibrium using any direct approach infeasible. To overcome this computational challenge and achieve a scalable implementation, we introduce a hierarchical scheme that approximates the underlying joint distribution over the recorded activities, such as system-calls, across the entire system, enabling system-level activities to be assessed at varying degrees of granularity, similar to state summarization in big data processing. Based on this hierarchical scheme, we show that the detection cost has a nested structure, enabling decomposition of the entire game into nested local sub-games. In this way, it is possible to compute the equilibrium efficiently via a dynamic program similar to backward induction in extensive-form games. The chapter also provides a complexity analysis of the proposed solution and an analysis of the behavior of the players at equilibrium over an illustrative example.

Chapter 8: Sensor Manipulation Games in Cybersecurity, *João P. Hespanha*
This chapter addresses sensor manipulation problems in which an attacker tries to induce a cyber defense system to make wrong decisions, such as denying access to a sensitive resource, flagging a computer as compromised, deauthorizing a user, or closing a firewall. The chapter presents a game-theoretical treatment of sensor manipulation and considers two attack models, each focusing on a different aspect of sensor manipulation: Measurement-manipulation games mostly explore the challenges that arise from not knowing which sensors have been manipulated and how this impairs the defender's ability to make correct decisions. In sensor-reveal games,

the defender actually knows which sensor has been revealed, but the challenge is to explore the information that the attacker implicitly provides by selecting a particular sensor.

Chapter 9: Adversarial Gaussian Process Regression in Sensor Networks, *Yi Li, Xenofon Koutsoukos, Yevgeniy Vorobeychik*

Autonomous cyber-physical systems typically feature a control loop that maps sensor measurements to control decisions. Often, the goal of the controllers is to keep particular system state variables in a safe range, with anomaly detection employed to ensure that malicious corruptions to the observed state do not cause catastrophic failures. We study a novel stealthy attack on anomaly detection using Gaussian process regression (GPR), in which an attacker's goal is to maximally distort the observed values of target state variables. Additionally, we study the problem of robust GPR for anomaly detection by modeling it as a Stackelberg game and present a novel algorithmic approach for solving it. Our experiments demonstrate the effectiveness of our approaches.

Chapter 10: Moving Target Defense Games for Cybersecurity: Theory and Applications, *Abdelrahman Eldosouky, Shamik Sengupta*

The recent widespread of using Internet-connected devices in cyber-physical systems and the IoT has brought forward many security threats and attacks. Attackers can exploit Internet connectivity to perform remote attacks that can cause severe damage to the systems. Therefore, it is imperative for cyber systems to adopt suitable security mechanisms to protect their assets and information. One recent effective security mechanism, shown to enable cyber systems to thwart prospective attacks, is known as moving target defense (MTD). MTD enables a system to randomize its configurations to harden the attacker's mission and increase the cost of a successful attack. To this end, this chapter aims at providing an in-depth study of the recent theory and applications of MTD. In particular, the chapter studies a number of theoretical frameworks that enable one to evaluate the effectiveness of MTD implementations, study the timing issue of MTD, and minimize the costs associated with MTD. This is followed by analyzing and comparing some game-theoretic models from literature that are proposed to study MTD. Game-theoretic models are chosen as they can best model the interactions between attackers and defenders in MTD scenarios. Next, a promising game-theoretic model, based on single controller stochastic games, is introduced as a general framework that can be used to study attacker–defender interactions in many MTD applications, in which the defender controls the changes in the system's configuration. A case study pertaining to wireless networks security is studied, in detail, to highlight the effectiveness of the proposed game-theoretic model. Within the case study, equilibrium analysis is analytically derived and simulation-based results are analyzed. Finally, the chapter covers some recent applications of MTD in both IoT and machine learning-related applications.

Chapter 11: Continuous Authentication Security Games, *Serkan Saritas, Ezzeldin Shereen, Henrik Sandberg, Gyorgy Dan*

Continuous authentication, as an extension to conventional authentication methods, is emerging as a promising technology for preventing and mitigating sophisticated identity theft and session hijacking attacks, which can cause catastrophic losses for individuals and companies, and possibly disasters for critical infrastructures.

Unlike traditional authentication techniques that make use of a single secret once, continuous authentication security schemes enhance the security level of the systems by passively, transparently, and continuously monitoring and testing the authenticity of users in real time and in ways that are difficult to replicate. In this chapter, the interaction between a system operator (defender), who maintains continuous authentication and an intrusion detection system (IDS) in its system comprising multiple resources, and an adversarial attacker, who tries to execute a rogue command on the system's resources, is modeled as a dynamic discrete stochastic game with imperfect information. In the considered game, if the attacker chooses to compromise the system, they observe the traffic generated by the user and may craft packets that appear to originate from that user, whereas the defender designs security measures to detect suspicious behavior and prevent unauthorized access while minimizing the monitoring expenses. Following common practice in game-theoretic models of security, the attacker is assumed to be aware of the strategy of the defender, whereas the defender is not aware of the attacker's strategy. Then, a corresponding optimal strategy of the attacker is characterized by formulating the attacker's utility as a backward dynamic programming recursion; the effects of the selection of the defender parameters on the attacker and defender utilities are illustrated to show the intrinsic trade-off between monitoring cost and security risk. Even though continuous authentication can be effective in minimizing security risk, the results indicate that on its own, it is not enough to secure the system and should be combined with appropriate incident detection, such as IDS, for optimal security risk minimization.

Chapter 12: Cyber Autonomy in Software Security: Techniques and Tactics, *Tiffany Bao, Yan Shoshitaishvili*

Software security in deployed systems is a dynamic interaction between attackers and defenders. These interactions can be formalized as computer security games between multiple parties, each of which interacts through actions such as finding a zero-day vulnerability from a software program, using an exploit, and deploying a patch. Game theory provides a framework to think through players' choices and consequences, as well as serves as a model of the components for optimizing real systems and responding to zero-day vulnerabilities. In this chapter, we consider both the techniques and strategy for discovering and dealing with software vulnerabilities. Furthermore, we explore the automation in both techniques and decision-making, called cyber autonomy in software security.

This chapter states cyber autonomy from both techniques and tactics aspects. In particular, we investigate concrete system instances of players as represented by the Cyber Reasoning Systems (CRS) found in the Defense Advanced Research Projects Agency (DARPA) Cyber Grand Challenge (CGC), such as Mayhem and Mechanical Phish. Nonetheless, cyber autonomy is a general concept and is also applicable for real-world scenarios such as zero-day vulnerability handling or even cyber warfare.

We address two main lines of research. First is studying player strategies based on game-theoretical models. We consider the interaction among multiple players, seek the optimal strategy corresponding to an equilibrium of the associated game, and explore the factors that affect the outcome of the game. Second, we study the critical components in the theoretical model and investigate the techniques that realize such actions in real systems. We are viewing the game from the system perspective, where we look into individual players and study the system components that support the actions under different strategies in real systems. In the end, we point out the current limitations in cyber autonomy with the aim to enlighten future work.

Part 3: Adversarial Machine Learning for Cybersecurity

The third part of the book covers several topics in the emerging area of adversarial machine learning (AML), which considers how machine learning methods perform in the presence of malicious adversaries (as in cybersecurity applications). This focus breaks many of the standard assumptions in machine learning and represents an area of overlap between the game-theoretic approaches that specifically model multiple players with different objectives and conventional machine learning approaches designed for single-agent contexts. This connection is made explicitly in Chapter 13, which develops game-theoretic models that help to understand AML problems in greater detail. Chapter 14 shows some specific applications of AML methods in the cybersecurity context, specifically developing attacks on 5G communications. Chapter 15 provides a broader analysis of the novel threat that AML methods pose in the cybersecurity domain, including ways that AML can change how attackers might execute attacks and evade detection, and a discussion of existing mitigation methods. Finally, Chapter 16 specifically focuses on deep learning methods that have been very popular and successful in many areas such as vision, but which also have problems with trust, resilience, and interpretability that must be addressed as we rely on these techniques for critical security applications.

Chapter 13: A Game-Theoretic Perspective on Adversarial Machine Learning and Related Cybersecurity Applications, *Yan Zhou, Murat Kantarcioglu, Bowei Xi*

Many machine learning applications need to work in an adversarial environment, where adversaries adapt their behavior in order to fool machine learning-based systems. Examples of such applications include malware detection, intrusion detection, credit card fraud detection, e-mail spam filtering, and web spam detection. To foil machine learning systems and evade detection, adversaries typically modify the data under their control so that the corrupted data invalidates the independent and identically distributed (i.i.d) assumption. In other words, at application time, machine learning systems are facing data that follow a different distribution other than the one assumed by the trained machine learning models. A large divergence of distribution in training and test data can seriously hamper the applicability of machine learning applications in practice.

We discuss adversarial challenges in the cybersecurity domain, where machine learning techniques are broadly desired. We explain why machine learning techniques have to adapt to these challenges in order to improve their resilience to adversarial attacks. We define an adversarial learning problem in which two or more opponents are playing games to maximize their gains and offer a game-theoretic perspective on the problem. We review the existing game-theoretic techniques developed for adversarial learning problems that represent a variety of conflicting interests. By modeling the adversarial learning problem as a game between players, we provide an extensive analysis on how each player's action and assumption would impact the other player's reaction. We discuss cases where the game can be modeled as simultaneous or sequential, depending on each player's motivation and objective. We further present different game-theoretic models that approach the problem from different angles and demonstrate two specific use cases in real-world applications.

Chapter 14: Adversarial Machine Learning in 5G Communications Security, *Yalin Sagduyu, Tugba Erpek, Yi Shi*

This chapter identifies vulnerabilities of machine learning-driven 5G communication systems, identifies the attack surface due to AML, and discusses the defense. While machine learning finds rich applications in wireless communications, its vulnerabilities have not been fully understood yet by accounting for unique differences from other data domains due to channel, interference, and radio hardware effects. This chapter discusses two stealth attacks against 5G, namely, attacks against (i) spectrum sharing of 5G with incumbent Citizens Broadband Radio Service (CBRS) users and (ii) physical layer authentication of 5G user equipment (UE) to support network slicing. For the first attack, the adversary relies on an exploratory attack to learn the 5G's transmit behavior, namely, when the 5G base station, gNodeB, successfully communicates with 5G UE. Then, the adversary launches an evasion attack over the air to manipulate the signal-level inputs to the deep learning classifier deployed at the Environmental Sensing Capability (ESC) to support the 5G system. The goal of the adversary is to decide when to jam 5G communications during data transmission or spectrum sensing to maximize its impact. For the second attack, the adversary applies the generative adversarial network (GAN) to spoof signals that cannot be reliably distinguished from intended 5G UE signals such that the deep learning-based physical layer authentication of the 5G gNodeB is fooled into granting access to the adversary. Both attacks operate with small footprints and can significantly reduce the 5G performance. We present a defense mechanism to slightly increase the uncertainty of the 5G communication behavior (by making a small number of wrong transmit decisions) such that the adversary cannot accurately learn this behavior and therefore cannot effectively attack 5G communications. As shown in this chapter, AML has emerged as a novel threat for 5G and raises the need for defense mechanisms.

Chapter 15: Machine Learning in the Hands of a Malicious Adversary: A Near Future If Not Reality, *Keywhan Chung, Xiao Li, Peicheng Tang, Zeran Zhu, Zbigniew T. Kalbarczyk, Thenkurussi Kesavadas, Ravishankar K. Iyer*
Targeted and sophisticated cyber threats are on the rise. To maximize efficiency, attackers spend significant amounts of time preparing attacks (i.e., reconnaissance, weaponization, and exploitation). However, an extended foothold into a system often leaves a trace that a well-prepared security team can use to track the attacker. In this chapter, we discuss the potential for threats to advance through the use of machine learning techniques. With machine learning-driven malware installed, an attacker might no longer need to maintain a foothold in a target system, but can let malware internally weaponize the attack while leaving minimal traces. In the context of two target applications (i.e., a smart building automation system and a surgical robot), we demonstrate how targeted attacks powered by machine learning techniques can probe the target system to infer actionable intelligence and customize the attack payload to increase the efficiency of the attack while reducing the time to success. The chapter discusses (i) why attackers might want to adopt machine learning techniques, despite the effort required to design and implement the key functional components and feature of machine learning-driven malware, and despite the computational resources required for such malware to infer actionable intelligence; (ii) the limitations of current mitigation methods; and (iii) ongoing research that can help prevent or detect such attacks. Recent research demonstrates the potential of machine learning-driven threats and shows that such threats might soon become a reality, if indeed they are not already active. Hence, it is important for cyber defenders to understand the mechanisms and implications of the malicious use of AI in order to stay ahead of threats and deploy defenses to prevent theft of data, system damage, or major disruption of critical infrastructure. This chapter should motivate further research on advanced offensive technologies, not to favor the adversaries, but to know them and be prepared.

Chapter 16: Trinity: Trust, Resilience, and Interpretability of Machine Learning Models, *Susmit Jha, Anirban Roy, Brian Jalaian, Gunjan Verma*
Deep learning methods have demonstrated impressive performance in many applications, such as image processing, speech processing, and natural language processing. Despite these remarkable strides over the last decade, the utilization of these techniques is typically limited to benign environments. The use of deep learning in applications such as biometric recognition, anomaly detection, malware detection, and intrusion detection requires them to operate in adversarial environments. But the overwhelming empirical studies and theoretical results have shown that these methods are extremely fragile and susceptible to adversarial attacks. The rationale for why these methods make the decisions they do is also notoriously difficult to interpret; understanding such rationales may be crucial for the applications mentioned. In this chapter, we summarize recent results on the connections between the related challenges of trust, resilience, and interpretability of deep learning models, and describe a novel integrated approach, Trinity (Trust, Resilience, and INterpretabilITY), for analyzing these models. We study the connections between the related

challenges of trust, resilience, and interpretability of deep learning models, which together hinder the adoption of these models in high-assurance, safety-critical applications. The Trinity framework has a suite of algorithms to establish trust over the safe behavior of systems with machine learning models, provide resilience against adversarial attacks, and simultaneously improve their interpretability. The neurosymbolic approach in Trinity combines symbolic methods developed for automated reasoning and formal analysis with connectionist deep learning methods that use deep neural networks. The symbolic representation of knowledge in Trinity is complemented by scalable generative methods useful for top-down inference and analysis by synthesis. This is a first step toward creating human-like learning approach that is trustworthy, resilient, and interpretable.

Part 4: Generative Models for Cybersecurity

The next section of the book focuses on generative models for producing deceptive objects using machine learning methods. This is a particular branch of AML that has great relevance for cybersecurity because the techniques can be used to produce highly effective decoys and deceptive objects. In turn, these can be used to evade existing IDSs, send concealed messaged, and produce effective honeypots or other decoy objects, and for many other purposes in cybersecurity. Chapter 17 shows how these adversarial learning methods can be used by attackers to evade many existing intrusion detection systems. Chapter 18 shows how similar methods can also be used to conceal communications from censors by evading censorship systems. While both of these use similar fundamental techniques, they are put to very different purposes showing the diversity of possible uses of these methods for both attack and defense.

Chapter 17: Evading Machine Learning-Based Network Intrusion Detection Systems with GANs, *Bolor-Erdene Zolbayar, Ryan Sheatsley, Patrick McDaniel, Mike Weisman*

Modern network intrusion detection systems (NIDS) are implemented with machine learning algorithms to detect unknown attacks and mutations of known attacks. However, the machine learning algorithms in NIDS are vulnerable to a class of countermeasures such as adversarial attacks. Attackers can evade machine learning-based NIDS by applying slight perturbations to malicious network traffic. Although adversarial examples have been extensively studied in the image domain, the associated threat model is unrealistic: it assumes adversaries have full control over the feature space, which is impractical in domains like network intrusion detection. In this chapter, we evaluate the threat of GAN-aided attacks on machine learning-based IDSs. We develop algorithms that extract and formulate the domain constraints and enforce these constraints on the adversarial examples during the training of our attack. In our threat model, an adversarial attacker, given access to a deep neural network (DNN) model for NIDS, adds a minimal perturbation to the feature values of attack traffic to change the DNN's prediction from "malicious" to "benign." Our experiments demonstrate that optimally trained DNN models of NIDS can be fooled by adversarial network traffic generated by our GAN-based algorithm. The effectiveness of the attack algorithm is validated by experiments performed on the state-of-the-art CICIDS-2017 dataset that includes the most-up-to-date attacks and the benchmark NSL-KDD dataset. Our algorithm achieved a high success rate of 100% on both datasets with an l^2 perturbation of 0.3.

Chapter 18: Concealment Charm (ConcealGAN): Automatic Generation of Steganographic Text Using Generative Models to Bypass Censorship, *Nurpeiis Baimukan, Quanyan Zhu*

Based on the recent observation of M. E. Roberts, digital censorship through information friction mechanism can effectively halt the spread of information, thus abridging the freedom of speech of online users. This work deconstructs digital censorship in the context of messenger and emailing services, where it has been shown that the confidentiality of the users is vulnerable when exchanging textual messages. The clear instance of the aforementioned censorship can be seen in the most popular Chinese application called WeChat. The research by the Citizen Lab of the University of Toronto has shown that there are specific words/phrases, which if written, will not be able to pass to the recipient.

We propose to use linguistic steganography to enhance the privacy of the users, but also ensure that the services are collecting inaccurate data on us. To exchange secret data in an open medium, such as social media, without detection by anyone, one can use a covert channel. Communication through a covert channel is hidden in a legitimate communication channel, which refers to steganography. Although steganographic techniques have been successfully employed when hiding the secret message in images and audio, it has been more challenging to show similar results when hiding information in text because test has a higher degree of information coding. In other words, text has less redundant information when compared to images. Simultaneously to avoid cover attacks, where given text is slightly modified, we propose to use to generate text from scratch using generative models, such as LeakGAN and recurrent neural networks (RNNs) in our system, which we call ConcealGAN.

ConcealGAN generates natural cover text without modifying known plaintext. It is done using a state-of-the-art natural language generation model that uses deep learning techniques. Firstly, it builds on RNN and LeakGAN to enhance imperceptibility or reduce detectability. Secondly, to enhance the difficulty of extraction, ConcealGAN proposes a new architecture of a double layer of encoding of the secret message.

Part 5: Reinforcement Learning for Cybersecurity
The next part of the book introduces the popular framework of reinforcement learning, in which agents can learn from their experience to optimize a policy in an unknown and uncertain environment. While this approach can be useful in many cybersecurity problems, most reinforcement learning techniques are not robust to adversarial conditions. Chapter 19 shows some of the problems with how reinforcement learning can be manipulated in an adversarial context. Chapter 20 provides a case study for how reinforcement learning (combined with deep learning methods) can be used to solve a specific cybersecurity problem.

Chapter 19: Manipulating Reinforcement Learning: Stealthy Attacks on Cost Signals,
Yunhan Huang, Quanyan Zhu
In this chapter, we have provided a basic introduction of reinforcement learning and discussed the potential threats in reinforcement learning. The security threats of reinforcement learning and its wide application require an urgent study of the security problems of reinforcement learning. However, only a limited number of studies have appeared in recent literature. Hence, we introduce in this chapter a general framework to study reinforcement learning under deceptive falsifications of cost signals where a number of attack models have been presented. These different attack models invoke a number of interesting security problems of reinforcement learning yet to be investigated. We have provided theoretical underpinnings for understanding the fundamental limits and performance bound on the attack and the defense in reinforcement learning systems. We have shown that in TD(λ), the approximation learned from the manipulated costs has an approximation error bounded by a discount factor-dependent constant times the magnitude of the attack. Further, the effect of the adversarial attacks does not depend on the choice of λ. In Q-learning, we have characterized conditions on deceptively falsified cost signals under which Q-factors learned by agents can produce the policy that adversaries aim for. The concept of robust region we introduce can be utilized by both offensive and defensive sides. A reinforcement learning agent can leverage the robust

region to evaluate the robustness to malicious falsifications. An adversary can use it to estimate whether certain objectives can be achieved or not. We characterize the conditions on attacks that provide a fundamental understanding of the possible strategic adversarial behavior of the adversary. These theoretical results are shown greatly help our understanding of the attainability of an adversary's objective. A numerical example is provided to illustrate the performance degradation of the reinforcement learning algorithms and corroborate the theoretical results given.

Chapter 20: Resource-Aware Intrusion Response Based on Deep Reinforcement Learning for Software-Defined Internet of Battle Things, *Seunghyun Yoon, Jin-Hee Cho, Gaurav Dixit, Ing-Ray Chen*

This chapter introduces deep reinforcement learning (DRL) to solve a cybersecurity problem. DRL is an interesting technique that combines deep learning (DL), which offers the capability to solve high-dimensional problems, with reinforcement learning (RL), which autonomously learns from experiences via trials and errors. In particular, we are interested in building a highly attack-resistant tactical network, namely, the Internet of Battle Things (IoBT), against both physical and cyberspace epidemic attacks. We aim to develop an autonomous intrusion response framework that can identify optimal response strategies based on DRL to deal with detected intrusions. In the context of resource-constrained tactical networks assigned with a time-sensitive mission, the responses to detected intrusions can vary depending on the importance of compromised assets (e.g., IoT devices or web servers) and the impact introduced by the compromised assets. Considering severe resource constraints in IoBT consisting of highly heterogeneous entities operating under high hostility and network dynamics, we take a resource-aware defense approach. First, we build a multi-layer defense network architecture that can construct a network topology based on the importance levels of nodes. The underlying idea is to provide more security protection for more important nodes while providing less protection for less important nodes, with the aim of maximizing both security (e.g., mean time to security failure) and performance (e.g., service availability). Second, we develop a resource-aware intrusion response framework that can identify the best action (destruction, repair, or replacement) in response to a detected failure/attack to maximize security/performance. In this chapter, we discuss the key findings and insights learned from our experiments in which we perform a comparative analysis of multiple state-of-the-art DRL algorithms with a baseline strategy (e.g., random) in selecting the optimal intrusion response strategy based on both system security and service availability metrics.

Part 6: Other Machine Learning Approaches to Cybersecurity

The applications of machine learning to cybersecurity are so broad that we cannot cover all of them in a single volume. In this final part, we include a few unique examples of machine learning approaches that do not fit neatly into other categories, but are useful for showing the breadth of applications in this area. Chapter 21 presents a machine learning approach for improving the critical task of network scanning. Chapter 22 provides a more abstract use of machine learning for parameter estimation that could potentially apply to many of the other theoretical models presented earlier in the book. Finally, Chapter 23 combines machine learning with another emerging technology in distributed systems (Blockchain) to provide a novel approach for adaptive cyber defense.

Chapter 21: Smart Internet Probing: Scanning Using Adaptive Machine Learning, *Armin Sarabi, Kun Jin, Mingyan Liu*

Network scanning is an extensively studied Internet measurement technique. Using modern scanners such as ZMap, one can probe for devices accessible on the public Internet across a wide range of ports/protocols (often over the entire IPv4 address space), and parse the acquired data to identify, profile, and capture the configuration of Internet-facing machines. Data resulting from Internet scans have many security and measurement applications, e.g., to identify misconfigured, vulnerable, and infected networked devices; gauge the attack surface of networks; and study trends in the Internet ecosystem. However, Internet scans generate large amounts of traffic for the probed networks and are not as effective over IPv6 networks, where exhaustive scanning is not feasible due to the vast number of addresses in this space. To address these shortcomings, in this study, we develop a machine learning-enabled framework for smart network scanning by preemptively predicting and refraining from sending probes to unresponsive IP addresses, thereby reducing their intrusiveness. To this end, we leverage a priori information including location and ownership (Autonomous System) information of networks to capture local patterns and combine that with cross-protocol information to predict ports that are possibly active on any given IP address. Our results evaluated on global snapshots of the Censys database across 20 ports suggest that we can reduce Internet scanning traffic by 20–80% while still discovering 90–99.9% of responsive IP addresses. Our proposed framework reduces the footprint of scans with negligible computational overhead and allows one to cover larger hitlists for IPv6 scanning, hence discovering more active devices for giving visibility into IPv6 networks. We also observe high coverage over small subpopulations of vulnerable and misconfigured devices, thus enabling efficient and accurate assessment of networks' security posture.

Chapter 22: Semi-automated Parameterization of a Probabilistic Model Using Logistic Regression—A Tutorial, *Stefan Rass, Sandra König, Stefan Schauer*

A major challenge when using a probabilistic model is to find reasonable values for the parameters that it contains. It is often convenient to "just assume" that some value p describes the likelihood for a certain event; however, the challenge of finding a reliable estimate for this variable p is a different and often untold story. The practical issue becomes worse if there are many parameters, as it is naturally the case when applying Markov chains to describe stochastic dynamics. The convenience of enjoying a rich theory and powerful tools to study and use Markov chains comes with the price of severe difficulties of instantiating the chain with real values in a practical scenario. As model parameterization may be a neglected issue in the literature, this chapter presents the concept of "parameterization by example": we consider a generic probability parameter p and show—in a tutorial way—how to apply logistic regression to find a value for p based on available data. Our exposition is in the R language for statistical computing, but straightforwardly translates to other programming languages. As this is a special instance of machine learning applied to the problem of parameter choice in stochastic models, we briefly compare the method to alternatives in the literature. We showcase the method on a model that simulates cascading effects of incidents in critical infrastructures, utilizing stochastic Mealy automata to capture the uncertainty. While this choice is arbitrary and only for the sake of illustration, the general application of logistic regression has wide applicability to instantiate even large lots of (conditional) probability parameters in a batch. Overall, the chapter brings the reader's attention to logistic regression and general machine learning to help not only construct, but equally importantly explore the problem of model parameterization in general.

Chapter 23: Resilient Distributed Adaptive Cyber Defense Using Blockchain, *George Cybenko, Roger A. Hallman*

Adaptive cyber defense systems change their behaviors as the environments in which they operate change. Those changes can be due to dynamic reconfigurations of the information infrastructure, such as the addition or removal of compute nodes, sensors, applications, and communications links. These defensive posture changes can be made by continuously monitoring the environment, learning its new characteristics, and implementing appropriate new control actions. The basis for making such adaptations in a stationary or slowly changing environment can be based on classical reinforcement learning and adaptive control ideas. However, if the operating environment changes because of adversary adaptations, existing mathematical and algorithmic principles for defensive adaptation do not apply. We study the problem of online learning using distributed Upper Confidence Bounds (UCB) algorithms in which cumulative regret (CR) is the performance criterion.

CR is an appropriate performance metric because it captures the overall rate at which the systems performance is approaching optimality, not just the asymptotic conditions under which optimality is reached. The faster a system can approach optimality, the more it will outperform an adversary that is also changing but at a slower rate. A key observation and technical contribution is the use of distributed systems to implement UCB-based learning. This entails the deliberate use of "suboptimal" actions in order to fully explore the values of the available actions. The system must sacrifice some of its capabilities to learn faster and perform better. However, because some agents within the distributed system are operating suboptimally, they will be compromised and so some subset of information in the distributed learning system will lack integrity. This requires adding technologies for making such systems robust even under untrusted agent operations. Accordingly, we discuss the introduction of Byzantine Fault tolerance and Blockchain technologies into a distributed adaptive cyber-defensive system to address.

Chapter 24: Summary and Future Work, *Quanyan Zhu, Fei Fang*
This book provides a collection of recent advances in machine learning and game theory for cybersecurity. Machine learning plays an important role in detecting anomalies, finding insider threats, and defending against malicious behaviors. Game theory has been used to understand the adversarial behaviors and design defense strategies to mitigate the impact of the attacks. We show that machine learning and game theory are two important AI techniques that are complementary to each other for cybersecurity applications. In particular, the incorporation of machine learning models into game theory will put game-theoretic algorithms into practical use. This book comprises foundational techniques and applied methods for developing quantitative cybersecurity models, mitigating risks of exploitable attack surfaces, and designing adaptive and proactive defenses for emerging applications.

References

T. Alpcan and T. Başar. *Network Security: A Decision And Game-Theoretic Approach.* Cambridge University Press, 2010.

C. Badue, R. Guidolini, R. V. Carneiro, P. Azevedo, V. B. Cardoso, A. Forechi, and L. Veronese. Self-driving cars: A survey. *arXiv preprint arXiv:1901.04407*, 2019.

D. S. Berman, A. L. Buczak, J. S. Chavis, and C. L. Corbett. A survey of deep learning methods for cyber security. *Information*, 10(4):122, 2019.

N. Brown and T. Sandholm. Superhuman AI for multiplayer poker. *Science*, 365(6456):885–890, 2019.

J. Ker, L. Wang, J. Rao, and T. Lim. Deep learning applications in medical image analysis. *IEEE Access*, 6:9375–9389, 2017.

J. Lewis. Economic impact of cybercrime, 2018. https://www.csis.org/analysis/economic-impact-cybercrime.

M. H. Manshaei, Q. Zhu, T. Alpcan, T. Başar, and J. P. Hubaux. Game theory meets network security and privacy. *ACM Computing Surveys (CSUR)*, 45(3):1–39, 2013.

J. Pawlick and Q. Zhu. *Game Theory for Cyber Deception: From Theory to Applications.* Springer, 2021.

S. Singh, A. Okun, and A. Jackson. Artificial intelligence: Learning to play Go from scratch. *Nature*, 550(7676):336–337, 2017.

Y. Xin, L. Kong, Z. Liu, Y. Chen, Y. Li, H. Zhu, … and C. Wang. Machine learning and deep learning methods for cybersecurity. *IEEE Access*, 6:35365–35381, 2018.

Part I

Game Theory for Cyber Deception

2

Introduction to Game Theory

Fei Fang[1], Shutian Liu[2], Anjon Basak[3], Quanyan Zhu[2], Christopher D. Kiekintveld[4], and Charles A. Kamhoua[3]

[1] *School of Computer Science and Institute for Software Research, Carnegie Mellon University, Pittsburgh, PA, USA*
[2] *Department of Electrical and Computer Engineering, NYU Tandon School of Engineering, New York University, Brooklyn, NY, USA*
[3] *Network Security Branch, Combat Capabilities Development Command, US Army Research Laboratory, Adelphi, MD, USA*
[4] *Department of Computer Science, The University of Texas at El Paso, El Paso, TX, USA*

2.1 Overview

Game theory is the study of mathematical models of conflict and cooperation between intelligent decision makers (Myerson 1991). Game theory stems from discussions of card games initially (Dimand and Dimand 1996), but it is not a collection of theories about entertaining games. Instead, game theory provides an essential class of models for complex systems with multiple agents or players. It has a profound impact in economics, with several Nobel Memorial Prize laureates in economics winning the prize for their foundational work in game theory. In addition to economics, game theory has wide applications in sociology, psychology, political science, biology, and other fields. Part 2 of this book introduces how game theory can be used to analyze cybersecurity problems and improve cyber defense effectiveness.

A game consists of at least three elements: the set of players, the set of actions for players to take, the payoff function that describes how much payoff or utility each player can get under different states of the world and joint actions chosen by the players. Games can be categorized into different types based on their differences in these elements. Two-player games have been studied much more extensively than the games with three or more players. A game is zero-sum if the players' payoffs always sum up to zero. Simultaneous games are games where all players take actions simultaneously with no further actions to be taken afterward, i.e. the game ends immediately. Even if the players do not move simultaneously, the game can still be viewed as a simultaneous game if the players who move later are completely unaware of the earlier players' actions: they cannot even make any inference about the earlier players' actions from whatever they observe. In contrast, a game is sequential if some players take action after other players, and they have some knowledge about the earlier actions taken by other players.

A game is with complete information if the players have common knowledge of the game being played, including each player's action set and payoff structure. Some games are with incomplete information. For example, in a single-item auction, each player knows how much she values the item and thus her utility function, but not how much other players value it. A game is with perfect information if players take action sequentially, and all other players can observe all actions the players take. A game has perfect recall if each player does not forget any of their previous actions.

Game Theory and Machine Learning for Cyber Security, First Edition.
Edited by Charles A. Kamhoua, Christopher D. Kiekintveld, Fei Fang, and Quanyan Zhu.
© 2021 The Institute of Electrical and Electronics Engineers, Inc. Published 2021 by John Wiley & Sons, Inc.

A game can be described or represented in different forms. The normal form (also called strategic form) is the most fundamental way of representing games. In normal form, each player chooses one action from their action set. The utility function directly maps the joint actions of all players to an N-dimensional real-valued vector, representing the payoff values for each player in an N-player game. It is straightforward to use the normal form to represent simultaneous games when there is no randomness in the environment. For sequential games or games involving randomness in the environment, they can still be represented in normal form through Harsanyi's transformation approach, as we will detail later in Section 2.7. In contrast to normal form, the extensive form explicitly captures the sequencing of players' moves in the game, as well as the players' choices at every decision point, and the possibly imperfect information each player has about the other player's moves when they make a decision. The extensive form is naturally used to represent sequential games. Extensive-form games can also be transformed into normal-form games. However, the transformation loses some of the game structure and leads to a many-to-one matching, i.e. multiple extensive-form games can be transformed into the same normal-form game.

In this chapter, we will mainly discuss noncooperative game theory, a dominant branch of game theory that views individual players as the basic unit of modeling. The name "noncooperative game theory" does not mean that the players never cooperate. Instead, it means that we focus on modeling individual players' actions and payoffs and analyzing the players' joint actions without considering the possibility that some agents form alliances or coalitions and share or transfer their payoff within the coalition. The other branch of game theory that directly models such coalitions is called cooperative game theory or coalitional game theory (Chalkiadakis et al. 2011) and is not the main focus of this chapter.

For the rest of the chapter, we will first introduce two-player zero-sum normal-form games with finite action sets through concrete examples. We will then introduce the formal definition of normal-form games, the commonly used solution concepts such as minmax, maxmin, and Nash equilibrium, and the basic algorithms for computing them. We will turn to the extensive-form games and the corresponding solution concepts such as subgame perfect equilibrium. Following that, we will provide a discussion on Stackelberg games and their applications in security domains. At the end of this chapter, we will formally introduce repeated games, Bayesian games, and stochastic games. We will also discuss some existing works that applies game theory to cybersecurity as we introduce the game models (Do et al. 2017).

2.2 Example Two-Player Zero-Sum Games

We start by discussing the simplest type of games: two-player zero-sum normal-form games with finite action sets. They involve two decision-makers who will each take a single action among a finite set of actions at the same time, and get a payoff based on their joint moves. Furthermore, the two players' payoffs sum up to zero. Thus, one player's gain or loss is the same as the other player's loss or gain in terms of the absolute value. An example of such a game is the classic Rock-Paper-Scissors (RPS) game, whose payoff matrix is shown in Table 2.1. The game has two players. Let us refer to them as the row player (or Player 1) and the column player (or Player 2). Each player can choose among three actions: Rock, Paper, and Scissors. Rock beats Scissors; Scissors beats Paper, and Paper beats Rock. In this payoff matrix, each row shows a possible action for the row player, and each column shows a possible action for the column player. Each entry in the payoff matrix has two numbers, showing the payoff for the two players, respectively, when choosing the actions shown in the corresponding row and column. If the row player chooses Paper and the column player chooses Rock, the row player wins and gets a payoff of 1 while the column player

Table 2.1 Rock-paper-scissors.

	Rock	Paper	Scissors
Rock	0, 0	−1, 1	1, −1
Paper	1, −1	0, 0	−1, 1
Scissors	−1, 1	1, −1	0, 0

loses and gets a payoff of −1 as shown by the numbers in the second row, first column. Since this is a zero-sum game, the second number in each entry is always the negation of the first number. Therefore, in some cases, we only show the payoff value for the row player in each entry in the payoff matrix for two-player zero-sum games.

In the RPS game, it is easy to see that the payoff the row player gets depends on not only the action she chooses,[1] but also the action chosen by the column player. However, the column player's action is not known in advance. The column player will try to choose an action that can lead to a high payoff for him, which depends on the action the row player chooses. This is why we need game theory: we need a framework to reason about such interdependencies.

A natural and important question is, how should the players play the RPS game? Is there an "optimal strategy" for the players? Should we expect the players to play specific strategies? If one is asked to play an RPS game with another person only once, how would this person choose the action to play? One kind of strategy is to stick to a particular action, say Rock, and play it. This kind of strategy is called *pure strategy*. One may also introduce randomness in their choice, and choose an action according to a probability distribution over actions. We call such a strategy a *mixed strategy*. The most straightforward mixed strategy is a uniform random strategy. For RPS, this means one will randomly choose among Rock, Paper, and Scissors when they play the game, and they do not know themselves which action they will choose before they play the game.

To discuss which strategy is the best for a player, we need to consider the other player's strategy and define an evaluation metric for a strategy. If the other player's strategy is known in advance, then the best strategy can be defined as one that maximizes the current player's (expected) utility, which is the (expectation of the) payoff the player can get. Such a strategy is called a *best response* to the other player's strategy. In the RPS game, if the row player chooses a pure strategy of Rock, and this strategy is known to the column player, then the column player can best respond by choosing pure strategy Paper to ensure a payoff of 1, and this is the highest expected utility he can get. If the row player chooses a mixed strategy that selects Rock with probability 0.3 and Paper with probability 0.7, then the column player's best response is the pure strategy Scissors, which leads to an expected utility of $0.4 = 0.3 \times (-1) + 0.7 \times 1$. The −1 here is the second number in the first row, the third column in Table 2.1, the payoff for the column player when the row player chooses Rock (which occurs with probability 0.3) and the column player chooses Scissors. However, in a game like RPS, a player often does not know the other player's strategy ahead of time. One way to evaluate a player's strategy is to discuss its performance in the worst case: what is the expected utility a player can get if the other player chooses an action or strategy that leads to the worst outcome for the first player? When we are concerned about such worse cases, the best strategy would be one that maximizes the player's expected utility in the worst case, and such a strategy is also called the *maxmin strategy*. Assume the row player in RPS chooses Rock,

---

1 For the expository purpose, we will refer to the row player (or the first player) in two-player games with female pronouns and the column player (or the second player) with male pronouns.

Paper, and Scissors with probability p, q, and r respectively ($p + q + r = 1$). She will get an expected utility of $p \times 0 + q \times 1 + r \times (-1) = q - r$ if the column player chooses Rock. Similarly, she will get an expected utility of $r - p$ and $p - q$ if the column player chooses Paper and Scissors, respectively. So the worst-case expected utility for the column player is $\underline{u}_1 = \min\{q - r, r - p, p - q\}$. The maxmin strategy maximized $\underline{u}_1$. Clearly $\underline{u}_1 \leq 0$ and $\underline{u}_1 = 0$ if and only if $p = q = r = 1/3$. Therefore, the row player's maxmin strategy in the RPS game is just the uniform random strategy, choosing each action with probability $1/3$. In addition to the maxmin strategy, one may also consider the *minmax strategy*, a strategy that minimizes the best-case expected utility for the other player. That is, if a player just wants to harm her opponent, what strategy she should choose. In a two-player zero-sum game, one player's loss is the other player's gain, so finding the minmax strategy is equivalent to finding the maxmin strategy. In RPS, the column player's expected utility is $r - q$, $p - r$, and $q - p$ if the column player chooses Rock, Paper, and Scissors, respectively. The minmax strategy aims to minimize $\overline{u}_2 = \max\{r - q, p - r, q - p\}$, and again the minimum value is 0 and is achieved when the row player chooses the uniform random strategy.

The maxmin strategy and minmax strategy describe what strategy one player should play. In game theory, there are other solution concepts that involve strategies of all players. The most commonly used solution concept is Nash equilibrium (NE), named after Dr. John Nash. A strategy profile, i.e. the collection of strategies chosen by each player in a game, is an NE if each player's strategy is a best response to other players' strategies. That means no player can gain from unilateral deviation to a different strategy and has no incentive to change their strategy if other players stick to their strategies. Thus, the strategy profile forms an "equilibrium." In RPS, if the row player chooses a pure strategy Rock and the column player chooses a pure strategy Scissors, it is not a Nash equilibrium since for the column player, his best response to the row player's strategy Rock is Paper. Assuming the row player sticks to the Rock strategy, the column player's utility can increase from -1 to 1 by switching from Scissors to Paper. Therefore, the column player has the incentive to deviate, and the current strategy profile is not an NE. So no pure strategy profile in RPS is an NE since at least one player can deviate from the current strategy to win the game. In contrast, both players choosing a uniform random strategy is an NE in RPS, and both players get an expected utility of 0. If the row player sticks to this uniform random strategy, even if the column player deviates to playing a pure strategy Rock, Paper, or Scissors, his expected utility is still 0 as he wins, loses, and ties each with probability $1/3$. Even if he changes to a mixed strategy other than uniform random, his expected utility is still 0 as the expected utility of playing a mixed strategy is just the weighted average of the expected utilities of playing the pure strategies. Thus, he cannot benefit from changing to another strategy. It is the same for the column player.

In the RPS example, it happens to be the case that the minmax strategy, maxmin strategy, and the strategies in the NE coincide. We will show in the next section that this conclusion generalizes to all finite two-player zero-sum games.

Another example of a two-player zero-sum game is Matching Pennies, whose payoff matrix is shown in Table 2.2 with $d = 1$. If two players both choose Heads or both choose Tails, the row player wins; otherwise, the column player wins. Following a reasoning similar to that in the RPS game, we will see that the minmax, maxmin, and NE strategies of the Matching Pennies game are again

Table 2.2 Matching pennies.

	Heads	Tails
Heads	$d, -d$	$-1, 1$
Tails	$-1, 1$	$1, -1$

the uniform random strategy. Clearly, uniform random strategy is not always the best when $d \neq 1$. If $d = 2$, the maxmin strategy for the row player is the one that choosing Heads with probability $2/5$ and Tails with probability $3/5$.

2.3 Normal-Form Games

In this section, we introduce normal-form games (NFGs) and formally introduce the solution concepts we mentioned in Section 2.2. Most of the notations and definitions are adapted from Leyton-Brown and Shoham 2008. The normal form is the most basic form of games to represent players' interactions and strategy space. An NFG captures all possible combinations of actions or strategies for the players and their corresponding payoffs in a matrix, or multiple matrices for more than two players. A player can choose either a pure strategy that deterministically selects a single strategy or play a mixed strategy that specifies a probability distribution over the pure strategies. The goal for all players is to maximize their expected utility. Formally, a finite, N-person NFG is described by a tuple $(\mathcal{N}, \mathcal{A}, u)$, where:

- $\mathcal{N} = \{1, \ldots, N\}$ is a finite set of N players, indexed by i.
- $\mathcal{A} = \mathcal{A}_1 \times \cdots \times \mathcal{A}_N$ is a set of joint actions of the players, where $\mathcal{A}_i$ is a finite set of actions available to player i. $a = (a_1, \ldots, a_N) \in \mathcal{A}$ is called an action profile with $a_i \in \mathcal{A}_i$.
- $u = (u_1, \ldots, u_N)$ where $u_i : \mathcal{A} \mapsto \mathbb{R}$ is a utility (or payoff) function for player i. It maps an action profile a to a real value. An important characteristic of a game is that player i's utility depends on not only his own action a_i but also the actions taken by other players; thus, the utility function is defined over the space of $\mathcal{A}$ instead of $\mathcal{A}_i$.

A player can choose a mixed strategy. We use $S_i = \Delta^{|\mathcal{A}_i|}$ to denote the set of mixed strategies for player i, which is the probability simplex with dimension $|\mathcal{A}_i|$. Similarly, $S = S_1 \times \cdots \times S_N$ is the set of joint strategies and $s = (s_1, \ldots, s_N) \in S$ is called a strategy profile. The support of a mixed strategy is defined as the set of actions that are chosen with a nonzero probability. An action of player i is a pure strategy and can be represented by a probability distribution with support size 1 (of value 1 in one dimension and 0 in all other dimensions). The utility function can be extended to mixed strategies by using expected utility. That is, if we use $s_i(a_i)$ to represent the probability of choosing action a_i in strategy s_i, the expected utility for player i given strategy profile s is $u_i(s) = \sum_{a \in \mathcal{A}} u_i(a) \prod_{i'=1}^{n} s_{i'}(a_{i'})$. A game is zero-sum if the utilities of all the players always sum up to zero, i.e. $\sum_i u_i(s) = 0, \forall s$ and is nonzero-sum or general-sum otherwise.

Many classic games can be represented in normal form. Table 2.3 shows the game Prisoner's Dilemma (PD). Each player can choose between two actions, Cooperate (C) and Defect (D). If they both choose C, they both suffer a small loss of -1. If they both choose D, they both suffer a big loss of -2. However, if one chooses C and the other chooses D, the one who chooses C suffers a huge loss while the other one does not suffer any loss. If the row player chooses a mixed strategy of playing C with probability 0.4 and D 0.6, while the column player chooses the uniform random strategy, then the row player's expected utility is $-1.4 = (-1) \cdot 0.4 \cdot 0.5 + (-3) \cdot 0.4 \cdot 0.5 + (-2) \cdot 0.6 \cdot 0.5$.

Now we provide the formal definition of best response. We use $-i$ to denote all players but i.

Table 2.3 Prisoner's dilemma.

	Cooperate	Defect
Cooperate	$-1, -1$	$-3, 0$
Defect	$0, -3$	$-2, -2$

Definition 2.1 *(Best Response)*: Player i's best response to other players $(-i)$'s strategy profile s_{-i} is a strategy $s_i^* \in S_i$ such that $u_i(s_i^*, s_{-i}) \geq u_i(s_i, s_{-i})$ for all strategies $s_i \in S_i$.

The best response can be a pure strategy or a mixed strategy. There can be more than one best response to a strategy profile s_{-i}. We use $BR(s_{-i})$ to denote the set of best responses. For example, in the RPS game, if the row player chooses the uniform random strategy, the column player will get an expected utility of zero no matter what strategy he chooses. Thus, the set of best responses is the entire mixed strategy space for the column player, and it has infinite size. As the following lemma indicates, if there is more than one pure strategy in the best response set, then any probability distribution over these best responding pure strategies is still a best response.

Lemma 2.1 *(Adapted from Nash 1951)*: A mixed strategy is a best response to other players' strategy profile s_{-i} if and only if all actions in the support are best responses to s_{-i}, i.e. $s_i \in BR(s_{-i})$ if and only if $\forall a_i : s_i(a_i) > 0, a_i \in BR(s_{-i})$.

In the PD game, at first sight, it seems that both players should choose C and suffer a small loss together. However, when one player chooses C, the other player's best response is D. This is the unique best response. Interestingly, if the one player chooses D, the other player's best response is still D. In a strategy profile (C,C), both players have the incentive to deviate to strategy D to increase their utility.

2.3.1 Solution Concepts

We will now introduce the formal definition of maxmin, minmax, and Nash equilibrium.

Definition 2.2 *(Maxmin Strategy)*: The maxmin strategy for player i is a strategy that maximizes the worst-case expected utility for player i, i.e. $\arg\max_{s_i} \min_{s_{-i}} u_i(s_i, s_{-i})$.

Definition 2.3 *(Minmax Strategy)*: The minmax strategy for player i against player i' is the strategy player i uses when she coordinates with all other players except player i' to minimize the best possible expected utility player i's can get, i.e. player i's component in the strategy profile $\arg\min_{s_{-i'}} \max_{s_{i'}} u_{i'}(s_{-i'}, s_{i'})$.

The minmax strategy is defined in this way because when there are more than two players, a player's utility depends on the strategy used by all other players. When there are only two players, the minmax strategy defined in Definition 2.3 becomes $\arg\min_{s_i} \max_{s_{-i}} u_{-i}(s_i, s_{-i})$. The maximum and minimum value achieved by maxmin and minmax strategies is called the maxmin value and minmax value for the player. Similar to best responses, there can be more than one maxmin and minmax strategy for a player.

Definition 2.4 *(Nash Equilibrium)*: A strategy profile $s = (s_1, \ldots, s_N)$ is a Nash equilibrium if $s_i \in BR(s_{-i}), \forall i$.

There can be multiple NEs in a game, and the strategies in an NE can either be a pure strategy or a mixed strategy. If all players use pure strategies in an NE, we call it a pure strategy NE (PSNE). If at least one player randomly chooses between multiple actions, i.e. the support set of their strategy has a size larger than 1, we call it a mixed strategy NE.

In the PD game, the strategy profile (D,D) is an NE because when one player chooses D, the other player's best response is D. Thus, both players are best responding to the other player's strategy.

Table 2.4 Football vs. concert.

	Football	Concert
Football	2, 1	0, 0
Concert	0, 0	1, 2

This NE may be counterintuitive as the players' rational choices lead to a bad outcome. This is also why the PD game becomes an important example in game theory.

In the game shown in Table 2.4, when both players choose Football (F) or Concert (C), they both get positive utility. The row player prefers F while the column player prefers C. There are two PSNEs in this game, (F,F) and (C,C). Besides, there is a mixed strategy NE, where the row player chooses F with probability 2/3 and C with probability 1/3, and the column player chooses F with probability 1/3 and C with probability 2/3. It is easy to check that these are indeed NEs. For (F,F) and (C,C), if any player deviates to a different pure strategy, they will get a utility of 0, and thus it is not profitable to deviate. For the mixed strategy NE, given the row player's mixed strategy, the column player's expected utility for choosing a pure strategy F is $2/3 = 2/3 \cdot 1 + 1/3 \cdot 0$ and the expected utility for the column player choosing C is again $2/3 = 1/3 \cdot 2 + 2/3 \cdot 0$. Thus, when the column player randomly chooses between F and C, his expected utility is still 2/3, and any strategy of the column player is a best response to the row player's strategy. Similarly, the row player's best response set to the column player's mixed strategy is also the entire mixed strategy space. So these two mixed strategies form an NE.

PSNEs are not guaranteed to exist, but there exists at least one NE in any finite game, as stated in the following theorem.

Theorem 2.1 *(Theorem 1 in (Nash 1951))*: Every finite game (i.e. a game with $N < \infty$ and $|A| < \infty$) has an NE.

The proof uses Brouwer's fixed point theorem, which states that there is a fixed point for any continuous function mapping a compact convex set to itself. We refer the interested readers to the original paper (Nash 1951).

Importantly, for two-player zero-sum games with finite action sets, the minmax strategy and maxmin strategy of individual players coincide with the players' strategies in NE, as stated by the following Minimax theorem.

Theorem 2.2 *(Minimax theorem, Adapted from Neumann 1928 and Nash 1951)*: In two-player zero-sum games, $\max_{s_i} \min_{s_{-i}} u_i(s_i, s_{-i}) = \min_{s_{-i}} \max_{s_i} u_i(s_i, s_{-i})$. Further, denote this optimal value as v^* and let S_i^* and S_{-i}^* be the set of strategies that achieves v^* on the left-hand side and right-hand side of the equation, respectively. The set of NE is $\{(s_i^*, s_{-i}^*) : s_i^* \in S_i^*, s_{-i}^* \in S_{-i}^*\}$, and all NEs lead to the same utility profile $(v^*, -v^*)$.

v^* is also called the value of the game.

There are many other solution concepts, and we briefly introduce some of them here. Dominant strategy solution prescribes that each player i plays a dominant strategy, which is a strategy s_i that is better than any other strategy of player i regardless of other players' strategies, i.e. $u_i(s_i, s_{-i}) > u_i(s_i', s_{-i}), \forall s_{-i} \in S_{-i}, s_i' \in S_i \backslash s_i$. In the PD game, (D,D) is a dominant strategy solution. However, the dominant strategy solution does not always exist. There is no dominant strategy solution in the

RPS game or the Football vs. Concert game. Approximate Nash equilibrium (also called ε-Nash equilibrium) is a generalization of NE. It relaxes the requirement on the best response to that any player cannot gain more than ε if they deviate. Quantal response equilibrium (QRE) (McKelvey and Palfrey 1995) is a smoothed version of NE and is defined based on a noisy best-response function for the players. Instead of requiring that every player best responds to other players' strategy, QRE requires that every player's strategy satisfies that the probability of choosing the jth pure strategy is proportional to the exponential of λ times its corresponding expected utility given the other players' strategies:

$$s_i(j) = \frac{e^{\lambda u_i(j, s_{-i})}}{\sum_k e^{\lambda u_i(k, s_{-i})}} \tag{2.1}$$

This means each player chooses actions following a softmax function on the expected utilities after applying a multiplier λ. The definition describes that the player chooses an action with higher expected utility with higher probability. Still, every single action will be chosen with some nonzero probability, no matter how bad a utility the action leads to. This can be viewed as a noisy version of the best response. When λ gets very large, the probability of choosing nonbest-response actions gets very small, and the QRE gets closer to an NE.

The computation of NEs is challenging. It has been proved that finding an NE is PPAD (Polynomial Parity Arguments on Directed graphs)-complete (Chen and Deng 2006, 2005; Daskalakis et al. 2009). This complexity result means that the problem belongs to a complexity class that is a superset of class P (whose solution can be found in polynomial time) and a subset of class NP (whose solution can be verified in polynomial time). This indicates that it is not promising to find a polynomial-time algorithm to find NEs in general, even in two-player games.

There are subclasses of games in which the computation of NE is computationally tractable. For example, the NE in two-player zero-sum games can be computed in polynomial time through linear programming (LP), i.e. formulating an optimization problem with linear objective and linear constraints (Bradley et al. 1977). As mentioned in Theorem 2.2, NE strategies coincide with minmax and maxmin strategies in these games. So we will first introduce the LP-based approach to compute minmax strategy. Let U_{jk}^i be the utility for Player i when Player 1 chooses the jth action in her action set and Player 2 chooses the kth action in his action set, which can be read from the payoff matrix directly. Let $x \in \mathbb{R}^{|A_1|}$ with the value in the jth dimension x_j being the probability that Player 1 chooses the jth action in a minmax strategy. To determine the value of x_j, we build the following LP:

$$\min_{x, v} \quad v \tag{2.2}$$

$$v \geq \sum_{j=1}^{|A_1|} x_j U_{jk}^2, \forall k \in \{1 \ldots |A_2|\} \tag{2.3}$$

$$\sum_{j=1}^{|A_1|} x_j = 1 \tag{2.4}$$

$$x_j \in [0, 1], \forall j \in \{1 \ldots |A_1|\} \tag{2.5}$$

v is an auxiliary variable. Constraints (2.4)–(2.5) ensure that x is a probability distribution. Constraint (2.3) restricts v to be greater than or equal to the expected utility for Player 2 when he is taking any pure strategy. Therefore, a feasible v will be greater than or equal to the maximum expected utility Player 2 can get when Player 1 is using a mixed strategy defined by x. Since the objective is to minimize v, the optimal value of v is achieved when $v = \max_k \sum_{j=1}^{|A_1|} x_j U_{jk}^2$, which is exactly Player 2s maximum expected utility. Thus, the optimal solution of this LP (x^*, v^*) provides the minmax strategy for Player 1 against Player 2 (x^*) as well as the value of the game $(-v^*)$. The LP can be solved in polynomial time in theory using the ellipsoid method (Bland et al. 1981) although other algorithms such as the simplex method (Nelder and Mead 1965) and primal-dual interior point method (Mehrotra 1992) often have superior empirical performance. In practice, commercial solvers such as IBM ILOG CPLEX and Gurobi have integrated many advanced techniques, and can solve large-scale LPs efficiently. The maxmin strategy of Player 1 can be found in a similar way by solving the following LP.

$$\max_{x,v} \quad v \tag{2.6}$$

$$v \le \sum_{j=1}^{|A_1|} x_j U^1_{jk}, \forall k \in \{1 \dots |A_2|\} \tag{2.7}$$

$$\sum_{j=1}^{|A_1|} x_j = 1 \tag{2.8}$$

$$x_j \in [0,1], \forall j \in \{1 \dots |A_1|\} \tag{2.9}$$

Here, the optimal value of v is equal to the expected utility Player 1 can get in the worst case.

For general NFGs, PSNEs can be found through enumerating all the action profiles and checking if the individual player's potential gain is positive if they deviate to a different action. For mixed strategy NE, a straightforward approach is to enumerate the support of the players' strategies in the NE. According to the definition of NE, each player's strategy is a best response to other players' strategies. The support of a strategy consists of all the actions chosen with nonzero probability. Thus, according to Lemma 2.1, every single action in the support has to be a best response. Therefore, given the other player's strategy, taking any of the actions in the support deterministically should lead to the same expected utility since otherwise, they cannot all be best responses. Also, the expected utility of taking an action in the support set should not be lower than the expected utility of taking any action that is not in the support. These two claims give us a set of linear equations and inequality constraints. The NE can be found by first solving the system of linear equations and then checking if the solution satisfies the inequality constraints. Alternatively, we can again take into account all the equations and inequalities through LP.

We illustrate this method by analyzing the Football vs. Concert game shown in Table 2.4. We enumerate the support of the players. First, assume the support of each player's strategy contains only one element, i.e. both players are playing pure strategies. There are four possible support combinations. It is easy to check that both players playing F or C are NEs as the utility for taking the action in the support is positive and unilaterally changing to the other action, which is not in the support, will lead to a utility of zero. Second, assume the support of both players to be of size 2, meaning both players choose F and C with nonzero probability. In this case, we can find the NE by solving the LP shown below.

$$\max_{x,y} \quad 1 \tag{2.10}$$

$$2 \cdot y_F + 0 \cdot y_C = 0 \cdot y_F + 1 \cdot y_C \tag{2.11}$$

$$1 \cdot x_F + 0 \cdot x_C = 0 \cdot x_F + 2 \cdot x_C \tag{2.12}$$

$$x_F + x_C = 1 \tag{2.13}$$

$$y_F + y_C = 1 \tag{2.14}$$

$$x_F, x_C, y_F, y_C \in [0,1] \tag{2.15}$$

x_F, x_C, y_F, y_C are the decision variables describing the probability of Players 1 and 2 choosing action Football and Concert, respectively. Constraint (2.11) means the expected utility for Player 1 taking F and C should be equal given that Player 2 is playing a mixed strategy y, i.e. $u_1(F, y) = u_1(C, y)$. Otherwise, if $u_1(F, y) > u_1(C, y)$, C will not be a best response, and should not be chosen at all in an NE, violating our assumption that Player 1 chooses both F and C with nonzero probability. Similarly, Constraint (2.12) means $u_2(x, F) = u_2(x, C)$. The next two constraints ensure that x and y are indeed a probability distribution. There are no inequalities in this LP because, in our assumption, the support is the whole action set, and there is no action outside of the support. In larger games, when the support we assume does not contain all the actions, we need to add inequalities to ensure $u_1(j, y) \geq u_1(k, y)$ if j is an action in the support and k is not. The solution to this LP is $x_F = y_C = 2/3$, $x_C = y_F = 1/3$, and this is a mixed strategy NE for the game. It is impossible to have different support sizes for the two players in this game because if one player chooses a pure strategy, the other player's utilities when choosing different actions cannot be the same according to the payoff matrix. Thus, this game only has three NEs, including two PSNEs and one mixed strategy NE.

The support enumeration method is very inefficient in finding the NEs. A better algorithm is the Lemke-Howson algorithm (Lemke and Howson 1964; Rosenmüller 1971), which relies on a Linear Complementarity formulation.

In practice, the NE of a small game can be computed using the software package Gambit (McKelvey et al. 2006). There are several different solvers for finding NEs in this toolkit. For example, one solver is based on the QRE and gradually reduces the "noise" part in the quantal response equation (Eq. 2.1).

2.4 Extensive-Form Games

Extensive-form games (EFGs) represent the sequential interaction of players using a rooted game tree. Figure 2.1 shows a simple example game tree. Each node in the tree belongs to one of the players and corresponds to a decision point for that player. Outgoing edges from a node represent actions that the corresponding player can take. The game starts from the root node, with the player corresponding to the root node taking an action first. The chosen action brings the game to the child node, and the corresponding player at the child node takes an action. The game continues until it reaches a leaf node (also called a terminal node), i.e. each leaf node in the game tree is a possible end state of the game. Each leaf node is associated with a tuple of utilities or payoffs that the players will receive when the game ends in that state. In the example in Figure 2.1, there are three nodes. Node 1 belongs to Player 1 (P1), and nodes 2 and 3 belong to Player 2 (P2). Player 1 first chooses between action L and R, and then Player 2 chooses between action l and r. Player 1's highest utility is achieved when Player 1 chooses L, and Player 2 chooses l.

There is sometimes a special fictitious player called Chance (or Nature), who takes an action according to a predefined probability distribution. This player represents the stochasticity in many problems. For example, in the game of Poker, each player gets a few cards that are randomly dealt. This can be represented by having a Chance player taking an action of dealing cards. Unlike the

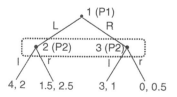

Figure 2.1 An example extensive-form game.

other real players, the Chance player does not rationally choose an action to maximize his utility since he does not have a utility function. In the special case where Nature only takes an action at the very beginning of the game, i.e. the root of the game tree, the game is essentially a Bayesian game, as we will detail in Section 2.7.

In a perfect information game, every player can perfectly observe the action taken by players in the previous decision points. For example, when two players are playing the classic board game of Go or Tic-Tac-Toe, each player can observe the other players' previous moves before she decides her move. However, it is not the case in many other problems. An EFG can also capture imperfect information, i.e. a game where players are sometimes uncertain about the actions taken by other players and thus do not know which node they are at exactly when they take actions. The set of nodes belonging to each player is partitioned into several information sets. The nodes in the same information set cannot be distinguished by the player that owns those nodes. In other words, the player knows that she is at one of the nodes that belong to the same information set, but does not know which one exactly. For example, in a game of Poker where each player has private cards, a player cannot distinguish between certain nodes that only differ in the other players' private cards. Nodes in the same information set must have the same set of actions since otherwise, a player can distinguish them by checking the action set. It is possible that an information set only contains one node, i.e. a singleton. If all information sets are singletons, the game is a perfect information game. The strategy of a player specifies what action to take at each information set. In the example game in Figure 2.1, the dashed box indicates that nodes 2 and 3 are in the same information set, and Player 2 cannot distinguish between them. Thus, nodes 2 and 3 have the same action set. This information set effectively makes the example game a simultaneous game as Player 2 has no information about Player 1s previous actions when he makes a move.

Given the definition of the information set, we can interpret the perfect-recall property as that for any two nodes in a same information set, the paths from the root to them consist of the same sequence of actions for the acting player at the information set. In a game with perfect information and perfect recall, every player is aware of which node she is at accurately and can recover the whole path from the root to the current node when it is her turn to take an action.

It is possible to convert an EFG to an NFG. In the constructed NFG, a pure strategy for Player i specifies an action for every information set that belongs to her. The resulting payoff at one entry of the payoff matrix of the constructed NFG can be found by checking at which terminal node the game will end, given all the players' pure strategies. It is possible that some of the information sets will not be reached at all although the players have specified an action for them. This simple representation often has many redundancies: some information sets are not reachable due to the actions taken by the acting player at a node higher up in the tree. Omitting these redundancies leads to the reduced-normal-form strategies, which is equivalent to the constructed normal-form strategies (Shoham and Leyton-Brown 2008) in terms of expected utility and equilibrium. The constructed game is very large, with the number of rows and columns exponential in the number of nodes in the game tree of the EFG.

A player can also play a mixed strategy in EFGs. There are two ways to represent a mixed strategy, and they are equivalent in terms of expected utility in games with perfect recall. The first way is to describe a probability distribution over actions for the actions at each information set. The players randomly sample an action at each information set independently. This is also called the behavioral strategy. The second way is to describe the mixed strategy as a probability distribution over pure strategies in the corresponding NFG.

Formally, an EFG is defined by a tuple $(\mathcal{N}, \mathcal{A}, \mathcal{H}, \mathcal{Z}, \chi, \rho, \sigma, u)$ where

- $\mathcal{N}$ is the set of players.
- $\mathcal{A}$ is the set of actions.

- $\mathcal{H}$ is the set of nonterminal nodes in the game tree, also known as histories as each node corresponds to a history of actions taken by the players.
- $\chi : \mathcal{H} \mapsto \{0,1\}^{|\mathcal{A}|}$. It is a function that specifies the set of actions available at each node.
- $\rho : \mathcal{H} \mapsto \mathcal{N}$. It is a function that specifies the acting player at each node.
- $\mathcal{Z}$ is the set of terminal nodes.
- $\sigma : \mathcal{H} \times \mathcal{A} \mapsto \mathcal{H} \cup \mathcal{Z}$. It is the successor function that specifies the successor node when the acting player takes an action at a node.
- $u = (u_1, \dots, u_N)$ where $u_i : \mathcal{Z} \mapsto \mathbb{R}$ is the utility function for player i.

We use $u_i(z)$ to denote the utility for player i if the game terminates at $z \in \mathcal{Z}$.

2.4.1 Solution Concepts

Since EFGs can be converted into NFGs, the solution concepts introduced in Section 2.3.1 still apply. Also, there are other solution concepts specifically for EFGs. We briefly introduce some of them at a high level.

One solution concept is subgame perfect equilibrium (or subgame perfect Nash equilibrium, SPE in short). It is a refinement of NE. This means the set of SPE is a subset of NE. A strategy profile is an SPE equilibrium if, in every subgame of the original game, the strategies remain an NE. A subgame is defined as the partial tree consisting of a node and all its successors, with the requirement that the root node of the partial tree is the only node in its information set. This means if the players play a small game that is part of the original game, they still have no incentive to deviate from their current strategy. A subgame can be a subgame of another subgame rooted at its ancestor node. A finite EFG with perfect recall always has an SPE. To find an SPE in a finite game with perfect information, one can use backward induction as all the information sets are singletons. First, consider the smallest subgames rooted at a parent node of a terminal node. After solving these subgames by determining the action to maximize the acting player's expected utility, one can solve a slightly larger subgame whose subgames have already been solved, assuming the player will play according to the smaller subgames' solution. This process continues until the original game is solved. The resulting strategy is an SPE. For games of imperfect or incomplete information, backward induction cannot be applied as it will require reasoning about the information sets with more than one node. The SPE concept can also be defined through the one-shot deviation principle: a strategy profile is an SPE if and only if no player can gain any utility by deviating from their strategy for just one decision point and then reverting back to their strategy in any subgame.

In the example game in Figure 2.1, if we remove the dashed box, the game becomes one with perfect information, and it has two subgames in addition to the original game, with the root nodes being nodes 2 and 3 respectively. In this new game, Player 1 choosing L and Player 2 choosing r at both nodes 2 and 3 is an NE but not an SPE, because, in the subgame rooted at node 3, Player 2 choosing r is not the best action. In contrast, Player 1 choosing R and Player 2 choosing r at node 2, l at node 3 is an SPE.

Another solution concept is sequential equilibrium (Kreps and Wilson 1982), which is a further refinement of SPE. A sequential equilibrium consists of not only the players' strategies but also the players' beliefs for each information set. The belief describes a probability distribution on the nodes in the information set since the acting player cannot distinguish them when they play the game. Unlike SPE which only considers the subgames rooted from a node that is the only element in its information set, the sequential equilibrium requires that the players are sequentially rational and takes the best action in terms of expected utility concerning the belief at every information set even if it is not a singleton.

2.5   Stackelberg Game

Consider a two-player NFG with finite actions. The two players choose an action in their action set simultaneously without knowing the other player's strategy ahead of time. In an NE of the game, each player's strategy is a best response to the other player's strategy. But what if one player's strategy is always known to other players ahead of time? This kind of role asymmetry is depicted in a Stackelberg game.

In a Stackelberg game, one player is the leader and chooses her strategy first, and the other players are followers who observe the leader's strategy and then choose their strategies. The leader has to commit to the strategy she chooses, i.e. she cannot announce a strategy to the followers and play a different strategy when the game is actually played. She can commit to either a pure strategy or a mixed strategy. If she commits to a mixed strategy, the followers can observe her mixed strategy but not the realization of sampled action from this mixed strategy. A Stackelberg game can be described in the same way as before, with a tuple $(\mathcal{N}, \mathcal{A}, u)$ for games in normal form and $(\mathcal{N}, \mathcal{A}, \mathcal{H}, \mathcal{Z}, \chi, \rho, \sigma, u)$ for games in extensive form, with the only difference that one of the players (usually Player 1) is assumed to be the leader.

At first sight, one may think that the followers have an advantage as they have more information than the leader when they choose their strategies. However, it is not the case. Consider the Football vs. Concert game again. If the row player is the leader and commits to choosing the action F, a rational column player will choose action F to ensure a positive payoff. Thus, the row player guarantees herself a utility of 2, the highest utility she can get in this game, through committing to playing pure strategy F. In the real world, it is common to see similar scenarios. When a couple is deciding what to do during the weekend, if one of them has a stronger opinion and insists on going to the activity he or she prefers, they will likely choose that activity in the end. In this example game, the leader ensures a utility that equals the best possible utility she can get in an NE. In some other games, the leader can get a utility higher than any NE in the game by committing to a good strategy. In the example game in Table 2.5 (Up, Right) is the only NE, with the row player's utility being 4. However, if the row player is the leader and commits to a uniform random strategy, the column player will choose Left as a best response, yielding an expected utility of 5 for the row player. This higher expected utility for the row players shows the power of commitment.

2.5.1   Solution Concept

As shown in the example, NEs are no longer well suited to prescribe the behavior of players in a Stackelberg game. Instead, the concept of Stackelberg equilibrium is used. A Stackelberg equilibrium in a two-player Stackelberg game is a strategy profile where the follower best responds to the leader's strategy, and the leader commits to a strategy that can maximize her expected utility, knowing that the follower will best respond. One subtle part of this definition is the tie-breaking rule. There can be multiple best responses. Although they lead to the same expected utility for the follower, they can result in different expected utility for the leader. Therefore, to define a Stackelberg equilibrium, we need to specify a best response function that maps a leader's strategy to the follower's best response strategy.

Table 2.5   Stackelberg game example.

	Left	Right
Up	6, 0	4, 1
Down	4, 3	1, 0

Definition 2.5 *(Best Response Function):* $f : S_1 \mapsto S_2$ is a best response function if and only if $u_2(s_1, f(s_1)) \geq u_2(s_1, s_2), \forall s_1 \in S_1, s_2 \in S_2$

Since the expected utility of playing a mixed strategy is a linear combination of the expected utility of playing the pure strategies in its support, the constraint is equivalent to $u_2(s_1, f(s_1)) \geq u_2(s_1, a_2), \forall s_1 \in S_1, a_2 \in A_2$. The Stackelberg equilibrium is defined based on the best response function definition.

Definition 2.6 *(Stackelberg Equilibrium):* A strategy profile $s = (s_1, f(s_1))$ is a Stackelberg equilibrium if f is a best response function and $u_1(s_1, f(s_1)) \geq u_1(s_1', f(s_1')), \forall s_1' \in S_1$.

For some best response functions, the corresponding Stackelberg equilibrium may not exist. A commonly used best response function is one that breaks ties in favor of the leader. The resulting equilibrium is called the strong Stackelberg equilibrium (SSE).

Definition 2.7 *(Strong Stackelberg Equilibrium (SSE)):* A strategy profile $s = (s_1, f(s_1))$ is a strong Stackelberg equilibrium if

- $u_2(s_1, f(s_1)) \geq u_2(s_1, s_2), \forall s_1 \in S_1, s_2 \in S_2$ (attacker best responds)
- $u_1(s_1, f(s_1)) \geq u_1(s_1, s_2), \forall s_1 \in S_1, s_2 \in BR(s_1)$ (tie-breaking is in favor of the leader)
- $u_1(s_1, f(s_1)) \geq u_1(s_1', f(s_1')), \forall s_1' \in S_1$ (leader maximizes her expected utility)

SSE is guaranteed to exist in two-player finite games. The weak Stackelberg equilibrium (WSE) can be defined in a similar way, with a tie-breaking rule against the leader. However, WSE may not exist in some games. SSE in two-player normal-form Stackelberg games can be computed in polynomial time through solving multiple linear programs, each corresponding to a possible best response of the follower (Conitzer and Sandholm 2006).

2.5.2 Stackelberg Security Games

A special class of Stackelberg games is the Stackelberg security game (SSG). The SSG model has wide applications in security-related domains. It has led to several application tools successfully deployed in the field, including ARMOR for setting up checkpoints and scheduling canine patrols at the Los Angeles International Airport, PROTECT for protecting ports of New York and Boston (Tambe 2011), and PAWS for protecting wildlife from poaching (Fang et al. 2016).

An SSG (Kiekintveld et al. 2009; Paruchuri et al. 2008) is a two-player game between a defender and an attacker. The defender must protect a set of T targets from the attacker. The defender tries to prevent attacks using M defender resources. In the simplest case, each defender resource (which corresponds to a patroller, a checkpoint, a canine team, or a ranger in security-related domains) can be allocated to protect one target and $M < T$. Thus, a pure strategy for the defender is an assignment of the M resources to M targets and a pure strategy for the adversary is a target $k \in \{1 \dots T\}$ to be attacked. In a slightly more complex setting, each defender resource can be assigned to a patrol schedule, which covers a subset of targets, and it is impossible to cover all the targets with limited resources. Denote the jth defender pure strategy as B_j, an assignment of all the security resources. B_j is represented as a column vector $B_j = \langle B_{jk} \rangle^T$, where $B_{jk} \in \{0, 1\}$ indicates whether target k is covered by B_j. For example, in a game with four targets and two resources, each of which can cover one target, $B_j = \langle 1, 1, 0, 0 \rangle$ represents the pure strategy of assigning one resource to target 1 and another to target 2. Each target $k \in \{1 \dots T\}$ is assigned a set of payoffs $\{P_k^a, R_k^a, P_k^d, R_k^d\}$: If an attacker attacks target k and it is protected by a defender resource, the attacker gets utility P_k^a and the defender gets utility R_k^d. If target k is not protected, the attacker gets utility R_k^a and the defender gets utility P_k^d. In order to be a valid security game, it must hold that $R_k^a > P_k^d$ and $R_k^d > P_k^d$, i.e. when the attacked target is covered, the defender gets a utility higher than that when it is not covered, and the opposite holds for the attacker. The defender is the leader and commits to a strategy first. The attacker can observe the defender's strategy before selecting their strategy. This assumption reflects what happens in the real world in many cases. For example, in the airport security domain, the airport security team sends canine teams to protect the terminals every day. A dedicated attacker may be able to conduct surveillance to learn the defender's strategy.

Many works in SSGs focus on the SSE. In an SSE, the defender selects an optimal strategy based on the assumption that the attacker will choose a best response, breaking ties in favor of the defender. The defender's strategy in an SSE is the optimal resource allocation strategy for the defender, given that the attackers are perfectly rational and best respond to the defender's strategy. The optimal strategy is usually a mixed strategy, which is a distribution over the set of pure defender strategies, as the adversary would easily exploit any deterministic defender strategy. The defender's mixed strategy can be represented as a vector $b = \langle b_j \rangle$, where $b_j \in [0, 1]$ is the probability of choosing pure strategy B_j. There is also a more compact marginal representation for defender strategies. Let $x = \langle x_k \rangle$ be the marginal strategy, where $x_k = \sum_j b_j B_{jk}$ is the probability that target k is covered. The defender's strategy in the SSE can be represented using either a mixed strategy b or a marginal strategy x.

There have been many algorithms and models developed to solve SSGs, including DOBSS (Paruchuri et al. 2008), which solves SSGs through mixed-integer linear programming, ORIGAMI (Kiekintveld et al. 2009), which provides a polynomial-time algorithm for SSGs with each patrol schedule only covering one target.

One of the strongest assumptions in classic game theory is that the players are perfectly rational utility maximizers. However, it is well-understood that human beings are boundedly rational. Thus, the attackers may not always choose to attack the target with the highest expected utility, i.e. best respond to the defender's strategy. Incorporating human behavioral models such as quantal response (QR) as defined in Eq. (2.1) into security games has been demonstrated to improve the performance of defender patrol strategies in both simulations and human subject experiments (Nguyen et al. 2013; Pita et al. 2010; Yang et al. 2012). Instead of choosing the action with the highest utility, the QR model predicts a probability distribution over adversary actions where actions with higher utility have a greater chance of being chosen. Nguyen et al. (2013) extended the QR model by proposing that humans use "subjective utility," a weighted linear combination of features (such as defender coverage, adversary reward, and adversary penalty) to make decisions. This subjective utility quantal response (SUQR) model was shown to outperform QR in human subject experiments. In this model, an attacker's choice is based on an evaluation of key properties of each target, including the coverage probability, the reward, and the penalty, represented by the parameter vector $\omega = (\omega_1, \omega_2, \omega_3)$. Given the defender's marginal strategy x, the probability that an attacker with parameter ω attacks target k is

$$s_2(k) = \frac{e^{\omega_1 x_k + \omega_2 R_k^a + \omega_3 P_k^a}}{\sum_j e^{\omega_1 x_k + \omega_2 R_k^a + \omega_3 P_k^a}} \tag{2.16}$$

When there are multiple attackers, the Bayesian SUQR model Yang et al. (2014) is proposed, which is an extension of the SUQR model. It captures the heterogeneity of a group of attackers and assumes that different attackers have different parameters.

2.5.3 Applications in Cybersecurity

Several works have used the Stackelberg game model to formulate problems in cybersecurity. Schlenker et al. (2018) and Thakoor et al. (2019) model the cyber deception problem as a Stackelberg game between the defender and the attacker. Similar to SSGs, the defender is protecting a set of targets from the attacker. In contrast to SSGs where the defender allocates resources to stop any potential attack on a subset of targets, in this work, the defender can choose to set up honeypots or provide a camouflage of existing machines in a network. By such deception techniques, the defender can make the honeypots appear to be important, or value targets appear unimportant. Thus, she can induce the attacker to attack a unimportant to the defender. Cranford et al. (2018) focuses on human behavior in cyber deception games. Schlenker et al. (2017) proposes a Stackelberg game model for the problem of allocating limited human limited human resources to investigate cybersecurity alerts. Stackelberg game models have also been used to model defense in cyber-physical systems (Wang et al. 2018; Yuan et al. 2016; Zhu and Martinez 2011). Li and Vorobeychik (2014, 2015) considers the detection and filtering of spam and phishing emails and formulate the problem as a Stackelberg game where an intelligent attacker will try to adapt to the defense strategy.

2.6 Repeated Games

A repeated game is an EFG consisting of repetitions of a stage game. The stage game can be an NFG. For example, the repeated PD game is a game where the players repeatedly play the stage game of a PD game for multiple rounds. Also, we often see people playing the RPS game for three or five consecutive rounds in practice. A finitely repeated game has a finite number of rounds, and a player's final payoff is the sum of their payoffs in each round. Infinitely repeated games last forever; there can be a discount factor $\delta \in (0, 1)$ for future rounds when computing the final payoff.

The repetition of the stage game enables a much larger strategy space for the players comparing with the strategy space of the stage game. A player's action in a round can be based on the players' actions in the previous rounds. For example, in the repeated PD game, a player can use a Tit-for-Tat strategy, meaning that the player always chooses the same action as the one taken by the other player in the previous round of the game except for the first round where the player chooses Cooperate. In other words, if her opponent chose Cooperate in the last round, she will reward the other player's kindness by choosing Cooperate in this round. Otherwise, she will penalize the other player by choosing Defect. A player can also choose a grim trigger strategy: she chooses Cooperate in the first round and continues to choose Cooperate if the other player never chooses Defect, but penalizes the other player by always choosing Defect for the rest of the game right after the other player chooses Defect for the first time.

In fact, both players playing the grim trigger strategy is an NE in an infinitely repeated PD game, as a player who deviates from it will get a utility of 0 in the round he deviates but at most −2 for each round in the rest of the game as the other player will always choose Defect. Both players choosing Tit-for-Tat is also an NE with a large enough δ. If the player deviates by choosing Defect for $k - 1$ consecutive rounds starting from round t and then chooses Cooperate again, he can get at most a total utility of $0 + \delta \times (-2) + \delta^2 \times (-2) + \cdots + \delta^{k-1} \times (-2) + \delta^k \times (-3)$ for round t to $t + k$. In comparison, not deviating leads to a total utility of $(-1) + \delta \cdot (-1) + \cdots + \delta^k \cdot (-1)$ during this period.

2.6.1 Solution Concepts

There are many NEs in infinitely repeated games. Let v be an arbitrary N-dimensional vector such that there exists a strategy profile whose corresponding utility profile is v. If δ is large enough and the ith value of v is higher than the ith player's minmax payoff, there exists an NE whose utility profile is v.

Since a repeated game can be viewed as an extensive-form game, the solution concepts such as SPE can also be applied. In finitely repeated games, if the stage game has a unique equilibrium, the repeated game has a unique SPE, which is that all players play the stage game equilibrium in each round.

2.6.2 Applications in Cybersecurity

Given that the agents' interaction in cybersecurity is often repeated in nature, the repeated game model has been used to model various cybersecurity problems, including in packet forwarding (Ji et al. 2010), cyber-physical power systems (Paul and Ni 2019), and network security with multiple defenders (Jiang et al. 2010). Paul and Ni (2019) consider smart attackers who launch attacks to gain access to and manipulate a power system. The attacks are not one-shot as the same set of players may show up again and again. Therefore, they use the repeated game framework to model the interaction.

2.7 Bayesian Games

Bayesian games are games where the players have incomplete information about the game settings. In this section, we introduce Harsanyi's approach to incomplete information (Harsanyi 1967, 1968), which uses random variables, known as "types," to capture uncertainties in the game.

The setting of incomplete information refers to where at least one player does not know the payoff of at least one of the other players. It differs from the complete information setting where the set of players $\mathcal{N}$, the action space for each player $\mathcal{A}_i$, and the payoff function for each player u_i are completely known to all the players. The set of players $\mathcal{N}$ and the action spaces $\mathcal{A}_i, i \in \mathcal{N}$ are considered as common knowledge in the games of incomplete information.

Harsanyi has introduced an approach to capture a game with incomplete information using the notion of types of players. The type of player can be considered as an aggregation of her private information. Thus, the player's type is only known to herself. The type is also called an *epistemic type* since it can involve the player's beliefs of different levels about the other players' payoffs. By considering the assignments of types as prior moves of the fictitious player, Nature, we can transform a game with incomplete information into a game with imperfect information about Nature's moves, and use standard techniques to analyze equilibrium behaviors.

To define Bayesian games formally, we let $\Theta = (\Theta_1, \ldots, \Theta_N)$ denote the type spaces of players in $\mathcal{N}$. Associate the type space with a prior distribution $p : \Theta \rightarrow [0, 1]$, which is common knowledge. The payoffs of the players are defined by functions of their types and the actions of all the players as $u_i : \mathcal{A} \times \Theta \rightarrow \mathbb{R}$. Then, a Bayesian game is defined as the tuple $(\mathcal{N}, \mathcal{A}, \Theta, p, u)$, where $u = (u_1, \ldots, u_N)$ and $\mathcal{A} = (\mathcal{A}_1, \ldots, \mathcal{A}_N)$. A pure strategy of player i is defined as a mapping $s_i : \Theta_i \rightarrow \mathcal{A}_i$. Mixed strategies extend pure strategies to probability distributions on the action set of the player. Note that the strategies and payoffs are mappings of the types.

A Bayesian game can be naturally transformed into an NFG by enumerating all the pure strategies, each of which specifies what action the player should take for each of the possible types. For example, in a two-player game, each row or column of the payoff matrix represents a vector of $|\Theta_i|$ actions specifying a pure strategy of the player.

2.7.1 Solution Concepts

The equilibrium solutions depend on the type space and the prior distributions. The goal of each player in the game is to find strategies to maximize her expected payoff. There are three types of equilibrium strategies and equilibrium payoff that are distinguished by the timing of commitment: ex-ante, interim, and ex-post ones. An ex-ante expected utility is obtained when the players commit their strategies prior to the observation or the knowledge of her type. An interim expected utility is obtained when the players commit their strategies after observing or knowing their own type. The expected utility becomes ex-post when the players commit their strategies after observing or knowing the types of all players. The ex-post game is equivalent to a complete information game.

Bayesian Nash equilibrium (BNE) is often used as a solution concept for Bayesian games under incomplete information. An BNE in a game of incomplete information is a strategy profile $s(\cdot) = (s_1(\cdot), \ldots, s_n(\cdot))$ that satisfies best response constraints in Bayesian setting. Formally, $s(\cdot)$ is a Bayesian Nash equilibrium under the ex-ante expected utility if, for each player i,

$$s_i(\cdot) \in \arg\max_{s'_i} \sum_{\theta_i}\sum_{\theta_{-i}} p(\theta_i, \theta_{-i}) u_i(s'_i(\theta_i), s_{-i}(\theta_{-i}), (\theta_i, \theta_{-i})). \tag{2.17}$$

Similarly, given the type θ_i, $s_i(\theta_i)$ is a Bayesian Nash equilibrium under the interim expected utility if, for every player i and every type θ_i,

$$s_i(\theta_i) \in \arg\max_{s'_i} \sum_{\theta_{-i}} p(\theta_{-i}|\theta_i) u_i(s'_i(\theta_i), s_{-i}(\theta_{-i}), (\theta_i, \theta_{-i})). \tag{2.18}$$

2.7.2 Applications in Cybersecurity

Bayesian games are useful to model many cybersecurity applications since they often involve unknown or unobservable information, as we have seen in man-in-the-middle attacks in remotely controlled robotic systems (Xu and Zhu 2015), spoofing attacks (Zhang and Zhu 2017), compliance control (Casey et al. 2015), deception over social networks (Mohammadi et al. 2016), and denial of service (Pawlick and Zhu 2017). The attackers often know more about the defenders than what defenders know about the attackers. The information asymmetry naturally creates an attacker's advantage. Cyber deception is an essential mechanism (Pawlick et al. 2018, 2019) to reverse the information asymmetry and create additional uncertainties to deter the attackers. In addition to the later chapters in this book, readers can also refer to Pawlick et al. (2019) for a recent survey of game-theoretic methods for defensive deception and Zhang et al. (2020) for a summary of Bayesian game frameworks for cyber deception. The Bayesian games can also be extended to dynamic ones to model multistage deception. Interested readers can see Huang and Zhu (2019) for more details.

Another important application of Bayesian games is mechanism design. According to the fundamental *revelation principle* (Myerson 1989, 1981), an equivalent revenue-maximizing mechanism can be designed under the scenario where players reveal their private types. Mechanism design has been applied to understanding security as a service (Chen and Zhu 2016, 2017), and pricing in IoT networks (Farooq and Zhu 2018).

2.8 Stochastic Games

Stochastic games are a class of games where the players interact for multiple rounds in different games. This section first presents the basic concepts of stochastic games along with their solution concepts. We will discuss the Markov perfect equilibria and the applications of stochastic games in cybersecurity.

The attribute that makes stochastic games is the concept of "states." At each state, players interact in a state-dependent game whose outcome determines the state and the associated game in the next round. Let Q denote a finite state space. Each state has an associated strategic game. The state transition is captured by the kernel function $P : Q \times A \times Q \to [0, 1]$, with $P(q_1, a, q_2)$ being the probability of transitioning to state q_2 from state q_1 when the players play action profile (a_1, a_2). Since the game played at each stage is determined by its state, the payoff functions also depend on the state. They are defined as $r = (r_1, \ldots, r_N)$ and $r_i : Q \times \mathcal{A} \to \mathbb{R}$. That is, each player gets a payoff in each state and the payoff is dependent on the current state and the joint actions of the players. A stochastic game is thus defined as the tuple $(Q, \mathcal{N}, \mathcal{A}, P, r)$. We assume the game at each stage is finite. It is clear that repeated games are special cases of stochastic games when the state space is a singleton.

The record of multiround interactions up to round t, consisting of actions and states, is called history denoted by $h_t = (q^0, a^0, \ldots, a^{t-1}, q^t)$, where the superscripts denote the rounds or stages. Let $s_i(h_t, a_i)$ be a behavioral strategy of player i, which indicates the probability of playing a_i under history h_t. Let q_t and q'_t denote the final states of histories h_t and h'_t, respectively. The behavioral strategy is a Markov strategy if, for each t, $s_i(h_t, a_i) = s_i(h'_t, a_i)$ whenever $q_t = q'_t$. In other words, the dependency on the history is only through the last state. It is analogous to the Markov property of stochastic processes. A Markov strategy is stationary if $s_i(h_{t_1}, a_i) = s_i(h'_{t_2}, a_i)$ whenever $q_{t_1} = q'_{t_2}$.

Stochastic games can also be viewed as a generalization of Markov decision processes (MDPs) to multiagent scenarios. Stochastic games are also referred to as Markov games. Here we briefly introduce the MDP formulation and explain how stochastic games generalize it. An MDP is defined by the set of states, the set of actions, the state transition probability and the reward or payoff function. Let q^0 be the initial state of the MDP and $p(q^{t+1}|q^t, a^t)$ be probability of transitioning from state q^t to state q^{t+1} under action a^t. The objective of a single-player MDP under discounted payoff aims to find a control policy u that maps from the states to actions to maximize the following infinite-horizon payoffs.

$$\max_u \sum_{t=t_0}^{\infty} \beta^t \mathbb{E}[r^t(q^t, u(q^t))], \tag{2.19}$$

Here, r^t is the scalar payoff under state q^t and action $u(q^t)$ at stage t. Dynamic programming can be used to find the optimal solutions to (2.19). Let $v^*(q^0)$ denote the optimal cost-to-go, which is the optimal value of the problem (2.19) given an initial state q^0. The Bellman equations state that the optimal cost-to-go at the stage t is the optimized current reward plus the discounted expected future reward at the next stage. More formally, the Bellman equations are given by

$$v^*(q^t) = \max_a r^t(q^t, a) + \beta \sum_{q^{t+1} \in Q} p(q^{t+1}|q^t, a)v(q^{t+1}). \tag{2.20}$$

The optimal payoff given an initial state can be further formulated as an LP problem by transforming (2.20) into a set of constraints as:

$$\min_v v(q)$$
$$\text{s.t.} \quad v(q^t) \geq r^t(q^t, a) + \beta \sum_{q^{t+1} \in Q} p(q^{t+1}|q^t, a)v(q^{t+1}), \forall a, \forall q^t. \tag{2.21}$$

Therefore, the problem in (2.19) can be solved by LP using a similar technique by optimizing a linear combination of $v(\cdot)$ functions of every state subject to all the corresponding constraints.

Stochastic games generalize the problem settings in (2.19) by introducing multiple agents into the process. Hence, we can use similar approaches to analyze all the agents' behaviors in their optimal policies. One key difference is that in stochastic games, we are not looking for utility-maximizing control policy for an individual agent but equilibrium solutions where no agents can benefit by unilateral deviations from the equilibrium. Interested readers can refer to Filar and Vrieze (2012) for more details.

2.8.1 Solution Concepts

The counterpart of SPE in stochastic games is Markov perfect equilibria. A Markov perfect equilibrium requires the strategy profile to be Markov and sub-game perfect. Let the total payoff of player i

be defined as the accumulative payoff with discount, $\sum_{j=1}^{\infty} \beta^j r_i^j$, where r_i^j is the stage payoff to player i at stage j. A Markov perfect equilibrium always exists in a finite stochastic game with discounted payoffs. When the total payoff is defined in the time-average sense as $\lim_{k \to \infty} \frac{1}{k} \sum_{j=1}^{k} r_i^j$, the existence of Nash equilibrium for a two-player, general-sum, irreducible game is guaranteed. Folk theorem (Dutta 1995; Fudenberg et al. 1994) characterizes the feasible regions of Nash equilibria when a sufficient number of stages are played, and the two players are sufficiently far-sighted.

2.8.2 Applications in Cybersecurity

The stochastic games have been widely used in cybersecurity, including network configurations (Zhu and Başar 2009; Zhu et al. 2010b), wireless security (Altman et al. 2007; Zhu et al. 2010a), cyber-physical systems (Miao and Zhu 2014; Xu and Zhu 2018), and cyber deception (Horák et al. 2017b; Huang and Zhu 2020). One important extension of stochastic games is to the ones under partial information. For example, the one-side partially observable stochastic games have been successfully used to capture the scenarios where one player can perfectly observe the state while the other player can only observe the state through partial observations. This class of games is particularly useful for cyber deception where the one-sided, partial observation naturally captures the information asymmetry between the defender and the attacker. See Horák et al. (2017a,b) for more details.

References

Eitan Altman, Konstantin Avratchenkov, Nicolas Bonneau, Mérouane Debbah, Rachid El-Azouzi, and Daniel Sadoc Menasché. Constrained stochastic games in wireless networks. In *IEEE GLOBECOM 2007-IEEE Global Telecommunications Conference*, pp. 315–320. IEEE, 2007.

Robert G. Bland, Donald Goldfarb, and Michael J. Todd. The ellipsoid method: A survey. *Operations Research*, 29(6):1039–1091, 1981.

Stephen P. Bradley, Arnoldo C. Hax, and Thomas L. Magnanti. *Applied Mathematical Programming*. Addison-Wesley, 1977.

William Austin Casey, Quanyan Zhu, Jose Andre Morales, and Bud Mishra. Compliance control: Managed vulnerability surface in social-technological systems via signaling games. In *Proceedings of the 7th ACM CCS International Workshop on Managing Insider Security Threats*, pp. 53–62. ACM, 2015.

Georgios Chalkiadakis, Edith Elkind, and Michael Wooldridge. Computational aspects of cooperative game theory. *Synthesis Lectures on Artificial Intelligence and Machine Learning*, 5(6):1–168, 2011.

Juntao Chen and Quanyan Zhu. Optimal contract design under asymmetric information for cloud-enabled internet of controlled things. In *International Conference on Decision and Game Theory for Security*, pp. 329–348. Springer, 2016.

Juntao Chen and Quanyan Zhu. Security as a service for cloud-enabled internet of controlled things under advanced persistent threats: A contract design approach. *IEEE Transactions on Information Forensics and Security*, 12(11): 2736–2750, 2017.

Xi Chen and Xiaotie Deng. 3-Nash is ppad-complete. *Electronic Colloquium on Computational Complexity*, 134: 2–29. Citeseer, 2005.

Xi Chen and Xiaotie Deng. Settling the complexity of two-player Nash equilibrium. In *2006 47th Annual IEEE Symposium on Foundations of Computer Science (FOCS'06)*, pp. 261–272. IEEE, 2006.

Vincent Conitzer and Tuomas Sandholm. Computing the optimal strategy to commit to. In *Proceedings of the 7th ACM Conference on Electronic Commerce*, pp. 82–90. ACM, 2006.

Edward A. Cranford, Christian Lebiere, Cleotilde Gonzalez, Sarah Cooney, Phebe Vayanos, and Milind Tambe. Learning about cyber deception through simulations: Predictions of human decision making with deceptive signals in Stackelberg security games. In *Proceedings of the 40th Annual Meeting of the Cognitive Science Society, CogSci 2018, Madison, WI, USA, 25–28 July 2018*, 2018.

Constantinos Daskalakis, Paul W. Goldberg, and Christos H. Papadimitriou. The complexity of computing a Nash equilibrium. *SIAM Journal on Computing*, 39(1):195–259, 2009.

Mary-Ann Dimand and Robert W. Dimand. *The History of Game Theory, Volume 1: From the Beginnings to 1945*, Volume 1. Routledge, 1996.

Cuong T. Do, Nguyen H. Tran, Choongseon Hong, Charles A. Kamhoua, Kevin A. Kwiat, Erik Blasch, Shaolei Ren, Niki Pissinou, and Sundaraja Sitharama Iyengar. Game theory for cyber security and privacy. *ACM Computing Surveys (CSUR)*, 50(2):1–37, 2017.

Prajit K. Dutta. A folk theorem for stochastic games. *Journal of Economic Theory*, 66(1):1–32, 1995.

Fei Fang, Thanh H. Nguyen, Rob Pickles, Wai Y. Lam, Gopalasamy R. Clements, Bo An, Amandeep Singh, Milind Tambe, and Andrew Lemieux. Deploying paws: Field optimization of the protection assistant for wildlife security. In *Proceedings of the Twenty-Eighth Innovative Applications of Artificial Intelligence Conference (IAAI 2016)*, pp. 3996–3973. AAAI Press, 2016.

Muhammad Junaid Farooq and Quanyan Zhu. Optimal dynamic contract for spectrum reservation in mission-critical UNB-IoT systems. In *2018 16th International Symposium on Modeling and Optimization in Mobile, Ad Hoc, and Wireless Networks (WiOpt)*, pp. 1–6. IEEE, 2018.

Jerzy Filar and Koos Vrieze. *Competitive Markov Decision Processes*. Springer Science & Business Media, 2012.

Drew Fudenberg, David Levine, and Eric Maskin. The folk theorem with imperfect public information. *Econometrica*, 62(5):997–1039, 1994.

John C. Harsanyi. Games with incomplete information played by Bayesian players, I–III: Part I. The basic model. *Management Science*, 14(3):159–182, 1967.

John C. Harsanyi. Games with incomplete information played by bayesian players, Part II. Bayesian equilibrium points. *Management Science*, 14(5): 320–334, 1968.

Karel Horák, Branislav Bošanský, and Michal Pěchouček. Heuristic search value iteration for one-sided partially observable stochastic games. In *Thirty-First AAAI Conference on Artificial Intelligence*. AAAI Press, 2017a.

Karel Horák, Quanyan Zhu, and Branislav Bošanský. Manipulating adversary's belief: A dynamic game approach to deception by design for proactive network security. In *International Conference on Decision and Game Theory for Security*, pp. 273–294. Springer, 2017b.

Linan Huang and Quanyan Zhu. Dynamic Bayesian games for adversarial and defensive cyber deception. In *Autonomous Cyber Deception*, pp. 75–97. Springer, 2019.

Linan Huang and Quanyan Zhu. Dynamic games of asymmetric information for deceptive autonomous vehicles. 2020.

Zhu Ji, Wei Yu, and K. J. Ray Liu. A belief evaluation framework in autonomous MANETs under noisy and imperfect observation: Vulnerability analysis and cooperation enforcement. *IEEE Transactions on Mobile Computing*, 9(9): 1242–1254, 2010.

Libin Jiang, Venkat Anantharam, and Jean Walrand. How bad are selfish investments in network security? *IEEE/ACM Transactions on Networking*, 19(2):549–560, 2010.

Christopher Kiekintveld, Manish Jain, Jason Tsai, James Pita, Fernando Ordóñez, and Milind Tambe. Computing optimal randomized resource allocations for massive security games. In *Proceedings of the 8th International Conference on Autonomous Agents and Multiagent Systems, Volume 1, AAMAS '09*, pp. 689–696. International Foundation for Autonomous Agents and Multiagent Systems, Richland, SC, 2009.

David M. Kreps and Robert Wilson. Sequential equilibria. *Econometrica: Journal of the Econometric Society*, 50(4):863–894, 1982.

Carlton E. Lemke and Joseph T. Howson, Jr. Equilibrium points of bimatrix games. *Journal of the Society for industrial and Applied Mathematics*, 12(2): 413–423, 1964.

Kevin Leyton-Brown and Yoav Shoham. Essentials of game theory: A concise multidisciplinary introduction. *Synthesis Lectures on Artificial Intelligence and Machine Learning*, 2(1):1–88, 2008.

Bo Li and Yevgeniy Vorobeychik. Feature cross-substitution in adversarial classification. In *Advances in Neural Information Processing Systems 27: Annual Conference on Neural Information Processing Systems 2014, 8–13 December 2014, Montreal, Quebec, Canada*, pp. 2087–2095, 2014.

Bo Li and Yevgeniy Vorobeychik. Scalable optimization of randomized operational decisions in adversarial classification settings. In *Artificial Intelligence and Statistics*, pp. 599–607, 2015.

Richard D. McKelvey and Thomas R. Palfrey. Quantal response equilibria for normal form games. *Games and Economic Behavior*, 10(1):6–38, 1995.

Richard D. McKelvey, Andrew M. McLennan, and Theodore L. Turocy. Gambit: Software tools for game theory. 2006. www.gambit-project.org (accessed 18 March 2021).

Sanjay Mehrotra. On the implementation of a primal-dual interior point method. *SIAM Journal on Optimization*, 2(4):575–601, 1992.

Fei Miao and Quanyan Zhu. A moving-horizon hybrid stochastic game for secure control of cyber-physical systems. In *Decision and Control (CDC), 2014 IEEE 53rd Annual Conference on*, pp. 517–522. IEEE, 2014.

Amin Mohammadi, Mohammad Hossein Manshaei, Monireh Mohebbi Moghaddam, and Quanyan Zhu. A game-theoretic analysis of deception over social networks using fake avatars. In *International Conference on Decision and Game Theory for Security*, pp. 382–394. Springer, 2016.

Roger B. Myerson. Optimal auction design. *Mathematics of Operations Research*, 6(1):58–73, 1981.

Roger B. Myerson. Mechanism design. In *Allocation, Information and Markets*, pp. 191–206. Springer, 1989.

Roger B. Myerson. *Game Theory: Analysis of Conflict*. Harvard University Press, 1991. http://www.jstor.org/stable/j.ctvjsf522.

John Nash. Non-cooperative games. *Annals of Mathematics*, 54(2):286–295, 1951.

John A. Nelder and Roger Mead. A simplex method for function minimization. *The Computer Journal*, 7(4):308–313, 1965.

J. v. Neumann. Zur theorie der gesellschaftsspiele. *Mathematische Annalen*, 100 (1):295–320, 1928.

Thanh Hong Nguyen, Rong Yang, Amos Azaria, Sarit Kraus, and Milind Tambe. Analyzing the effectiveness of adversary modeling in security games. *Proceedings of the AAAI Conference on Artificial Intelligence*, Vol. 27, No. 1. AAAI, 2013.

Praveen Paruchuri, Jonathan P. Pearce, Janusz Marecki, Milind Tambe, Fernando Ordonez, and Sarit Kraus. Playing games for security: An efficient exact algorithm for solving Bayesian Stackelberg games. In *Proceedings of the 7th International Joint Conference on Autonomous Agents and Multiagent Systems, Volume 2, AAMAS '08*, pp. 895–902. International Foundation for Autonomous Agents and Multiagent Systems, Richland, SC, 2008.

Shuva Paul and Zhen Ni. A strategic analysis of attacker-defender repeated game in smart grid security. In *2019 IEEE Power & Energy Society Innovative Smart Grid Technologies Conference (ISGT)*, pp. 1–5. IEEE, 2019.

Jeffrey Pawlick and Quanyan Zhu. Proactive defense against physical denial of service attacks using poisson signaling games. In *International Conference on Decision and Game Theory for Security*, pp. 336–356. Springer, 2017.

Jeffrey Pawlick, Edward Colbert, and Quanyan Zhu. Modeling and analysis of leaky deception using signaling games with evidence. *IEEE Transactions on Information Forensics and Security*, 14(7):1871–1886, 2018.

Jeffrey Pawlick, Edward Colbert, and Quanyan Zhu. A game-theoretic taxonomy and survey of defensive deception for cybersecurity and privacy. *ACM Computing Surveys (CSUR)*, 52(4):1–28, 2019.

James Pita, Manish Jain, Fernando Ordonez, Milind Tambe, and Sarit Kraus. Robust solutions to Stackelberg games: Addressing bounded rationality and limited observations in human cognition. *Artificial Intelligence Journal*, 174(15):1142–1171, 2010.

J. Rosenmüller. On a generalization of the Lemke–Howson algorithm to noncooperative n-person games. *SIAM Journal on Applied Mathematics*, 21 (1):73–79, 1971.

Aaron Schlenker, Haifeng Xu, Mina Guirguis, Christopher Kiekintveld, Arunesh Sinha, Milind Tambe, Solomon Sonya, Darryl Balderas, and Noah Dunstatter. Don't bury your head in warnings: A game-theoretic approach for intelligent allocation of cyber-security alerts. In *Proceedings of the Twenty-Sixth International Joint Conference on Artificial Intelligence, IJCAI 2017, Melbourne, Australia, 19–25 August 2017*, pp. 381–387, 2017. https://doi.org/10.24963/ijcai.2017/54.

Aaron Schlenker, Omkar Thakoor, Haifeng Xu, Milind Tambe, Phebe Vayanos, Fei Fang, Long Tran-Thanh, and Yevgeniy Vorobeychik. Deceiving cyber adversaries: A game theoretic approach. In *International Conference on Autonomous Agents and Multiagent Systems, Stockholm, Sweden. 10–15 July*, 2018.

Yoav Shoham and Kevin Leyton-Brown. *Multiagent Systems: Algorithmic, Game-Theoretic, and Logical Foundations*. Cambridge University Press, 2008.

Milind Tambe. *Security and Game Theory: Algorithms, Deployed Systems, Lessons Learned*. Cambridge University Press, 2011.

Omkar Thakoor, Milind Tambe, Phebe Vayanos, Haifeng Xu, Christopher Kiekintveld, and Fei Fang. Cyber camouflage games for strategic deception. In *International Conference on Decision and Game Theory for Security*, pp. 525–541. Springer, 2019.

Kun Wang, Li Yuan, Toshiaki Miyazaki, Yuanfang Chen, and Yan Zhang. Jamming and eavesdropping defense in green cyber–physical transportation systems using a Stackelberg game. *IEEE Transactions on Industrial Informatics*, 14(9):4232–4242, 2018.

Zhiheng Xu and Quanyan Zhu. A cyber-physical game framework for secure and resilient multi-agent autonomous systems. In *Decision and Control (CDC), 2015 IEEE 54th Annual Conference on*, pp. 5156–5161. IEEE, 2015.

Zhiheng Xu and Quanyan Zhu. Cross-layer secure and resilient control of delay-sensitive networked robot operating systems. In *2018 IEEE Conference on Control Technology and Applications (CCTA)*, pp. 1712–1717. IEEE, 2018.

Rong Yang, Fernando Ordonez, and Milind Tambe. Computing optimal strategy against quantal response in security games. In *Proceedings of the 11th International Conference on Autonomous Agents and Multiagent Systems, Volume 2*, pp. 847–854. International Foundation for Autonomous Agents and Multiagent Systems, 2012.

Rong Yang, Benjamin Ford, Milind Tambe, and Andrew Lemieux. Adaptive resource allocation for wildlife protection against illegal poachers. In *International Conference on Autonomous Agents and Multiagent Systems (AAMAS), Paris, France, 5–9 May*, 2014.

Yuan Yuan, Fuchun Sun, and Huaping Liu. Resilient control of cyber-physical systems against intelligent attacker: A hierarchal Stackelberg game approach. *International Journal of Systems Science*, 47(9):2067–2077, 2016.

Tao Zhang and Quanyan Zhu. Strategic defense against deceptive civilian gps spoofing of unmanned aerial vehicles. In *International Conference on Decision and Game Theory for Security*, pp. 213–233. Springer, 2017.

Tao Zhang, Linan Huang, Jeffrey Pawlick, and Quanyan Zhu. Game-theoretic analysis of cyber deception: Evidence-based strategies and dynamic risk mitigation. In Charles A. Kamhoua, Laurent L. Njilla, Alexander Kott, and Sachin Shetty, editors. *Modeling and Design of Secure Internet of Things*, pp. 27–58. Wiley Online Library, 2020.

Minghui Zhu and Sonia Martinez. Stackelberg-game analysis of correlated attacks in cyber-physical systems. In *Proceedings of the 2011 American Control Conference*, pp. 4063–4068. IEEE, 2011.

Quanyan Zhu and Tamer Başar. Dynamic policy-based IDS configuration. In *Proceedings of the 48h IEEE Conference on Decision and Control (CDC) held jointly with 2009 28th Chinese Control Conference*, pp. 8600–8605. IEEE, 2009.

Quanyan Zhu, Husheng Li, Zhu Han, and Tamer Başar. A stochastic game model for jamming in multi-channel cognitive radio systems. In *2010 IEEE International Conference on Communications*, pp. 1–6. IEEE, 2010a.

Quanyan Zhu, Hamidou Tembine, and Tamer Başar. Network security configurations: A nonzero-sum stochastic game approach. In *Proceedings of the 2010 American Control Conference*, pp. 1059–1064. IEEE, 2010b.

3

Scalable Algorithms for Identifying Stealthy Attackers in a Game-Theoretic Framework Using Deception

Anjon Basak[1], Charles A. Kamhoua[1], Sridhar Venkatesan[2], Marcus Gutierrez[3], Ahmed H. Anwar[1], and Christopher D. Kiekintveld[3]

[1] *Network Security Branch, Combat Capabilities Development Command, US Army Research Laboratory, Adelphi, MD, USA*
[2] *Perspecta Labs Inc., Basking Ridge, NJ, USA*
[3] *Department of Computer Science, The University of Texas at El Paso, El Paso, TX, USA*

3.1 Introduction

Cyber attackers pose a serious threat to economies, national defense, critical infrastructure, and financial sectors apt (APT37 2017; APT38 2018; M-Trends 2019). Early detection and identification of a cyber attacker can help a defender to make better decisions to mitigate the attack. However, identifying the characteristics of an attacker is challenging as they may employ many defense evasion techniques def. Adversaries may also mask their operations by leveraging white-listed tools and protocols (Wheeler and Larsen 2003; Tsagourias 2012).

There are many existing intrusion detection methods to detect attacks (Wheeler and Larsen 2003; Nicholson et al. 2012). We focus here on using honeypot (HP) for detection and identification, though our models could be extended to other types of defensive actions. HP are systems that are designed to attract adversaries (Spitzner 2003; Kreibich and Crowcroft 2004) and to monitor (Nicholson et al. 2012) attacker activity so that the attack can be analyzed (usually manually by experts). There are works on automating the analysis process (Raynal et al. 2004a) using host-based and network-based data for correlation to identify a pattern. Deductive reasoning can be used to conclude the attacker (Raynal et al. 2004b).

Most current approaches focus on detection during an ongoing attack, possibly with some effort to categorize different types of detections. More detailed identification is done during later forensic analysis. Here we focus on formally modeling what types of actions the defender can take to support more detailed attacker identification early in the attack chain, allowing more information to target specific defensive responses. This can be challenging; for example, many different attackers may use the same core malware toolkits and common tactics apt (APT33 2017; APT38 2018a; APT37 2017; APT38 2018b). Attackers may intentionally try to look similar to other actors, and may even change during an attack (e.g., when compromised resources are sold to other groups), leading to a complete change in focus apt (APT40 2009; APT34 2017; M-Trends 2019).

We focus our model on detecting an attacker type early. An attacker type is defined by an Attack Graph (AG) specific to that attacker that describes his possible actions and goals in planning an attack campaign. An AG represents the set of attack paths that an attacker can take to achieve their goal in the target network (Durkota et al. 2015). Depending on the observed network and the exploits they have available, the attacker tries to choose the optimal sequence of attack actions to achieve their particular goal. Identification of different attack types in our model corresponds

Game Theory and Machine Learning for Cyber Security, First Edition.
Edited by Charles A. Kamhoua, Christopher D. Kiekintveld, Fei Fang, and Quanyan Zhu.
© 2021 The Institute of Electrical and Electronics Engineers, Inc. Published 2021 by John Wiley & Sons, Inc.

to identification of possible AGs this particular attacker is using in a particular interaction. The defender chooses deception actions that try to force the attackers into choices that will reveal which type they are early on, which can then inform later defenses. First, we propose a multi-stage game-theoretic model to strategically deploy HPs to force the attacker to reveal his type early. We present case studies to show how this model works for realistic scenarios, and to demonstrate how attackers can be identified using deception. Then, we present a general model, a basic solution algorithm, and simulation results that show that if the defender uses proactive deception the attacker type can be detected early compared to if he only observes the activity of an attacker. However, the optimal algorithm considers every possible action space for the defender and as a result, does not scale very well. Finally, we use some heuristics considering domain knowledge to reduce the action space to a reasonable amount to make the algorithm more scalable. Our experiments show that using heuristics improves the scalability of the algorithm by reasonable margins.

3.2 Background

AGs represent sequential attacks by an attacker to compromise a network or a particular computer (Durkota et al. 2015). AGs can be automatically generated using a known vulnerabilities database (Ingols et al. 2006; Ou et al. 2006). Due to resource limitations, the automatically generated AGs are often used to identify high priority vulnerabilities to fix (Sheyner et al. 2002; Noel and Jajodia 2008). We use AGs as a library of attack plans that can represent different attackers. The optimal plan for any attacker will depend on his particular options and goals, reflected in the AG. The AG for each attacker will also change depending on the network, including changes made by the defender (e.g., introducing HPs). We model a multistage Stackelberg Security game (SSG) with a leader and a follower. The defender commits to a strategy considering the attacker's strategy. The attacker observes the strategy of the leader and chooses an optimal attack strategy using the AG. The AG of the attackers we considered is defined by initial access and lateral movement actions (Enterprise Tactics 2020) to reach their corresponding goal node g_i as shown in Figure 3.1.

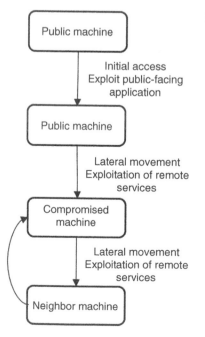

Figure 3.1 An attack graph.

Initial access represents the vectors an attacker uses to gain an initial foothold in the network. We consider the technique called *Exploit Public-Facing Application* mit (Initial Access 2018) for initial access, where an attacker uses tools to exploit the weakness of the public-facing systems. Lateral movement is a tactic to achieve greater control over network assets. We consider the example of *Exploitation of Remote Services* technique for lateral movement mit (Lateral Movement 2018) where an attacker exploits the vulnerabilities of a program. While there may be similarities in the AG for different attackers, having different goals and options available mean that the plans may eventually diverge. The overlap between multiple attack plans is the number of actions that are identical at the start of the plan.

3.3 Case Studies

We present three case studies that consider different types of attackers. We look at different pairs of attackers based on what exploits are shared between them and whether their final objective is the same or not. We use the network shown in Figure 3.2, where a router is R_i, a host is H_i, a firewall is F_i, a switch is S. An exploit $\phi_i(c)$ with cost c on an edge allows an attacker to move laterally if an attacker a_i has exploit $\phi_i(c)$. The cost of using an exploit represents both the time and effort as well as the risk of detection that attackers want to minimize. An attacker tries to reach a goal by making lateral movements using an attack plan with the minimum cost. If there is more than one minimum cost plan, attackers choose the ones that maximize the overlap with other attackers. We assume that when an attacker reaches his goal the game ends. He also has complete knowledge (e.g. vulnerabilities) about the network but does not know which nodes are HPs. For each case study, we first analyze the attack plans based on the AGs. Then we analyze what proactive deceptive action a defender can take to detect the attacker type earlier. We assume that the attacks are fully observable. Since we are interested in scenarios where attackers have common attack plans in their AG, we assume that host H_1 is where all attackers initially enter the network.

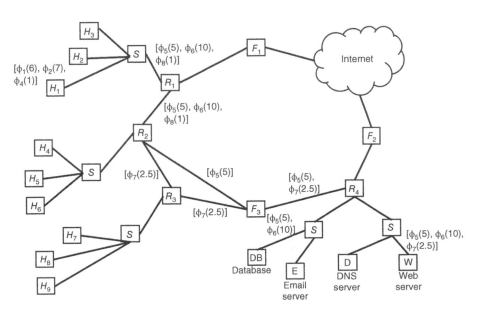

Figure 3.2 Network for case study.

3.3.1 Case Study 1: Attackers with Same Exploits but Different Goals

We consider two attackers a_1 and a_2 with the set of exploits $\phi_1, \phi_2, \phi_5, \phi_6$. Goal nodes for the attackers are defined by $g(a_1) = $ DB and $g(a_2) = $ W. The attack plans are shown in Figure 3.3a. The defender cannot distinguish which attacker he is facing until the attacker reaches to his goal node. Here, the defender can use a decoy node of either the DB or the W to reduce the overlap in

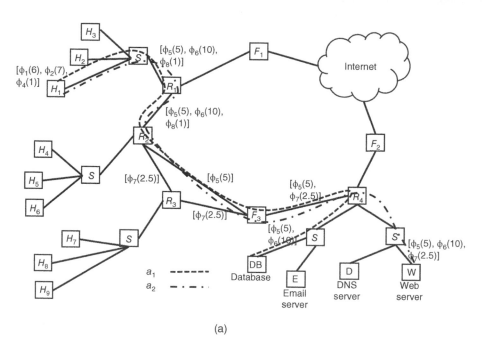

(a)

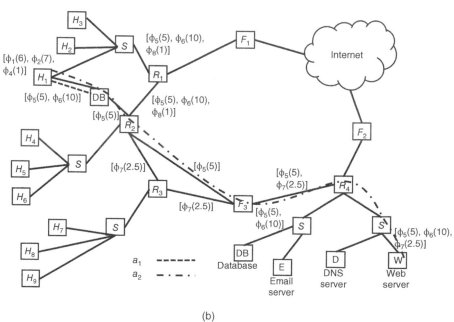

(b)

Figure 3.3 Case study 1. Attack plans of the attackers a_1 and a_2 before (a) and after deception (b).

the attack plans by a_1 and a_2. Since both of the attackers have the same set of exploits a defender cannot use any decoy node with vulnerabilities. The attackers have different goals and the defender can take advantage of that situation.

Figure 3.3b shows the use of a decoy DB (with dotted line) between H_1 and R_2 with unique exploits on the edges to force only the targeted attacker to go through the decoy. Since other attackers will not use that plan, this creates a unique attack plan that can be identified. In Figure 3.3b, we notice that if the acting attacker is a_1, then he will go for the decoy DB. Attacker a_1 does not take the longer path to compromise the DB because he chooses an attack plan with minimum cost. To maximize the overlapping attack plan attacker a_2 will choose the plan through the decoy DB instead of R_1 even though the two plans cost the same. In the next case studies, the defender avoids using the goal node as a decoy to reduce the cost associated with the decoy server.

3.3.2 Case Study 2: Attackers with Shared Exploits and Different Goals

Now we consider attacker a_1 with exploits $(\phi_1, \phi_2, \phi_5, \phi_6)$ and $g(a_1) =$ DB. Attacker a_3 has exploits $\phi_1, \phi_2, \phi_6, \phi_7$ and $g(a_3) = W$. The attackers a_1 and a_3 compute attack plans as shown in the Figure 3.4a. If the defender just observes, he will be able to detect the acting attacker after R_2. Since the attackers have shared and unique exploits in their possession, these create a nonoverlapping path in the middle of their attack sequence, that causes the defender to detect the differentiation between the attackers' attack plans.

However, if the defender uses a HP as shown in Figure 3.4b the attack sequence of the attacker's changes. Attacker a_1 cannot go through the HP due to his lack of exploiting $\phi_7(2.5)$ that causes the attackers to diverge at an earlier stage facilitating identification by the defender.

3.3.3 Case Study 3: Attackers with Shared Exploits but Same Goals

Now we consider attacker a_4 with exploits $(\phi_1, \phi_2, \phi_5, \phi_6, \phi_7)$ and $g(a_4) = W$ and for a_5 exploits $(\phi_1, \phi_4, \phi_7, \phi_8), g(a_5) = W$. The attack plans are shown in Figure 3.5a. Attacker a_4 and a_5 executes the exact same attack plan. Even though attacker a_4 could go directly to F_3 from R_2 he chooses the plan to through R_3 minimize the chance of getting type detected. Here, the defender has zero probability to differentiate the attackers. The defender cannot use any HP at the initial stage as in case study 1 or 2 since those will not make any difference in the attackers plan. However, if the defender deploys a HP as shown in the Figure 3.5b, the attackers choose the plans with the minimum costs that leads to an earlier identification for the defender.

3.4 Game Model

We now present a general game-theoretic framework to model the interactions between the defender and the attacker. There is a network with nodes $t_i \in T$ similar to an enterprise network shown in Figure 3.2. Each node $t_i \in T$ has a value v_{t_i} and a cost c_{t_i}. Some nodes are public nodes that can be used to access the network. There are goal nodes $g_j \subset T$ (e.g. DB) that are high valued nodes. Each node $t_i \in T$ has some vulnerabilities that can be exploited using $\phi_{t_i} \in \Phi$ on the edges. A node t_i can be compromised from a node t_j using the exploit ϕ_{t_i} if there is an edge from t_j to node t_i and if node t_j allows the use of the exploit ϕ_{t_i} and if an attacker has any of the ϕ_{t_i} exploits. The game has N attacker types a_i and one defender d. The attackers have complete knowledge about the network and have single deterministic policies $\pi(a_i)$ to reach their goals and some exploit $\phi_{a_i} \in \Phi$. However, he does not know whether a node is a HP or not. The defender can

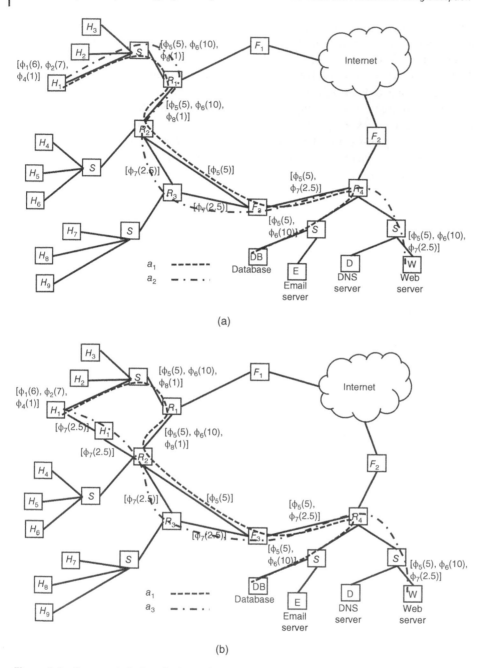

Figure 3.4 Case study 2. Attack plans of the attackers a_1 and a_3 before (a) and after deception (b).

deploy deception by allocating HP $h \in H$ in the network. The configurations of the HPs are chosen randomly from real nodes T in the network. The game starts by choosing an attacker type a_i randomly. In round r, the defender knows the history up to round $r - 1$, that is used to update the defender's belief about attacker type. Next, the defender allocates k HPs to force the attacker to reveal his type earlier. The defense actions also may thwart the plan of the attacker of reaching the goal if the attacker moves laterally to a HP or if the attacker does not find any plan to move laterally. The network change is visible to the attacker in the current round. This can be justified

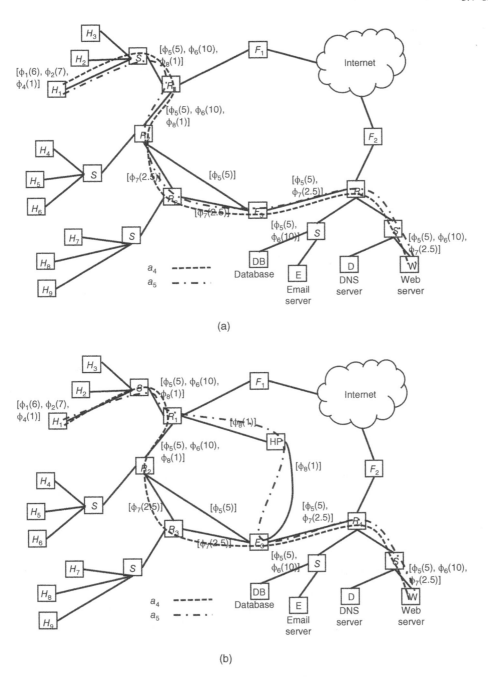

Figure 3.5 Case study 3. Attack plans of the attackers a_4 and a_5 before (a) and after deception (b).

because an APT usually performs reconnaissance again and monitors defender activity after making a lateral movement. The attacker recomputes his attack plan using his AG and chooses an optimal attack plan with the minimum cost from his current position to reach his goal. If there are multiple plans with the same cost, we break the tie in favor of the attacker where the attackers have the maximum common overlapping length in their attack plans. The game ends if the attacker gets caught by the deception set up by the defender or if he reaches his goal node.

3.5 Defender Decision Making

In each round of the game the defender updates his beliefs about the attacker types and the attackers' goals. According to Bayes' theorem, given the sequence of lateral movement seq(t) up to node t from the starting node t_p, the probability that the defender is facing attacker a_i is:

$$p(a_i|\text{seq}(t)) = \frac{p(\text{seq}(t)|a_i)p(a_i)}{\sum_{j=0}^{N}p(\text{seq}(t)|a_j)p(a_j)}$$

where $p(a_i)$ is the prior probability of facing the attacker a_i and $p(\text{seq}(t)|a_i)$ is the likelihood of the observation seq(t) given that we are facing the attacker a_i. Similarly, belief about goal can be computed. The probability of the plan of the attacker is P_g from the start node t_p is:

$$p(P_g|\text{seq}(t), a_i) = \frac{p(\text{seq}(t)|P_g, a_i)p(P_g, a_i)}{\sum_{\forall g \in G} p(\text{seq}(t)|P_g, a_i)p(P_g, a_i)}$$

Next, the defender considers all the possible deception deployments $c \in C$ where there are edges $t_m \rightarrow t_n$ from the attacker's last observed position t_{lp} where t_m can be reached from node t_{lp}. Without affecting the existing connections of the network deceptions are deployed between two nodes. The defender has a library of AGs for each of the attackers that he can use to optimize the decision making. We consider three possible objectives the defender uses to make this decision. In *Minimizing Maximum Overlapping Length*, the defender chooses his deception deployment by minimizing the sum of the attackers' overlapping actions. Another variation would be to minimize the attacker's maximum overlapping length with other attackers by considering each of the attackers. Minimizing the maximum overlapping length of attack plans may not always focus on all the attackers' attack plans, e.g. if all the attackers have high overlapping (of attack plans) with each other except the acting attacker. To overcome the issue the defender can compute the expected overlapping length of the attack plans: *Minimizing Expected Overlapping Length*. According to information theory, one way to reduce the anonymity between the attacker types is to deploy deception in such a way that will minimize entropy. If $X_1 = p(a_0)$, $X_2 = p(a_1)$ and $X_3 = p(a_2)$ are three random variables for the attacker types where $X_1 + X_2 + X_3 = 1$, then entropy can be written as follows: $H(X) = -\sum_{i=0}^{i=1} p(a_i)\log_b p(a_i)$ where $p(a_i)$ is the posterior probability for the attacker a_i. In *Minimizing Entropy*, the defender chooses the deception deployment that results in the minimum entropy for all the attackers A.

3.6 Attacker Decision Making

Now we present the mixed-integer program (MIP) for an attacker, where he chooses the minimum cost plan to reach his goal. Ties are broken by choosing a plan that maximizes the sum of the common overlapping length of the attack plans.

$$\max \sum_{ij} d_{ij} \tag{3.1}$$

$$0 \leq d_{ij} - \sum_{0}^{me} d_{ijme} \leq 0 \quad \forall i, j \tag{3.2}$$

$$d_{ijme} \leq e_{iem} \text{ AND } d_{ijme} \leq e_{jem} \quad \forall i, j, m, e \tag{3.3}$$

$$d_{ijme} \leq d_{ij(m-1)e} \quad \forall i, j, e, m = 1, 2, \ldots, M \tag{3.4}$$

$$\sum_{m} \sum_{e} e_{iem} c_e \leq C_i \quad \forall i \tag{3.5}$$

$$\sum_m e_{iem} - \sum_m e_{ie'm} = \begin{cases} 1 & \text{if } s \in e \\ -1 & \text{if } g \in e \qquad \forall i, \forall t \\ 0 & \text{otherwise} \end{cases} \tag{3.6}$$

$$\sum_m e_{ie(m-1)} - \sum_m e_{ie'm} = 0 \quad \forall i, \forall t, t \neq s, t \neq g \tag{3.7}$$

Equation 3.1 is the objective function where the attacker computes the maximum sum of the overlapping length of attack plans among all the attackers. Constraint 3.2 assigns the sum of the overlapping length between attacker i,j up to move m into d_{ij}. In d_{ijme}, e is the edge identifier. Constraint 3.3 computes the overlapping length between the attacker plans where e_{iem} is a binary variable for attacker i representing an edge for edge e (subscript) at mth move. Constraint 3.4 makes sure that the overlapping starts from the beginning of the plans and not in the middle; if two plans start differently but merge in the middle somewhere. Constraint 3.5 ensures that each attacker chooses a minimum cost plan to reach the goal node. Constraint 3.6 and 3.7 are path flow constraints for attackers.

3.7 Simulation Results

We want to show that using proactive deception a defender can reveal the attacker type earlier than otherwise. We define an early identification as to when the defender can use proactive deception to determine the attacker type in an earlier round compared to when the defender just observes. Vulnerabilities in the nodes are (indirectly) represented by exploits on the edge between two nodes. We randomly generated 20 networks, somewhat similar to Figure 3.2, with 18 nodes with values and costs chosen randomly from the range [0, 10] including one public node. We considered exploits $\phi_0, \phi_1, \phi_2, \phi_3, \phi_4, \phi_5$ with cost chosen randomly from the range [0, 10]. Next, we assign exploits to the edges in such a way that it allows the attackers to have unique attack plans to their goals except for the starting node. Depending on the edge density (number of edges from a node) and shared vulnerability parameters we randomly connect edges with exploits between nodes where the two nodes are in different attack plans of different attackers. We used six HPs and the vulnerabilities are picked from randomly chosen nodes existing in the network so that the HPs can act as decoys. The games are limited to five rounds. In each round r, the defender d deploys $0 \leq k \leq 2$ decoys. In the attack plan library, we considered three attacker types, a, b, c with different goals.

In the first experiment, we show that depending on different density of edges and shared vulnerabilities between nodes, how early a defender can identify the attacker type he is facing varies. Each of the three attackers, a, b, c has some unique and some shared exploits in their possession. Attacker a has exploits ϕ_0, ϕ_1, ϕ_2. Attacker b has ϕ_2, ϕ_3, ϕ_1. Attacker c has exploits ϕ_4, ϕ_5, ϕ_2. We picked the attacker b as the acting attacker.

Figure 3.6 shows the results. In the first row in Figure 3.6a-c, defender just observes the attacker actions. As the density and shared vulnerabilities increases, it takes more rounds for the defender to identify the attacker type b. In the second row in Figure 3.6e and f, the defender deploys HPs. If we compare the figures of the same edge density and shared vulnerabilities from the two rows, it is easy to notice that the use of deception facilitates early identification except Figure 3.6d where it was the same. However, an increase in edge density and shared vulnerabilities between nodes harms performance.

The second experiment is the same as the first except that we kept the edge density and shared vulnerabilities between nodes fixed to 40% and we varied the shared exploits between the attackers. We chose the attacker c as the acting attacker. We can observe in Figure 3.7a-c that as we increase

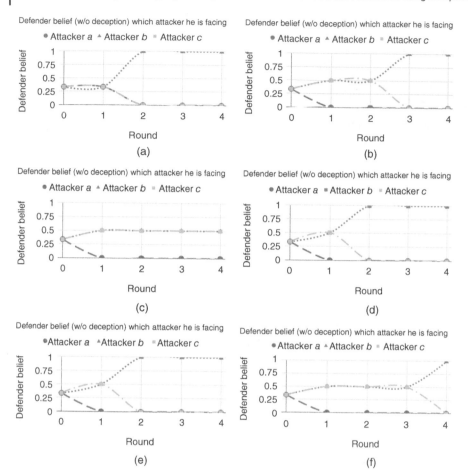

Figure 3.6 Comparison between just observation (first row) and use of proactive deception (second row). The shared exploits are fixed to 40% between attackers. (a) Edge density and shared vulnerabilities 20%. (b) Edge density and shared vulnerabilities 40%. (c) Edge density and shared vulnerabilities 80%. (d) Edge density and shared vulnerabilities 20%. (e) Edge density and shared vulnerabilities 40%. (f) Edge density and shared vulnerabilities 80%.

the sharing of exploits between the attackers it takes longer for the defender to identify the attacker type c. When all the attackers have the same exploits the defender was unable to identify the attacker type even at round 4 without using any deception. However, in the second row in Figure 3.7d–f as the defender strategically uses deception, identification of attacker c happens earlier. The performance of the early identification decreases as the shared exploits between the attacker's increases. Another observation is noticeable in Figure 3.7d: the defender was not able to identify the attacker type, however, the attacker did not find any policy to continue its attack. This shows that the use of strategical deception can also act as a deterrent for the attackers.

For our last experiment, we compared different techniques a defender can use to facilitate early identification; minimizing maximum overlap: *min-max-overlap*, minimizing maximum expected overlap: *min-max-exp-overlap* and minimizing entropy: *min-entropy*. Data are averaged for all the attacker types we considered. Figure 3.8a and b shows on average how many rounds it took for the defender to identify the attackers using different techniques. In both figures, we can see that using deception facilitates earlier identification. We did not notice any particular difference between different techniques except when all the attackers have the same exploits, and in that case

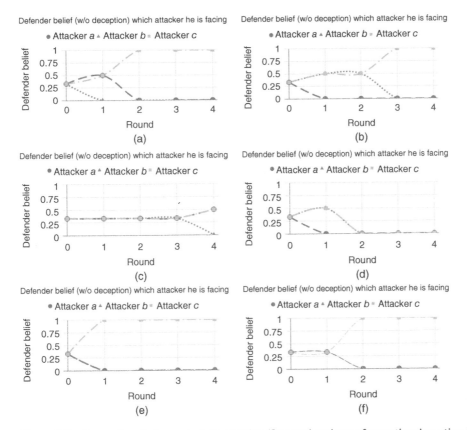

Figure 3.7 Comparison between just observation (first row) and use of proactive deception (second row). We increase shared exploits between the attackers. The edge density and shared vulnerabilities between nodes are fixed to 40%. (a) Unique exploits for the attackers. (b) 40% shared exploits between attackers. (c) All attackers have the same set of exploits. (d) Unique exploits for the attackers. (e) 40% shared exploits between attackers. (f) All attackers have the same set of exploits.

min-max-overlap performed better. From all the experiments, it is clear that the use of deception will speed the identification of an attacker that is very important in cybersecurity scenarios as different real-world attackers, e.g. APTs can lay low for a long time and detecting the attacker early can facilitate informed decision making for the defender.

3.8 Scalability

In our current game model, the defender makes a move after considering the attacker's last known position, and every possible way k HPs can be allocated between two real nodes in the network. Let's define $\sigma(i,j)$ as a slot where a HP can be allocated between node i and node j. The current algorithm makes sure that node i is always one hop distance away from the attacker's last known position and node j does not include the goal nodes to exclude trivial cases.

Using this approach the number of slots of $\sigma(i,j)$ where k HPs can be allocated increases very quickly with both network size (branching factors) and k that makes the algorithm not very scalable. For our initial experiments, we used a network of size 18. Now we will present a simple heuristics that makes sure that the algorithm can handle larger instances of networks compared to our initial approach.

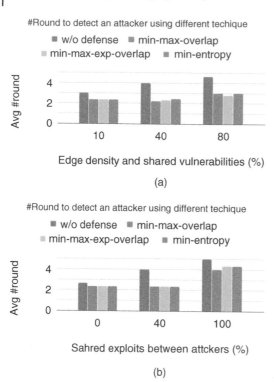

#Round to detect an attacker using different techique

■ w/o defense ■ min-max-overlap
■ min-max-exp-overlap ■ min-entropy

Edge density and shared vulnerabilities (%)

(a)

#Round to detect an attacker using different techique

■ w/o defense ■ min-max-overlap
■ min-max-exp-overlap ■ min-entropy

Sahred exploits between attckers (%)

(b)

Figure 3.8 Comparison between different techniques used by the defender.

3.8.1 Heuristics

In the real world, sensors, intrusion detection systems, and forensic analysis are used to collect data and to analyze the alerts and to understand the Techniques, Tactics and Procedures (TTP) used by an attacker. Utilizing the same tools and analytics it is also possible to form a belief on the approximate time interval of the attacks of an attacker and his preferences towards different features of the network, for example, OS, application, ports, hardware, etc. Rather than just focusing on TTP used by an attacker the defender can try to understand the attacker's behavior that drives the attacks and the attacker's propagation throughout the network. If the defender knows the last position of the attacker in the network, the time interval and preferences and attacker behavior can be used to get an estimate on his future attacks in the network to form a radius from the last known position of the attacker to consider the slots for HP allocation. Using these estimations on the attacker's future positions in the network, many of the unnecessary *slots* also can be filtered out.

In our game model, we do not capture this complex behavior of attacker preferences, behaviors, and intervals of attacks. We simplify it by assuming that the defender knows the attacker's last position in each round and the defender also knows that the attacker only moves one hop per round. Using these assumptions the defender can consider *slots* $\sigma(i,j)$ where the distance between node i and node j is always two hops and node i is always one hop away from the last known position of the attacker. It is also unnecessary to consider *slots* $\sigma(i,j)$ where node i and node j are only one hop away since it will only introduce costs to the paths and we assume that the attacker chooses a path that minimizes his costs. More unnecessary slots can be eliminated from considerations utilizing domain knowledge; for example, the defender knows that the attacker only moves forward (which is not realistic) however, this is one of our assumptions of the game model. As a result, it is not necessary to consider *slots* that are behind the attacker's last known position.

3.9   Evaluation of Heuristics

Our goal is to evaluate the heuristics we used to reduce the number of possible ways HPs can be allocated in the network to make the algorithm more scalable compared to our previous experiments. In the first experiment, we evaluate the solution quality of the algorithm with and without heuristics by comparing how early the defender can identify an attacker. We used different sizes of networks with different limits on the number of rounds the players can play. Each of the three attackers we considered, a, b, c, has some unique and some shared exploits in their possession. Attacker a has exploits ϕ_0, ϕ_1, ϕ_2. Attacker b has exploits ϕ_2, ϕ_3, ϕ_1. Attacker c has exploits ϕ_4, ϕ_5, ϕ_2. We picked the attacker b as the acting attacker. We kept the edge density and shared vulnerabilities between nodes to 40%. Results are averaged over 20 game instances. In the Figure 3.9, the first row, Figure 3.9a and b computes the solution without using the heuristics and the second row, Figure 3.9c and d computes the solution using the heuristics to reduce the action space. As we can see, the solution quality did not degrade for the experiments we conducted. However, currently, we have no proof to guarantee an optimal solution with the use of heuristics.

In the next experiment, we compare the run time between the algorithms using without and with heuristics for different sizes of networks and the same setup as the previous experiment. Run times are averaged over all the attackers for a particular size of games. As shown in Figure 3.10, for smaller instances the algorithm with heuristics can compute the solution much quicker than if we do not use any heuristics. For larger instances of games if we do not heuristics, the algorithm runs out of memory very quickly (as shown by "x"), whereas using the heuristic we can compute the solutions.

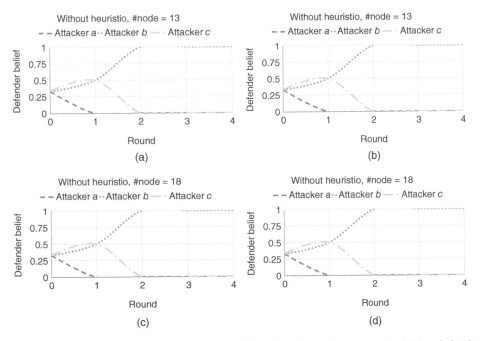

Figure 3.9 Comparison between algorithms without heuristics (first row) and with heuristics (second row). Edge density and shared vulnerabilities between nodes 40%. (a) Solution quality without heuristics #nodes= 13. (b) Solution quality without heuristics #nodes= 18. (c) Solution quality with heuristics #nodes= 13. (d) Solution quality with heuristics #nodes= 18.

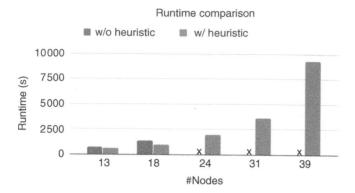

Figure 3.10 Comparison between run times of the algorithms for different sizes of networks. Edge density and shared vulnerabilities between nodes 40%.

The experiments above clearly show that if we reduce the action space, it is possible to make the algorithms faster. However, there can be other ways to increase the scalability performance of the algorithms. For example, we can reduce the graph itself to construct an abstracted version of the game. However, we will consider that as our future work.

3.10 Conclusions and Future Direction

Identification of an attacker is an even harder problem in many ways than detection, especially when many attackers use similar TTP in the early stages of attacks. However, any information that can help to narrow down the intention and likely TTP of an attacker can also be of immense value to the defender if it is available early on.

We present several case studies and a formal game model showing how we can use deception techniques to identify different types of attackers represented by the different AGs they use in planning optimal attacks based on their individual goals and capabilities. We show that strategically using deception can facilitate significantly earlier identification by leading attackers to take different actions early in the attack that can be observed by the defender. Our simulation results show this in a more general setting. However, the optimal algorithm does not scale very well since it considers all the possible action space to allocate HPs. So, we presented a scalable version of the optimal algorithm that reduces the action space by a reasonable margin to handle larger instances of games. In future work, we plan to explore how this model can be extended to different types of deception strategies, integration with other IDS techniques, as well as larger and more diverse sets of possible attacker types to make it operational in the real world.

References

APT33 2017. https://www.fireeye.com/blog/threat-research/2017/09/apt33-insights-into-iranian-cyber-espionage.html.

APT34 2017. https://www.fireeye.com/blog/threat-research/2017/12/targeted-attack-in-middle-east-by-apt34.html.

APT37(REAPER) 2017. https://www2.fireeye.com/rs/848-DID-242/images/rpt_APT37.pdf.

APT38 2018a. https://www.fireeye.com/blog/threat-research/2018/10/apt38-details-on-new-north-korean-regime-backed-threat-group.html.

APT38 2018b. Un-usual Suspects. https://content.fireeye.com/apt/rpt-apt38.

APT40 2019. https://www.fireeye.com/blog/threat-research/2019/03/apt40-examining-a-china-nexus-espionage-actor.html.

M-Trends 2019. https://content.fireeye.com/m-trends.

Defense Evasion 2018. https://attack.mitre.org/tactics/TA0005/.

Initial Access 2018. https://attack.mitre.org/tactics/TA0001/.

Lateral Movement 2018 https://attack.mitre.org/tactics/TA0008/.

Enterprise Tactics 2020. https://attack.mitre.org/tactics/enterprise/.

Karel Durkota, Viliam Lisỳ, Branislav Bošanskỳ, and Christopher Kiekintveld. Optimal network security hardening using attack graph games. In *Twenty-Fourth International Joint Conference on Artificial Intelligence, Buenos Aires, Argentina, 25–31 July 2015*, 2015.

Kyle Ingols, Richard Lippmann, and Keith Piwowarski. Practical attack graph generation for network defense. In *2006 22nd Annual Computer Security Applications Conference (ACSAC'06), Miami, FL, 11–15 December 2006*, pp 121–130. IEEE, 2006.

Christian Kreibich and Jon Crowcroft. Honeycomb: Creating intrusion detection signatures using honeypots. *ACM SIGCOMM Computer Communication Review*, 34(1):51–56, 2004.

Andrew Nicholson, Tim Watson, Peter Norris, Alistair Duffy, and Roy Isbell. A taxonomy of technical attribution techniques for cyber attacks. In *European Conference on Information Warfare and Security, Laval, France, 5–6 July 2012*, pp 188. Academic Conferences International Limited, 2012.

Steven Noel and Sushil Jajodia. Optimal IDs sensor placement and alert prioritization using attack graphs. *Journal of Network and Systems Management*, 16(3):259–275, 2008.

Xinming Ou, Wayne F. Boyer, and Miles A. McQueen. A scalable approach to attack graph generation. In *Proceedings of the 13th ACM Conference on Computer and Communications Security, Alexandria, VA, USA, 30 October–3 November 2006*, pp 336–345. ACM, 2006.

Frederic Raynal, Yann Berthier, Philippe Biondi, and Danielle Kaminsky. Honeypot forensics. In *Proceedings from the Fifth Annual IEEE SMC Information Assurance Workshop, New York, 10–11 June 2004*, pp 22–29. IEEE, 2004a.

Frederic Raynal, Yann Berthier, Philippe Biondi, and Danielle Kaminsky. Honeypot forensics, part II: Analyzing the compromised host. *IEEE Security & Privacy*, 2(5):77–80, 2004b.

Oleg Sheyner, Joshua Haines, Somesh Jha, Richard Lippmann, and Jeannette M Wing. Automated generation and analysis of attack graphs. In *Proceedings of the 2002 IEEE Symposium on Security and Privacy, Berkeley, California, 12–15 May 2002*, pp 273–284. IEEE, 2002.

Lance Spitzner. *Honeypots: Tracking Hackers*, volume 1. Addison-Wesley, Reading, 2003.

Nicholas Tsagourias. Cyber attacks, self-defense and the problem of attribution. *Journal of Conflict and Security Law*, 17(2):229–244, 2012.

David A. Wheeler and Gregory N. Larsen. Techniques for cyber attack attribution. *Technical Report*. Institute for Defense Analyses, 2003.

4

Honeypot Allocation Games over Attack Graphs for Cyber Deception

Ahmed H. Anwar[1], Charles A. Kamhoua[1], Nandi Leslie[1], and Christopher D. Kiekintveld[2]

[1] *US Army Research Laboratory, Adelphi, MD, USA*
[2] *Department of Computer Science, The University of Texas at El Paso, El Paso, TX, USA*

4.1 Introduction

The increasing attention in computer security research has lead to an increasing dependence on proactive defense approaches. Moreover, the need for hiding the network structure and vulnerabilities from adversaries made cyber deception a hot area for research and development. The huge growth of networks' sizes through added wireless-enabled devices. For instance, the Internet of Things (IoT), robots, sensors, etc., causes an ever-increasing networks' size (Forecast, 2019). Hence, such complex heterogeneous networks are more vulnerable to different types of attacks. Military tactical networks also face similar security challenges due to the increasing use of soldiers' wearable devices and the Internet of battlefield things (Russell and Abdelzaher, 2018).

The IoT network structure has been deployed in battlefield contexts, where it is known as the Internet of Battlefield Things (IoBT) (Kott et al., 2016; Kamhoua, 2018). In a broader sense, the IoBT also refers to devices useful for military battles that may communicate over tactical networks other than the Internet. Accordingly, securing the resilience and robustness of critical nodes of such a network on a battlefield is crucial. In this chapter, we propose an approach for optimizing cyber deception to prevent possible attackers from characterizing effective attack strategies. In fact, attackers spend time and effort collecting information about the networks before launching their attacks (Rowe and Goh, 2007). Information gathering (also called the reconnaissance stage) is the stage of attack in which an attacker collects inside information about the targeted network using a set of tools and scanning techniques (Engebretson, 2013). In some other scenarios, adversaries can gain intelligence information that maps the internal structure of the system/network. Attackers use software scanning tools like Nmap (Lyon, 2009) to map out the network structure. On the defender side, the network administrator aims to protect the network during this reconnaissance stage using cyber deception techniques to manipulate the network interfaces to disguise the true state of the network. The scope of this chapter is to investigate a realistic attack scenario in which the defender protects critical nodes and important system components via introducing false information (i.e. deception) to disrupt the attacker's decision-making, providing him with a false sense of certainty in a game-theoretic framework.

Cyber deception is a proactive defense approach in which the network administrators aim to expend adversaries' resources and time and gather information about the adversaries' strategies, tools, techniques, and targets through deploying decoy assets such as honeypots and fake nodes. Moreover, the network defender may choose to hide a subset of the network true node through

Game Theory and Machine Learning for Cyber Security, First Edition.
Edited by Charles A. Kamhoua, Christopher D. Kiekintveld, Fei Fang, and Quanyan Zhu.

camouflaging techniques. This lead to the development of a new class of security games, which is referred to as cyber deception games (CDGs). In this class of games, the defender manipulates the true state of the network using decoys and misleading signals as in Carroll and Grosu (2011), Li et al. (2018), Xu et al. (2005), Clark et al. (2012), Jajodia et al. (2015).

Recent work in literature formulated deception games using honeypots as a strategic game Çeker et al. (2016); Bilinski et al. (2018); Zhang and Zhu (2018). Schlenker et al. (2018) developed a novel defense scheme using a Stackelberg game framework that magnifies the uncertainty of the attacker. Hence, the adversary reconnaissance outcome is destroyed through misguiding the network scanning means. Hence, the attacker gets false information regarding the installed operating system on different nodes, port status, the number of different services, and the names of subnetworks and active users. Letchford and Vorobeychik (2013) proposed a Stackelberg game where the network defender optimizes over the type of mitigating technique to reduce the capability of the attacker to achieve his/her goals. The authors also investigated the complexity of the developed game model.

The importance of the developed game-theoretic models is that they shape our judgment of robustness for networked devices in a tactical network like in IoBT. Over the past years, a growing concern has developed over resilience from both industries and government. Since security breaches are becoming wider and more pervasive, hence cyber deception approaches are seen as an effective proactive in this regard. For instance, in 2019, the US Department of the Army issued a report on its strategy for military deception for multi-domain operations (ARMY, 2019), including network security, that addresses deception as a mechanism to prevent network attacks.

To this end, we formulate two-person zero-sum strategic form games to optimally allocate a set of honeypots and obtain practical deception strategy. Although, we have introduced this framework in part in Anwar et al. (2020), in this chapter, we characterize Nash equilibrium solution to the formulated game. Moreover, we extend the deception game model to consider ℓ levels of deception in the network. The complexity of the extended model is investigated and a scalable algorithm to overcome the exponential complexity of the game is proposed. We present numerical results that validate the proposed defense approach and the efficiency of the scalable solution as well.

The rest of this chapter is organized as follows: in Section 4.2, we present our model for the wireless network, along with the attacker and defender goals. We present our game formulation in Section 4.3 and present our scalable approach. We present our numerical results in Section 4.5 and conclude in Section 4.6 and discuss ongoing and future research to this work.

4.2 System and Game Model

In this section, we describe our system model, formulate the game between the two players, define the control variables of each player, and finally discuss the reward function and the game parameters.

4.2.1 Attack Graph

We consider an attack graph of N nodes represented as a graph $G(\mathcal{V}, \mathcal{E})$, where $N = |\mathcal{V}|$. Each node represents a vulnerability associated with a host or a machine in the network. An edge $e_{u,v} \in \mathcal{E}$ that is connecting two nodes u and v indicates reachability to exploit a vulnerability at node v through a vulnerability at node u. Further discussion on software vulnerabilities is presented in Chapter 12.

Each node in the graph has a value, w_v, that reflects its importance to the network administrator, and hence nodes with a higher value are considered a valuable asset to the network that contains important databases and critical information to the military group. Therefore, it is practical

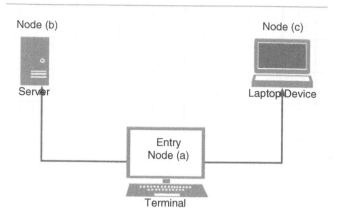

Figure 4.1 System model illustration.

to assume that nodes of high value are very attractive to adversaries and network attackers. The attacker wants to maximize his/her expected reward through wisely selecting a victim node among the set of all nodes, $\mathcal{V}$. We start our game formulation and analysis by assuming that the attacker knows the values associated with each node. This assumption is justified since attackers usually can obtain some internal information regarding the network structure and can probe nodes with network scanning tools (Lyon, 2009).

Figure 4.1 illustrates our system model and the associated game to be played between the network admin and the attacker. In this figure, we have three nodes, let for instance node $a \in \mathcal{V}$ represent an entry node to the network. Node a has two edges, $e_{a,b}$ and $e_{a,c}$ connecting it to node b and c, respectively. We consider a practical threat model for the attacker. More specifically, we assume that the defender does not know the exact location of the attacker. The defender knows that attackers can penetrate the network through a set of entry points. Since the network records can easily provide a distribution $f_a(.)$ over the set of entry points, where $\mathcal{V}_e \subset \mathcal{V}$ denotes the set of entry points. In other words, the defender knows the probability that an attacker will penetrate the network through an entry point $u \in \mathcal{V}_e$ is $f_a(u)$, such that $\sum_{u \in \mathcal{V}_e} f_a(u) = 1$. We can now readily define the game played at each of the entry points.

4.2.2 General Game Formulation

As introduced in Chapter 2, a two-player zero-sum strategic game is defined as a triple $(\mathcal{N}, \mathcal{A}, \mathcal{R})$, where,

- $\mathcal{N} = \{1, 2\}$ is the set of players, where player 1 is the network defender and player 2 is the attacker.
- $\mathcal{A} = \mathcal{A}_1 \times \mathcal{A}_2$ is the game action space, where $\mathcal{A}_1$ and $\mathcal{A}_2$ is the defender and attacker action spaces, respectively. An action profile that denotes a joint action taken by the two opponents (a_1, a_2) determines the reward received by both players.
- $\mathcal{R} = \{\mathcal{R}_1, \mathcal{R}_2\}$, where $\mathcal{R}_1 + \mathcal{R}_2 = 0$. $\mathcal{R} : \mathcal{A} \rightarrow \mathbb{R}^2$ is the reward function for the defender and $\mathcal{R}_2$ is the reward function for the attacker.

4.2.2.1 Defender Action

Considering a proactive deception approach as we did in Chapter 3, the defender is assumed to allocate a set of k honeypots along the attacker path to deceive the attacker. The placed attractive honeypots along his/her way will deviate the attacker from reaching real nodes. Honeypots are allocated as fake services along the different edges connecting between network vulnerabilities. If the attacker exploited such fake service (honeypot) placed over $e_{u,v}$ to reach node v from u, the defender tracks the new location of the attacker and the fake service will reveal important information about the attacker hacking techniques, etc. This tracking is extremely important for the defender to make his/her future defense actions. However, in this section, we focus on maximizing the defender reward due to tracking the attacker at every possible game. Therefore, the proposed solution represents the core of the solution to the dynamic game formulation that will be studied in Section 4.4.

Assuming that the attacker is attacking from entry node a as shown in Figure 4.1. The defender is allocating one honeypot and he needs to decide which edge to allocate the honeypot on. Recall that the defender is not certain about that the attacker is at node a, he also considers a no-allocation action as a possible choice to avoid the allocation cost as discussed below. Overall, the defender action in this case is to place the honeypot on $e_{a,b}$ or $e_{a,c}$ or no-allocation.

4.2.2.2 Attacker Action

On the attacker side, his/her action is to decide which node to attack next. Given the current example in Figure 4.1, the attacker can decide to exploit the vulnerability at node b or the one at node c. Since the attacker wants to remain stealthy, there is a cost per attack. Hence, the attacker may choose to back off completely in some circumstances to avoid such cost which we consider to abstract the risk of getting caught associated with the attack action. Finally, the attack action space is to either attack node b or c or to back off.

4.2.3 Reward Function

As introduced earlier, there is a cost associated with each action. The network defender incurs a fixed cost for placing a new honeypot on an edge in the network. Let P_c denote the honeypot placement cost. On the attacker side, there is a cost per attack denoted as A_c. The attack cost reflects the risk taken by the attacker as mentioned earlier.

If the defender placed honeypot on the same edge the attacker exploits, the defender gains a capturing reward. Otherwise, if the attacker exploited another safe edge, the attacker gains a successful attack reward. Let Cap and Esc denote the defender capture reward and the attacker successful reward, respectively.

To account for different nodes in the network, we adopt a reward function that takes into account the importance of the network nodes. Therefore, both capturing reward and the successful attack reward are weighted by the value of the secured or attacked node value, w_v. We start by expressing the reward matrix for the game illustrated in Figure 4.1 and present the general reward matrix afterward.

$$R_1 = \begin{bmatrix} -P_c + A_c + \text{Cap} * w_b & -P_c + A_c - \text{Esc} * w_c & -P_c \\ -P_c + A_c - \text{Esc} * w_b & -P_c + A_c + \text{Cap} * w_c & -P_c \\ A_c - \text{Esc} * w_b & A_c - \text{Esc} * w_c & 0 \end{bmatrix}. \tag{4.1}$$

The attacker reward matrix is $R_2 = -R_1$. The reward, $R_1(1, 1)$ and $R_1(2, 2)$, represents a captured attacker as the defender installed a honeypot at the attacked node. Therefore, the defender pays placement cost P_c and gains a capturing reward weighted by the value of the defended node. The attacker incurs an attack cost, A_c, which represents a reward for the defender in a zero-sum game. On the other hand, $R_1(1, 2)$ and $R_1(2, 1)$ represent a successful attack, where the honeypot is allocated at a different node. Therefore, there is a Esc loss weighted by the compromised node value. In $R_1(3, 1)$ and $R_1(3, 2)$, the defender decides not to place any honeypot to save placement cost P_c. Similarly, in $R_1(1, 3)$, $R_1(2, 3)$, and $R_1(3, 3)$, the attacker backs off to either avoid capture cost or attack cost A_c.

The reward function can easily be generalized to an arbitrary number of possible edges as follows.

$$R_1(a_1, a_2) = \begin{cases} -P_c + A_c + \text{Cap} * w_v; & a_1 = e_{a,v}, a_2 = v & \forall v \in \mathcal{V} \\ -P_c + A_c - \text{Esc} * w_u; & a_1 = e_{a,v}, a_2 = u & \forall u \neq v \in \mathcal{V} \\ -P_c; & a_1 = e_{a,v}, a_2 = 0 & \forall v \in \mathcal{V} \\ 0; & a_1 = 0, a_2 = 0 \end{cases}$$ (4.2)

where $a_1 = 0$ denotes the defender is not allocating any new honeypot. Similarly, $a_2 = 0$ denotes the attacker decided to back off.

4.2.4 Mixed Strategy

We have defined the actions available to each player in the game, i.e. $\mathcal{A}_1$ and $\mathcal{A}_2$. A pure strategy is a strategy that selects one of these actions. Alternatively, a player may choose to use a randomized (mixed) strategy defined through a probability distribution over these actions. Given the set of actions of player 1, $\mathcal{A}_1$, let $\Pi(\mathcal{A}_1)$ denote the set of all probability distributions over $\mathcal{A}_1$. Then, the set of mixed strategies for player 1 is $\Pi(\mathcal{A}_1)$, denoted by $\mathcal{X}_1$. Therefore, in a mixed strategy $\mathbf{X} \in \mathcal{X}_1$, action a_1^i is played with probability x_i such that,

$$\mathbf{X} = [x_1, x_2, \dots, x_n]^{\mathrm{T}},$$ (4.3)

where $n = |\mathcal{A}_1|$. Similarly, the attacker may also play a randomized strategy, $\mathbf{Y} = [y_1, y_2, \dots, y_m]^{\mathrm{T}}$, where $m = |\mathcal{A}_2|$.

Hence, the expected admin reward, denoted U_1, can be expressed as:

$$U_1 = \mathbf{X}^{\mathrm{T}} \mathbf{R}_1 \mathbf{Y}$$ (4.4)

Each player aims to maximize his/her own reward. In a zero-sum game, this implies minimizing the other player's reward. The expected utility for player 1 in equilibrium is $U_1 = -U_2$. The minimax theorem implies that U_1 holds constant in all equilibria and is the same value that player 1 achieves under a minimax strategy by player 2. Using this result, we can construct the optimization problem of player 1 as a linear program (LP) as follows:

$$\begin{aligned} \underset{\mathbf{X}}{\text{maximize}} \quad & U_1 \\ \text{subject to} \quad & \sum_{a_1 \in \mathcal{A}_1} r_1(a_1, a_2) x_{a_1} \geq U_1, \qquad \forall a_2 \in \mathcal{A}_2. \\ & \sum_{i=1}^{n} x_i = 1, \quad x_i \geq 0, \ i = 1, \dots, n \ . \end{aligned}$$ (4.5)

Similarly, the attacker solves the counter minimization problem as follows:

$$\begin{aligned} \underset{\mathbf{Y}}{\text{minimize}} \quad & U_1 \\ \text{subject to} \quad & \sum_{a_2 \in \mathcal{A}_2} r_1(a_1, a_2) y_{a_2} \leq U_1, \qquad \forall a_1 \in \mathcal{A}_1. \\ & \sum_{i=1}^{m} y_i = 1, \quad y_i \geq 0, \ i = 1, \dots, n \ . \end{aligned}$$ (4.6)

The first constraint follows from the definition of the Nash Equilibrium. Therefore, the expected reward is greater than the value of the game. Since the value of the game depends on the mixed strategy played by player 2 (the attacker), the admin should constrain his/her response to the best response set that guarantees a higher reward. The remaining two constraints ensure that $\mathbf{X}$ is a valid probability distribution. The attacker solves a corresponding LP that can be characterized along the same lines to satisfy that the optimal mixed strategy $\mathbf{Y}$ is the best response for every possible action played by the defender. The resulting mixed strategies forms a Nash equilibrium for the two players.

4.2.5 System Parameters

In our model, we introduced four different parameters each affects the attacker and defender actions. To better understand the effect and role of each of these parameters, we analyze the game NE theoretically.

For the sake of simplicity and to avoid distractions, we consider a toy graph. In this scenario, we assume two nodes, u, v, connected through a single edge. The attacker is currently entering the network through node u. The attacker has to decide whether to attack the following node v or to back off. On the other side, the defender decides to allocate a honeypot over the existing edge or to save the cost of honeypot allocation. To derive the game equilibrium in mixed strategies, let the attacker attack node v with probability y and the defender allocate a honeypot with probability x. The reward matrix can be extracted from (4.4) by removing the second row and second column and replacing w_b by w_v. Therefore, the defender expected reward can be expressed as:

$$u_d(x, y) = (-P_c - A_c + Cap.w_v)xy - P_c(1 - y)x + (A_c - Esc.w_v)y(1 - x). \tag{4.7}$$

For every attack strategy y adopted by the attacker, the defender reward:

$$u_d(x) = (-2A_c y + (Cap + Esc)w_v y - P_c)x + (A_c - Escw_v)y; \quad \forall y \in [0, 1]. \tag{4.8}$$

As shown in (4.8), the defender expected reward forms a line equation, i.e. $u_d(x) = mx + b$. If the slope of this line is negative, the defender best response for the attacker strategy y is to play $x^* = 0$. On the other hand, if $m > 0$, the defender best response is $x^* = 1$, i.e. the defender allocates the honeypot through edge $e_{u,v}$. If the slope $m = 0$, this makes the defender action indifferent to the played strategy y. Hence, it satisfies the equilibrium point of the game.

Lemma 4.1 For the formulated game, the defender surely allocates a honeypot if the allocation cost A_c satisfies the following condition:

$$A_c < \frac{(Cap + Esc)w_v y - P_c}{2y}; \quad \forall y \in [0, 1], \tag{4.9}$$

Lemma 4.2 Similarly, the attacker decides to back off if the cost upon captured Cap satisfies the following condition:

$$Cap > \frac{Esc(1 - x) - A_c(1 - 2x)}{w_v}; \quad \forall x \in [0, 1]. \tag{4.10}$$

Theorem 4.1 The formulated game has a Nash equilibrium in mixed strategies at $x^* = \frac{Escw_v - A_c}{(Cap + Esc)w_v - 2A_c}$ and $y^* = \frac{P_c}{(Cap + Esc)w_v - 2A_c}$. Moreover, the game admits a Nash equilibrium in pure strategies such that the defender allocates a honeypot and the attacker backs off if Lemma 4.1 and $Cap > A_c/w_v$.

Proof: The proof of the theorem follows directly from NE definition. The mixed strategy x^* stated in the theorem ensures that the defender reward, 4.8, is indifferent to the attacker action; hence, the defender has no incentive to deviate from this point. Similarly, y^* makes the attacker reward indifferent to the defender strategy. In the second part of the theorem, the equilibrium characterized in pure strategies follows directly from Lemmas 4.1 and 4.2.

Game Complexity: Although, the run-time complexity of the LPs in (4.5) grows linearly with the starting node degree. Therefore, the complexity of a single allocation is independent of the network size which can be larger number than node degrees. Hence, one can efficiently solve the optimization in (4.5) and find the Nash equilibrium strategy in mixed strategies for each players. If the defender is allocating multiple honeypots to cover a set of edges at a time, the complexity of the game will be exponentially growing with the number of allocated honeypots. To over come this prohibitive complexity, in the following section, we propose a progressive decomposition algorithm.

4.3 Allocating ℓ Honeypots Model

Instead of securing immediate neighboring nodes, in this section, we extend the game model where the defender is trying to protect a larger subset of nodes. In such a model, the defender is securing a subset of nodes that are located within ℓ-hops away from the entry node. Practically, the attacker follows an attack path to reach a targeted node. Therefore, the defense strategy is optimized to allocate honeypots that cover a specific path. A path is represented by a collection of connected edges.

Following the same game formulation, to allocate ℓ honeypots, both players extend their action spaces to consider all paths in the network. However, we need to enumerate all the possible pure actions for each player as in (4.5). Obviously, the complexity of the extended optimization problem is now growing exponentially in ℓ. To overcome such problem, we propose a progressive decomposition-based algorithm.

4.3.1 The Algorithm

The reward function of allocating ℓ honeypots can be expressed as a sum over ℓ-steps reward functions as follows:

$$\bar{R}_1 = \sum_{h=1}^{\ell} R_1(a_1^h, a_2^h); \ (a_1^h, a_2^h) \in \mathcal{A}_1^h \times \mathcal{A}_2^h. \tag{4.11}$$

Note that the action spaces for both players in each the hth hop depends on their action history. Therefore, in order to redefine the action spaces, one needs to keep track of the previous actions taken by both players. This implicit dependency is coupling the objective function of the game LP.

To overcome this issue, we start from the network entry point where all the information of the game is available to the players. After the one-hop depth game is played, the resulting mixed strategies are passed forward to the next set of nodes. Hence, we know the probability of being at each of the nodes that are located two hops away from the entry node and the new game is played with a known probability, since we are passing the beliefs from one hop to the other. The new game is now properly defined. In other words, the new action spaces for both players are known. After repeating this procedure ℓ times, we multiply the mixed strategies of allocating a honeypot at every edge over all the possible paths and normalize the resulting strategies to have a well-defined probability distribution.

Algorithm 4.1 Progressive decomposition-based algorithm.

Result: **Normalize X, Y**
$h = 1$;
while $h \leq \ell$ **do**
   Define all games $h - 1$ away from entry;
   Calculate probability of each game: P_g using $\mathbf{X}^{h-1}, \mathbf{Y}^{h-1}$;
   if $P_g > 0$ **then**
      Solve LP associated with game i;
      Find $\mathbf{X}^h_i, Y^h_i$;
      Move to the next game;
   else
      Move to the next game;
   end
   $h + +$;
   Forward $\mathbf{X}^h, \mathbf{Y}^h$.
end

In the following section, we present our numerical results that show the effectiveness of the presented game model, defense strategies, as well as the scalability of the decomposition algorithm.

4.4 Dynamic Honeypot Allocation

In the above section, the formulated game for a one-shot case, in this section, we formulate a dynamic game between the network defender and the attacker. In a dynamic environment, both players play the game in different settings every time. For example, the two players experience different network connectivity at different instances. Also, the network entry point can change, as well as the attacker location. Therefore, to understand the development of this game between the two players and its evolution, we model it as a dynamic game to capture the state of the game after the actions are played by both the defender and the attacker. Later in Chapter 6, the theory of hypergames on graphs is discussed to investigate the case of a defender with private information about the attacker.

To model the game state at each instant, we introduce a state variable s_k. Hence, the dynamic game Γ is defined as the tuple $(\mathcal{N}, \mathcal{A}, \mathcal{S}, \mathcal{P}, \mathcal{R})$, where $\mathcal{S}$ is a finite set of states and $\mathcal{P} : \mathcal{S} \times \mathcal{A} \times \mathcal{S} \rightarrow [0, 1]$ is the transition probability function between states, i.e. $P(s, a, s')$ is the probability of transitioning to a future state s' from the current state s after action profile $a = (a_1, a_2)$. The action space $\mathcal{A}$ and reward function $\mathcal{R}$ are as defined in Section 4.2.

A state $s \in \mathcal{S}$ is a normal form game played by two players, where $\mathcal{S} = \{s^1, \ldots, s^n\}$ and $n = |\mathcal{S}|$. We generally use s to denote an arbitrary state. A state at time t is denoted $s_t = (\mathcal{E}_t, \mathcal{I}_t, \mathcal{T}_t)$, where $\mathcal{E}_t$ is the set of vulnerabilities on the attack graph at state s_t, which is observable to both players. The set $\mathcal{I}_t$ denotes the identity set of each vulnerability. Finally, the set $\mathcal{T}_s$ is the set of exploited vulnerabilities up to the current stage s.

4.4.1 Mixed Strategy, State Evolution, and Objective Function

Unlike MDPs, the optimal stationary policy for dynamic games need not be deterministic. Since we will be searching for an optimal mixed (randomized) stationary policy, we start with a definition of stationary mixed strategies for both players.

Let $x^i(s), i = 1, \dots, |\mathcal{A}^1|$, and $y^j(s), j = 1, \dots, |\mathcal{A}^2|$, denote the probability that the defender and attacker play the pure action $d^i \in \mathcal{A}^d$ and $a^j \in \mathcal{A}^a$ while at state s, respectively. A mixed strategy is then defined as a probability distribution over the whole action space. Specifically, given a state s, $\mathbf{x}(s) = [x^1(s), \dots, x^{|\mathcal{A}_d|}(s)]^T$ and $\mathbf{y}(s) = [y^1(s), \dots, y^{|\mathcal{A}_a|}(s)]^T$ are the mixed strategies for the defender and the attacker at state s, respectively. A stochastic stationary policy is readily defined as $\pi = \{\mathbf{x}(s^1), \dots, \mathbf{x}(s^n)\}$ for the defender, and $\theta = \{\mathbf{y}(s^1), \dots, \mathbf{y}(s^n)\}$ for the attacker, where $n = |S|$ is the total number of states.

In general setup, the game evolves from state s at time t to state s' at time $t + 1$ according to the evolution equation:

$$s' = f(s, a^{(d)}, a^{(a)}, \zeta_t) \tag{4.12}$$

for some deterministic function f, where ζ_t is a random variable representing an exogenous control variable based on the environment or the system randomness. However, in our game, such variable does not exists. Equation (4.12) defines the transition probability $P(s, a, s')$.

The defender goal is to maximize the expected sum of discounted rewards. Under randomized stationary defense and attack policies π and θ, the expected sum of discounted rewards starting from state $s \in S$ at time $t = 0$ is,

$$V(s, \pi, \theta) = \mathbb{E}\left[\sum_{t=0}^{\infty} \gamma^t R\left(s_t, a_t^{(d)}, a_t^{(a)}, s_{t+1}\right) | s_0 = s, \pi, \theta\right], \tag{4.13}$$

where the expectation is over the randomness of the players' actions (recalling that π and θ are randomized policies) and the state evolution (the exogenous variables). Subscript t denotes the tth stage and $0 < \gamma < 1$ is a discount factor. The goal is to solve for the optimal stationary policies for both players, i.e. obtain the solution to

$$V^*(s) := V(s, \pi^*, \theta^*) = \max_\pi \min_\theta V(s, \pi, \theta). \tag{4.14}$$

The optimal randomized stationary policies $\mathbf{x}^*(s)$ and $\mathbf{y}^*(s)$ for state s are the solutions to

$$V^*(s) = \max_{\mathbf{x}(s)} \min_{\mathbf{y}(s)} \mathbb{E}\left[R(s, a^{(d)}, a^{(a)}, s') + \gamma V^*(s') | \mathbf{x}(s), \mathbf{y}(s)\right] \tag{4.15}$$

where the immediate reward term is given by

$$\mathbb{E}\left[R(s, a^{(d)}, a^{(a)}, s') | \mathbf{x}(s), \mathbf{y}(s)\right] = \sum_{s' \in S} \sum_{i=1}^{|\mathcal{A}^d|} \sum_{j=1}^{|\mathcal{A}^a|} R(s, d^i, a^j, s') x^i(s) y^j(s) p(s' | s, d^i, a^j). \tag{4.16}$$

Thus, the optimal stationary policies are $\pi^* = \{\mathbf{x}^*(s^1), \dots, \mathbf{x}^*(s^n)\}$, $\theta^* = \{\mathbf{y}^*(s^1), \dots, \mathbf{y}^*(s^n)\}$.

4.4.2 Q-Minmax Algorithm

MDPs (single-player Markov game) can be solved using value iteration (Bertsekas, 1995). There, the total expected discounted reward $V^*(s)$ is termed the value of state s since larger rewards are collected from states with larger values of V^*. The value of a state $V^*(s)$ satisfies

$$V(s) = \max_{a' \in \mathcal{A}} Q(s, a'), \tag{4.17}$$

where $Q(s, a)$ is the quality of the state-action pair (s, a) defined as the total expected discounted reward attained by the nonstationary policy that takes action a then follows with the optimal policy from this point on. Given two players, this notion can be extended so that

$$Q(s, a^{(d)}, a^{(a)}) = \mathbb{E}\left[R(s, a^{(d)}, a^{(a)}, s') + \gamma V^*(s') | a^{(d)}, a^{(a)}\right] \tag{4.18}$$

To solve (4.14) for the optimal policy, we adopt the Q-minmax value-iteration-based algorithm of Littman (1994) extended to zero-sum stochastic games, which replaces the maximization in (4.17) with a maxmin operator to account for the actions of the second player.

$$V^*(s) = \max_{x(s)\in\chi(\mathcal{A}^d)} \min_{a^{(a)}\in\mathcal{A}^a} \sum_{i=1}^{|A^d|} Q(s, d^i, a^{(a)})x^i(s), \tag{4.19}$$

where $\chi(\mathcal{A}^d)$ is the space of probability distributions defined over $\mathcal{A}^d$.

4.5 Numerical Results

In this section, we present our numerical results and discuss the effect of game parameters as defined in Section 4.2.

In Figure 4.2, we plot the defender reward at different attack costs. It is obvious that the defender reward increases as the attack cost increases. As the attack cost increases, the rational attacker tends to back off more often (i.e. with higher probability). Therefore, the reward of the defender increases. We also compare the reward when the defender adopts NE deception strategy to random allocation strategy to show the effectiveness of our game formalism. The shown comparison is for the network shown in Figure 4.1, of only two nodes. As the number of nodes increases, the gap between the optimal reward and random deception reward will increase dramatically. These results are obtained at $Esc = 2, P_c = 5$ and $Cap = 2$.

In Figure 4.3, we plot the defender reward versus the placement cost, P_c. As shown in the figure, the defender reward decreases as the cost per allocation increases. However, the NE deception strategy is yielding higher reward for the network defender.

To show the effectiveness of the proposed decomposition algorithm, we compare the run time required to obtain exact random strategy to the decomposition-based randomized strategy in Figure 4.4. In Figure 4.5, we plot the defender reward on a seven-node network as plotted in Figure 4.6. For this network, $\ell = 2$. The algorithm solved one game in the first round and two other games at $h = 2$. The proposed algorithm is yielding a better reward for the defender that is

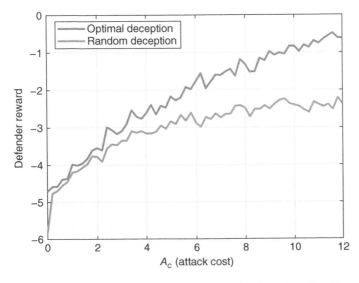

Figure 4.2 Comparing defender reward at NE and random allocation at different attack cost.

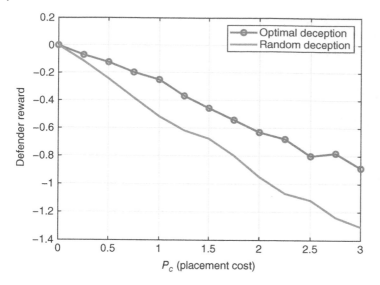

Figure 4.3 Comparing defender reward at NE and random allocation at different placement cost.

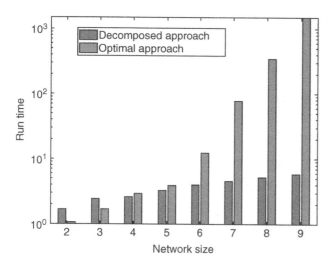

Figure 4.4 Run-time comparison between decomposition-based algorithm and optimal approach algorithm.

higher than simply randomized the allocation policy blindly. It is also shown in Figure 4.5 the defender reward when the attacker is having stronger computational capabilities and can solve the full game while the defender is using the decomposition-based algorithm solely.

In order to show the effectiveness of the proposed deception algorithm and understand the dynamics of the game played between the network defender and the attacker, next we plot the results of Q-minmax algorithm described in Section 4.4. Figure 4.7 illustrates the convergence of the state value function defined in 4.19. The shown value function corresponds to state 10 and it converges after 15 steps. Moreover, in Figure 4.8, we plot the learning curve for the defender and the attacker converging to their optimal strategies. For the shown state, both the attacker and the defender have two possible actions; hence, the optimal strategy for the defender can be characterized as p and $1 - p$, and for the attacker as q and $1 - q$.

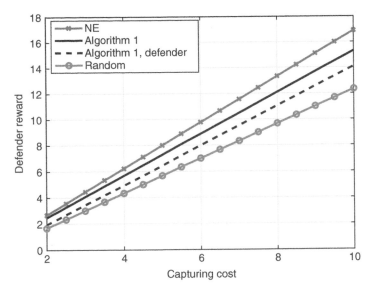

Figure 4.5 Algorithm is yielding a comparable reward to the full-game Nash equilibrium.

Figure 4.6 A 7-node tree network topology.

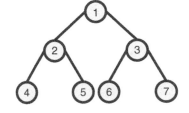

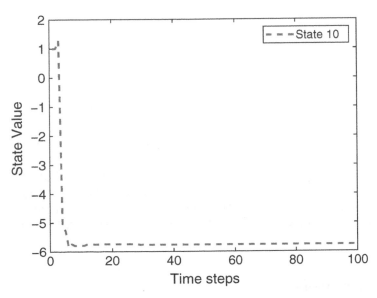

Figure 4.7 State Value function converges over time using Q-minmax algorithm.

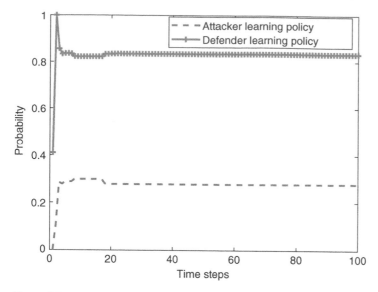

Figure 4.8 Players strategies learned over time.

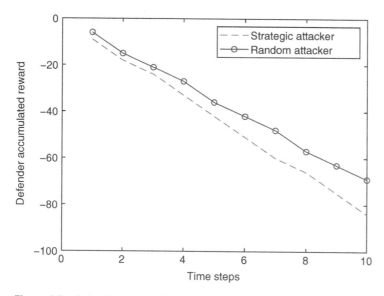

Figure 4.9 Defender-accumulated reward comparing NE reward and random attack strategy.

Finally, in Figure 4.9, we compare the defender-accumulated reward over time when both players adopt NE strategies and when the attacker is less informative and adopts a random attack strategy. As shown deviating from the NE strategy, the attack reward can never yield a higher reward. Hence, for the zero-sum game, the defender reward is higher when the attacker deviates from NE strategy and attacks randomly.

4.6 Conclusion and Future Work

In this chapter, a cyber deception approach has been proposed through allocating honeypots over the network topology in a game-theoretic framework. To this end, several game models were developed, and a scalable allocation algorithm has been developed to over computationally prohibitive games. This novel game formulation captured the graph structure as well as different node values. The game model investigated the trade-off between security cost and deception reward for the network administrator. On the other side, the attacker is assumed to maximize her reward through compromising the network without revealing being captured in honeypots. Nash equilibrium strategies have been analyzed. The general game complexity has been discussed, and a scalable algorithm is developed to ensure that the defense approach applies to large-scale networks. Moreover, a dynamic version of the game model was investigated to study the evolution of the game over time.

Acknowledgment

Research was sponsored by the Army Research Laboratory and was accomplished under Cooperative Agreement Number W911NF-19-2-0150. The views and conclusions contained in this document are those of the authors and should not be interpreted as representing the official policies, either expressed or implied, of the Army Research Laboratory or the US Government. The US Government is authorized to reproduce and distribute reprints for government purposes notwithstanding any copyright notation herein.

References

Ahmed H. Anwar, Charles Kamhoua, and Nandi Leslie. Honeypot allocation over attack graphs in cyber deception games. In *2020 International Conference on Computing, Networking and Communications (ICNC)*, pp. 502–506. IEEE, 2020.

Dimitri P. Bertsekas. *Dynamic Programming and Optimal Control*, volume 1. Athena Scientific Belmont, MA, 1995.

Mark Bilinski, Ryan Gabrys, and Justin Mauger. Optimal placement of honeypots for network defense. In *International Conference on Decision and Game Theory for Security*, pp. 115–126. Springer, 2018.

Thomas E. Carroll and Daniel Grosu. A game theoretic investigation of deception in network security. *Security and Communication Networks*, 4(10): 1162–1172, 2011.

Hayreddin Çeker, Jun Zhuang, Shambhu Upadhyaya, Quang Duy La, and Boon-Hee Soong. Deception-based game theoretical approach to mitigate dos attacks. In *International Conference on Decision and Game Theory for Security*, pp. 18–38. Springer, 2016.

Andrew Clark, Quanyan Zhu, Radha Poovendran, and Tamer Başar. Deceptive routing in relay networks. In *International Conference on Decision and Game Theory for Security*, pp. 171–185. Springer, 2012.

Department of the Army. Army support to military deception. 2019. https://fas.org/irp/doddir/army/fm3-13-4.pdf.

Patrick Engebretson. *The Basics of Hacking and Penetration Testing: Ethical Hacking and Penetration Testing Made Easy*. Elsevier, 2013.

Global Mobile Data Traffic Forecast. Cisco visual networking index: Global mobile data traffic forecast update, 2017–2022. *Update*, 2017:2022, 2019.

Sushil Jajodia, Paulo Shakarian, V.S. Subrahmanian, Vipin Swarup, and Cliff Wang. *Cyber Warfare: Building the Scientific Foundation*, volume 56. Springer, 2015.

Charles A. Kamhoua. Game theoretic modeling of cyber deception in the internet of battlefield things. In *2018 56th Annual Allerton Conference on Communication, Control, and Computing (Allerton)*, pp. 862–862. IEEE, 2018.

Alexander Kott, Ananthram Swami, and Bruce J West. The internet of battle things. *Computer*, 49(12):70–75, 2016.

Joshua Letchford and Yevgeniy Vorobeychik. Optimal interdiction of attack plans. In *Proceedings of the 2013 International Conference on Autonomous Agents and Multi-agent Systems*, pages 199–206. International Foundation for Autonomous Agents and Multiagent Systems, 2013.

Yapeng Li, Yu Xiao, Yong Li, and Jun Wu. Which targets to protect in critical infrastructures-a game-theoretic solution from a network science perspective. *IEEE Access*, 6:56214–56221, 2018.

Michael L. Littman. Markov games as a framework for multi-agent reinforcement learning. In *Proceedings of the Eleventh International Conference on Machine Learning*, volume 157, pp. 157–163, 1994.

Gordon Fyodor Lyon. *Nmap Network Scanning: The Official Nmap Project Guide to Network Discovery and Security Scanning*. Insecure, 2009.

Neil C. Rowe and Han C. Goh. Thwarting cyber-attack reconnaissance with inconsistency and deception. In *Information Assurance and Security Workshop, 2007. IAW'07. IEEE SMC*, pp. 151–158. IEEE, 2007.

Stephen Russell and Tarek Abdelzaher. The internet of battlefield things: The next generation of command, control, communications and intelligence (c3i) decision-making. In *MILCOM 2018-2018 IEEE Military Communications Conference (MILCOM)*, pp. 737–742. IEEE, 2018.

Aaron Schlenker, Omkar Thakoor, Haifeng Xu, Fei Fang, Milind Tambe, Long Tran-Thanh, Phebe Vayanos, and Yevgeniy Vorobeychik. Deceiving cyber adversaries: A game theoretic approach. In *Proceedings of the 17th International Conference on Autonomous Agents and MultiAgent Systems*, pp. 892–900. International Foundation for Autonomous Agents and Multiagent Systems, 2018.

Wenyuan Xu, Wade Trappe, Yanyong Zhang, and Timothy Wood. The feasibility of launching and detecting jamming attacks in wireless networks. In *Proceedings of the 6th ACM International Symposium on Mobile ad hoc Networking and Computing*, pp. 46–57. ACM, 2005.

Tao Zhang and Quanyan Zhu. Hypothesis testing game for cyber deception. In *International Conference on Decision and Game Theory for Security*, pp. 540–555. Springer, 2018.

5

Evaluating Adaptive Deception Strategies for Cyber Defense with Human Adversaries

Palvi Aggarwal[1], Marcus Gutierrez[2], Christopher D. Kiekintveld[2], Branislav Bošanský[3], and Cleotilde Gonzalez[1]

[1] Department of Social and Decision Sciences, Carnegie Mellon University, Pittsburgh, PA, USA
[2] Department of Computer Science, The University of Texas at El Paso, El Paso, TX, USA
[3] Department of Computer Science, Czech Technical University in Prague, Prague, Czechia

5.1 Introduction

More than two thousand years ago, the Chinese military strategist Sun Tzu defined **deception** as one of the most important principles of war (as explained in the 2005 translated text; Giles et al. 1910). Deception has indeed been an important military strategy in many naturalistic conflicts. Notably, in World War II, decoy equipment and vehicles were used to appear as actual military equipment; while they were made of plywood and inflatable materials, they were able to give the aerial impression of real equipment (Garber 2013).

While techniques for deception may appear obvious in the physical world, they are less obvious in cyberspace. This is because of the multidimensional dynamics of the cyberworld, where geography and physical proximity do not matter; where the adversary has a strong advantage over the defender; and where the effects of malicious actions appear less concerning given the "intangible" items (e.g., information) that one deals with (Geers 2010). Despite these difficulties, deception is a common technique for cyber defense. In fact, deception is a powerful technique; the use of it causes adversaries to be confused, waste their time and leverage the wrong information to conduct their crimes (Ferguson-Walter et al. 2018; Gonzalez et al. 2020). Some of the common techniques of deception in cybersecurity include masking of information to make the real objects more difficult to identify and creating decoys to present false objects that grab the attention of the human adversaries.

Honeypots are perhaps the most common example of decoys, and they have been used for decades (Spitzner 2003). Honeypots are used for detection to catch illicit interactions; in prevention, to assist in slowing adversaries down; and many other defense strategies (Almeshekah and Spafford 2016). However, to correctly implement honeypots, one must not only have a solid understanding of the technologies behind honeypots, but also a solid understanding of the human adversary behaviors. Honeypots are effective when they can manipulate their features in a way that can make them more believable to adversaries (i.e., successfully deceive adversaries) because the value of honeypots depends on the adversary's ability to detect and avoid honeypots. In Chapter 3, Basak et al. (2020) demonstrate the strategic use of deception in the form of honeypots for early detection and identification of attacker type represented by the attack graphs.

Game Theory and Machine Learning for Cyber Security, First Edition.
Edited by Charles A. Kamhoua, Christopher D. Kiekintveld, Fei Fang, and Quanyan Zhu.
© 2021 The Institute of Electrical and Electronics Engineers, Inc. Published 2021 by John Wiley & Sons, Inc.

A significant issue with current techniques to design and allocate honeypots is that the algorithms of defense are often static (Ferguson-Walter et al. 2019). Chapter 4, proposed the scalable honeypot allocation algorithms over the attack graphs. Human adversaries, in turn, are excellent at learning with little or no information and adapting to changing conditions of an environment (Cheyette et al. 2016). As a result, many of the current static defense allocation techniques are ineffective.

In this chapter, we study the effectiveness of a set of defense algorithms against human adversaries. Importantly, these algorithms vary in three main dimensions: determinism, adaptivity, and customization. These algorithms are studied in the context of an abstract "HoneyGame" in human experiments. Before we consider the defense algorithms, we first introduce the HoneyGame, which will serve to illustrate the mechanics of the algorithms. Next, we present the human experiments and the results comparing the effectiveness of these defense algorithms. We conclude with a discussion of the importance of developing adaptive algorithms that are informed by models that accurately represent human behavior.

5.1.1 HoneyGame: An Abstract Interactive Game to Study Deceptive Cyber Defense

The HoneyGame is an abstract representation of a common cybersecurity problem. A defender (i.e., the defense algorithm) assigns decoys to protect network resources and an adversary (i.e., a human) aims to capture those resources. This game has been reported in past studies, which report partial findings of the work we present in the current chapter (Gutierrez et al. 2019).

A screenshot of the user interface of the HoneyGame is shown in Figure 5.1. In this game, a network has five nodes. Each node in the network is assigned a v_i value of node i; a c_i^a cost of attacking node i, and a c_i^d cost of defending node i. The reward $v_i - c_i^a$ of attacking a nonhoneypot appears as a positive number on top of each node. The cost of attacking a honeypot $-c_i^a$ appears as a negative number at the bottom of the node.

Table 5.1 shows the specific values used in the HoneyGame in the experiments reported in this chapter. We designed the node values to fit common risk-reward archetypes (e.g., low-risk/low-reward, high-risk/high-reward, low-risk/high-reward). The explicit values shown in each node give an adversary the possibility of making informed decisions that will be combined with experiential decisions by Lejarraga et al. (2012). However, the adversary never knows the policy of the defender (i.e., the probabilities of a node being a honeypot).

At the beginning of each round, the defender spends a budget D to turn some subset of the nodes into honeypots, such that the total cost is $\leq D$. Once the defender deploys honeypots, the adversary selects a node to attack. If the adversary's chosen node i is not a honeypot, the adversary receives

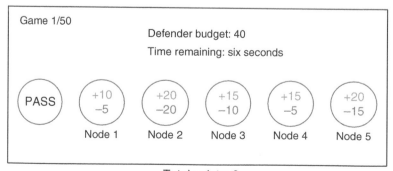

Total points: 0

Figure 5.1 User interface for the HoneyGame.

Table 5.1 Node parameters for the online human experiment.

	Pass	Node 1	Node 2	Node 3	Node 4	Node 5
v_i	0	15	40	35	20	35
c_i^a	0	5	20	10	5	15
c_i^d	0	10	20	15	15	20

the reward $v_i - c_i^a$, and the defender receives a reward of 0. If the adversary's chosen node i was a honeypot, the adversary receives the negative reward $-c_i^a$, and the defender receives the positive reward v_i.[1] However, the adversary may also decide to "pass", which is a "safe" choice, where an adversary and the defender receive a reward of 0.

Each round is composed of a set of 50 rounds. At the end of each round, the game resets, so that at the beginning of each round a new action from the defense algorithm is drawn (i.e., a combination of nodes are selected to be honeypots). The adversaries are only informed of the rewards they receive after each action, and they do not directly observe all the selected honeypots in the defense action.

5.2 An Ecology of Defense Algorithms

The importance of advancing defense strategies from static to dynamic has been widely recognized (Cranford et al. 2020; Ferguson-Walter et al. 2018, 2019; Gonzalez et al. 2020). Dynamic defenses should be able to actively change a defense strategy according to the actions taken by adversaries during an attack. Previous work has not tested the benefits of the ability for defense strategies to dynamically adapt to the actions taken by adversaries. Table 5.2 presents six defense algorithms of defense that vary in three dimensions: determinism, adaptivity, and customization.

The algorithms are said to be **nondeterministic** if, in each round, the defense strategy is not fixed, but rather it is sampled from a distribution. An algorithm is **adaptive** if it considers its past decisions and the history of rewards actively to change the defense strategy in the future rounds. An algorithm is said to be **customized** to the adversary when it relies on an active adversarial algorithm to modify the defense strategy in each round.

Table 5.2 Features of the six algorithms presented in this chapter.

	Deterministic	Adaptive	Customization
Pure	yes	No	rationality
Equilibrium	No	No	rationality
LLR	Yes	Yes	None
BR-TS	Yes	Yes	sampling
PBR-TS	No	Yes	sampling
FTRL	No	Yes	regularizer

---

1 We assume $v_i \geq c_i^a$ and $\sum_{i \in N} c_i^d > D$.

Table 5.2 highlights the high-level feature differences between our various defenders which will each be detailed in the coming sections. Here, we would like to note the differences in their customizability. Pure Static and Static Equilibrium both come from basic game-theory and utilize the Nash equilibrium concept to calculate their strategies. Nash equilibrium assumes both players are fully rational and are attempting to maximize their individual outcomes. This rationality assumption can be swapped out for any other type of player (e.g., a punishing attacker or a max-min attacker). LLR has no room for customization as only the exploration constant, L, may be changed, but even this constant is kept to the original authors' recommendation. The two Best Response algorithms utilize random sampling to simulate the next move of the attacker over thousands of samples. In this paper, we leverage Thompson sampling as the Best Response's attacker model; however, this could naturally be swapped out for other predictor models. Lastly, FTRL utilizes function regularizers to balance over-fitting with quick learning. If FTRL learns too quickly it can be "trained" to play in a desired way by the attacker then exploited in later rounds. Conversely, the defender needs to learn quickly otherwise be exploited early on for quick gains. In summary, each algorithm has some degree of customizability in its assumptions about the attacker's decision-making.

5.2.1 Static Pure Defender

The *Static Pure Defender* employs a "set and forget" defense that implements an unchanging, greedy strategy that spends its defense budget to protect the highest valued nodes. We derived this defender from the game-theoretic process of finding a nonchanging Pure Strategy Nash Equilibrium. For the scenario in Figure 5.1, the defender always sets nodes 2 and 5 as honeypots, leading to nodes 3 and 4 being the optimal ones to attack, which will always yield 15 points each to the adversary every round. Against this defender, the adversary can gain a maximum of 750 total points in this specific scenario by always attacking nodes 3 or 4 for all 50 rounds.

This defender acts as our baseline for how quickly humans learn static defense strategies. Although this defender is trivial in nature and should perform poorly as a competitive defender, it is worth the analysis when comparing predictor algorithms that attempt to emulate human behavior; more specifically, when predicting how quickly humans can adapt to static defenses.

5.2.2 Static Equilibrium Defender

The *Static Equilibrium Defender* plays according to a fixed probability distribution over the possible combinations of honeypot assignments to nodes. A new combination is selected randomly each round according to the distribution shown in Table 5.3. We derived the distribution from a game-theoretic Mixed Strategy Nash Equilibrium that optimizes the defender's expected utility by assuming a single, nonrepeated interaction (i.e., the defender assumes only a single round game) against a fully rational adversary. For further reading on how to derive a Mixed Strategy Nash Equilibrium, we direct the reader to chapter 13 of Microeconomics and Behavior (Frank et al. 2008).

This defense takes an improved step in defense over the Static Pure Defender by adding randomness; however, this defense maintains the same probability distribution throughout the duration

Table 5.3 Static equilibrium defender probabilistic strategy.

Defended nodes	{1, 3, 4}	{2, 3}	{2, 5}	{3, 5}
Probability	≈ 0.303	≈ 0.095	≈ 0.557	≈ 0.0448

of the (multiround) game. In a real cyber defense setting, this defense would naïvely assume an adversary would only attempt to infiltrate a network once.

The optimal strategy for the adversary against this strategy is to attack node 4, with an expected total value of ≈ 447 points for the adversary.

5.2.3 Learning with Linear Rewards (LLR)

The *Adaptive Learning with Linear Rewards Defender (LLR)* (Gai et al. 2012) plays an adaptive, learning defense strategy that tries to maximize reward by balancing exploration and exploitation using an approach designed for Multi-Armed Bandit (MAB) learning. A_a in LLR is the set of all individual actions (nodes to defend). In the scenario from Figure 5.1, A_a is the set containing all five nodes. LLR uses a learning constant L, which we set to $L = 3$, since this is the maximum number of nodes we can play in a defense from the scenario described in Section 5.1.1. In their original description of LLR, Gai et al. recommends setting L to the maximum number of arms on the bandit (i.e., nodes to play). LLR has an initialization phase for the first $N = 5$ rounds where it plays random defense actions, but guarantees playing each node at least once. This initialization is to guarantee there is some information known about each node before the main decision-making loop takes over. $\left(\hat{\theta}_i\right)_{1 \times N}$ is the vector containing the mean observed reward $\hat{\theta}_i$ for all nodes i. $\left(m_i\right)_{1 \times N}$ is the vector containing m_i, or number of times node i has been played. The vectors are updated after each round.

After the initialization phase, LLR solves the maximization problem in Eq. 5.1, and deterministically selects the subset of nodes that maximizes the equation each round until the end of the game. The algorithm tries to balance between nodes with high observed means (i.e., have captured the adversary often in the past) and exploring less frequently played nodes (which the adversary may move to avoid capture). While LLR makes no assumptions that it is facing an adaptive opponent, it indirectly adapts to an adversary based on the observations of previous rewards that depend on the adversary's strategy. If it fails to capture the adversary on certain nodes, it will explore other, less-played nodes to try to obtain a higher expected reward.

We utilized this algorithm as LLR is a well-known solution to the Multi-Armed Bandit problem in online learning and it has numerous properties that set it apart from the other defenses in this chapter. As mentioned, this defense is deterministic and if the algorithm is recognized as the selected defense, one might assume that it would be easy for an adversary to optimize over to never be caught. However, its randomness in the initialization stage and the adversary's limited observation (i.e., the adversary can only observe what she attacks and can never view all the defended nodes in a round) prevent the adversary from fully exploiting this defense. There are a total of 432 opening defense sequences in this game for LLR.

5.2.4 Best Response with Thompson sampling (BR-TS)

The Static Equilibrium Defender plays optimally under two major assumptions: the game is played only in a single round and that the adversary is a perfectly rational agent that tries to maximize her expected reward. Game theoretically speaking, we would want to use a perfect model for human decision-making (as opposed to a full rationality model) and evaluate every possible state in this game to develop a decision tree that will effectively solve the game.

One major issue is that with the six possible adversary decisions and eight possible defense combinations, there are 10^{80} possible games (assuming the defender does not purposefully leave a node it can afford undefended, such as leaving all nodes undefended with an unspent budget). The second issue is that while humans are capable of significant strategic reasoning, they are typically

---

Algorithm 5.1 Learning with Linear Rewards (LLR)

If $max|\mathcal{A}_a|$ is known, let $L = max|\mathcal{A}_a|$; else, $L = N$

for $t = 1$ to N **do**

  Play any action a such that $t \in \mathcal{A}_a$

  Update $(\hat{\theta}_i)_{1 \times N}$, $(m_i)_{1 \times N}$ accordingly

end for

for $t = N + 1$ to ∞ **do**

  Play an action a which solves the maximization:

$$a = \underset{a \in F}{\mathrm{argmax}} \sum_{i \in \mathcal{A}_a} a_i \left(\hat{\theta}_i + \sqrt{\frac{(L+1)\ln n}{m_i}} \right), \tag{5.1}$$

  Update $(\hat{\theta}_i)_{1 \times N}$, $(m_i)_{1 \times N}$ accordingly

end for

---

resource-constrained and make suboptimal or heuristic decisions, so it is often a challenging problem to predict human behavior with high accuracy.

To address these issues, we designed an adaptive Best Response (BR) defender that uses Thompson sampling (BR-TS) as a human predictor model and only looks ahead a single round to simulate what the adversary might do to provide an effective defense (Agrawal and Goyal 2012). We find that evaluating all possible states of this game is unnecessary as human participants are often only considering their next move or two. This limited computation also saves time as to not burden the human participants with long wait times in-between rounds for the experiment.

Because the defender only knows status of what was defended, the defender is forced to infer which actions the adversary may have chosen when the adversary went undetected. As an example, assume the players are currently on round 5 and the defender detected the adversary in all previous rounds except for the second round, where the defender covered nodes 1, 3, and 4. The defender would then have full information about the adversary's history except for round 2 where the adversary could have passed, attacked node 2, or attacked node 5. The BR defender then runs thousands of simulations, inferring what the adversary might have done (i.e., choosing a random branching path) then seeing what decision is made on round 5 and selecting an optimal defense based on the simulations.

The key to building a strong "Best Response" defender here is to utilize an accurate history inference model (e.g., uniform random, static attack probability distribution, changing attack probability distribution, etc.) and an accurate adversary predictor model for the current round in question. Naturally, a uniform random history inference model might perform poorly in this game as the node values are advertised, and naturally, participants will "pass" less than they will attack a high-valued node.

For this BR defender, we use participant attack distributions collected in early trials to act as our history inference model. For our adversary predictor, we utilized Thompson sampling which has shown success in emulating human behavior in bandit settings (Agrawal and Goyal 2012). As we see later, Thompson sampling ends up performing poorly as a predictor model for this defender and we leave exploring better predictors (such as Instanced-Based Learning) for future work. We limited BR-TS to 1500 simulations per round and only predicted the next round instead of future rounds primarily to prevent burdening the human participants with unreasonable computation time, but still maintaining defensive quality.

5.2.5 Probabilistic Best Response with Thompson Sampling (PBR-TS)

Our previously described BR-TS defender deterministically selects the highest performing defense over the sampled adversary simulations. Here we consider a BR-TS defender that polls randomly from the distribution of the predicted expected rewards of each defense.

5.2.6 Follow the Regularized Leader (FTRL)

A shortcoming for LLR is that it naïvely assumes that the expected payoffs of each node are fixed, independent, and identically distributed. This may hold for some human adversaries, but certainly not all. As human adversaries learn, they will likely adapt to the defenses they encounter and, in turn, update the attack distributions they follow. To address this, we look to **Follow the Perturbed Leader**, an adversarial MAB algorithm that does assume an adversary (e.g., human adversary) is actively adjusting the expected rewards for each node.

The normal Follow the Leader algorithm falls victim to adversarial training in that it can be "taught" to play as the adversary wishes. Follow the Perturbed Leader directly addresses this by allowing some suboptimal actions. FTRL takes this a step further by using regularizers to further avoid over-fitting and adversarial training. We adopt the FTRL algorithm described by Zimmert et al. Zimmert et al. (2019).

The authors of the algorithm use a hybrid regularizer that combines Shannon entropy and Tsallis entropy (with power $1/2$). Early on, when the defender has little information about the nodes and attack information, Tsallis entropy guides the decision-making, whereas Shannon entropy takes over near the end of the game when the defender has more knowledge.

Algorithm 5.2 FTRL with hybrid regularizer

   Parameter: $\gamma < 1$
   Initialize: $\hat{L}_0 = (0, \dots, 0), \eta_t = 1/\sqrt{t}$
   for $t = 1, 2, \dots$ **do**
      compute
$$x_t = \underset{x \in Conv(\mathcal{X})}{\arg\min} \langle x, \hat{L}_{t-1}\rangle + \eta_t^{-1}\Psi(x)$$
      where $\Psi(\cdot)$ is defined in Eq. (1)
      sample $X_t \sim P(x_t)$
      observe $o_t = X_t \circ \ell_t$
      construct estimator $\hat{\ell}_t : \hat{\ell}_{ti} = \dfrac{(o_{ti} + 1)\mathbb{1}_t(i)}{x_{ti}} - 1, \forall i$
      update $\hat{L}_t = \hat{L}_{t-1} + \hat{\ell}_t$
   end for

Each round, FTRL computes the regularized leader x_t as seen in Algorithm 5.2 such that $Conv(\mathcal{X})$ is the convex hull of $\mathcal{X}$, $\hat{L}_{t-1}$ is the cumulative estimated loss, η_t is the defined learning rate, and $\Psi(x)$ is some chosen regularizer, which we define in Eq. 5.2. For a further description of the algorithm, we direct the reader to its original description by Zimmert et al. (2019).

$$\Psi(\S) = \sum_{i=1}^{d} -\sqrt{\S_i} + \gamma(1 - \S_i)\log(1 - \S_i) \tag{5.2}$$

5.3 Experiments

Using the *HoneyGame* we conducted two experiments. Experiment 1 includes two static defense algorithms i.e. Static Pure, Static Equilibrium and one dynamic but deterministic algorithm, LLR, that constantly tries to explore the best defenses.

From our observations in Experiment 1, we designed another Experiment 2 that includes FTRL, BR-TS, and PBR-TS algorithms which use Thomson sampling as a predictor of the adversary's actions to create adaptive defenses. The Thomson Sampling used the data collected in Experiment 1 to predict the adversary's actions. For both experiments, we used the following measures for data analysis.

5.3.1 Measures

For each experimental condition (i.e., defense algorithm), we calculated the average reward, proportion of attacks on honeypots, switching and attack distribution.

Average rewards represent the average score of adversaries. The proportion of attacks on honeypots indicates the number of times adversaries selected a honeypot. The switching behavior is a common measure of exploration used in human decision-making and learning studies (Gonzalez and Dutt 2016; Todd and Gigerenzer 2000). This measure indicates the proportion of switches from one node selection to a different one in the next trial. High switching indicates high exploration and low switching indicates exploitation in the case of a static defender and static environment. We measure different switching types: *Win-Stay*, *Win-Shift*, *Lose-Stay*, and *Lose-Shift*. Win-Stay measures the proportion of selections of the same node in the current trial after a successful attack of that node the previous trial (i.e., it was a real node, receiving a positive reward). Lose-Shift measures the proportion of switches in the current trial after attacking a honeypot in the previous (i.e., receiving a negative reward). These two strategies correspond to the "Win-Stay" and "Lose-Shift" (WSLS) heuristic, which is very commonly observed in experimental economics (Biele et al. 2009; Robbins 1985). Win-Shift measures the proportion of switches in the current trial after attacking a real node in the previous trial. Lose-Stay measures the proportion of selections of the same node in the current trial after attacking a honeypot node in the previous trial.

5.4 Experiment 1

5.4.1 Participants

We recruited 303 human participants on Amazon's Mechanical Turk (AMT). After excluding 4 participants with partial responses, the final data included 299 human participants (128 female, 169 male, and 2 participants reporting as other). All participants were above the age of 18 (range 19–73), and the median age was 32 (mean = 34.87). Participants interacted with one of the 3 defense algorithms for 50 rounds. 100 participants played against the Static Pure Defender; 99 played against the Static Equilibrium Defender, and 100 played against the LLR defender. Participants took roughly 10 minutes from start to finish. They were paid US\$ 1.00 for completing the experiment and were given a bonus payment proportional to their performance in the 50 round game, ranging from US\$0 to an extra US\$3.25.

5.4.2 Procedure

Participants were asked for informed consent according to the protocol approved by the Institutional Review Board at Carnegie Mellon University. Then, participants were given detailed instructions and definitions of the concepts needed to perform the task (i.e., HoneyGame). Participants were told that the defender has a budget $D = 40$ that limits the number of honeypot configurations (i.e., combinations of defended nodes). In each round, the participant attacks a node and receives either a positive reward $v_i - c_i^a$ or a negative reward $-c_i^a$ depending on the defender's action. The setup in Figure 5.1 was the same for every participant. Once they finished the experiment, they were directed to fill in a brief demographic survey about their age and gender.

5.4.3 Results

5.4.3.1 Average Rewards

Figure 5.2 shows the average rewards of the human adversaries against each algorithm. The human adversaries performed best against the Static Pure defender, yielding an average score of 610.35 points. The human adversaries performed significantly worse against the Static Equilibrium defender, yielding an average of 255.91 points. Finally, the adversaries performed the worst against the most resilient defender LLR yielding an average of 172.6 points.

We conducted a one-way ANOVA to compare the effect of defense algorithm on average rewards. There was a significant effect of the defense algorithm ($F(2,296) = 253.1, p < 0.05, \eta^2 = 0.63$). Post hoc comparisons using the Tukey HSD test indicated a significant difference in the rewards between the Static Equilibrium Defender ($M = 255.91$, SD $= 142.49$) and the Pure Static Defender ($M = 610.35$, SD $= 169.16$) conditions. Similarly, there was a significant difference in the rewards between the LLR ($M = 172.6$, SD $= 123.02$) and the Pure Static ($M = 610.35$, SD $= 169.16$) conditions. The rewards for the LLR ($M = 172.6$, SD $= 123.02$) and rewards for the Static Equilibrium defenders ($M = 255.91$, SD $= 142.49$) were also significantly different. The results suggest the best defense algorithm is the LLR, followed by the Static Equilibrium algorithm and finally the Pure Static algorithm. The more adaptive algorithm was substantially better.

5.4.3.2 Attacks on Honeypots

Figure 5.3 shows the proportion of attacks on honeypots per block (a consecutive group of 10 rounds) for each of the defense algorithms. The Static Pure defender predictably resulted in

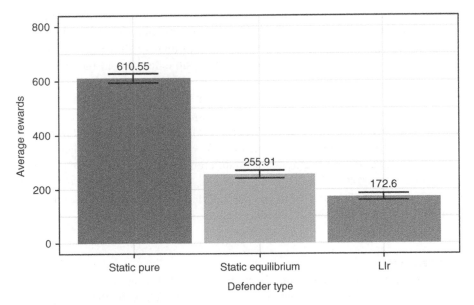

Figure 5.2 The average rewards in different algorithms.

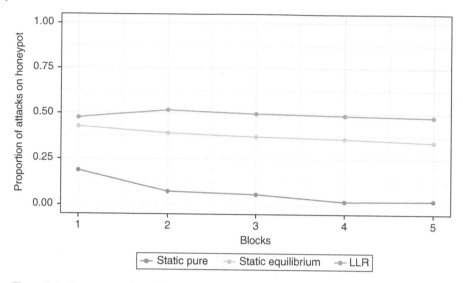

Figure 5.3 The proportions of attacks on honeypots.

Table 5.4 Repeated measures ANOVA.

Effect	df	F-ratio	p	η^2
Defender	2, 296	574.31	<0.001	0.79
Block	4, 1184	14.84	<0.001	0.05
Defender: block	8, 1184	6.48	<0.001	0.04

the lowest proportion of attacks to honeypots, followed by the Static Equilibrium defender, and the LLR.

A repeated measures ANOVA was conducted to compare the effect of defense algorithm and blocks on the proportion of honeypot attacks. There was a significant effect of defense algorithm, block, and their interaction was also significant. The results are summarized in Table 5.4. The proportion of attacks to honeypots decreased, mostly due to the visible decrease in the Static Pure algorithm.

Post hoc comparisons using the Tukey HSD test indicated that there was a significant difference in the proportion of honeypot attacks for the Static Equilibrium defender ($M = 0.38$, SD $= 0.09$) and the Pure Static defender ($M = 0.07$, SD $= 0.08$). The proportion of honeypot attacks for the LLR defender ($M = 0.49$, SD $= 0.10$) and the Static Pure defender ($M = 0.07$, SD $= 0.08$) was significantly different. Similarly, we found that there was a significant difference in the proportion of honeypot attacks for LLR defender ($M = 0.49$, SD $= 0.10$) and the Static Equilibrium defender ($M = 0.38$, SD $= 0.09$) conditions. Again, the best algorithm was the LLR, followed by the Static Equilibrium defender and then the Pure Static defender. The more adaptive the algorithm, the more attacks to honeypots it elicited.

5.4.3.3 Switching Behavior
We computed the contingency matrix including the probabilities of staying or switching after a gain (selection of a real node) or a loss (selection of a honeypot), for each of the three algorithms. The contingency matrix based on their attack in the previous trial and the strategy in the current

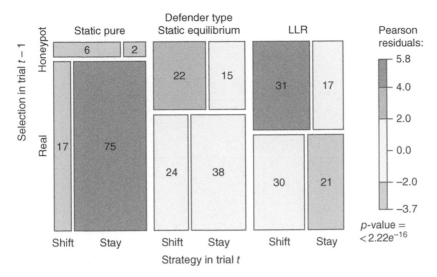

Figure 5.4 Transition probability matrix.

trial is shown in Figure 5.4 and demonstrates the Win-Stay, Win-Shift, Lose-Stay, and Lose-Shift strategies as define in measures section. The Red tiles indicate significant negative residuals, where the frequency is less than expected and the Blue tiles indicate significant positive residuals, where the frequency is greater than expected. The intensity of the color represents the magnitude of the residual.

Results show that participants playing against Static Pure defender had a 75% tendency to select the same node in trial t after selecting a real node in trial $t - 1$, compared to shifting to other nodes. Similarly, for the participants playing against the Static Equilibrium defender, the proportion of stay actions at trial t was higher after an attack to a real node in trial $t - 1$ (38%) compared to the proportion of shifts. In contrast, for the participants playing against the LLR defense, the proportion of stay at trial t after selecting a real node in trial $t - 1$ was lower (21%) compared to the proportion of shifts. Furthermore, with the LLR defense, attacks on honeypots were higher than in the Pure Static and the Static Equilibrium defenders.

To understand the behavior of the Win-Stay, Win-Shift, Lose-Stay, and Lose-Shift over the 50 rounds, we computed the average proportions of switching actions per block, where a block consisted of a contiguous sequence of 10-rounds. Figure 5.5a shows the dynamics of the Lose-Shift, Figure 5.5b shows the Lose-Stay, Figure 5.5c shows the Win-Shift, and Figure 5.5d shows the Win-Stay strategies over blocks for each of the three defense algorithms.

As summarized in Table 5.5, the proportion of Lose-Shift strategies were significantly different over the blocks. The Static Pure defender triggered fewer switches after attacking a honeypot and the proportion of switching decreased over time. Adversaries against the LLR algorithm comparatively encounter more honeypots and more switching compared to static defense algorithms. The proportion of lose-stay strategies was significantly different over the blocks and the proportion of stay strategy was different across different defense algorithms. The static pure defender triggered less proportion of honeypots, thus the proportion of stay after triggering honeypot is the lowest. Similarly, the proportion of Win-Shift strategy decreased significantly over blocks and this behavior was different across different defense algorithms.

The most interesting patterns appeared in the Win-Stay strategy. For the static pure defender, the proportion of staying after an attack on a real node in the previous round increases and is higher compare to the Static Equilibrium and the LLR algorithms. Suggesting that participants learned to

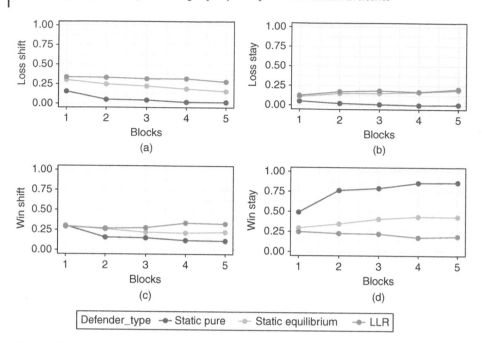

Figure 5.5 Switching behavior.

Table 5.5 Repeated measures ANOVA for shifting–staying behavior in Experiment 1.

Dependent Measure	Effect	*df*	*F*-ratio	*p*	η^2
Lose Shift	Defender	2, 296	132.8	<0.001	0.47
	Block	4, 1184	28.57	<0.001	0.09
	Defender: block	8, 1184	3.26	<0.001	0.02
Lose Stay	Defender	2, 296	77.45	<0.001	0.34
	Block	4, 1184	4.97	0.0010	0.017
	Defender: block	8, 1184	6.32	<0.001	0.041
Win Shift	Defender	2, 296	503.28	<0.001	0.63
	Block	4, 1184	13.38	<0.001	0.043
	Defender type: block	8, 1184	9.15	<0.001	0.058
Win Stay	Defender	2, 296	1380.37	<0.001	0.82
	Block	4, 1184	39.02	<0.001	0.116
	Defender: block	8, 1184	28.20	<0.001	0.160

recognize the nodes that were honeypots and attacked more on real nodes. In contrast, the LLR algorithm has the lowest proportion of staying when the participant attacked a real node in the previous round.

5.4.3.4 Attack Distribution

To investigate the attack preferences of human participants, we looked at the proportion of attacks on different nodes sorted according to their Expected Values (EVs). The expected values of each

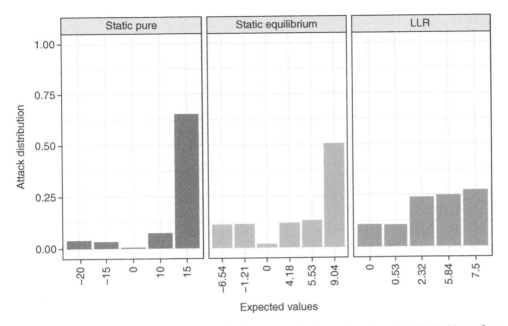

Figure 5.6 Attack distributions organized by the expected value of the five nodes in the HoneyGame.

node are the sum of the proportion of each node acting as honeypot ($P(H)$) multiplied by its penalty and the proportion of each node acting as real ($P(R)$) multiplied by its reward.

$$EV = P(R) * \text{rewards} + P(H) * \text{penalty} \tag{5.3}$$

The attack distributions shown in Figure 5.6 suggest that participants were able to identify and choose the nodes with the highest EV in the Static Pure and Static Equilibrium algorithms. However, in the LLR defense algorithm, participants were unable to choose purely based on EVs which explains the best performance of these algorithms. A chi-square test of independence confirmed that the attack distributions were significantly different across the three defense algorithms ($\chi^2(10) = 7922, p < 0.001$).

5.5 Experiment 2

Using the same procedures in Experiment 1, we conducted a second experiment, where human adversaries played against three defense algorithms: BR-TS, PBR-TS, and FTRL.

5.5.1 Participants

We recruited 303 human participants on Amazon's Mechanical Turk (AMT) where 6 participants were excluded due to incomplete data. Out of 297 participants, 112 reported female and 183 reported male with 2 participants reporting as other. All participants were above the age of 18 (range 20–72), and the median age was 33 (mean = 35.39). Participants interacted with one of the 3 defense algorithms for 50 rounds. 96 participants played against the BR-TS Defender, 98 played against the PBR-TS Defender, and 103 played against the FTRL defender. Participants took roughly 10 minutes from start to finish. They were paid US$1.00 for completing the experiment and were given a bonus payment proportional to their performance in the 50 round game, ranging from US$0 to an extra US$3.45.

5.5.2 Results

5.5.2.1 Average Rewards
Figure 5.7 shows the average rewards of the human adversaries in each algorithm. The human adversaries against the PBR-TS algorithm performed the best, yielding an average score of 306.73 points. The adversaries against the BR-TS defense algorithm performed similar to the PBR-TS, yielding an average of 289.95 points. Finally, the adversaries playing against the FTRL defense algorithm performed the worst for the human adversaries, yielding an average score of 198.11 points.

A one-way ANOVA was conducted to compare the effect of defense algorithm on average rewards. There was a significant effect of defense algorithm ($F(2,294) = 16.03, p < 0.05, \eta^2 = 0.098$). Post hoc analyses using the Tukey HSD test indicated no significant difference in the rewards between BR-TS ($M = 289.73, \text{SD} = 158.79$) and PBR-TS ($M = 306.73, \text{SD} = 170.44$) algorithms. However, the average rewards in FTRL ($M = 198.11, \text{SD} = 104.86$) and the PBR-TS ($M = 306.73, \text{SD} = 170.44$) were significantly different. The rewards for FTRL ($M = 198.11, \text{SD} = 104.86$) and the BR-TS ($M = 289.73, \text{SD} = 158.79$) algorithms were also significantly different. Results suggest the best defense algorithm is FTRL, followed by BR-TS and PBR-TS algorithms.

5.5.2.2 Attacks on Honeypots
Figure 5.8 shows the proportion of attacks on honeypot by block and defense algorithm. The PBR-TS algorithm resulted in the lowest proportion of attacks on honeypots, followed by the BR-TS and the FTRL algorithms. A repeated ANOVA was conducted to compare the effect of defense algorithm and block on the proportion of honeypot attacks. There was a significant effect of defense algorithm, block and their interaction summarized in Table 5.6. The proportion of attacks on honeypot decreased significantly over the blocks in the PBR-TS and BR-TS algorithms. However, the proportion of attacks on honeypots increased over the blocks for the FTRL algorithm. Post hoc analyses using the Tukey HSD test indicated no significant difference in the proportion of attacks on honeypots for the BR-TS ($M = 0.37, \text{SD} = 0.104$) and PBR-TS ($M = 0.36, \text{SD} = 0.113$). However, the average attacks on honeypot were significantly higher in the FTRL algorithm ($M = 0.45, \text{SD} = 0.085$) compared to the PBR-TS ($M = 0.36, \text{SD} = 0.113$) and the

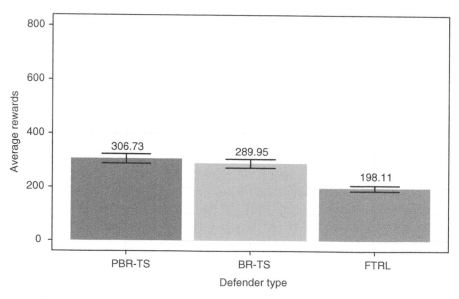

Figure 5.7 Average rewards.

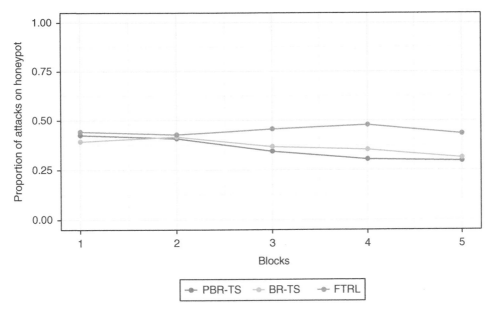

Figure 5.8 The proportions of attacks on honeypots by human adversaries when playing against three defense algorithms: PBR-TS, BR-TS, and FTRL.

Table 5.6 Repeated measures ANOVA.

Effect	df	F-ratio	p	η^2
Defender	2, 294	24.49	<0.001	0.051
Block	4, 1176	10.295	<0.001	0.021
Defender: block	8, 1176	5.381	<0.001	0.022

BR-TS ($M = 0.37$, SD $= 0.104$) algorithms. Again, the best algorithm for defense was FTRL, followed by the PBR-TS and the BR-TS algorithms.

5.5.2.3 Switching Behavior

We calculated the contingency matrix including the probabilities of staying or switching after a gain or a loss, for each of the three algorithms. The contingency matrix based on their attack in the previous trial and the strategy in the current trial is shown in Figure 5.9. Results show that participants in the PBR-TS algorithm had a 39% tendency to select the same node in trial t after selecting a real node in trial $t - 1$, compared to shifting to another node. Similarly, in the BR-TS algorithm, the proportion of stay actions at trial t was higher after an attack to a real node in trial $t - 1$ (35%) compared to the proportion of the shifts. In contrast, in the FTRL defense, the proportion of stay at trial t after selecting a real node in trial $t - 1$ was similar to the proportion of shifts ($29{\sim}28\%$). Furthermore, with the FTRL algorithm, attacks on honeypots were higher than in the BR-TS and PRB-TS algorithms.

To understand the overtime behavior of the Win-Stay, Win-Shift, Lose-Stay, and Lose-Shift, we computed the average proportions of switching actions per block, where a block consisted of a contiguous sequence of 10-rounds. Figure 5.10a shows the dynamics of the Lose-Shift, Figure 5.10b shows the Lose-Stay, Figure 5.10c shows the Win-Shift, and Figure 5.10d shows the Win-Stay strategies over blocks for each of the three defense algorithms.

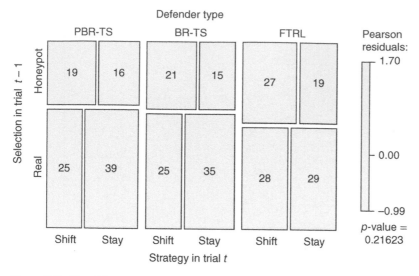

Figure 5.9 Transition probability matrix.

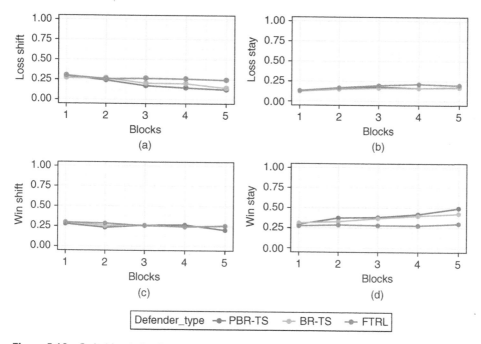

Figure 5.10 Switching behavior.

As summarized in Table 5.7, the proportion of lose-shift strategies significantly differs over the blocks. The PBR-TS defender triggered fewer switches after attacking a honeypot and the proportion of switching decreased over time. Adversaries that faced the FTRL algorithm comparatively encounter more honeypots and more switching compared to PBR-TS and BR-TS algorithms. The proportion of lose-stay strategies was significantly different over the blocks. However, the proportion of stay strategy was similar across different defense algorithms. Similarly, the proportion of win-shift strategy was only significantly different over the blocks; there was no significant difference across different defense algorithms. The most interesting patterns appeared in the Win-Stay strategy. For the PBR-TS defender, the proportion of staying after an attack on a real node in the previous trial increased and it was higher compared to BR-TS and FTRL algorithms. These results

Table 5.7 Repeated measures ANOVA.

Dependent measure	Effect	df	F-ratio	p	η^2
Lose Shift	Defender	2, 294	5.83	<0.01	0.020
	Block	4, 1176	26.46	<0.001	0.038
	Defender: block	8, 1176	3.94	<0.001	0.011
Lose Stay	Defender	2, 294	1.81	0.17	0.006
	Block	4, 1176	11.27	<0.001	0.017
	Defender: block	8, 1176	0.76	0.64	0.002
Win Shift	Defender	2, 294	0.28	0.76	0.001
	Block	4, 1176	4.02	<0.003	0.005
	Defender: block	8, 1176	1.43	0.17	0.004
Win Stay	Defender	2, 294	6.94	<0.001	0.029
	Block	4, 1176	17.23	<0.001	0.019
	Defender: block	8, 1176	3.81	<0.001	0.008

suggest that participants learned to recognize the nodes that were honeypots and attacked more normal nodes. In contrast, the FTRL algorithm had the lowest proportion of staying when an adversary did not attack a honeypot in the previous round.

5.5.2.4 Attack Distribution

To investigate the attack preferences of human participants, we looked at the proportion of attacks on different nodes sorted according to their expected value. The expected values of each node were calculated using the proportion of times each node acting as a honeypot. The attack distributions shown in Figure 5.11 suggest that participants were unable to identify and choose the nodes with the highest EV. In the BR-TS and PBR-TS algorithms, participants seems to stick with a "good

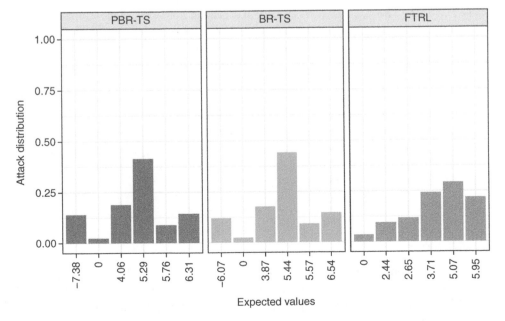

Figure 5.11 Attack distribution.

enough" node. Participants tended to select the nodes with higher EVs more often in the FTRL, however; they were unable to choose purely based on EVs, which explains the reason for the best performance of this algorithm. A chi-square test of independence confirmed that the attack distributions were significantly different across the three algorithms ($\chi^2(10) = 4386, p < 0.001$).

5.6 Towards Adaptive and Personalized Defense

Across the two experiments, we found that participants (in the role of adversaries) performed worst when playing against the LLR defender in Experiment 1, and when playing against the FTRL defender in Experiment 2. With repeated interactions in the HoneyGame and against static algorithms (Static Pure and Static Equilibrium), participants were able to learn the probabilities and patterns of the static algorithms of defense, which they exploited to their benefit.

The BR-TS and PBR-TS algorithms were less successful than we expected. These algorithms used Thompson sampling to predict human actions to customize the defense strategy. Thompson sampling has shown success in emulating human behavior in bandit settings (Agrawal and Goyal 2012), and we expected Thompson sampling would be adequate for the predictions of human actions and improve the allocation of defenses in these algorithms. For both, the BR-TS and PBR-TS algorithms, we used the attack distributions collected in Experiment 1 to act as the history inference model. In retrospect, this is perhaps not the best way to predict human behavior in Experiment 2. As the results in experiment 2 suggest, both best response algorithms performed poorly. Interestingly, they did move participants away from attacking the highest valued nodes, yet the overall actions of exploiting the nodes after attacking a real node were high compared to the FTRL. The best algorithm out of the six ones tested was LLR.

Another possible explanation for the poor performance of the BR-TS and PBR-TS algorithms is that Thompson sampling is not an accurate representation of human attack behavior. A better strategy might be to use cognitive models to emulate human attack behavior and use the predictions of those models to adapt the defense strategies. Recent research (Gonzalez et al. 2020) provides a research framework for generating dynamic, adaptive, and personalized strategies of defense using Instance-Based Learning Theory (IBLT) i.e., a theory of decisions from experience (Gonzalez et al. 2003). Furthermore, results from a demonstration in the use of IBL models to generate adaptive and personalized defenses (Cranford et al. 2020) suggests that adapting game-theory models using the IBL models resulted in effective defense strategies, compared to models that used other Machine Learning approaches.

As an initial test of the use of IBL models in the context of the HoneyGame, Gutierrez et al. (2019) developed Thompson sampling and IBL models to predict human actions. These models were used to adapt the LLR algorithm presented in Experiment 1. The IBL model was a better predictor of human adversaries compared to the Thompson sampling algorithm. In the future, it will be important to pursue this kind of investigation, where adaptive and personalized algorithms of defense are informed by cognitive models.

5.7 Conclusions

In this chapter, we report on the effectiveness of six deceptive algorithms for cyber defense by pitting them against human participants (in the role of adversaries). These six algorithms differ particularly in three dimensions: determinism, adaptivity, and customization to the adversary's actions.

The results clearly illustrate how human participants were able to learn the patterns of the more static algorithms and suggest how challenging it is to design algorithms that are truly adaptive to human actions. The nondeterministic, adaptive, and customizable algorithms in our set (BR-TS and PBR-TS), were less effective than the deterministic and noncustomizable, LLR, algorithm. The reason, it turns out, might be in the model used to emulate the human adversary's actions (i.e., Thompson sampling), or on the data, we used to construct the Thompson sampling model.

Based on new directions in the design of defense algorithms (Gonzalez et al. 2020), we propose to use Instance-Based Learning Theory (Gonzalez et al. 2003), to develop models that emulate human behavior and inform the design of dynamic, adaptive, and personalized algorithms of defense.

Acknowledgements

Research was sponsored by the Army Research Laboratory and was accomplished under Cooperative Agreement Number W911NF-13-2-0045 (ARL Cyber Security CRA). The views and conclusions contained in this document are those of the authors and should not be interpreted as representing the official policies, either expressed or implied, of the Army Research Laboratory or the U.S. Government. The U.S. Government is authorized to reproduce and distribute reprints for Government purposes notwithstanding any copyright notation here on. The authors thank Orsolya Kovacs from the Dynamic Decision Making Laboratory at Carnegie Mellon University for her help with data collection.

References

Shipra Agrawal and Navin Goyal. Analysis of thompson sampling for the multi-armed bandit problem. In *Conference on Learning Theory, Edinburgh, Scotland, 25–27 June 2012*, pp 39–41, 2012.

Mohammed H. Almeshekah and Eugene H. Spafford. Cyber security deception. In *Cyber Deception*, pp 23–50. Springer, 2016.

Guido Biele, Ido Erev, and Eyal Ert. Learning, risk attitude and hot stoves in restless bandit problems. *Journal of Mathematical Psychology*, 53(3):155–167, 2009.

Samuel Cheyette, Emmanouil Konstantinidis, Jason L. Harman, and Cleotilde Gonzalez. Choice adaptation to increasing and decreasing event probabilities. In *CogSci, Philadelphia, USA, 10–13 August*, 2016.

Edward A. Cranford, Christian Lebiere, Palvi Aggarwal, Cleotilde Gonzalez, and Milind Tambe. Adaptive cyber deception: Cognitively-informed signaling for cyber defense. In *Proceedings of the 53rd Hawaii International Conference on System Sciences, Maui, Hawaii, 7–10 January 2020*. IEEE, 2020.

Kimberly Ferguson-Walter, Temmie Shade, Andrew Rogers, Michael Christopher Stefan Trumbo, Kevin S. Nauer, Kristin Marie Divis, Aaron Jones, Angela Combs, and Robert G. Abbott. The tularosa study: An experimental design and implementation to quantify the effectiveness of cyber deception. Technical report, Sandia National Lab. (SNL-NM), Albuquerque, NM, USA, 2018.

Kimberly Ferguson-Walter, Sunny Fugate, Justin Mauger, and Maxine Major. Game theory for adaptive defensive cyber deception. In *Proceedings of the 6th Annual Symposium on Hot Topics in the Science of Security, Nashville TN, USA, 1–3 April 2019*, pp 1–8, 2019.

Robert H. Frank. *Microeconomics and Behavior*, 7 edition. McGraw-Hill Irwin, Boston, 2008.

Yi Gai, Bhaskar Krishnamachari, and Rahul Jain. Combinatorial network optimization with unknown variables: Multi-armed bandits with linear rewards and individual observations. *IEEE/ACM Transactions on Networking (TON)*, 20(5):1466–1478, 2012.

Megan Garber. Ghost army: The inflatable tanks that fooled hitler, May 2013. https://www.theatlantic.com/technology/archive/2013/05/ghost-army-the-inflatable-tanks-that-fooled-hitler/276137/?utm_source=share&utm_campaign=share.

Kenneth Geers. *Sun Tzu and Cyber War*. Cooperative Cyber Defence Centre of Excellence (CCD COE), 2010.

Lionel Giles. *Sun Tzŭ on the Art of War: The Oldest Military Treatise in the World*. Luzac & Company, 1910.

Cleotilde Gonzalez and Varun Dutt. Exploration and exploitation during information search and experimential choice. *Journal of Dynamic Decision Making*, 2(1): Article 2 (1–8), 2016.

Cleotilde Gonzalez, Javier F. Lerch, and Christian Lebiere. Instance-based learning in dynamic decision making. *Cognitive Science*, 27(4):591–635, 2003.

Cleotilde Gonzalez, Palvi Aggarwal, Christian Lebiere, and Edward Cranford. Design of dynamic and personalized deception: A research framework and new insights. In *Proceedings of the 53rd Hawaii International Conference on System Sciences, Maui, Hawaii, 7–10 January 2020*, 2020.

Marcus Gutierrez, Noam Ben-Asher, Efrat Aharonov, Branislav Bošanský, Christopher Kiekintveld, and Cleotilde Gonzalez. Evaluating models of human adversarial behavior against defense algorithms in a contextual multi-armed bandit task. In *41st Annual Meeting of the Cognitive Science Society (CogSci 2019)*, Montreal, QC (2019 (in press)), 2019.

Tomás Lejarraga, Varun Dutt, and Cleotilde Gonzalez. Instance-based learning: A general model of repeated binary choice. *Journal of Behavioral Decision Making*, 25(2):143–153, 2012.

Herbert Robbins. Some aspects of the sequential design of experiments. In *Herbert Robbins Selected Papers*, pp 169–177. Springer, 1985.

Lance Spitzner. *Honeypots: Tracking Hackers*, volume 1. Addison-Wesley Reading, 2003.

Peter M. Todd and Gerd Gigerenzer. Précis of simple heuristics that make us smart. *Behavioral and Brain Sciences*, 23(5):727–741, 2000.

Julian Zimmert, Haipeng Luo, and Chen-Yu Wei. Beating stochastic and adversarial semi-bandits optimally and simultaneously. *arXiv preprint arXiv:1901.08779*, 2019.

6

A Theory of Hypergames on Graphs for Synthesizing Dynamic Cyber Defense with Deception

Abhishek N. Kulkarni[1] and Jie Fu[2]

[1]*Robotics Engineering Program, Worcester Polytechnic Institute, Worcester, MA, USA*
[2]*Department of Electrical and Computer Engineering, Robotics Engineering Program, Worcester Polytechnic Institute, Worcester, MA, USA*

6.1 Introduction

Cyber deception is a key technique in network defense. With cyber deception, the defender creates uncertainties and unknowns for the attacker. By doing so, the attacker's strategy in exploiting the system becomes less effective, thus, resulting in improved security and safety of the network. In this chapter, we investigate a formal methods approach for synthesizing defensive strategies in cyber network systems with cyber deception. We employ formal security specifications to express a rich class of desired properties. For example, a defender may need to satisfy a safety property in terms of preventing the attacker from reaching critical data server. He may also need to satisfy a liveness property stating that a service should eventually be provided to the user after being made temporarily unavailable. Given formal security specifications, formal synthesis is to compute a defense strategy, if exists, with which the defender can provably satisfy his specification against all possible actions from the attacker.

Formal methods have been employed to verify the security of network systems. Formal graphical security models such as attack graphs (Jha et al. 2002) and attack trees (Schneier 2007) are used in model-based verification of system security. An attack graph captures multiple paths that an attacker can carry out by exploiting vulnerabilities and their dependencies in a network to reach the attack goal. Given an attack graph, the formal security specification can be verified using model checking algorithms for transition systems (Baier and Katoen 2008). An attack tree builds a tree structure that describes how the attacker can achieve his goal by achieving a set of subgoals. The root of the tree is the main attack goal and the leaves of the tree are elementary attack subgoals. The internal tree nodes show the logical dependency between subgoals at different levels of the tree. To incorporate defender's countermeasures, attack-defense trees (Kordy et al. 2010, 2014) are proposed to capture the dependencies between actions and subgoals for both attacker and defender. These models are used in verifying quantitative security properties in temporal logic (Aslanyan et al. 2016; Hansen et al. 2017; Kordy and Wideł 2018). The major limitation of attack trees is that it does not characterize network status changes under the attack actions and thus may fail to generate some attack scenarios. It is also noted that these formal graphical models do not capture the asymmetric information between the attacker and the defender due to cyber deception. Specifically, these models assume both defender and attacker knows the game they are playing, while as with cyberdeception, the defender intentionally introduces incorrect or uncertain information about the game to the attacker.

Game Theory and Machine Learning for Cyber Security, First Edition.
Edited by Charles A. Kamhoua, Christopher D. Kiekintveld, Fei Fang, and Quanyan Zhu.

Active deception (Jajodia et al. 2016) employs decoy systems and other defenses, including access control and online network reconfiguration, to conduct deceptive planning against the intrusion of malicious users who have been detected and confirmed by sensing systems. To design defense strategies with deception, game theory has been employed (Cohen 2006; Horák (2017); Zhu and Rass 2018; Al-Shaer et al. 2019; Horak et al. 2019). These game-theoretic models express the attacker and defender's objectives using reward-loss functions. In Horak et al. (2019), a partially observable stochastic game is formulated to capture the interaction between an attacker and a defender with one-sided partial observations. The attacker is to exploit and compromise the system without being detected and has complete observation. The defender is to detect the attacker and reconfigure the honeypots. Horák (2017) consider the case when the attacker has incomplete information and forms a belief about the defender's unit. Players employ Bayesian rules to update the belief about the state in the game. Leveraging the attacker's incomplete information, the defender may mislead the attacker's belief and thus his actions to minimize the damage to the network measured by a state-dependent loss function. However, reward and loss functions are not expressive enough to capture more complex qualitative defense/attack objectives studied in attack graph, such as safety and temporally extended attack goals. These objectives can be captured succinctly using temporal logic (Manna and Pnueli 1992). When formal specification is used in specifying defense objectives, there is a lack of formal synthesis methods which employ cyber deception to ensure the security goals are met.

We study the problem of formal synthesis of secured network systems with active cyber deception. We view the interactions between the defender and the attacker as a two-player game played on a finite graph. Combining the game graph abstraction with the logical security specifications, we construct a model of an attack-defend game as a game on a graph with temporal logic objectives (Pnueli and Rosner 1989; Chatterjee and Henzinger 2012). This game includes both the controllable and uncontrollable actions to represent the actions and exploits by the defender and the attacker, respectively.

In such a game between a defender and an attacker, the attacker plays with incomplete information, if he does not know the locations of honeypots. Furthermore, if the attacker mistakes a honeypot as a critical host, then we say that he has a misperception about the game. We extend the theory of hypergame to reason about the asymmetric incomplete information between players and to enable synthesis of deceptive strategies. A hypergame (Bennett 1980; Vane 2000; Kovach et al. 2015) is a game of perceptual games, i.e. games perceived by individual players given the information available to them and the higher-order information known to them, i.e. what the player knows about the information known to the opponent. Based on the hypergame modeling, the key questions are: how will the attacker carry out his attack mission, given his incomplete or incorrect information? And, how to synthesize effective defense strategies, which leverage the defender's private information to ensure that the defender's logical security specifications are satisfied?

Our insight is that deception with honeypots can create a misperception about the labeling function of the attack-defend game. A labeling function relates an outcome-a sequence of states in the game graph-to the properties specified in logic. When honeypots are introduced, an attacker might mislabel a honeypot as a critical host and pursue to reach it. Under this formulation, our main algorithmic contribution is the solution of hypergames under labeling misperception and linear temporal logic objectives. Our solution approach includes two steps: the first step is to synthesize the rational attack strategy perceived by the attacker using solutions of omega-regular games (Zielonka 1998; Chatterjee and Henzinger 2012). The synthesized strategy serves as a predictive model of the attacker's rational behavior, which is then used to refine the original game graph to eliminate actions perceived to be irrational from the attacker's perspective. In the second step, a level-2 hypergame is solved, yielding a deceptive defense strategy, if one exists, that ensures a specification is satisfied with probability one, given the misperception of the attacker. A case study is employed to illustrate how to apply the game-theoretic reasoning to synthesize deceptive strategies.

We structure the remainder of the chapter so as to provide rigorous mathematical treatment of the topic for a reader familiar with formal methods and support it with elaborate descriptions, discussions, and examples to illustrate our approach to a reader new to the area.

6.2 Attack-Defend Games on Graph

In this section, we introduce a model, called *An Attack-Defend (AD) Game on a Graph*, that augments the attack graph model with the defense actions available to the defender. Our AD game on graph model resembles the game on graph model, which is commonly used in reactive synthesis (Pnueli and Rosner 1989).

Formally, an AD game on graph can be written as a tuple $\mathcal{G} = \langle G, \varphi \rangle$ where the two main components are (i) G: a game arena and (ii) φ: the Boolean payoff function (Linear Temporal Logic (LTL) specification) of the defender. Let us understand each of the component in more detail.

6.2.1 Game Arena

A game arena is a transition system with labels assigned to the states. It captures different configurations of the network and the actions that the attacker and the defender may use to change the current configuration. A configuration of system is a set of state variables that jointly define the current state of the system. For instance, a state variable may be a collection of the current host compromised by the attacker, IP addresses of different hosts over the network, an enumeration of services running over each host, or a list of users currently accessing the hosts with their privileges (root, user, and none). Suppose that there are n state variables and we denote the ith state variable as X_i, then the domain of a state space can be given by $S = X_1 \times X_2 \times \cdots \times X_n$. Given this notion of state, we formally define a game arena as follows:

Definition 6.1 *(Arena)* A turn-based, deterministic game arena between two players P1 (defender, pronoun "he") and P2 (attacker, pronoun "she") is a tuple

$$G = \langle S, A, T, \mathcal{AP}, L \rangle,$$

whose components are defined as follows:

- $S = S_1 \cup S_2$ is a finite set of states partitioned into two sets S_1 and S_2. At a state in S_1, P1 chooses an action and at a state in S_2, P2 selects an action.
- $A = A_1 \cup A_2$ is the set of actions. A_1 (resp., A_2) is the set of actions for P1 (resp., P2);
- $T : (S_1 \times A_1) \cup (S_2 \times A_2) \to S$ is a *deterministic* transition function that maps a state-action pair to the next state.
- $\mathcal{AP}$ is the set of atomic propositions.
- $L : S \to 2^{\mathcal{AP}}$ is the labeling function that maps each state $s \in S$ to a set $L(s) \subseteq \mathcal{AP}$ of atomic propositions that evaluate to true at that state.

The last two components of the game arena are related to the security specifications.
We discuss an example to illustrate the above concept.

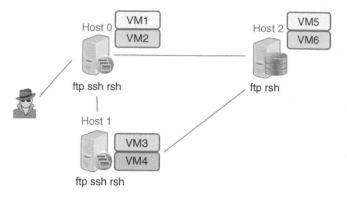

Figure 6.1 Configuration of the system.

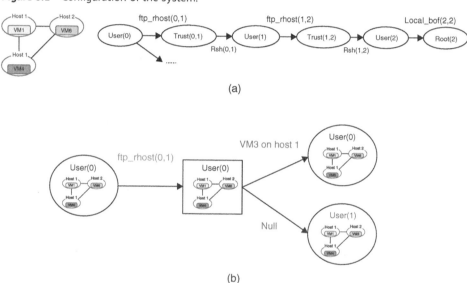

Figure 6.2 The comparison between the attack graph and the attack-defend game arena. (a) A fragment of the attack graph corresponding to one fixed configuration. (b) A fragment of the attack-defend game arena.

Example 6.1 Consider the system as shown in Figure 6.1 consisting of three hosts with platform diversity. Each host can hold up to two Virtual Machine (VM)s with different operation systems and services. For a fixed configuration of VMs, a fragment of the attack graph can be generated based on the set of known vulnerabilities, as shown in Figure 6.2a.

In reactive defense, the defender has detected the attacker is in host 0 and can exploit the vulnerability to gain user access to host 1. In that case, the defender can change the platform in host 1 to be VM3, for which the attack action will not be effective. The interaction is then captured in the game on graph, as shown in Figure 6.2. The action of the defender can also be stopping or running a service at the managed endpoints, which are omitted. We distinguish the set of states into square states at which the defender makes a move and circle states at which the attacker makes a move.

6.2.2 Specifying the Security Properties in LTL

We consider qualitative formal specifications for defender and attacker objectives. Different from quantitative utility functions in terms of costs, qualitative logic formulas capture hard security constraints that the network defense system must satisfy.

The defender has two types of goals, namely (i) operational objectives, such as the services should eventually be available to the legitimate users, and (ii) defense objectives, such as the attacker should never be able to compromise servers with sensitive information. However, the intention of attacker is often unknown. Thus, we consider the worst-case scenario where the attacker's objective is to violate the security goal of the defender.

We choose to express the security goal of the defender using LTL (Manna and Pnueli 1992). LTL allows us to express the security properties of system with respect to time. We shall now present the formal syntax and semantics of LTL and then discuss several examples.

Let $\mathcal{AP}$ be a set of atomic propositions. Linear Temporal Logic (LTL) has the following syntax,

$$\varphi := \top \mid \bot \mid p \mid \varphi \mid \neg\varphi \mid \varphi_1 \wedge \varphi_2 \mid \bigcirc\varphi \mid \varphi_1 \, \mathsf{U} \, \varphi_2,$$

where

- $\top, \bot$ represent universally true and false, respectively.
- $p \in \mathcal{AP}$ is an atomic proposition.
- $\bigcirc$ is a temporal operator called the "next" operator (see semantics below).
- U is a temporal operator called the "until" operator (see semantics below).

Let $\Sigma := 2^{\mathcal{AP}}$ be the finite alphabet. Given a word $w \in \Sigma^\omega$, let $w[i]$ be the ith element in the word and $w[i \ldots]$ be the subsequence of w starting from the ith element. For example, $w = abc$, $w[0] = a$, and $w[1 \ldots] = bc$. Formally, we have the following definition of the semantics:

- $w \vDash p$ if $p \in w[0]$;
- $w \vDash \neg p$ if $p \notin w[0]$;
- $w \vDash \varphi_1 \wedge \varphi_2$ if $w \vDash \varphi_1$ and $w \vDash \varphi_2$.
- $w \vDash \bigcirc\varphi$ if $w[1 \ldots] \vDash \varphi$.
- $w \vDash \varphi \, \mathsf{U} \, \psi$ if $\exists i \geq 0$, $w[i \ldots] \vDash \psi$ and $\forall 0 \leq j < i$, $w[j \ldots] \vDash \varphi$.

From these two temporal operators ($\bigcirc$, U), we define two additional temporal operators: $\Diamond$ "eventually" and $\square$ "always." Formally, $\Diamond \varphi = \top \, \mathsf{U} \, \varphi$ and $\square \varphi = \neg\Diamond \neg\varphi$. For details about the syntax and semantics of LTL, the readers are referred to (Manna and Pnueli 1992).

Here we present some examples. Suppose *root*(2) is an atomic proposition that the attacker has root privilege on host 2, then a safety property that attacker never has root privilege on host 2 can be written in LTL as a formula $\varphi_1 = \square \neg root(2)$, which is read as "proposition *root*(2) is *always* false." Similarly, a property that the attacker first gains a user privilege on host 1 and then a root privilege on host 2 can be expressed using an LTL formula $\varphi_2 = \Diamond (user(1) \wedge \Diamond root(2))$, which is read as "*eventually* proposition *user*(1) becomes true and then the proposition *root*(2) becomes true." In general, it is also possible to express properties such as recurrence (some event occurs infinitely often) or persistence (some property eventually becomes true and remains true thereafter) using LTL. However, in this chapter, we restrict ourselves to a subclass of LTL called syntactically co-safe LTL (scLTL) (Kupferman and Vardi 2001). Using scLTL, we can reason about the reachability and safety[1] properties.

This concludes a brief introduction to the concept of AD games on graphs, which do not model the asymmetric incomplete information available with the players. In the next section, we extend the notion of hypergames that incorporate the different perceptions that players may have due to incomplete information available to them.

---

1 Safety and reachability are dual problems. Hence, reasoning about the safety objectives can be done by reasoning about the dual reachability problem.

6.3 Hypergames on Graphs

A hypergame models the situation where different players perceive their interaction with other players differently and consequently play different games in their own minds depending on their perception. We consider the case where the difference in perception arises because of incomplete and potentially incorrect information. For instance, suppose a subset of nodes in the network are honeypots, the attacker may mistake these to be true hosts. We formulate a hypergame to model the interaction between the defender and the attacker given asymmetric information.

First, let's review the definition of hypergames.

Definition 6.2 *(Hypergame; Bennett 1980; Vane 2000)* Given two players, a game perceived by player 1 is denoted by G_1, and a game perceived by player 2 is denoted by G_2. A level-1 hypergame is defined as a tuple:

$$HG^1 = \langle G_1, G_2 \rangle,$$

In a level-1 hypergame, none of the player's is aware of other player's perception.

When one player becomes aware of the other player's (mis)perception, the interaction is captured by a level-2 two-player hypergame, defined as a tuple:

$$HG^2 = \langle HG^1, G_2 \rangle.$$

where P1 perceives the interaction as a level-1 hypergame (as P1 is aware of P2's game G_2 in addition to his own) and P2 perceives the interaction as the game G_2.

We refer to the games G_1 (resp., G_2) as P1's (resp., P2's) perceptual game in level-1 hypergame, and HG^1 as P1's perceptual game in level-2 hypergame. As P2 is not aware that she might be misperceiving the game, her perceptual game in level-2 hypergame is still G_2.

In general, if P1 computes his strategy by solving an $(m - 1)$th level hypergame and P2 computes her strategy using an nth level hypergame with $n < m$, then the resulting hypergame is said to be a level-m hypergame given as:

$$HG^m = \langle HG_1^{m-1}, HG_2^n \rangle.$$

Next, we show that by introducing honeypots, the attacker's perceptual game deviates from the actual game. This mismatch occurs in the labeling function. Recall that a labeling function L assigns every state in the game arena with a subset of atomic propositions that are true at that state. Let us consider a network with decoys where attacker is not aware of which hosts are decoys. Suppose p is a proposition that a host h is a decoy. Then, defender's labeling function, say L_1, labels h correctly as a decoy. However, the attacker's labeling function, say L_2, will incorrectly label h as a regular host. Given a path $\rho \in S^*$ in the game arena, this path may satisfy the security specification as $L_1(\rho) \vDash \varphi$, in which case the defender obtains payoff of 1 and the attacker obtains payoff of 0. However, due to misperception in the labeling, the attacker may have $L_2(\rho) \vDash \neg\varphi$ and thus have a misperception of the payoff of the path. We capture this misperception and asymmetric information using the new class of hypergames, defined as follows:

Definition 6.3 *(A Hypergame on a Graph with One-sided Misperception of Labeling Function)* Let $G_1 = \langle S, A, T, AP, L_1 \rangle$ be the game arena as constructed by P1. Similarly, let $G_2 = \langle S, A, T, AP, L_2 \rangle$ be a game arena as constructed by P2 based on her perception. Let φ be the defense objective of P1. Then, we construct two games $G_1 = \langle G_1, \varphi \rangle$ and $G_2 = \langle G_2, \varphi \rangle$. When

P1 is aware of P2's misperception, i.e. P1 knows L_2 and, therefore, $\mathcal{G}_2$, we have the model of their interaction as a hypergame of level-2,

$$\mathcal{HG}^2 = \langle \mathcal{HG}^1, \mathcal{G}_2 \rangle,$$

where $\mathcal{HG}^1 = \langle \mathcal{G}_1, \mathcal{G}_2 \rangle$ is a hypergame of level-1 and is P1's perceptual game. P2's perceptual game is $\mathcal{G}_2$. We say $\mathcal{HG}^2$ to be a hypergame on a graph with one-sided misperception when the labeling function of P1 coincides with the ground-truth labeling function, i.e. $L_1 = L$.

6.4 Synthesis of Provably Secure Defense Strategies Using Hypergames on Graphs

Given the hypergame model, we present a solution approach to automatically synthesize a strategy for defender such that, for every possible action of attacker, the strategy ensures that the security goals (i.e. φ) of defender are satisfied. In order to understand the synthesis approach for hypergame, we first look at the conventional solution approach used for a game on graph (McNaughton 1993; Zielonka 1998).

6.4.1 Synthesis of Reactive Defense Strategies

Recall that in an AD game on graph model; $\mathcal{G} = \langle G, \varphi \rangle$, we assume that the information available to both players is complete and symmetric. Under this assumption, the solution for game on graph can be computed by constructing a game transition system and then using an algorithm to identify the winning regions for the attacker and the defender.

Before we introduce the game transition system, let us visit the equivalence of an scLTL specification with a Deterministic Finite-State Automaton (DFA).

Definition 6.4 *(Specification DFA)*   A DFA is a tuple,

$$\mathcal{A} = \langle Q, \Sigma, \delta, I, F \rangle,$$

where

- Q is a finite set of DFA states.
- $\Sigma = 2^{AP}$ is an alphabet.
- $\delta : Q \times \Sigma \to Q$ is a deterministic transition function. The transition function can be extended recursively as: $\delta(q, uv) = \delta(\delta(q, u), v)$ for some $u, v \in \Sigma^*$.
- $I \in Q$ is a unique initial state.
- $F \subseteq Q$ is a set of final states.

A word $w = \sigma_0 \sigma_1 \ldots \sigma_n$ is accepted by the DFA if and only if $\delta(q_0, w) \in F$. Given an scLTL specification φ, a DFA $\mathcal{A}$ is called a specification DFA when every word w defined over the alphabet Σ that satisfies $w \vDash \varphi$ is accepted by DFA $\mathcal{A}$.

Using this notion of equivalence between a scLTL formula and DFA, we define the game transition system as follows:

Definition 6.5 *(Game Transition System)*   Let $\mathcal{A} = \langle Q, \Sigma, \delta, I, F \rangle$ be a DFA equivalent to the specification φ. Then, given $\mathcal{G} = \langle G, \varphi \rangle$, the game transition system, represented as $G \otimes \mathcal{A}$, is the following tuple:

$$G \otimes \mathcal{A} = \langle S \times Q, A, \Delta, (s_0, q_0), S \times F \rangle,$$

where

- $S \times Q$ is a set of states partitioned into P1's states $S_1 \times Q$ and P2's states $S_2 \times Q$;
- $A = A_1 \cup A_2$ is the same set of actions as labeled transition system G;
- $\Delta : (S_1 \times Q \times A_1) \cup (S_2 \times Q \times A_2) \to S \times Q$ is a *deterministic* transition function that maps a game state (s, q) and an action a to the next state (s', q') where $s' = T(s, a, s')$ and $q' = \delta(q, L(s'))$.
- $(s_0, q_0) \in S \times Q$ where $q_0 = \delta(I, L(s_0))$ is an initial state of the game transition system and
- $S \times F \subseteq S \times Q$ is a set of final states.

The following theorem is a well-known result in game theory (McNaughton 1993; Zielonka 1998).

Theorem 6.1 *(Determinacy of Game on Graph)* All two-player zero-sum deterministic turn-based games on graph are determined.

Theorem 6.1 is a very important result because it provides us with a characterization of the game state space. It states that, *at any state in the game transition system, either the defender or the attacker has a winning strategy*. In other words, the state space of game transition system is divided into two sets, one consisting of states from which the defender is guaranteed to satisfy his security objectives, and the second consisting of states from which attacker has a strategy to violate the defender's objectives.

Example 6.2 We illustrate the game on graph using a toy example. Consider a network system where the defender can switch between two network topologies, giving rise to two attack graph under two network topologies (shown in Figure 6.3). For simplicity, as the graph is deterministic, we omit the attack action labels on the graph. Incorporating defender's actions into the attack graph, we obtain the arena of the game, shown in Figure 6.4. A circle state $(0, A)$ can be understood as the attacker is at node 0, the network configuration is A, and it is attacker's turn to make a transition. A square state $(1, A)$ can be understood as the attacker is at node 1, the network configuration is A, and it is defender's turn to make a switch. A transition from circle $(0, A)$ to square $(1, A)$ means that the attacker exploits a vulnerability on node 1 and reach node 1. The goal of the attacker is node 3. That is, if the attacker can reach any of the square states $(3, A)$ or $(3, B)$, then she wins the game. The goal of the defender is to prevent the attacker from reaching the goal. In this game, we can compute the attacker's strategy shown in Figure 6.4 where dashed edges indicate the choice of attacker. For example, if the attacker is at host 1 given topology B, she reaches host 2. If the defender switches to A, then she will take action to reach square $(3, A)$. If the defender switches to B, then she will take action to reach square $(3, B)$. In this game, there is no winning strategy for the defender given the initial state of the game. In fact, the winning region of the defender is empty.

(a)   (b)

Figure 6.3 The attack graphs under different network topologies. (a) Attack graph under topology A. (b) Attack graph under topology B.

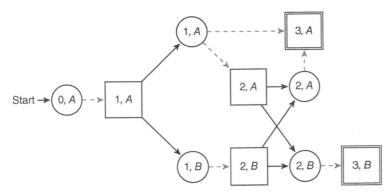

Figure 6.4 The game transition system given topology switching with simple attacker's reachability objective.

Figure 6.5 The automaton representing the attacker's objective

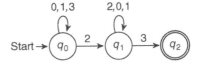

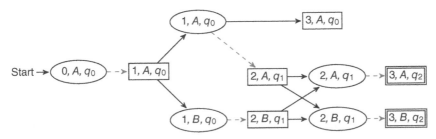

Figure 6.6 The game transition system for LTL co-safe formula $\Diamond$ ($2 \wedge \Diamond$ 3). The dashed edges are the attacker's strategy.

Now, let's consider a different specification of the attacker: $\Diamond$ ($2 \wedge \Diamond$ 3). That is, the attacker must reach node 2 first and then node 3. The LTL formula translates to DFA in Figure 6.5. Given the new specification, we construct the game transition system in Figure 6.6. An example of transition $(1, A, q_0, \text{circle}) \rightarrow (2, A, q_1, \text{square})$, where circle, square indicate the shapes of the nodes, is defined jointly by $(1, A, \text{circle}) \rightarrow (2, A, \text{square})$, and $q_0 \xrightarrow{2} q_1$. Given this LTL task, the winning strategy of the attacker is indicated with dashed edges. It is noted that when the attacker is at node 1, she will not choose to reach 3 but to reach 2, required by the new specification.

6.4.2 Synthesis of Reactive Defense Strategies with Cyber Deception

When the attacker has a one-sided misperception of labeling function, as defined in Definition 6.3, the defender *might* strategically utilize this misperception to deceive the attacker into choosing a strategy that is advantageous to the defender. To understand when the defender might have such a deceptive strategy and how to compute it, we study the solution concept of hypergame.

Solution Approach A hypergame $\mathcal{HG}^2 = \langle \mathcal{HG}^1, \mathcal{G}_2 \rangle$ is defined using two games, namely $\mathcal{G}_1$ and $\mathcal{G}_2$. Under one-sided misperception of labeling function, defender is aware of both games. Therefore, to synthesize a deceptive strategy, the defender must take into account the strategy that the attacker

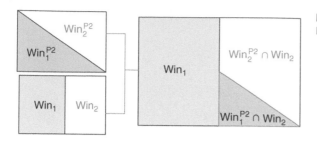

Figure 6.7 Illustration of the partition given by different perceptual game.

will use, based on her misperception. That is, the defender must solve two games: game G_2 to identify the set of states in the game transition system $G_2 \otimes A$ that the attacker perceives as winning for her under labeling function L_2 and game G_1 to identify the set of states in the game transition system $G_1 \otimes A$ that are winning for the defender under (ground-truth) labeling function $L = L_1$. After solving the two games, the defender can integrate the solutions to obtain a set of states, at which the attacker makes mistakes due to the difference between L_2 and L. Let us introduce a notation to denote these sets of winning states.

- G_1: P1's winning region is $\text{Win}_1 \subseteq S \times Q$ and P2's winning region is $\text{Win}_2 \subseteq S \times Q$.
- G_2: P1's winning region is $\text{Win}_1^{P2} \subseteq S \times Q$ and P2's winning region is $\text{Win}_2^{P2} \subseteq S \times Q$.

Figure 6.7 provides a conceptual representation partitions of the state space of a game transition system. Due to misperception, the set of states are partitioned into the following regions:

- Win_1: is a set of states from which P1 can ensure satisfaction of security objectives, even if P2 has complete and correct information. Thus, P1 can take the winning strategy π_1.
- $\text{Win}_1^{P2} \cap \text{Win}_2$ is a set of states where P2 is truly winning but perceives the states to be losing for her due to misperception. Thus, P2 may either give up the attack mission or play randomly.
- $\text{Win}_2^{P2} \cap \text{Win}_2$ is a set of states in which P2 is truly winning and perceives those states to be winning. In this scenario, she will carry out the winning strategy π_2^{P2}. However, this strategy can be different from the true winning strategy π_2 that P2 should have played if she had complete and correct information. This difference creates unique opportunities for P1 to enforce security of the system.

To compute deceptive strategy, the defender must reason about (a) how the attacker responds given her perception? and (b) how does the attacker expect the defender to respond, given her perception? It is noted that given the winning regions Win_1^{P2} (resp. Win_2^{P2}), there exists more than one strategies that P2 perceives to be winning for P2 (resp. for P1). The problem is to compute a strategy for the defender π_1^*, if exists, such that *no matter which* perceptual winning strategy that P2 selects, P1 can ensure the security specification is satisfied surely, without contradicting the perception of P2.

Given P2's perceptual game $G_2 = (S \times Q, A, \Delta, (s_0, q_{0,2}), S \times F)$, there can be infinitely many such almost-sure-winning mixed strategies for P2 (Bernet et al. 2002), we take an approximation of the *set* of almost-sure-winning strategy as a *memoryless set-based strategy* as follows.

$$\pi_2^{P2}(s, q) = \{a \mid \Delta_2((s, q), a) \in \text{Win}_2^{P2}\}. \tag{6.1}$$

In other words, P2 can select any action as long as she can stay within her perceived winning region Win_2^{P2}. Given the rational player 2, for a given state (s, q), an action a that is not in $\pi_2^{P2}(s, q)$ is *irrational* as it drives P2 from the perceived sure-winning region to the perceived losing region.

For a state $(s, q) \in \text{Win}_1^{P2} \cap \text{Win}_2$, P2 perceives P1 to be winning under labeling function L_2, when she is truly winning under ground-truth labeling function L. In this case, P1's deceptive strategy should conform to P2's perceptual winning strategy for P1. Otherwise, P2 would know that she is misperceiving the game when she observes P1 deviating from his rational behavior in the perceptual game of P2. When P1's action is inconsistent from what P2 perceives P1 should do, then P2 knows that she have misperception about the game.

Again, we take an approximation of the *set* of P1's almost-sure-winning strategy perceived by P2 as a *memoryless set-based strategy* as follows.

$$\pi_1^{P2}((s,q)) = \{a \mid \Delta_2((s,q),a) \in \text{Win}_1^{P2}\}. \tag{6.2}$$

Next, by removing P2's actions from G_1 that P2 perceives to be irrational as well as P1's actions that contradicts P2's perception, we obtain a different game. Now, we incorporate the knowledge of P1's actions that P2 would perceive to be irrational into the hypergame model. This results in a modified hypergame model as defined below.

Definition 6.6 Given the games $G_1 = G_1 \otimes A$ constructed using the true labeling function L and $G_2 = G_2 \otimes A$ with P2's misperceived labeling function L_2, the *deceptive sure-winning strategy of P1* is the sure-winning strategy of the following game:

$$HG = (S \times Q \times Q, A, \bar{\Delta}, (s_0, q_0, p_0), \text{Win}_1 \times Q),$$

where the transition function $\bar{\Delta}$ is defined such that

- For $(s, q, p) \in S_1 \times Q \times Q \backslash (\text{Win}_1 \times Q)$, if $(s, p) \in \text{Win}_1^{P2}$, then actions in $\pi_1^{P2}(s, p)$ are enabled. Otherwise, all actions $a \in A_1$ are enabled. For each enabled action a, let $\bar{\Delta}((s, q, p), a) = (s', q', p')$ where $s' = T(s, a)$, $q' = \delta(q, L(s'))$ and $p' = \delta(p, L_2(s'))$.
- For $(s, q, p) \in S_2 \times Q \times Q \backslash (\text{Win}_1 \times Q)$, if $(s, p) \in \text{Win}_2^{P2}$, then actions in $\pi_2^{P2}(s, p)$ are enabled. Otherwise, all actions $a \in A_2$ are enabled. For each enabled action a, let $\bar{\Delta}((s, q, p), a) = (s', q', p')$ where $s' = T(s, a)$, $q' = \delta(q, L(s'))$ and $p' = \delta(p, L_2(s'))$.
- The initial state is (s_0, q_0, p_0) where (s, q_0) is the initial state in G_1 and (s_0, p_0) is the initial state in G_2.

The transition function can be understood as follows. At a P1 state, when P2 perceives a state (s, q, p) to be winning for P1, the permissive actions in π_1^{P2} of P1 are enabled at that state. Otherwise, P2 would assume that P1 may choose any action from A_1. Similarly, at a P2 state, when P2 perceives a state to be winning for herself, she might choose any action from her permissive action set π_2^{P2}. Whereas, if P2 perceives the current state (s, q, p) to be losing for her, given her perception, she would choose any action from A_2.

Lemma 6.1 The sure-winning strategy π_1^* of the game HG in Definition 6.6 is stealthy for any state (s, q, p) where $(s, q) \in \text{Win}_2$ as it does not reveal any information with which P2 can deduce that some misperception exists.

Proof: In P2's perceived winning region Win_2^{P2} for herself, any strategy of P1 is losing. Thus, π_1^* will not contradict P2's perception. In P2's perceived winning region Win_1^{P2} for P1, an action, which is selected by π_1^* (if defined for that state), ensures that P1 to stay within Win_1^{P2} and thus will not contradict the perception of P2. In both cases, P2 will not deduce the fact that there is a misperception.

Example 6.3 *(Continued)*
Let us continue with the toy example and the simple reachability objective $\Diamond 3$ (eventually reach node 3). Suppose the node 2 is a decoy. Then the attacker's labeling function L_2 differs from the

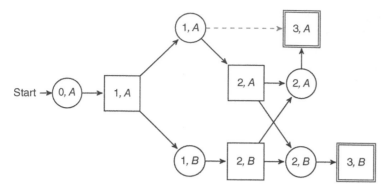

Figure 6.8 The game transition system and the attacker's winning strategy given that she is to reach node 3 and knows that node 2 is decoy.

defender's labeling function L: $L_2((2, X)) = \emptyset$ and $L((2, X)) =$ decoy where $X \in \{A, B\}$. As the attacker is to avoid reaching decoys, if she knows the true labeling L, then at the attacker's circle state $(1, A)$ in Figure 6.8, she will not choose to reach node 2 (and thus the square state $(2, A)$). The attacker's strategy given the true labeling function is given in Figure 6.8. In this game, the attacker has no winning strategy to reach 3 from 0 in the true game as the defender can choose to switch to topology B $((1, A, \text{square}) \rightarrow (1, B, \text{circle}))$. However, with the misperception on the labeling function, the attacker believes that she has a winning strategy from node 0 (see the attacker's strategy in Figure 6.4).

Consider now the initial state is $(1, A, \text{circle})$, that is, the attacker has compromised node 1 and the current network topology is A. If the attacker knows 2 is a decoy, then she will choose to reach 3 deterministically. If the attacker does not know 2 is a decoy, then she is indifferent to reaching node 2 or node 3, because in her perception, these two actions ensures that with probability one, she can reach node 3 in finitely many steps. Given this analysis, it is not difficult to see that $(1, A, \text{circle})$ is sure-winning for the attacker given the true game, but positive winning for the defender given the perceptual game of the attacker, with the incorrect labeling. Here, positive winning means that the defender wins with a positive probability, that is, the probability when the attacker makes mistakes (visiting decoy node 2) due to her misperception.

Finally, we construct the hypergame $\mathcal{HG}$ in Figure 6.9 where the defender's objective is $\neg p \ U$ decoy, where p is an atomic proposition that evaluates true when node 3 is compromised. The labeling functions are for $X \in \{A, B\}$, $L(1, X) = L_2(1, X) = \emptyset$, $L(3, X) = L_2(3, X) = \{p\}$ and $L(2, X) =$ decoy but $L_2(2, X) = \emptyset$. The states (shaded) are when the attacker's perceived automaton state differs from the defender's automaton state. For example, $((2, A), q_2, q_0, \text{circle})$ means that the defender knows that the attacker reached a decoy, but the attacker is unaware of this fact.

6.5 Case Study

We consider a simple network system illustrated in Figure 6.10.

In this network, each host runs a subset Servs $= \{0, 1, 2\}$ of services. A user in the network can have one of the three login credentials credentials $= \{0, 1, 2\}$ standing for "no access" (0), "user" (1), and "root" (2). There are a set of vulnerabilities in the network, each of which is defined by a pre-condition and a post-condition. The pre-condition is a Boolean formula that specifies the set of logical properties to be satisfied for an attacker to exploit the vulnerability instance. The post-condition is a Boolean formula that specifies the logical properties that can be achieved after

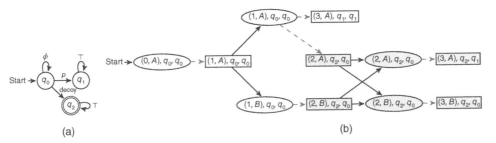

Figure 6.9 (a)The DFA for the defender's objective $\neg p$ U decoy. (b) The game that P1 uses to compute deceptive sure-winning strategy. The dashed edges are perceived winning actions of the attacker.

Figure 6.10 A example of network system.

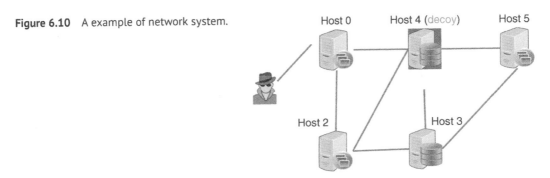

Table 6.1 The pre- and post-conditions of vulnerabilities.

Vulnerability ID	Pre- and post-conditions
0	Pre: $c \geq 1$, service 0 running on target host,
	Post: $c = 2$, stop service 0 on the target, reach
	target host.
1	Pre: $c \geq 1$, service 1 running on the target host,
	Post: reach target host.
2	Pre: $c \geq 1$, service 2 running on the target host
	Post: $c = 2$, reach target host.

the attacker has exploited that vulnerability. The set of vulnerabilities are given in Table 6.1 and are generated based on the vulnerabilities described in (Jha et al. 2002).

The defender can temporally suspend noncritical services from servers. To incorporate this defense mechanism, we assign each host a set of noncritical services that can be suspended from the host. In Table 6.2, we list the set of services running on each host, and a set of noncritical services that can be suspended by the defender. Other defenses can also be considered. For example, if the network topology can be reconfigured online, then the state in the game arena should keep track of the current topology configuration of the network. In our experiment, we consider simple defense actions. However, our method extends to more complex defense mechanisms.

The attacker, at a given attacker's state, can exploit any existing vulnerability on the current host. The defender, at a defender's state, can choose to suspend a noncritical service on any host in the

Table 6.2 The defender's options.

Host ID	Services	Non-critical services
0	$\{0, 1, 2\}$	$\emptyset$
1	$\{1, 0\}$	$\emptyset$
2	$\{1, 2\}$	$\emptyset$
3	$\{0, 1, 2\}$	$\{0, 1\}$
4	$\{0, 1\}$	$\emptyset$
5	$\{0, 1, 2\}$	$\emptyset$

network. The attacker's objective is expressed using scLTL formula

$$\varphi_2 = \neg\text{decoy } \mathsf{U} \ p_2 \wedge \neg\text{decoy } \mathsf{U} \ p_5,$$

where p_i means that the attacker has compromised host i and gained user or root access on that machine. However, the attacker does not know the location of the decoys. In this system, decoy is host 4.

The following result is obtained from hypergame analysis: at the initial state, the attacker is at host 0 with user access.

- The initial state is perceived to be winning by the attacker.
- Assuming complete, symmetric information, the size of winning region for the defender is 131.
- With asymmetric information, when the attacker plays a perceived winning strategy, the size of winning region for the defender is 193 which is greater than that with symmetric information.
- The defender has a winning strategy to prevent the attacker from achieving her objective in the network, using the solution of the hypergame.

It is noted that, using the winning region of the defender, we know for a given initial state, whether the security specification is satisfied. For a different initial state, for example, the attacker visited host 2 with user privileges, we can directly examine whether the security is ensured by checking if the new initial state is in the winning region for the defender. This set provides us important insight to understand the weak points in the network and can be used for guide the decoy allocation.

6.6 Conclusion

The goal of formal synthesis is to design dynamic defense with guarantees on critical security specifications in a cyber network. In this chapter, we introduced a game on graph model for capturing the attack-defend interactions in a cyber network for reactive defense, subject to security specifications in temporal logic formulas. In reactive defense, the defender can take actions in response to the exploit actions of the attacker. We introduced a hypergame for games on graphs to capture payoff misperception for the attacker caused by the decoy systems. The solution concept of hypergames enables us to synthesize effective defense strategy given the attacker's misperception of the game, without contradicting the belief of the attacker. There are multiple extensions from this study: the framework assumes asymmetric information but complete observations for both defender and attacker. It is possible to extend for defense design with a partially observable defender/attacker. By examining the winning region, it provides important insights for the resource allocation of decoy systems.

References

Ehab Al-Shaer, Jinpeng Wei, Kevin W Hamlen, and Cliff Wang. Dynamic bayesian games for adversarial and defensive cyber deception. In *Autonomous Cyber Deception*, pp. 75–97. Springer, 2019.

Zaruhi Aslanyan, Flemming Nielson, and David Parker. Quantitative verification and synthesis of attack-defence scenarios. In *2016 IEEE 29th Computer Security Foundations Symposium (CSF)*, pp. 105–119, 2016. doi: 10.1109/CSF.2016.15.

Christel Baier and Joost-Pieter Katoen. *Principles of Model Checking (Representation and Mind Series)*. The MIT Press, 2008.

Peter G. Bennett. Hypergames: Developing a model of conflict. *Futures*, 12(6): 489–507, 1980.

Julien Bernet, David Janin, and Igor Walukiewicz. Permissive strategies: From parity games to safety games. *RAIRO-Theoretical Informatics and Applications*, 36(3):261–275, 2002. doi: 10.1051/ita:2002013.

Krishnendu Chatterjee and Thomas A. Henzinger. A survey of stochastic omega-regular games. *Journal of Computer and System Sciences*, 78(2): 394–413, 2012. doi: http://dx.doi.org/10.1016/j.jcss.2011.05.002. Games in Verification.

Fred Cohen. The use of deception techniques: Honeypots and decoys. In *Handbook of Information Security 3.1*. Wiley, 2006.

René Rydhof Hansen, Peter Gjøl Jensen, Kim Guldstrand Larsen, Axel Legay, and Danny Bøgsted Poulsen. Quantitative evaluation of attack defense trees using stochastic timed automata. In *International Workshop on Graphical Models for Security*, pp. 75–90. Springer, 2017.

Karel Horák, Quanyan Zhu, and Branislav Bošanský. Manipulating adversary's belief: A dynamic game approach to deception by design for proactive network security. In *International Conference on Decision and Game Theory for Security,* pp. 273–294. Springer, Cham, 2017.

Karel Horak, Branislav Bošanský, Christopher Kiekintveld, and Charles Kamhoua. Compact representation of value function in partially observable stochastic games. In *International Joint Conferences on Artificial Intelligence Organization*, pp. 350–356, 2019. doi: 10.24963/ijcai.2019/50.

Sushil Jajodia, V. S. Subrahmanian, Vipin Swarup, and Cliff Wang. *Cyber Deception: Building the Scientific Foundation*. Springer, 2016. doi: 10.1007/978-3-319-32699-3.

S. Jha, O. Sheyner, and J. Wing. Two formal analyses of attack graphs. *Proceedings of the Computer Security Foundations Workshop*, 2002:49–63, 2002. doi: 10.1109/CSFW.2002.1021806.

Barbara Kordy and Wojciech Wideł. On quantitative analysis of attack–defense trees with repeated labels. In Lujo Bauer and Ralf Küsters, editors, *Principles of Security and Trust, Lecture Notes in Computer Science*, pp. 325–346. Cham, 2018. Springer International Publishing. doi: 10.1007/978-3-319-89722-6_14.

Barbara Kordy, Sjouke Mauw, Saša Radomirović, and Patrick Schweitzer. Foundations of attack–defense trees. In *International Workshop on Formal Aspects in Security and Trust*, pp. 80–95. Springer, 2010.

Barbara Kordy, Ludovic Piètre-Cambacédès, and Patrick Schweitzer. DAG-based attack and defense modeling: Don't miss the forest for the attack trees. *Computer Science Review*, 13–14:1–38, 2014. doi: 10.1016/j.cosrev.2014.07.001.

Nicholas S. Kovach, Alan S. Gibson, and Gary B. Lamont. Hypergame theory: A model for conflict, misperception, and deception. *Game Theory*, 2015:1–20, 2015. doi: 10.1155/2015/570639.

Orna Kupferman and Moshe Y. Vardi. Model checking of safety properties. *Formal Methods in System Design*, 19(3):291–314, 2001.

Zohar Manna and Amir Pnueli. *The Temporal Logic of Reactive and Concurrent Systems: Specification*. Springer-Verlag, New York, 1992. doi: 10.1007/978-3-4612-0931-7.

Robert McNaughton. Infinite games played on finite graphs. *Annals of Pure and Applied Logic,* 65(2):149–184, 1993.

Amir Pnueli and Roni Rosner. On the synthesis of an asynchronous reactive module. In *International Colloquium on Automata, Languages, and Programming*, vol. 372, pp. 652–671, 1989.

Bruce Schneier. Attack trees. http://www.schneier.com/paper-attacktrees-ddj-ft.html, August 2007.

Russell Richardson III Vane. Using hypergames to select plans in competitive environments. PhD thesis, George Mason University, 2000.

Quanyan Zhu and Stefan Rass. On multi-phase and multi-stage game-theoretic modeling of advanced persistent threats. *IEEE Access*, 6:13958–13971, 2018. doi: 10.1109/ACCESS.2018.2814481.

Wies?aw Zielonka. Infinite games on finitely coloured graphs with applications to automata on infinite trees. *Theoretical Computer Science*, 200(1-2): 135–183, 1998. doi: 10.1016/S0304-3975(98)00009-7.

Part II

Game Theory for Cyber Security

7

Minimax Detection (MAD) for Computer Security: A Dynamic Program Characterization

Muhammed O. Sayin[1], Dinuka Sahabandu[2], Muhammad Aneeq uz Zaman[3], Radha Poovendran[2], and Tamer Başar[3]

[1] *Laboratory for Information and Decision Systems, Massachusetts Institute of Technology, Cambridge, MA, USA*
[2] *Department of Electrical and Computer Engineering, University of Washington, Seattle, WA, USA*
[3] *Coordinated Science Laboratory, University of Illinois at Urbana-Champaign, Urbana, IL, USA*

7.1 Introduction

For the security of digital systems, defensive measures and adversaries have been combating in a nonending game. In that game, either side could have temporal successes or failures over the course of time. However, recently, the odds of successful attacks have been boosted at an accelerating pace in favor of adversaries. Data breach news are being received at an unprecedented frequency (Yao et al. 2017). This is a bothersome indication of ineffectiveness of existing detection mechanisms.

A classical and widely deployed approach is **signature-based detection**. After conducting a cumbersome analysis of an attack detected, security experts extract a signature distinct to its system-level behavior and develop a detection mechanism specifically searching for that signature across the entire system activities. Such a detection mechanism is remarkably successful at detecting that specific attack when it behaves exactly the way in its extracted signature. However, there is no guarantee beyond this. For example, this scheme cannot detect any attack for which no signature is extracted. Indeed, it is not only vulnerable to new attacks but also vulnerable to modified versions of old attacks, where the signatures extracted are no longer applicable. Apart from these vulnerabilities, it is also becoming computationally infeasible to check enormously many (and growing number of) different attack signatures in a computer system (Yao et al. 2017).

Another approach is **anomaly detection**, although it is not that widely deployed. Quite contrary to signature-based detection that is specialized on attacks, anomaly detection is specialized on systems and looks for any activity that is anomalous with respect to their normal behavior. Its main premise is that system activities under an adversarial attack are distinguishable from the system's normal activities. However, this is not entirely accurate. An adversarial intervention can behave similar to normal system activities (or can be designed to mimic the normal behavior; Wagner and Soto 2002) while conducting its malicious task. Therefore such attacks could undermine anomaly detection mechanisms. Furthermore these schemes cause too many false alarms due to rare system activities. Enhancement in powerful machine learning techniques has boosted applications of anomaly detection also in computer security (e.g., see Manzoor et al. 2016 and Du et al. 2017 and the references therein).

Recall that signature-based detection builds a defense based only on old attacks analyzed, which makes it vulnerable to new attacks, whereas anomaly detection builds a defense based only on the

Game Theory and Machine Learning for Cyber Security, First Edition.
Edited by Charles A. Kamhoua, Christopher D. Kiekintveld, Fei Fang, and Quanyan Zhu.
© 2021 The Institute of Electrical and Electronics Engineers, Inc. Published 2021 by John Wiley & Sons, Inc.

system's normal behavior, which makes it vulnerable to attacks that behave like the system's normal behavior even when they are known and well-analyzed. A more effective defense measure would be to combine these approaches together. In this chapter, we propose a new defense layer on top of these defense measures to evaluate their assessments in a *cohesive* and *strategic* way. Particularly, the proposed defense measure will evaluate system activities in terms of their likeliness (similar to anomaly detection) and riskiness (similar to signature-based detection).

7.1.1 Need for Cohesive Detection

A modern computing system can conduct a complex task by distributing it across multiple processes or multiple applications or even across multiple computers. Similarly an adversarial intervention can seek to conduct its malicious task in such a distributed manner. As benign examples, we can list multiprocess applications such as Internet browser (Elisan 2013) or parallel computing over a computer network (Almasi and Gottlieb 1994). As malicious ones, we can list modular malware equipped with multiple components such as rootkits and bot agents (Elisan 2013) or distributed denial-of-service (DDoS) attacks via a botnet across a network (Lau et al. 2000). Therefore, different from the previous approaches on program anomaly detection,[1] a cohesive approach is a future trend in adversarial intervention detection since advanced attackers can conduct the malicious task in a distributed manner over multiple processes or even multiple computers (Liu et al. 2019). However, it is a challenge for detection mechanisms to process an enormous amount of data collected across the entire system.

7.1.2 Need for Strategic Detection

A detection mechanism designed without taking into account how adversaries would react to it makes it vulnerable once it is publicly deployed and the adversaries have the chance to learn and adapt to it. Therefore it is important to design detection mechanisms by taking adversaries' reactions into account, e.g., designing it under the solution concept of Nash equilibrium.[2] Even though existence or uniqueness of an equilibrium is not guaranteed in general, when there exists a unique equilibrium, both strategic players end up playing the actions that attain the equilibrium, which leads to a theoretical guarantee over the outcome of the game for both players (Başar and Olsder 1999). We note that computing the strategies that attain an equilibrium can be computationally demanding if the players have large strategy spaces (Daskalakis et al. 2009).

In the literature, there are several studies that propose detection mechanisms under the solution concept of game theoretical equilibrium. In the inaugural study (Dalvi et al. 2004), the authors have introduced a nonzero-sum game between an attacker and a classifier; however, the proposed defense does not attain any notion of equilibrium. In Brückner and Scheffer (2009, 2011), the authors have addressed adversarial prediction problems for a certain class of learners, e.g., support-vector-machines, under the solution concepts of Nash and Stackelberg equilibria, respectively. In Dritsoula et al. (2017), the authors have analyzed adversarial binary-classification as a nonzero-sum game, where the attacker seeks to maximize his reward at the expense of additive detection cost while the detector seeks to minimize the attacker's reward and maximize his detection cost at the expense of additive false alarm cost. Reference Sayin et al. (2018) has proposed and analyzed a two-level game theoretical framework for inter-process adversarial intervention detection. Sayin et al. (2020) has proposed and analyzed a game theoretical adversarial

---

1 Interested reader can refer to Shu et al. (2015), Yao et al. (2017) and Liu et al. (2018) for a detailed overview of these studies.
2 We say that a pair of actions attains a Nash equilibrium if neither player has an incentive to take any other action since any unilateral deviation could not benefit him/her further (Başar and Olsder 1999).

intervention detection mechanism for smart road signs that incorporate smart codes (e.g., visible at infrared) on their surface to provide more detailed information to smart vehicles in addition to robustness against adversarial examples (e.g., see Eykholt et al. 2018) that can deceive the state of the art image classification algorithms.

Remark 7.1 Recall that it is already a computational challenge for detection mechanisms to process data collected across the entire system even in a nonstrategic manner. When we seek to design a detection mechanism in a strategic manner, due to the reasons explained above, the corresponding game possesses an enormous size, which makes it very challenging to compute its equilibrium through a direct approach.

7.1.3 Minimax Detection (MAD)

In a modern computing system, we consider the scenarios where an adversary seeks to intervene the system. For example, when there is an adversarial intervention, any system activity might be generated by the adversary out of the system's or the user's intention. As a counter measure, there exists a detector that can monitor certain system activities, e.g., system-calls of processes, and seeks to detect adversarial intervention by assessing these activities in terms of two metrics: their likeliness and riskiness. These metrics are common knowledge of the detector and the adversary. We model the interaction between the detector and the adversary as a *zero-sum game*. We name it "minimax detection" (MAD) since MAD leads to minimum (detection) cost to the system against an adversary seeking to maximize it.

Remark 7.2 Players have very large, yet finite, action spaces. We can convert the computation of the mixed strategy equilibrium of this zero-sum game into a linear program (LP) and then apply existing powerful computational tools, e.g., ellipsoid methods or interior point techniques, available to solve LPs (Boyd and Vandenberghe 2004). However, computational complexity of an LP is weakly polynomial in the number of constraints and variables (Boyd and Vandenberghe 2004), which makes such a direct approach infeasible for any existing computational resource.

To mitigate this computational challenge, MAD evaluates system activities within the following hierarchical scheme. Consider activities monitored across the entire system as a composition of[3] "building blocks" (the smallest units in the scheme), similar to the atomic units in Gao et al. (2004). For example, a building block can be a trace of system-calls of a process over a certain time interval. We say that a building block gets "infected" if any of the activities within that block is generated by the adversary. We view activities realized at a block as a "word" drawn from a finite-size "dictionary." Correspondingly, the associated dictionary is the set of all possible system activities that can be realized at that block.

MAD assesses each block whether it is infected or not according to the likeliness and riskiness of the words realized. However, a word coming from a single block might appear benign (or malicious) in itself while a combination of words coming from different blocks might reveal it maliciousness (or benignness). Correspondingly, a detection mechanism would be cohesive by evaluating all words together, which is, however, not feasible computationally.

Example 7.1 Analogously this is similar to detecting whether there are any malicious words in a book by evaluating the text across the entire book in a cohesive way. On the other hand, a book

---

3 We introduce our own terminology while modeling the hierarchical scheme for ease of reference.

consists of multiple paragraphs, which are compositions of multiple sentences, which are compositions of multiple words. After assessing each word in itself in terms of its metrics, we can assign metrics for each sentence composed of the words assessed. Then we can assess each sentence in itself in terms of its metrics assigned and continue in this hierarchical scheme until we reach the top layer where we assess the entire book in itself in terms of its metrics assigned.

Inspired by this analogy, we can pack certain multiple blocks into a larger block and assign the metrics of this larger block accordingly. We can continue to pack larger blocks into larger and larger ones as desired. Eventually we obtain blocks formed within a hierarchical scheme, i.e., level by level. MAD assesses each of these blocks (building blocks or the packed ones) by designing a *randomized* detection rule that depends on the metrics of the word realized.

We emphasize that the detector and the adversary are playing a single game across the entire system. However, this hierarchical scheme brings this normal form game into a structure similar to the extensive form games that model sequential interactions between players. Correspondingly, we prove that this entire game can be decomposed into certain nested local subgames played by the detector and the adversary at each level. These local subgames are also zero-sum games with compact strategy spaces, which are finite and convex polytopes, while the game objective is linear in the players' strategies. Equilibrium at each local subgame could be computed via an LP. Based on this nested structure, we provide a dynamic program to compute minimax detection rules across all blocks and levels similar to backward induction in extensive form games.

We can list the main novelties of MAD as follows:

- MAD is a detection scheme that incorporates signature-based detection and anomaly detection together for more effective defense
- MAD is a detection scheme that guarantees minimum detection cost against adversarial interventions seeking to maximize it
- MAD is a detection scheme that assesses activities across the entire system in a cohesive way through a hierarchical scheme
- MAD is a detection scheme that can be designed computationally efficiently via a dynamic programming approach

Organization of the chapter is as follows: In Section 7.2, we describe the hierarchical scheme and provide the game formulation. In Section 7.3, we analyze the equilibrium and compute the best randomized detection rule. We examine the equilibrium behavior of the players over an illustrative example in Section 7.4. We conclude the chapter in Section 7.5 with several remarks and identifying possible research directions. Appendix 7.A provides the technical details for a generalization of the results.

Notation throughout the chapter is as follows: Superscripts denote the level while subscripts denote the node index. Random variables are denoted by bold letters, e.g., $\boldsymbol{\omega}$.

7.2 Problem Formulation

We consider a computing system in which certain system-level activities, e.g., reading/writing or sending/receiving a file, take place while a recording mechanism monitors these activities at the level of kernel. Recent enhancements in transparent computing have mitigated the visibility issues in modern computing systems and there exist powerful monitoring resources that can record system activities at large scales and even at the resolution of system-calls efficiently (Zhang and Zhou

2006; Ma et al. 2016; Ji et al. 2017). Therefore, the underlying system could correspond to a spectrum of computing systems varying from a single process to a network of multiple computers.

We seek to detect any adversarial intervention in the system activities. In other words, our goal is to detect whether any of the activities being recorded are generated by an adversary. To this end, we consider two rational decision makers: a detector (MAD) and an adversary (MAX).[4] In the following section, we describe the details of the underlying system before modeling MAD and MAX.

7.2.1 System Model

As explained in Section 7.1, MAD evaluates system activities across the entire system in a cohesive way within a hierarchical scheme. Formally, we consider a hierarchically ordered tree graph whose vertices/nodes correspond to blocks. We say that a level precedes (respectively, succeeds) another one if the former one is hierarchically below (above) the latter. At a level, except the lowest level, each node is connected to certain nodes from the preceding level according to the composition of the associated block from the blocks at the preceding level. In these connections, the node at the succeeding level is called the "parent" while the node at the preceding level is called its "child." The nodes at the lowest level are not parents of any nodes and we call them "(local) elementary nodes" of the system.

Note that blocks at a level do not overlap with each other, i.e., their detection/monitoring zones do not overlap. As we will show in Section 7.3, this hierarchical tree model enables us to decompose the interactions between MAD and MAX into an entirely nested structure, which leads to efficient and effective detection algorithms. It can be shown that a directed acyclic graph scheme that allows overlaps in detection/monitoring zones is not advantageous in terms of either computational complexity or detection performance compared to a tree scheme, where we consider a hyper node corresponding to the nodes with overlapping detection/monitoring zones.

The framework provides flexibility on how to design the hierarchical structure. For example, we can prefer to evaluate building blocks within a process across certain time interval together or we can prefer to evaluate building blocks, from different processes, being executed at a certain time interval together. We also emphasize that this framework gives flexibility beyond models that consider sliding window, forgetting factors or Markovian behavior over dynamic execution flow of processes since it can be shown that they are special cases of the proposed framework.

Suppose that the graph has depth κ and has a single root. Let n_i^k, for $i \in \mathcal{I}^k$, denote the ith node at level k, where $\mathcal{I}^k$ denotes the index set of all the nodes at level k. At each n_i^k, MAD observes a word ω_i^k from a finite dictionary $\mathcal{D}_i^k$, i.e., $\omega_i^k \in \mathcal{D}_i^k$. At the elementary nodes, the words are certain sampling of system activities, e.g., recorded system-call traces. However, the words at nonelementary nodes, are constructed based on certain compressed/sampled/summarized versions of the words at their children. Let $\mathcal{M}_i^k$ denote the index set for children of n_i^k. For each n_i^k, except the top level, there exists a compression/encoding mapping $\phi_i^k : \mathcal{D}_i^k \to \bar{\mathcal{D}}_i^k$, where $\bar{\mathcal{D}}_i^k$ denotes the set of compressed words. By the nature of compression, we can expect $|\bar{\mathcal{D}}_i^k| \ll |\mathcal{D}_i^k|$; however, the proposed framework provides flexibility to design the encoding mapping in any desired way, including also the identity map. We denote the compressed version of $\omega_i^k \in \mathcal{D}_i^k$ by $\bar{\omega}_i^k \in \bar{\mathcal{D}}_i^k$, i.e.,

$$\bar{\omega}_i^k = \phi_i^k(\omega_i^k). \tag{7.1}$$

Then, at the parent n_j^{k+1}, MAD observes

$$\omega_j^{k+1} = \{\bar{\omega}_i^k\}_{i \in \mathcal{M}_j^{k+1}}, \tag{7.2}$$

---

4 We refer to the detector as "MAD" since it attains minimax detection performance and refer to the adversary as "MAX" since it maximizes the detection cost of the system.

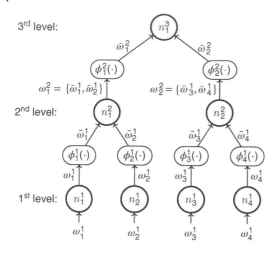

Figure 7.1 An illustration of the model with three levels and $4 + 2 + 1$ nodes. Edges are directed in the information flow rather than the hierarchical order.

as illustrated in Figure 7.1. The dictionary of the parent n_j^{k+1} is given by

$$\mathcal{D}_j^{k+1} = \underset{i \in \mathcal{M}_j^{k+1}}{\times} \bar{\mathcal{D}}_i^k. \tag{7.3}$$

In the following sections, we introduce defense and threat models, respectively.

7.2.2 Defense Model

At each n_i^k, MAD observes a word $\omega_i^k \in \mathcal{D}_i^k$, which may reveal certain information about whether there is any adversarial intervention or not. MAD assesses the word ω_i^k observed in terms of the following two metrics:

- a *likeliness* metric $l_i^k : \mathcal{D}_i^k \to [0, 1]$ where $l_i^k(\omega)$ corresponds to the probability that $\omega \in \mathcal{D}_i^k$ would be observed if there were no adversarial intervention
- a *riskiness* metric $r_i^k : \mathcal{D}_i^k \to [0, \infty)$ where $r_i^k(\omega)$ corresponds to the measure of damage induced by $\omega \in \mathcal{D}_i^k$ if n_i^k is infected

MAD selects a randomized detection policy β_i^k, where $\beta_i^k(\omega) \in [0, 1]$ is the probability that MAD triggers an alert for the word $\omega \in \mathcal{D}_i^k$. MAD seeks to minimize the total expected risk[5] across the entire system under a hard constraint on the false alarm rate for words that lead to the same compressed version. Correspondingly, the strategy space of MAD is given by $\mathcal{B} = \times_{i,k} \mathcal{B}_i^k$, where

$$\mathcal{B}_i^k := \left\{ \beta_i^k \in [0, 1]^{|\mathcal{D}_i^k|} \mid \mathbb{E}_o\{\beta_i^k(\omega) | \phi_i^k(\omega) = \bar{\omega}\} \le \tau_i^k, \ \forall \bar{\omega} \in \bar{\mathcal{D}}_i^k \right\} \tag{7.4}$$

where $\tau_i^k \in [0, 1]$ is a threshold on the false alarm rate and the expectation operator $\mathbb{E}_o$ takes expectation with respect to the underlying distribution over the normal behavior of the system, i.e., the likeliness metric $\{l_i^k\}$.

Remark 7.3 Note that a tolerable false alarm rate that is set to zero at a node can be viewed as if MAD does not allocate resources for detection at that node. This aspect of the proposed machinery provides freedom to allocate resources accordingly if there is any limitation on them. □

---

5 We will describe the cost function later in (7.6) after we introduce the threat model.

7.2.3 Threat Model

We seek to design the detection mechanism against adversaries that would maximize the detection cost. To this end, we model MAX as knowing the likeliness and riskiness metrics of each word at each node. MAX seeks to maximize the cost of MAD, by injecting words into any elementary node. For example, if n_i^1 is infected, then MAX selects a mixed strategy over $\mathcal{D}_i^1$, denoted by $\alpha_i^1 \in \Delta(\mathcal{D}_i^1)$, where $\alpha_i^1(\omega)$ is the (selected) probability that MAX injects $\omega \in \mathcal{D}_i^1$ into the node n_i^1. Correspondingly, the strategy space of MAX is given by

$$\mathcal{A} := \underset{i}{\times}\Delta(\mathcal{D}_i^1). \tag{7.5}$$

Although MAX only infects elementary nodes, we say that a nonelementary node is infected if any of its descendant (elementary) nodes is infected. We consider the scenarios where there is a Bernoulli random variable θ_i^k for all n_i^k and $\theta_i^k = 1$ means that there is an infection at n_i^k. Joint distribution of these random variables is a common knowledge of the players.

7.2.4 Game Model

Detection cost $U(\cdot)$, minimized by MAD (and maximized by MAX), is given by

$$U(\alpha, \beta) = \sum_{k=1}^{K}\sum_{i \in \mathcal{I}^k} \mathbb{E}\left\{r_i^k(\omega_i^k) - c_i^k\beta_i^k(\omega_i^k)|\theta_i^k = 1\right\}, \tag{7.6}$$

where $c_i^k \geq 0$ corresponds to the detection gain of MAD and the expectation is taken with respect to all the randomness (including MAX's mixed strategy). In other words, MAD seeks to minimize the total expected damage MAX can cause, with an additional reward for each detection.

We model the interaction between MAD and MAX as a zero-sum game. MAX selects a mixed strategy α from the strategy space $\mathcal{A}$ to maximize (7.6) while MAD selects a randomized detection rule β from the strategy space $\mathcal{B}$ to minimize (7.6). A pair of strategies (α^*, β^*) attains an equilibrium (i.e., a saddle point equilibrium) provided that

$$\beta^* \in \underset{\beta \in \mathcal{B}}{\operatorname{argmin}}\, U(\alpha^*, \beta), \tag{7.7a}$$

$$\alpha^* \in \underset{\alpha \in \mathcal{A}}{\operatorname{argmax}}\, U(\alpha, \beta^*). \tag{7.7b}$$

Note that the cost function (7.6) is an affine/linear function of the players' strategies while the players have compact and convex strategy spaces. Correspondingly, Minimax Theorem (Başar and Olsder 1999) implies that there exists an equilibrium and the value of the game is unique. In other words, we have

$$U(\alpha^*, \beta^*) = \min_{\beta \in \mathcal{B}}\max_{\alpha \in \mathcal{A}} U(\alpha, \beta) = \max_{\alpha \in \mathcal{A}}\min_{\beta \in \mathcal{B}} U(\alpha, \beta) \tag{7.8}$$

and β^* is a minimax detection strategy satisfying

$$\beta^* \in \underset{\beta \in \mathcal{B}}{\operatorname{argmin}}\, \max_{\alpha \in \mathcal{A}} U(\alpha, \beta). \tag{7.9}$$

Therefore it would not make any difference if the game is analyzed instead under the solution concept of Stackelberg equilibrium, where either MAD or MAX is the leader.

7.3 Main Result

In this section, we seek to compute minimax detection rules efficiently. To this end, we show that the game can be viewed as nested local subgames played at each node of the hierarchical scheme proposed. Each of these subgames is also a zero-sum game and their unique outcomes determine the configuration of the local subgames at higher levels. Hence, an equilibrium for the entire game can be computed through a dynamic program similar to backward induction in extensive form games. This mitigates computational challenges induced by the size of the entire game. We now derive the results over a simple example illustrated in Figure 7.2 for ease of representation without delving into any complicated notation. We, however, also provide a generalization of the results in Appendix 7.A for the reader's reference. Additionally at the end of this section, we analyze the complexity of the proposed dynamic program characterization compared to the direct approaches.

Consider the example hierarchical scheme described in Figure 7.2. In this example, n_1^2 is parent of n_1^1 and n_2^1. They, respectively, have the dictionaries $\mathcal{D}_1^2 = \{\omega_{1j}^2\}_{j=1}^4$, $\mathcal{D}_1^1 = \{\omega_{1j}^1\}_{j=1}^4$, and $\mathcal{D}_2^1 = \{\omega_{2j}^1\}_{j=1}^4$. Note that in Figure 7.2, vertices of the graph correspond to words rather than the nodes. And directed edges from dictionaries $\mathcal{D}_i^k$ to $\bar{\mathcal{D}}_i^k$ represent the encoding/compression mapping. For example, at n_1^1, words $\omega_{1,1}^1$ and $\omega_{1,2}^1$ are compressed into $\bar{\omega}_{1,1}^1$ while $\omega_{1,3}^1$ and $\omega_{1,4}^1$ are compressed into $\bar{\omega}_{1,2}^1$. On the other hand, directed edges from compressed dictionaries to dictionaries represent how the words at higher levels are formed as a composition of compressed words from lower levels. For example, at n_1^2, one of the words in its dictionary is $\omega_{1,1}^2 = (\bar{\omega}_{1,1}^1, \bar{\omega}_{2,1}^1)$.

Note that the words ω_1^1, ω_2^1, and ω_1^2 cannot be completely arbitrary. For example, if word $\omega_{1,2}^2 = (\bar{\omega}_{1,1}^1, \bar{\omega}_{2,2}^1)$ is observed at n_1^2, then at n_1^1, the word observed cannot be $\omega_{1,3}^1$ or $\omega_{1,4}^1$ since they entail a compressed word other than $\bar{\omega}_{1,1}^1$. However, given that at n_1^1, the word $\omega_{1,2}^2$ is observed, at n_1^1, any of the words $\omega_{1,1}^1$ and $\omega_{1,2}^1$ might have been observed. This yields that MAX has flexibility on selecting

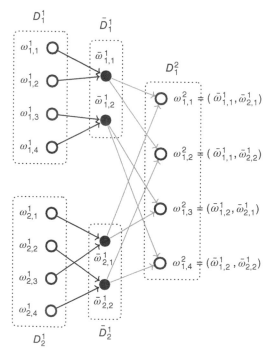

Figure 7.2 An illustrative example where n_1^2 is the root node while n_1^1 and n_2^1 are its children. Vertices correspond either to words or to their compressed versions, rather than the nodes. Moving from left to right, the first set of edges corresponds to compression whereas the second set corresponds to the Cartesian product.

a mixed strategy over the words that lead to the same compressed version while seeking to evade detection across the elementary nodes.

Example 7.2   Coming back to Example 7.1, suppose that there are two levels, e.g., the root node corresponds to the entire book, and its children correspond to paragraphs in it. Within the hierarchical scheme, MAD assesses short summaries of all paragraphs together at the root node and assesses each paragraph in itself at the bottom level. On the other hand, MAX decides the content of the paragraphs it infected. Since only the coupling between the compressed versions of the paragraphs is assessed by MAD, MAX can view the game as being played level by level in a top-down direction. For example, at the top level, MAX decides summarized contents of paragraphs infected and MAD assesses the correlation among the summaries. At the bottom level, MAX decides contents of the paragraphs infected under the constraint that their summarized versions lead to the summaries decided at the top level and MAD assesses the contents of these paragraphs in itself.

Inspired by this analogy, we seek to decompose the game into a nested structure played level by level. The question, however, is that how the decisions made by the players at a level impact their decisions at other levels. To this end, let us take a closer look at the simple example, described in Figure 7.2. For this example, the cost function is composed of three parts:

$$\mathbb{E}\{r_i^1(\omega_i^1) - c_i^1\beta_i^1(\omega_i^1)|\theta_i^1 = 1\} = \mathbb{P}(\theta_i^1 = 1) \sum_{\omega_i^1 \in D_i^1} \alpha_i^1(\omega_i^1)(r_i^1(\omega_i^1) - c_i^1\beta_i^1(\omega_i^1)) \tag{7.10}$$

for $i = 1, 2$, corresponding to the cost for MAD attained at n_i^1, $i = 1, 2$, and

$$\mathbb{E}\{r_1^2(\omega_1^2) - c_1^2\beta_1^2(\omega_1^2)|\theta_1^2 = 1\} \tag{7.11}$$

$$= \mathbb{P}(\theta_1^1 = 1, \theta_2^1 = 1) \sum_{(\omega_1^1, \omega_2^1)} \alpha_1^1(\omega_1^1)\alpha_2^1(\omega_2^1)(r_1^2(\omega_1^2) - c_1^2\beta_1^2(\omega_1^2)) \tag{7.12}$$

$$+ \sum_{i=1,2} \mathbb{P}(\theta_i^1 = 1, \theta_j^1 = 0) \sum_{\omega_i^1} \alpha_i^1(\omega_i^1)\mathbb{E}_o\{r_1^2(\omega_1^2) - c_1^2\beta_1^2(\omega_1^2)|\omega_i^1 = \omega_i^1\}, \tag{7.13}$$

where $j \in \{1, 2\}$ and $j \neq i$, and the first term in the right hand side (7.12) corresponds to the cost for MAD attained at n_1^2 for the scenario where there are infections at n_1^1 and n_2^1 while the second term (7.13) corresponds to the cost for MAD attained at n_1^2 for the scenarios where there is an infection either at n_1^1 or at n_2^1.

We let $\bar{\alpha}_i^1$ be a probability distribution over the compressed dictionary $\bar{D}_i^1$ such that $\bar{\alpha}_i^1(\bar{\omega})$ corresponds to the probability that $\bar{\omega} \in \bar{D}_i^1$ would have been observed if there were an infection at n_i^1, and it is given by

$$\bar{\alpha}_i^1(\bar{\omega}) = \sum_{\omega \in \mathcal{E}_i^1(\bar{\omega})} \alpha_i^1(\omega), \tag{7.14}$$

where $\mathcal{E}_i^1(\bar{\omega})$ is the inverse image of the compressed word $\bar{\omega}$, i.e.,

$$\mathcal{E}_i^1(\bar{\omega}) := \{\omega \in D_i^1 | \phi_i^1(\omega) = \bar{\omega}\}. \tag{7.15}$$

Furthermore, we let $\alpha_i^1(\cdot|\bar{\omega})$ be a probability distribution over $\mathcal{E}_i^1(\bar{\omega})$ such that $\alpha_i^1(\omega|\bar{\omega})$ corresponds to the probability that ω would have been observed if there was an infection at n_i^1 and the associated compressed version was $\bar{\omega} \in \bar{D}_i^1$. Correspondingly we have $\alpha_i^1(\omega) = \bar{\alpha}_i^1(\bar{\omega})\alpha_i^1(\omega|\bar{\omega})$.

By using these auxiliary variables, we can write the summation in (7.10) as

$$\sum_{\bar{\omega}_i^1 \in \bar{D}_i^1} \bar{\alpha}_i^1(\bar{\omega}_i^1)R_i^1(\bar{\omega}_i^1), \tag{7.16}$$

where

$$R_i^1(\bar{\omega}_i^1) := \sum_{\omega_i^1 \in \mathcal{E}_i^1(\bar{\omega}_i^1)} \alpha_i^1(\omega_i^1|\bar{\omega}_i^1)(r_i^1(\omega_i^1) - c_i^1 \beta_i^1(\omega_i^1)). \tag{7.17}$$

Furthermore the summations in (7.12) and (7.13), respectively, can be written as

$$\sum_{(\bar{\omega}_1^1, \bar{\omega}_2^1)} \bar{\alpha}_1^1(\bar{\omega}_1^1)\bar{\alpha}_2^1(\bar{\omega}_2^1)(r_1^2(\omega_1^2) - c_1^2 \beta_1^2(\omega_1^2)) \tag{7.18}$$

and

$$\sum_{\bar{\omega}_i^1 \in \bar{D}_i^1} \bar{\alpha}_i^1(\omega_i^1) \mathbb{E}_o\{r_1^2(\omega_1^2) - c_1^2 \beta_1^2(\omega_1^2)|\omega_i^1 = \omega_i^1\}. \tag{7.19}$$

Hence, for this simple example, we obtain that

$$U(\alpha, \beta) = \sum_{i=1,2} \mathbb{P}(\theta_i^1 = 1) \sum_{\bar{\omega}_i^1 \in \bar{D}_i^1} \bar{\alpha}_i^1(\bar{\omega}_i^1)R_i^1(\bar{\omega}_i^1) \tag{7.20}$$

$$+\mathbb{P}(\theta_1^1 = 1, \theta_2^1 = 1) \sum_{(\bar{\omega}_1^1, \bar{\omega}_2^1)} \bar{\alpha}_1^1(\bar{\omega}_1^1)\bar{\alpha}_2^1(\bar{\omega}_2^1)(r_1^2(\omega_1^2) - c_1^2 \beta_1^2(\omega_1^2)) \tag{7.21}$$

$$+\sum_{i=1,2} \mathbb{P}(\theta_i^1 = 1, \theta_j^1 = 0) \sum_{\bar{\omega}_i^1 \in \bar{D}_i^1} \bar{\alpha}_i^1(\omega_i^1) \mathbb{E}_o\{r_1^2(\omega_1^2) - c_1^2 \beta_1^2(\omega_1^2)|\omega_i^1 = \omega_i^1\}. \tag{7.22}$$

Note that $\bar{\alpha}_i^1 \in \Delta(\bar{D}_i^1)$ is a probability distribution. Therefore, we can multiply the summations (7.20) and (7.22) by $\sum_{\bar{\omega}_j^1} \bar{\alpha}_j^1(\bar{\omega}_j^1) = 1$ without loss of generality, which yields that

$$U(\alpha, \beta) = \mathbb{P}(\theta_1^2 = 1) \sum_{\omega_1^2} \alpha_1^2(\omega_1^2)(\tilde{r}_1^2(\omega_1^2) - c_1^2 \tilde{\beta}_1^2(\omega_1^2)), \tag{7.23}$$

where we define

$$\alpha_1^2(\omega_1^2) := \bar{\alpha}_1^1(\bar{\omega}_1^1)\bar{\alpha}_2^1(\bar{\omega}_2^1), \ \forall \ \omega_1^2 = (\bar{\omega}_1^1, \bar{\omega}_2^1) \in D_1^2 \tag{7.24}$$

and

$$\tilde{r}_1^2(\omega_1^2) := \mathbb{P}(\theta_1^1 = 1, \theta_2^1 = 1|\theta_1^2 = 1)r_1^2(\omega_1^2) \tag{7.25}$$

$$+\sum_{i=1,2} \mathbb{P}(\theta_i^1 = 1, \theta_j^1 = 0|\theta_1^2 = 1)\mathbb{E}_o\{r_1^2(\omega_1^2)|\omega_i^1 = \omega_i^1\} \tag{7.26}$$

$$+\sum_{i=1,2} \mathbb{P}(\theta_i^1 = 1|\theta_1^2 = 1)R_i^1(\omega_i^1), \tag{7.27}$$

$$\tilde{\beta}_1^2(\omega_1^2) := \mathbb{P}(\theta_1^1 = 1, \theta_2^1 = 1|\theta_1^2 = 1)\beta_1^2(\omega_1^2) \tag{7.28}$$

$$+\sum_{i=1,2} \mathbb{P}(\theta_i^1 = 1, \theta_j^1 = 0|\theta_1^2 = 1)\mathbb{E}_o\{\beta_1^2(\omega_1^2)|\omega_i^1 = \omega_i^1\}. \tag{7.29}$$

The cost function written in the form of (7.23) resembles to the payoff function of a game where there is only the root note n_1^2. Furthermore, all the interaction at level 1 is confined into the terms $R_i^1(\omega_i^1)$, $i = 1, 2$, described in (7.17). This yields a nested structure where the inner interactions between the players (i.e., the ones at lower levels) are confined into single terms while the cost function is an affine function of these terms. The following lemma shows that such a nested structure can be decomposed into nested local subgames.

Lemma 7.1 Consider a zero-sum game between $\mathcal{P}_X$ and $\mathcal{P}_Y$, where the players select a pair of strategies from decoupled, nonempty and compact strategy spaces. Let $x = (x_1, x_2)$ denote $\mathcal{P}_X$'s pair selected from $\mathcal{X}_1 \times \mathcal{X}_2$ while let $y = (y_1, y_2)$ denote $\mathcal{P}_Y$'s pair selected from $\mathcal{Y}_1 \times \mathcal{Y}_2$. $\mathcal{P}_Y$ seeks to minimize a cost function that can be written in the following nested structure:

$$U(x, y) = U_1(x_1, y_1, U_2(x_2, y_2)), \tag{7.30}$$

where U_1 is linear/affine function of each argument when the others are fixed, it is also a monotonically increasing function of U_2 while U_2 is a linear/affine function of its arguments. Then (x^*, y^*) is a saddle point if, and only if, the pair (x_2^*, y_2^*) is a saddle point of $U_2(\cdot, \cdot)$ and given $u_2^* = U_2(x_2^*, y_2^*)$, the pair (x_1^*, y_1^*) is a saddle point of $U_1(\cdot, \cdot, u_2^*)$.

Proof: The proof follows by contradiction. To this end, suppose that (x^*, y^*) attains the saddle point equilibrium, but (x_2^*, y_2^*) is not a saddle point for $U_2(\cdot, \cdot)$. This implies that there exists a y_2 such that

$$U_2(x_2^*, y_2^*) > U_2(x_2^*, y_2) \tag{7.31}$$

or there exists an x_2 such that

$$U_2(x_2^*, y_2^*) < U_2(x_2, y_2^*). \tag{7.32}$$

Since U_1 is a monotonically increasing function of its third argument, the former case (7.31) yields that the pair of strategies $(x^*, \{y_1^*, y_2\})$ yields smaller game outcome for the minimizing $\mathcal{P}_Y$ and the latter case (7.32) yields that the pair of strategies $(\{x_1^*, x_2\}, y^*)$ yields larger game outcome for the maximizing $\mathcal{P}_X$. Both lead to contradiction and therefore (x_2^*, y_2^*) must be a saddle point of U_2.

Now suppose that (x_1^*, y_1^*) is not a saddle point for $U_1(\cdot, \cdot, U_2(x_2^*, y_2^*))$. This implies that there exists a y_1 that leads to smaller game outcome for the minimizing $\mathcal{P}_Y$, which is a contradiction, or there exists an x_1 that leads to larger game outcome for the maximizing $\mathcal{P}_X$, which is also a contradiction.

This lemma yields that the equilibrium for the simple example (α^*, β^*) satisfies the following conditions:

- For each $\bar{\omega}_i^1 \in \bar{D}_i^1$, $i = 1, 2$, the pair of strategies $(\alpha_i^{1*}(\cdot | \bar{\omega}_i^1), \beta_i^{1*})$ is a saddle point of $R_i^1(\bar{\omega}_i^1)$, described in (7.17), and it has a unique value
- Given the value of $R_i^1(\bar{\omega}_i^1)$, the pair of strategies $(\alpha_1^{2*}, \beta_1^{2*})$ is a saddle point of (7.23)

Coming back to the illustration Figure 7.2, at each compressed word (highlighted vertex), the players select randomized strategies over the words leading to the compressed word. We can also view the top level as compressed into a single word such that all belong to a single equivalence class. And we can compute the equilibrium of the game by traversing the directed acyclic graph illustrated in Figure 7.2.

We reemphasize that even though we have only shown the decomposition of the game into nested local subgames over the simple example, illustrated in Figure 7.2, the results can be generalized to arbitrary depth and number of nodes, as stated in the following theorem and as shown in Appendix 7.A.

Theorem 7.1 The game can be decomposed into nested local (zero-sum) subgames played at each node of the underlying hierarchical scheme over each equivalence class of words that lead to the same compressed version. The cost functions in these local subgames are linear/affine functions of the players' strategies and have a structure similar to the one described in (7.16). The unique outcomes of these local games determine the cost function of the local subgames at higher levels.

Furthermore, in a bottom-up dynamic program, we can compute a saddle point equilibrium of the game and obtain minimax detection rules.

Note that damage induced by words observed at lower levels, if they were coming from the adversary, accumulates when we consider multiple of them together at higher levels. However, at that accumulation, how the players select their strategies plays a deterministic role. Correspondingly, the detection rules at higher levels can focus on the discrepancy with respect to the coupled behavior of the blocks at the lower level instead of evaluating them one by one since that evaluation has already been conducted at the lower level by the associated detection rules. This is essential in order to avoid wasting the system's resources due to overlaps in nested detection zones.

7.3.1 Complexity Analysis

Let us denote necessary computational resources to compute mixed strategy equilibrium of a zero-sum game where players have finite action spaces of size d by $Q(d)$, which is known to be weakly polynomial. As an illustrative example, consider the scenario where there are n building blocks and each of them has dictionary of size d. Then, the computational complexity is given by $Q(d^n)$ without the proposed hierarchical scheme. In the hierarchical scheme, suppose that each virtual block composes m blocks from the preceding level and r denotes the compression ratio. Correspondingly, the complexity is given by

$$n \times \frac{d}{r} \times Q(r) + \frac{n}{m} \times \frac{1}{r}\left(\frac{d}{r}\right)^m \times Q(r)$$

$$+ \frac{n}{m^2} \times \frac{1}{r}\left(\frac{1}{r}\left(\frac{d}{r}\right)^m\right)^m \times Q(r) + \cdots , \tag{7.33}$$

where at each additive term, the first multiplicative term corresponds to the number of blocks at the associated level, the second multiplicative term corresponds to the number of compressed words, and the last term denotes the complexity of the local subgame. Depending on the compression ratio r, the proposed hierarchical structure can reduce the complexity of the game substantially. For example, suppose that r is selected such that the size of the dictionary at each node is d, i.e., $d = (d/r)^m$. Then, the overall complexity is given by

$$n \sum_{k=0}^{\lfloor \log_m n \rfloor} \left(\frac{1}{m}\right)^k Q(r) \leq \frac{nm}{m-1} Q(r), \tag{7.34}$$

which is substantially smaller than $Q(d^n)$.

7.4 Illustrative Examples

In this section we analyze the behavior of MAD and MAX over an illustrative example seen in Figure 7.3. It consists of three levels with $4 + 2 + 1$ nodes. At each node, 4 words could be realized. At each level except the top one, words are compressed into 2 words arbitrarily, as seen in Figure 7.3. We set likeliness and riskiness metrics arbitrarily as plotted in Figure 7.4. We consider the scenarios where uncertainty on whether there is an infection at an elementary node is independent of all other uncertainties. For example, we set infection probabilities for elementary nodes as $p_1^1 = 0.025$, $p_2^1 = 0.1$, $p_3^1 = 0.075$ and $p_4^1 = 0.05$.

In Figure 7.4, we plot equilibrium behavior of each player at each node across the tree in addition to the likeliness and riskiness metrics. Intuitively, it is expected that MAX would have a tendency to select words that have high likeliness and riskiness scores with small probability of alert. In

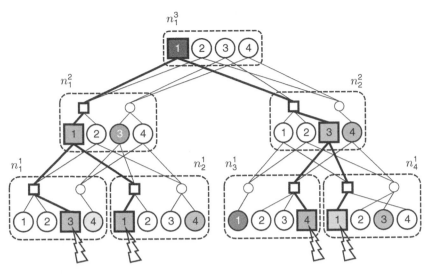

Figure 7.3 Hierarchical scheme (tree structure) of the example analyzed. Vertices of the graph, i.e., circles or squares, correspond either to a word or to a compressed word. Distinctively, we represent compressed words via smaller circles. Squares are the ones that would be selected by MAX at an equilibrium if it infected the associated node. Shaded words are the ones that MAD triggers an alert if they are realized. We have coded the probability of an alert by the intensity of the shade. In other words, the darker its shade is, the higher the probability of an alert when the word is realized. We emphasize that likeliness and riskiness metrics, plotted in Figure 7.4, play an important role on the players' decisions.

accordance with the intuition, at the root node n_1^3, MAX selects the word $\omega_{1,1}^3$ that has the highest likeliness and riskiness scores when there is an infection at the root node. We emphasize that MAX is not actually selecting the word realized at higher levels. MAX selects words at the elementary nodes that it infected and these words induce the words realized at higher levels. In this example, it turns out that at an equilibrium MAX selects the words $\omega_{1,3}^1, \omega_{2,1}^1, \omega_{3,4}^1$, and $\omega_{4,1}^1$ when it infects the elementary nodes n_1^1, n_2^1, n_3^1, and n_4^1, respectively.

Quite contrary to the intuition, the players can prefer to lose some battles to win the war. In other words, the players can select strategies that do not look optimal at a small scale, e.g., within the scope of a single node; however, those strategies could be more advantageous at the entire scale of the game. For example, at n_1^1, MAX selects $\omega_{1,3}^1$ instead of $\omega_{1,4}^1$, which has both more likeliness and riskiness scores in addition to smaller probability that MAD would trigger an alert. However, it becomes clear why it can be more advantageous for MAX to select $\omega_{1,3}^1$ instead of $\omega_{1,4}^1$ when we examine the parent node n_1^2. For example, if MAX selected $\omega_{1,4}^1$, this would yield that MAX selected $\omega_{1,3}^2$ instead of $\omega_{1,1}^2$ at n_1^2. However, $\omega_{1,3}^2$ has similar riskiness score with $\omega_{1,1}^2$ yet has lower likeliness score and there is higher probability of alert. Hence, this is an important indication of necessity for cohesive assessment of all activities across the entire system. In this chapter, minimax detection via a hierarchical scheme seeks to provide such a cohesive detection effectively and efficiently.

Over the same hierarchical scheme, let us compare performances of minimax detector and an anomaly detector that only considers the likeliness metric and ignores the riskiness metric in (7.6) when there is an adversary that seeks to maximize the detection cost. In Figure 7.5, we plot the detection cost for minimax detector and the one for anomaly detector for varying thresholds on false alarm rates. It shows that minimax detector outperforms the anomaly detector for any threshold on false alarm rates, as expected. This emphasizes the importance of accounting for the malicious content contributed by each defined word via a suitable metric in defense mechanisms against worst case adversarial interventions. Along the lines of what would be expected, the minimax detector also outperforms a signature-based detector that ignores the likelihood metric in (7.6) when there is an adversary that seeks to maximize the detection cost.

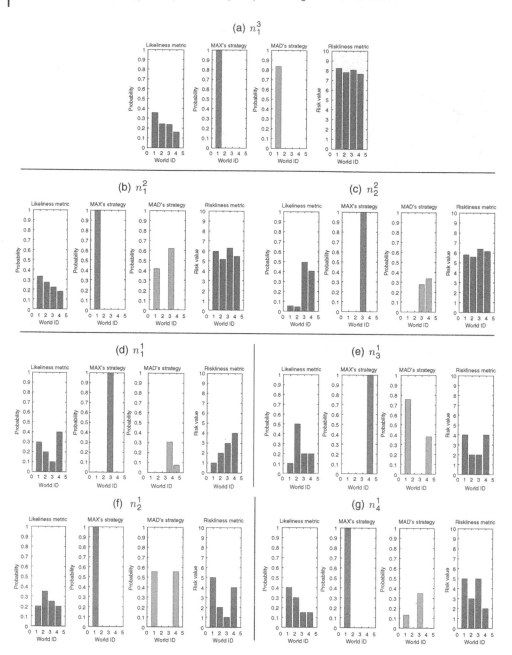

Figure 7.4 All subfigures (a)–(g) illustrate likeliness metric, equilibrium strategies of MAX and MAD, and riskiness metric (in this order from left to right) at the associated nodes of the hierarchical tree described in Figure 7.3.

7.5 Conclusion

In this chapter, we have proposed and analyzed minimax detectors that can assess system activities in terms of their riskiness (similar to signature-based detection) and likeliness (similar to anomaly detection) across a modern computing system in a cohesive and strategic way. Within the

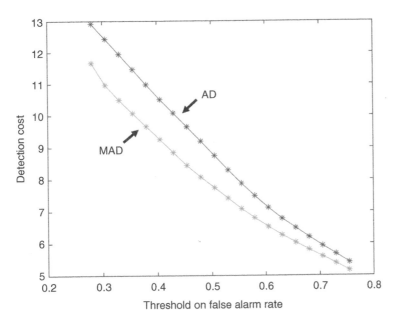

Figure 7.5 Comparison of the performances of the minimax detector (MAD) and an anomaly detector (AD) that does not take into account the riskiness metric while making its decision in terms of detection cost for varying thresholds on false alarm rates.

framework of zero-sum game, we have designed a detection mechanism that seeks to minimize the detection cost against an adversary maximizing it by intervening into the system activities. Since direct approaches to compute equilibrium of the game at that scale would not be feasible, we have introduced a hierarchical scheme where the system-level activities are assessed at varying granularity similar to state summarization used to handle big data applications, e.g., see Haider et al. (2019). Based on the hierarchical scheme, we have shown that the detection cost has a nested structure that would enable us to decompose the entire game into nested local subgames. Hence, the equilibrium could be computed efficiently via a dynamic program similar to the backward induction in extensive form games. Finally, we have provided complexity analysis of the proposed solution concept and analyzed the behavior of the players at an equilibrium over an illustrative example.

Some future research directions can be listed as:

- The proposed solution concept brings down the computation of the equilibrium into a dynamic program; however, it requires preprocessing to compute certain parameters. An interesting research direction would be to come up with efficient methods to compute these parameters either exactly or approximately
- Experimental evaluation/validation of the solution concept would play an important role towards deployment of such solution concepts in computer systems
- In this chapter we have supposed that a hierarchical scheme and the associated encoding/summarization mapping are given externally, e.g., based on expert knowledge. However, an interesting research challenge is to identify or characterize hierarchical schemes and summarization techniques that would lead to smaller detection cost for the system
- Another interesting research challenge is how to allocate the detector's resources, e.g., tolerable false alarm rate or multiplication factor for detection gain, across nodes and levels

Acknowledgements

This research was supported by the U.S. Office of Naval Research (ONR) MURI grant N00014-16-1-2710.

Appendix 7.A

Generalization of the Results

In order to generalize the result for the simple example to hierarchical schemes with arbitrary depth and number of nodes, we start by introducing a generalized version of the auxiliary variables introduced in Section 7.3 as

$$\alpha_i^k(\omega_i^k) := \prod_{j \in \mathcal{M}_i^k} \bar{\alpha}_j^{k-1}(\bar{\omega}_j^{k-1}), \tag{7.A.1}$$

where

$$\bar{\alpha}_i^k(\bar{\omega}) := \sum_{\omega \in \mathcal{E}_i^k(\bar{\omega})} \alpha_i^k(\omega), \tag{7.A.2}$$

and

$$\alpha_i^k(\omega|\bar{\omega}) := \frac{\alpha_i^k(\omega)}{\bar{\alpha}_i^k(\bar{\omega})} \text{ if } \bar{\alpha}_i^k(\bar{\omega}) > 0. \tag{7.A.3}$$

For ease of representation, we also introduce the following short-hand notations:

- $\mathcal{J}^k$ denotes a subset of indices at level k, i.e., $\mathcal{J}^k \subseteq \mathcal{I}^k$
- Given a subset $\mathcal{J}^k \subseteq \mathcal{I}^k$, we denote the indices of the parents of all nodes in $\mathcal{J}^k$ by $\mathcal{J}^{k+1} \subseteq \mathcal{I}^{k+1}$
- Given a subset $\mathcal{J}^k \subseteq \mathcal{I}^k$, we denote the set of all nodes at level k that share the same parents with the nodes in $\mathcal{J}^k$ by $\bar{\mathcal{J}}^k$, which can be viewed as its closure
- Given a node n_i^k, we denote its descendants at level $m \in \{1, \dots, k-1\}$ by $\mathcal{M}_{i,m}^k$
- We denote the probability that there exists an infection only at the nodes from an index set $\mathcal{J}$ by $\mathbb{P}(\mathcal{J})$
- We denote a product set of dictionaries (or compressed dictionaries) of nodes from an index set $\mathcal{J}$ by $\times(\mathcal{D}; \mathcal{J})$ (or $\times(\bar{\mathcal{D}}; \mathcal{J})$) instead of *** $\times_{j \in \mathcal{J}} \mathcal{D}_j$
- We define

$$\gamma_i^k(\omega_i^k) := r_i^k(\omega_i^k) - c_i^k \beta_i^k(\omega_i^k) \tag{7.A.4}$$

- We also define

$$\alpha^k(\omega^k; \mathcal{J}) := \prod_{j \in \mathcal{J}} \alpha_j^k(\omega_j^k) \text{ and } \bar{\alpha}^k(\bar{\omega}^k; \mathcal{J}) := \prod_{j \in \mathcal{J}} \bar{\alpha}_j^k(\bar{\omega}_j^k) \tag{7.A.5}$$

- We denote a tuple of words (or compressed words) from a level k by ω^k (or by $\bar{\omega}^k$) since the index of these words will be clear by the context
- We denote a set of words indexed by a set $\mathcal{J}$ by $\{\omega_j\}_{\mathcal{J}}$ instead of $\{\omega_j\}_{j \in \mathcal{J}}$
- In order to show that a set $\mathcal{J}$ is a nonempty subset of another set $\mathcal{I}$, we use $\mathcal{J} \subset' \mathcal{I}$
- Finally, we denote the probability that there is an infection at n_i^k by p_i^k

By using the notation introduced above, we can rewrite the cost function, described in (7.A.6), (by expanding the expectation operator) as

$$U(\alpha, \beta) = \sum_{\mathcal{J}^1 \subset' \mathcal{I}^1} \mathbb{P}(\mathcal{J}^1) \sum_{\omega^1 \in \times(\mathcal{D}^1; \mathcal{J}^1)} \alpha^1(\omega^1; \mathcal{J}^1) \sum_{k=1}^{K} \sum_{t \in \mathcal{J}^k} \mathbb{E}_o\{\gamma_t^k(\boldsymbol{\omega}_t^k) | \{\omega_j^1\}_{\mathcal{J}^1}\}, \tag{7.A.6}$$

in which we consider each infection scenario across elementary nodes individually. For example, given that only the elementary nodes indexed by a nonempty set $\mathcal{J}^1 \subset' \mathcal{I}^1$ are infected, the expected cost to MAD at node n_t^k is given by

$$\sum_{\omega^1 \in \times(\mathcal{D}^1; \mathcal{J}^1)} \alpha^1(\omega^1; \mathcal{J}^1) \mathbb{E}_o\{\gamma_t^k(\boldsymbol{\omega}_t^k) | \{\omega_j^1\}_{\mathcal{J}^1}\}, \tag{7.A.7}$$

where the expectation term corresponds to the expected detection cost to MAD if MAX has selected the words $\{\omega_j^1\}_{j \in \mathcal{J}^1}$. Note that even though the words selected by MAX are fixed, word realized at n_t^k is not deterministic since it also depends on the words at other elementary nodes that are not infected. Correspondingly the word realized at n_t^k also depends on the underlying distribution over the words across the nodes when there is no adversarial intervention. By taking expectation with respect to that distribution, we can compute the associated expected detection cost. These expected detection costs are averaged over all tuple of words from the product set of dictionaries at the nodes infected with respect to the probabilities that MAX would select these words. A weighted average of the cost for each infection scenario with respect to the probability of that scenario yields the cost (7.A.6) to MAD.

Our goal is to transform the cost function into a nested structure that would enable us to invoke Lemma 7.A.1 to decompose the entire game into nested local subgames. To this end, we evaluate the cost induced at elementary and nonelementary nodes separately by partitioning the cost function (7.A.6) into two parts Ⓐ and Ⓑ, given by

$$Ⓐ := \sum_{\mathcal{J}^1 \subset' \mathcal{I}^1} \mathbb{P}(\mathcal{J}^1) \sum_{\omega^1 \in \times(\mathcal{D}^1; \mathcal{J}^1)} \alpha^1(\omega^1; \mathcal{J}^1) \sum_{t \in \mathcal{J}^1} \mathbb{E}_o\{\gamma_t^1(\boldsymbol{\omega}_t^1) | \{\omega_j^1\}_{\mathcal{J}^1}\}, \tag{7.A.8}$$

$$Ⓑ := \sum_{\mathcal{J}^1 \subset' \mathcal{I}^1} \mathbb{P}(\mathcal{J}^1) \sum_{\omega^1 \in \times(\mathcal{D}^1; \mathcal{J}^1)} \alpha^1(\omega^1; \mathcal{J}^1) \sum_{k=2}^{K} \sum_{t \in \mathcal{J}^k} \mathbb{E}_o\{\gamma_t^k(\boldsymbol{\omega}_t^k) | \{\omega_j^1\}_{\mathcal{J}^1}\}. \tag{7.A.9}$$

Lemma 7.A.1 The first part Ⓐ can be written as

$$Ⓐ = \sum_{\mathcal{J}^2 \subset' \mathcal{I}^2} \mathbb{P}(\mathcal{J}^2) \sum_{\omega^2 \in \times(\mathcal{D}^2; \mathcal{J}^2)} \alpha^2(\omega^2; \mathcal{J}^2) \sum_{j \in \mathcal{J}^2} \bar{\delta}_{j,o}^2(\omega_j^2), \tag{7.A.10}$$

where

$$\bar{\delta}_{j,o}^2(\omega_j^2) := \frac{1}{p_j^2} \sum_{\mathcal{N}_j \subset' \mathcal{M}_{j,1}^2} p(\mathcal{N}_j) \sum_{t \in \mathcal{N}_j} R_t^1(\bar{\omega}_t^1) \tag{7.A.11}$$

and

$$R_j^1(\bar{\omega}) := \sum_{\omega \in \mathcal{E}_j^1(\bar{\omega})} \alpha_j^1(\omega | \bar{\omega}) \gamma_j^1(\omega). \tag{7.A.12}$$

Proof: Since $\mathbb{E}_o\{\gamma_t^1(\boldsymbol{\omega}_t^1) | \{\omega_j^1\}_{\mathcal{J}^1}\} = \gamma_t^1(\omega_t^1)$, we can write (7.A.8) as

$$Ⓐ = \sum_{\mathcal{J}^1 \subset' \mathcal{I}^1} \mathbb{P}(\mathcal{J}^1) \sum_{\omega^1 \in \times(\mathcal{D}^1; \mathcal{J}^1)} \alpha^1(\omega^1; \mathcal{J}^1) \sum_{t \in \mathcal{J}^1} \gamma_t^1(\omega_t^1) \tag{7.A.13}$$

and (7.A.3) yields that

$$\text{Ⓐ} = \sum_{\mathcal{J}^1 \subset' \mathcal{I}^1} \mathbb{P}(\mathcal{J}^1) \sum_{\bar{\omega}^1 \in \times(\bar{D}^1; \mathcal{J}^1)} \bar{\alpha}^1(\bar{\omega}^1; \mathcal{J}^1) \sum_{j \in \mathcal{J}^1} R_j^1(\bar{\omega}_j^1), \tag{7.A.14}$$

where $R_j^1(\bar{\omega}_j^1)$ is as described in (7.A.12).

We could confine the local interactions into the terms R_j^1's. For a given infection scenario $\mathcal{J}^1$, certain nodes at level 2 are infected and the associated set is denoted by $\mathcal{J}^2$. We now seek to write (7.A.14) in terms of $\mathcal{J}^2$ instead of $\mathcal{J}^1$. For an infection scenario $\mathcal{J}^1$, however, there might be elementary nodes that are not indexed by $\mathcal{J}^1$ but share the same parent with a node indexed by $\mathcal{J}^1$. To this end, we multiply the right hand side of (7.A.14) with $\sum_{\bar{\omega} \in \bar{D}_t^1} \bar{\alpha}_t^1(\bar{\omega}) = 1$ for each $t \in \mathcal{I}^1$ that satisfies $t \in \mathcal{M}_{i,1}^2$ for some $i \in \mathcal{J}^2$ yet $t \notin \mathcal{J}^1$. Then we obtain

$$\text{Ⓐ} = \sum_{\mathcal{J}^1 \subset' \mathcal{I}^1} \mathbb{P}(\mathcal{J}^1) \sum_{\bar{\omega}^1 \in \times(\bar{D}^1; \bar{\mathcal{J}}^1)} \bar{\alpha}^1(\bar{\omega}^1; \bar{\mathcal{J}}^1) \sum_{j \in \mathcal{J}^1} R_j^1(\bar{\omega}_j^1) \tag{7.A.15}$$

$$= \sum_{\mathcal{J}^1 \subset' \mathcal{I}^1} \mathbb{P}(\mathcal{J}^1) \sum_{\omega^2 \in \times(D^2; \mathcal{J}^2)} \alpha^2(\omega^2; \mathcal{J}^2) \sum_{j \in \mathcal{J}^1} R_j^1(\bar{\omega}_j^1). \tag{7.A.16}$$

Note that we can write the set of all infection scenarios across elementary nodes as

$$\{\mathcal{J}^1 \subset' \mathcal{I}^1\} = \left\{ \bigcup_{j \in \mathcal{J}^2} \mathcal{N}_j \mid \mathcal{N}_j \subset' \mathcal{M}_{j,1}^2, j \in \mathcal{J}^2, \text{ and } \mathcal{J}^2 \subset' \mathcal{I}^2 \right\}. \tag{7.A.17}$$

Therefore, (7.A.16) can be written as

$$\sum_{\mathcal{J}^2 \subset' \mathcal{I}^2} \sum_{\omega^2 \in \times(D^2; \mathcal{J}^2)} \alpha^2(\omega^2; \mathcal{J}^2) \sum_{j \in \mathcal{J}^2} \sum_{\mathcal{N}_j \subset' \mathcal{M}_{j,1}^2} p(\mathcal{N}_j, \mathcal{J}^2) \sum_{t \in \mathcal{N}_j} R_t^1(\bar{\omega}_t^1), \tag{7.A.18}$$

where $p(\mathcal{N}_j, \mathcal{J}^2)$ corresponds to the probability that infection scenario $\mathcal{J}^2$ occurs at level 2 and only the children of n_j^2 that are indexed by $\mathcal{N}_j$ are infected.

Given an infection scenario $\mathcal{J}^2$, we can multiply (7.A.18) with $\sum_{\bar{\omega} \in \bar{D}_t^1} \bar{\alpha}_t^1(\bar{\omega}) = 1$ for all $t \in \mathcal{I}^2 \backslash \mathcal{J}^2$ without loss of generality. Then we obtain

$$\text{Ⓐ} = \sum_{\omega^2 \in \times(D^2; \mathcal{I}^2)} \alpha^2(\omega^2; \mathcal{I}^2) \sum_{\mathcal{J}^2 \subset' \mathcal{I}^2} \sum_{j \in \mathcal{J}^2} \sum_{\mathcal{N}_j \subset' \mathcal{M}_{j,1}^2} p(\mathcal{N}_j, \mathcal{J}^2) \sum_{t \in \mathcal{N}_j} R_t^1(\bar{\omega}_t^1)$$

$$= \sum_{\omega^2 \in \times(D^2; \mathcal{I}^2)} \alpha^2(\omega^2; \mathcal{I}^2) \sum_{j \in \mathcal{I}^2} \sum_{\mathcal{N}_j \subset' \mathcal{M}_{j,1}^2} p(\mathcal{N}_j) \sum_{t \in \mathcal{N}_j} R_t^1(\bar{\omega}_t^1), \tag{7.A.19}$$

where $p(\mathcal{N}_j) = \mathbb{P}\left(\{\theta_t^1 = 1\}_{t \in \mathcal{N}_j}, \{\theta_{t'}^1 = 0\}_{t' \in \mathcal{M}_j^2 \backslash \mathcal{N}_j} \right)$ corresponds to the probability that among n_j^2's children only the ones that are indexed by $\mathcal{N}_j$ are infected. Note that (7.A.19) is not written in terms of infection scenarios at level 2, i.e., $\mathcal{J}^2 \subset' \mathcal{I}^2$. However, it can be written as[6]

$$\text{Ⓐ} = \sum_{\mathcal{J}^2 \subset' \mathcal{I}^2} \mathbb{P}(\mathcal{J}^2) \sum_{\omega^2 \in \times(D^2; \mathcal{I}^2)} \alpha^2(\omega^2; \mathcal{I}^2) \sum_{j \in \mathcal{J}^2} \frac{1}{P_j^2} \sum_{\mathcal{N}_j \subset' \mathcal{M}_{j,1}^2} p(\mathcal{N}_j) \sum_{t \in \mathcal{N}_j} R_t^1(\bar{\omega}_t^1)$$

since given $\omega^2 \in \times(D^2; \mathcal{I}^2)$, we have

$$\sum_{j \in \mathcal{I}^2} \lambda_j(\omega^2) = \sum_{\mathcal{J}^2 \subset' \mathcal{I}^2} \mathbb{P}(\mathcal{J}^2) \sum_{j \in \mathcal{J}^2} \frac{1}{P_j^2} \lambda_j(\omega^2), \tag{7.A.20}$$

---

6 We implicitly assume that there is a nonzero probability for the infection of any node.

where $\lambda_j(\omega^2) := \sum_{\mathcal{N}_j \subset' \mathcal{M}_{j,1}^2} p(\mathcal{N}_j) \sum_{t \in \mathcal{N}_j} R_t^1(\bar{\omega}_t^1)$. Since α_j^2 is a probability distribution, which implies that $\sum_{\omega \in D_j^2} \alpha_j^2(\omega) = 1$, we obtain (7.A.10). This completes the proof. $\square$

Let us next focus on the second part in the cost function (7.A.6).

Lemma 7.A.2 The second part Ⓑ can be written as

$$\text{Ⓑ} = \sum_{\mathcal{J}^2 \subset' \mathcal{I}^2} \mathbb{P}(\mathcal{J}^2) \sum_{\omega^2 \in \times(D^2;\mathcal{J}^2)} \alpha^2(\omega^2;\mathcal{J}^2) \tag{7.A.21}$$

$$\times \left(\sum_{j \in \mathcal{J}^2} \bar{\delta}_{j,n}^2(\omega_j^2) + \sum_{k=3}^{\kappa} \sum_{j \in \mathcal{J}^k} \delta_{j,2}^k(\{\omega_t^2\}_{\mathcal{J}^2}) \right), \tag{7.A.22}$$

where

$$\bar{\delta}_{j,n}^2(\omega_j^2) := \frac{1}{p_j^2} \sum_{\mathcal{N}_j \subset' \mathcal{M}_{j,1}^2} p(\mathcal{N}_j) \mathbb{E}_o\{\gamma_j^2(\omega_j^2)|\{\bar{\omega}_t^1\}_{\mathcal{N}_j}\} \tag{7.A.23}$$

$$\delta_{j,2}^k(\{\omega_t^2\}_{\mathcal{J}^2}) := \frac{1}{\mathbb{P}(\mathcal{J}^2)} \sum_{\substack{\mathcal{N}_j^1 \subset' \mathcal{M}_{j,1}^k \\ \ni \mathcal{N}_j^2 = \mathcal{J}^2}} p(\mathcal{N}_j^1) \mathbb{E}_o\{\gamma_j^k(\omega_j^k)|\{\bar{\omega}_t^1\}_{\mathcal{N}_j^1}\}. \tag{7.A.24}$$

Proof: Since $\alpha_j^1(\cdot|\bar{\omega})$ is a probability distribution satisfying $\mathbf{1}'\alpha_j^1(\cdot|\bar{\omega}) = 1$ and

$$\mathbb{E}_o\{\gamma_t^k(\omega_t^k)|\{\omega_j^1\}_{\mathcal{J}^1}\} = \mathbb{E}_o\{\gamma_t^k(\omega_t^k)|\{\bar{\omega}_j^1\}_{\mathcal{J}^1}\},$$

we can write (7.A.9) as

$$\text{Ⓑ} = \sum_{\mathcal{J}^1 \subset' \mathcal{I}^1} \mathbb{P}(\mathcal{J}^1) \sum_{\bar{\omega}^1 \in \times(\bar{D}^1;\mathcal{J}^1)} \bar{\alpha}^1(\bar{\omega}^1;\mathcal{J}^1) \sum_{k=2}^{\kappa} \sum_{t \in \mathcal{J}^k} \mathbb{E}_o\{\gamma_t^k(\omega_t^k)|\{\bar{\omega}_j^1\}_{\mathcal{J}^1}\}.$$

For each infection scenario $\mathcal{J}^1$, by multiplying it with $\sum_{\bar{\omega} \in \bar{D}_t^1} \bar{\alpha}_t^1(\bar{\omega}) = 1$ for $t \in \mathcal{I}^1$ that satisfies $t \in \mathcal{M}_{i,1}^2$ for some $i \in \mathcal{J}^2$ yet $t \notin \mathcal{J}^1$, we obtain

$$\text{Ⓑ} = \sum_{\mathcal{J}^1 \subset' \mathcal{I}^1} \mathbb{P}(\mathcal{J}^1) \sum_{\bar{\omega}^1 \in \times(\bar{D}^1;\bar{\mathcal{J}}^1)} \bar{\alpha}^1(\bar{\omega}^1;\bar{\mathcal{J}}^1) \sum_{k=2}^{\kappa} \sum_{t \in \mathcal{J}^k} \mathbb{E}_o\{\gamma_t^k(\omega_t^k)|\{\bar{\omega}_j^1\}_{\mathcal{J}^1}\}$$

$$= \sum_{\mathcal{J}^1 \subset' \mathcal{I}^1} \mathbb{P}(\mathcal{J}^1) \sum_{\omega^2 \in \times(D^2;\mathcal{J}^2)} \alpha^2(\omega^2;\mathcal{J}^2) \sum_{k=2}^{\kappa} \sum_{t \in \mathcal{J}^k} \mathbb{E}_o\{\gamma_t^k(\omega_t^k)|\{\bar{\omega}_j^1\}_{\mathcal{J}^1}\}, \tag{7.A.25}$$

which can also be written as

$$\sum_{\mathcal{J}^2 \subset' \mathcal{I}^2} \sum_{\omega^2 \in \times(D^2;\mathcal{J}^2)} \alpha^2(\omega^2;\mathcal{J}^2) \sum_{k=2}^{\kappa} \sum_{j \in \mathcal{J}^k} \sum_{\mathcal{N}_j \subset' \mathcal{M}_{j,1}^k} p(\mathcal{N}_j, \mathcal{J}^2) \mathbb{E}_o\{\gamma_j^k(\omega_j^k)|\{\bar{\omega}_t^1\}_{\mathcal{N}_j}\} \tag{7.A.26}$$

similar to (7.A.18). Given an infection scenario $\mathcal{J}^2$, we can multiply (7.A.26) with $\sum_{\bar{\omega} \in \bar{D}_t^1} \bar{\alpha}_t^1(\bar{\omega}) = 1$ for all $t \in \mathcal{I}^2 \backslash \mathcal{J}^2$. Then we obtain

$$\text{Ⓑ} = \sum_{\omega^2 \in \times(D^2;\mathcal{I}^2)} \alpha^2(\omega^2;\mathcal{I}^2) \sum_{k=2}^{\kappa} \sum_{j \in \mathcal{I}^k} \sum_{\mathcal{N}_j \subset' \mathcal{M}_{j,1}^k} p(\mathcal{N}_j) \mathbb{E}_o\{\gamma_j^k(\omega_j^k)|\{\bar{\omega}_t^1\}_{\mathcal{N}_j}\}. \tag{7.A.27}$$

Based on (7.A.20), similar to (7.A.10), we obtain (7.A.22). This completes the proof. $\square$

Combining Ⓐ and Ⓑ, as described in Lemmas 7.A.1 and 7.A.2, respectively, we obtain that (7.A.6) can be written as

$$U(\alpha, \beta) = \sum_{\mathcal{J}^2 \subset' \mathcal{I}^2} \mathbb{P}(\mathcal{J}^2) \sum_{\omega^2 \in \times (\mathcal{D}^2; \mathcal{J}^2)} \alpha^2(\omega^2; \mathcal{J}^2)$$

$$\times \left(\sum_{j \in \mathcal{J}^2} \bar{\delta}_{j,0}^2(\omega_j^2) + \bar{\delta}_{j,n}^2(\omega_j^2) + \sum_{k=3}^{\kappa} \sum_{j \in \mathcal{J}^k} \delta_{j,2}^k(\{\omega_t^2\}_{\mathcal{J}^2}) \right). \tag{7.A.28}$$

In order to write (7.A.6) and (7.A.28) in the same compact form, we define $\delta_{j,m}^k$ as follows:

- If $m = 1$ and $k \geq m$, then

$$\delta_{j,m}^k(\{\omega_t^m\}_{\mathcal{J}^m}) = \mathbb{E}_o\{\gamma_j^k(\omega_j^k)|\{\omega_t^m\}_{\mathcal{J}^m}\} \tag{7.A.29}$$

- If $m > 1$, $k = m$, and for all $j \in \mathcal{J}^m$, then

$$\delta_{j,m}^k(\{\omega_t^m\}_{\mathcal{J}^m}) = \sum_{\mathcal{N}_j \subset' \mathcal{M}_{j,m-1}^k} \frac{p(\mathcal{N}_j)}{p_j^k} \left(\delta_{j,m-1}^k(\{\bar{\omega}_t^{m-1}\}_{\mathcal{N}_j}) + \sum_{t \in \mathcal{N}_j} R_t^{m-1}(\bar{\omega}_t^{m-1}) \right),$$

where

$$R_t^{m-1}(\bar{\omega}_t^{m-1}) := \sum_{\omega \in \mathcal{E}_t^{m-1}(\bar{\omega}_t^{m-1})} \alpha_t^{m-1}(\omega | \bar{\omega}_t^{m-1}) \delta_{t,m-1}^{m-1}(\omega) \tag{7.A.30}$$

- If $m > 1$ and $k > m$, then

$$\delta_{j,m}^k(\{\omega_t^m\}_{\mathcal{J}^m}) = \sum_{\substack{\mathcal{N}_j^{m-1} \subset' \mathcal{M}_{j,m-1}^k \\ \ni \mathcal{N}_j^m = \mathcal{J}^m}} \frac{p(\mathcal{N}_j^{m-1})}{\mathbb{P}(\mathcal{J}^m)} \delta_{j,m-1}^k(\{\bar{\omega}_t^{m-1}\}_{\mathcal{N}_j^{m-1}}) \tag{7.A.31}$$

- Furthermore $\delta_{j,m}^k(\{\omega_t^m\}_{\mathcal{J}^m}) = \delta_{j,m}^k(\{\bar{\omega}_t^m\}_{\mathcal{J}^m})$ if $k > m$ since ω_j^k is independent of $\{\omega_t^m\}_{\mathcal{J}^m}$ conditioned on $\{\bar{\omega}_t^m\}_{\mathcal{J}^m}$

Then the cost function (7.A.6) and (7.A.28) can be written as

$$U(\alpha, \beta) = \sum_{\mathcal{J}^m \subset' \mathcal{I}^m} \mathbb{P}(\mathcal{J}^m) \sum_{\omega^m \in \times (\mathcal{D}^m; \mathcal{J}^m)} \alpha^m(\omega^m; \mathcal{J}^m) \sum_{k=m}^{\kappa} \sum_{j \in \mathcal{J}^m} \delta_{j,m}^k(\{\omega_t^m\}_{\mathcal{J}^m}) \tag{7.A.32}$$

for $m = 2$ and $m = 3$, respectively. Note that in (7.A.32), the interaction between the players at levels lower than the mth level are confined into the terms $\delta_{j,m}^k$. Therefore the cost function (7.A.32) has the nested structure described in Lemma 7.A.1 and correspondingly the game can be decomposed into nested local (zero-sum) subgames played for each equivalence class of words that lead to the same compressed version. For example, at n_j^k, the players play the subgame over $\mathcal{E}_j^k(\bar{\omega})$, where $\bar{\omega} \in \bar{\mathcal{D}}_j^k$, and the cost function of MAD is given by

$$\sum_{\omega \in \mathcal{E}_j^k(\bar{\omega})} \alpha_j^k(\omega | \bar{\omega}) \delta_{j,k}^k(\omega), \tag{7.A.33}$$

and $\delta_{j,k}^k(\cdot)$ is an affine function of MAD's detection rule at n_j^k, i.e., β_j^k.

References

G. S. Almasi and A. Gottlieb. *Highly Parallel Computing*. Benjamin-Cummings Publishing Co., Inc., Redwood City, CA, 1994.

T. Başar and G. J. Olsder. *Dynamic Noncooperative Game Theory*. Society for Industrial Mathematics (SIAM) Series in Classics in Applied Mathematics, 1999.

S. Boyd and L. Vandenberghe. *Convex Optimization*. Cambridge University Press, 2004.

M. Brückner and T. Scheffer. Nash equilibria of static prediction games. In *Proceedings of Advances in Neural Information Processing (NIPS), Vancouver, BC, Canada, December*, pp. 171–179, 2009.

M. Brückner and T. Scheffer. Stackelberg games for adversarial prediction problems. In *Proceedings of the 17th ACM SIGKDD International Conference on Knowledge Discovery and Data Mining*, 2011.

N. Dalvi, P. Domingos, Mausam, S. Sanghai, and D. Verma. Adversarial classification. In *Proceedings of the 10th ACM SIGKDD International Conference on Knowledge Discovery and Data Mining, Seattle WA, USA, August*, 99–108, 2004.

C. Daskalakis, P. W. Goldberg, and C. H. Papadimitriou. The complexity of computing a Nash equilibrium. *SIAM Journal on Computing*, 39(1):195–259, 2009.

L. Dritsoula, P. Loiseau, and J. Musacchio. A game-theoretic analysis of adversarial classification. *IEEE Transactions on Information Forensics and Security*, 12(12):3094–3109, 2017.

M. Du, F. Li, G. Zheng, and V. Srikumar. DeepLog: Anomaly detection and diagnosis from system logs through deep learning. In *Proceedings of the 24th ACM Conference on Computer and Communications Security, Dallas TX, USA*, 1285–1298, 2017.

C. C. Elisan. *Malware, Rootkits and Botnets: A Beginner's Guide*. Mc Graw Hill, 2013.

K. Eykholt, I. Evtimov, E. Fernandes, B. Li, A. Rahmati, C. Xiao, A. Prakash, T. Kohno, and D. Song. Robust physical-world attacks on deep learning visual classification. In *Proceedings of the IEEE Conference on Computer Vision and Pattern Recognition (CVPR), Salt Lake City, UT, USA, June*, 1625–1634, 2018.

D. Gao, M. K. Reiter, and D. Song. On gray-box program tracking for anomaly detection. In *Proceedings of the 13th Conference on USENIX Security Symposium, San Diego CA, USA, August*, 2004.

W. Haider, J. Hu, Y. Xie, X. Yu, and Q. Wu. Detecting anomalous behavior in cloud servers by nested arc hidden semi-markov model with state summarization. *IEEE Transactions on Big Data*, 5(3):305–316, 2019.

Y. Ji, S. Lee, E. Downing, W. Wang, M. Fazzini, T. Kim, A. Orso, and W. Lee. RAIN: Refinable attack investigation with on-demand inter-process information flow tracking. In *Proceedings of the 24th ACM Conference on Computer and Communications Security, Dallas TX, USA, October*, 377–390, 2017.

F. Lau, S. H. Rubin, M. H. Smith, and L. Trajkovic. Distributed denial of service attacks. In *IEEE International Conference on Systems, Man and Cybernetics, Nashville TN, USA, October*, pp. 2275–2280, 2000.

M. Liu, Z. Xue, X. Xu, C. Zhong, and J. Chen. Host-based intrusion detection system with system calls: Review and future trends. *ACM Computing Surveys*, 51(5):1–36, 2019.

Shiqing Ma, Xiangyu Zhang, and Dongyan Xu. ProTracer: Towards practical provenance tracing by alternating between logging and tainting. In *Annual Network and Distributed System Security Symposium (NDSS), San Diego, CA, USA, February*, 1–15, 2016.

E. Manzoor, S. M. Milajerdi, and L. Akoglu. Fast memory-efficient anomaly detection in streaming heterogeneous graphs. In *Proceedings of the 22nd ACM SIGKDD International Conference on Knowledge Discovery and Data Mining, San Francisco CA, USA, August*, 1035–1044, 2016.

M. O. Sayin, H. Hosseini, R. Poovendran, and T. Başar. A game theoretical framework for inter-process adversarial intervention detection. In T. Başar, R. Poovendran, L. Bushnell, E. Vorobeychik, and Q. Zhu, editors. *Proceedings of International Conference on Decision and Game Theory for Security on Lecture Notes in Computer Science*. Springer, Seattle, WA, 2018.

M. O. Sayin, C.-W. Lin, E. Kang, S. Shiraishi, and T. Başar. Reliable smart road signs. *IEEE Transactions on Intelligent Transportation Systems*, 21(12): 4995–5009, 2020.

X. Shu, D. Ye, and B. G. Ryder. A formal framework for program anomaly detection. In H. Bos, F. Monrose, G. Blanc, editors. *Research in Attacks, Intrusion, and Defenses. Lecture Notes in Computer Science*, volume 9404. Springer, Cham, 2015.

D. Wagner and P. Soto. Mimicry attacks on host-based intrusion detection systems. In *Proceedings of the 9th ACM Conference on Computer and Communications Security, Washington, DC, USA, November,* 2002.

D. Yao, X. Shu, L. Cheng, and S. J. Stolfo. Anomaly detection as a service: Challenges, advances, and opportunities. *Synthesis Lectures on Information Security, Privacy, and Thrust #22.* Morgan & Claypool Publishers, 2017.

Y. Zhang and Y. Zhou. Transparent computing: A new paradigm for pervasive computing. In J. Ma, H. Jin, L. T. Yang, and J. J. Tsai, editors. *Ubiquitous Intelligence and Computing on Lecture Notes in Computer Science.* Springer, Berlin, Heidelberg, 2006

8

Sensor Manipulation Games in Cyber Security

João P. Hespanha

Center for Control Dynamical-Systems, and Computation, University of California, Santa Barbara, CA, USA

8.1 Introduction

This chapter addresses the problem of making decisions based on sensor measurements that may have been manipulated by an adversary. For concreteness, we focus our attention on making a binary decision that, in the context of cybersecurity, could correspond to denying access to a sensitive resource, flagging a computer as compromised, deauthorizing a user, closing a firewall, etc. Such decisions are typically based on measurements collected by cybersecurity sensors that analyze records of events (logs) and provide recommendations on what the binary decision should be. Sophisticated cybersecurity systems typically rely on multiple such sensors to maximize their ability to catch attacks, while maintaining a small probability of false alarms. The use of multiple cyber sensors also provides protection against adversaries that may have compromised some of these sensors, either by disabling a sensor or actively manipulating its output.

Noncooperative game theory provides a mathematical framework to reason about decision making by a group of agents (players) in which the decision of one agent affects the costs/rewards incurred by the other agents. To apply ideas from game theory to cybersecurity, we regard the cyber defense system as one player and the attacker as the other player. In the simplest form of the game, the defender wants to minimize the probability of making the wrong decision, while the attacker wants to maximize this probability. However, we shall see below that the goals of the players may be more complex. While the defender's decision is assumed to be binary, we consider a much richer set of choices for the attacker, as its decision space involves selecting if and how to manipulate sensor measurements.

We shall consider two types of sensor manipulation games that differ on the type of sensor manipulation available to the attacker. In *measurement manipulation games*, the attacker is able to manipulate the measurements of M out of N sensors available to the defender, but the latter does not know which sensors have been manipulated. In *sensor-reveal games*, the attacker exposes to the defender the measurement of a single sensor out of N sensor possibilities, with the caveat that revealing data from noninformative sensors may be costly and interpreted by the defender as strong indication that an attack is afoot. These games cover different aspects of sensor manipulation: Measurement manipulation games mostly explore the challenges that arise from not knowing which sensors have been manipulated and how this impairs the defender's ability to make correct decisions. In sensor-reveal games, the defender actually knows which sensor has been revealed, but the challenge is to explore the information that the attacker implicitly provides by selecting a particular sensor. Several combinations of these games are possible (and relevant).

Game Theory and Machine Learning for Cyber Security, First Edition.
Edited by Charles A. Kamhoua, Christopher D. Kiekintveld, Fei Fang, and Quanyan Zhu.

The remainder of this chapter discusses these two types of games and is mostly based on results from Vamvoudakis et al. (2014) for measurement manipulation games and from Hespanha and Garagic (2019) for sensor reveal games. The goal of this chapter is to provide a tutorial view on the use of game theory to model and solve sensor manipulations problems that arise in the cybersecurity domain. We refer the reader to the following two recent surveys for excellent literature reviews on the application of game theory to the cybersecurity domain and to the detection of cyber attacks (Do et al. 2017; Giraldo et al. 2018). Chapter 7 of this book also considers the detection of cyber attacks (Kamhoua et al. 2020, Chapter 7), but the focus there in the detection of multistep dynamic attacks, rather than sensor manipulation.

8.2 Measurement Manipulation Games

One can model making a binary decision as estimating the value of a Bernoulli random variable $\theta \in \{0, 1\}$, with the understanding that if the defender's estimate $\hat{\theta}$ is equal to θ then the "right" decision has been made. The Bernoulli parameter

$$P(\theta = 1) = 1 - P(\theta = 0) = p \tag{8.1}$$

expresses any a-priori information that we may have on θ. To make its decision, the defender has available a vector $Y := (Y_1, Y_2, \ldots, Y_N)$ of N binary "noisy" sensor measurements that provide recommendations on how $\hat{\theta}$ should be selected. We assume that the measurements $Y_i \in \{0, 1\}$, $i \in \{1, 2, \ldots, N\}$ are conditionally independent, given θ. Specifically, $Y_i = 1$ means that sensor i recommends selecting $\hat{\theta} = 1$, with the understanding that

$$P(Y_i = 1 | \theta; Y_{j \neq i}) = \begin{cases} p_{\text{err}} & \theta = 0 \\ 1 - p_{\text{err}} & \theta = 1, \end{cases} \quad \forall i \in \{1, 2, \ldots, N\}, \tag{8.2}$$

where p_{err} denotes the sensor's error probability that, for simplicity, is assumed the same for every sensor.

Measurement manipulation arises because the defender must build its estimate $\hat{\theta}$ of θ based on a version $Z := (Z_1, Z_2, \ldots, Z_N)$ of the measurement vector Y that may have been "corrupted" by an attacker. Specifically, with a probability $p_{\text{att}} \in [0, 1]$, the attacker manipulated the values of $M \leq N$ entries of Y and therefore only $N - M$ of the entries of Z are guaranteed to match those of Y, but the defender does not know which. The probability p_{att} should be viewed as a design parameter that reflects how certain the estimator is that the measurements have been manipulated. For $p_{\text{att}} = 0$, there is no attack and the estimation of θ is a standard Bayesian estimation problem subject to stochastic measurement errors. However, for $p_{\text{att}} > 0$ the solution to this problem requires game theoretical tools.

Our problem can be viewed as a two-player partial information game: The *defender* must select an estimate $\hat{\theta}$ for θ based on the vector Z of possibly corrupted sensor measurements. Since the defender does not know which measurements have been manipulated and all sensors are assumed equally reliable, the defender's decision must be based solely on the total number of zeros and ones that appear in the N-vector Z. This means that the defender's policy is a function μ that maps integers in $\{0, 1, \ldots, N\}$ into the binary estimate:

$$\hat{\theta} = \mu \left(\sum_{i=1}^{N} Z_i \right). \tag{8.3}$$

Since the domain of μ has $N + 1$ elements and its codomain has 2 elements, the set $\mathcal{U}$ of all possible estimation policies contains 2^{N+1} policies.

The *attacker* is able to manipulate M out of the N entries of Z and must thus decide how many of these M measurements should be set to report a zero. We assume that the attacker knows the value of θ but not the measurements Y_i that it cannot manipulate. The attack policy δ is thus a function that maps the binary variable $\theta \in \{0, 1\}$ to the integer set $\{0, 1, \ldots, M\}$. Since the domain of δ has 2 elements and its codomain has $M + 1$ elements, the set D of all possible attack policies contains $(M + 1)^2$ policies.

In measurement manipulation games, we consider a simple zero-sum formulation in which the defender wants to minimize the probability of making the wrong decision

$$J_{\text{def}} := P(\hat{\theta} \neq \theta), \tag{8.4}$$

whereas the attacker wants to maximize it.

8.2.1 Saddle-Point Equilibria

The model described above for measurement manipulation defines a zero-sum noncooperative game in which the detector selects a policy $\mu \in \mathcal{U}$ and the attacker a policy $\delta \in D$ to minimize and maximize, respectively, the probability of an estimation error (8.4) (Kamhoua et al. 2020, Chapter 2). Since the sets of policies are finite, we have a (finite) matrix game defined by a 2^{n+1} by $(M + 1)^2$ matrix A, with each row corresponding to an estimation policy μ for the defender and each column to an attack policy δ. Straightforward computations show that the entry of A corresponding to a defender policy μ and an attack policy δ that sets to 0 and to 1 a number of sensors equal to $\delta(\theta)$ and $M - \delta(\theta)$, respectively, is given by

$$P(\hat{\theta} \neq \theta)$$

$$= (1 - p) \left(P_{\text{att}} \sum_{k=M-\delta(0)}^{N-\delta(0)} \mu(k) \binom{N - M}{k - M + \delta(0)} p_{\text{err}}^{k-M+\delta(0)} (1 - p_{\text{err}})^{N-k-\delta(0)} \right.$$

$$\left. + (1 - P_{\text{att}}) \sum_{k=0}^{N} \mu(k) \binom{n}{k} p_{\text{err}}^k (1 - p_{\text{err}})^{N-k} \right)$$

$$+ p \left(P_{\text{att}} \sum_{k=M-\delta(1)}^{N-\delta(1)} (1 - \mu(k)) \binom{N - M}{k - M + \delta(1)} (1 - p_{\text{err}})^{k-M+\delta(1)} p_{\text{err}}^{N-k-\delta(1)} \right.$$

$$\left. + (1 - P_{\text{att}}) \sum_{k=0}^{N} (1 - \mu(k)) \binom{n}{k} (1 - p_{\text{err}})^k p_{\text{err}}^{N-k} \right). \tag{8.5}$$

Often, this game does not have pure saddle-point equilibria and the players will need to seek for mixed policies, which correspond to selecting probability distributions over the sets of actions $\mathcal{U}$ and D (Başar and Olsder 1995; Hespanha 2017). While the number of rows of the matrix A defined above grows exponentially with the number of sensors, it turns out that a large number of rows can be ignored using policy domination and this game has mixed saddle-point policies that involve randomization only over a small number of pure policies that we define below:

1. The attacker's *deception rule* is the pure policy

$$\delta(\theta) = \delta_{\text{dec}}(\theta) := \begin{cases} 0 & \theta = 0, \\ M & \theta = 1, \end{cases}$$

which sets all M manipulated sensors equal to 1 when $\theta = 0$ and all M sensors equal to 0 when $\theta = 1$.

2. The attacker's *no-deception rule* is the pure policy

$$\delta(\theta) = \delta_{\text{no-dec}}(\theta) := \begin{cases} M & \theta = 0, \\ 0 & \theta = 1, \end{cases}$$

which sets all M manipulated sensors equal to θ.

3. The detector's *majority rule* is the pure policy

$$\mu\left(\sum_{i=1}^{N} Z_i\right) = \mu_{\text{maj}}\left(\sum_{i=1}^{N} Z_i\right) := \begin{cases} 0 & \sum_{i=1}^{N} Z_i \leq \frac{N-1}{2}, \\ 1 & \sum_{i=1}^{N} Z_i \geq \frac{N+1}{2}, \end{cases}$$

which corresponds to setting $\hat{\theta} = 0$ if more than half the sensors reported the value 0.

4. The detector's *no-consensus rule* is the pure policy

$$\mu\left(\sum_{i=1}^{N} Z_i\right) = \mu_{\text{no-cons}}\left(\sum_{i=1}^{N} Z_i\right)$$

$$:= \begin{cases} 0 & 0 < \sum_{i=1}^{N} Z_i \leq \frac{N-1}{2} \text{ or } \sum_{i=1}^{N} Z_i = N, \\ 1 & n > \sum_{i=1}^{N} Z_i \geq \frac{N+1}{2} \text{ or } \sum_{i=1}^{N} Z_i = 0. \end{cases}$$

This somewhat unexpected policy is like the majority rule, except that if all sensors agree on a particular value (i.e., $Z_i = 1$, $\forall i$ or $Z_i = 0$, $\forall i$), the estimate $\hat{\theta}$ should take the opposite value.

The deception rule seems intuitive, since the attacker should try to deceive the defender by reporting the opposite of θ. Similarly, the majority rule also seems reasonable, at least when the attacker can manipulate less than half of the sensors. To gain intuition on why the remaining policies need to be considered, suppose that the estimator observes that all sensors report the same value: Three options are possible: 1) there was no attack and all sensors are correct, 2) the attacker did not flip any bit and all sensors are correct, or 3) the attacker flipped M bits and the remaining $N - M$ sensors reported incorrect values. In cases 1) and 2), the defender should use the majority rule, but in case 3) the estimator should choose precisely what we called the nonconsensus rule. At a saddle point, the attacker will choose the probabilities of using the deception versus no-deception rules to keep the estimator guessing. Because of this, it is not surprising to discover that the optimal estimator's policy will sometimes select the majority rule (typically with high probability) and sometimes the no-consensus rule (typically with low probability).

The following result provides an explicit formula for a mixed saddle-point equilibrium for this game.

Theorem 8.1 Consider $p = 1/2$ in (8.5) and an odd number of sensors $N \geq 3$, for which

$$M \leq \min\left\{\frac{N-1}{2}, \frac{N+1}{2} - \frac{p_{\text{err}}}{1 - p_{\text{err}}} \frac{N-1}{2}\right\}, \tag{8.6}$$

$$p_{\text{err}} \leq \frac{2}{N+1}, \quad p_{\text{err}} \leq \left(1 + \frac{\frac{N-1}{2}!(N-M)!}{\frac{N-2M+1}{2}!}\right)^{-1}, \tag{8.7}$$

and, for the case $M \geq 2$, further assume that

$$p_{\text{att}} \leq \frac{1}{1 + \frac{1}{n}\binom{N-M}{M-1}\frac{p_{\text{err}}^{N-2M+1}(1-p_{\text{err}})^{M-1}}{p_{\text{err}}(1-p_{\text{err}})^{N-1} - p_{\text{err}}^{N-1}(1-p_{\text{err}})}}. \tag{8.8}$$

In this case, the value of the game is given by

$$v^* = \alpha + p_{\text{att}} \min\left\{\gamma, \gamma + \frac{\eta(\beta - p_{\text{err}}^{N-M})}{(1-p_{\text{err}})^{N-M} + p_{\text{err}}^{N-M}}\right\} \tag{8.9}$$

and a mixed saddle-point policy corresponds to selecting

$$
\begin{cases} \mu_{\text{maj}} & \text{w.p. } 1 - p_{\text{no-cons}}, \\ \mu_{\text{no-cons}} & \text{w.p. } p_{\text{no-cons}}, \end{cases} \qquad \begin{cases} \delta_{\text{dec}} & \text{w.p. } 1 - p_{\text{no-dec}}, \\ \delta_{\text{no-dec}} & \text{w.p. } p_{\text{no-dec}}, \end{cases}
\tag{8.10}
$$

where

$$
p_{\text{no-cons}} := \begin{cases} \Pi\left(\dfrac{\eta}{(1-p_{\text{err}})^{N-M} + p_{\text{err}}^{N-M}} \right) & p_{\text{att}} \geq \dfrac{(1-p_{\text{err}})^{N} - p_{\text{err}}^{N}}{p_{\text{err}}^{N-M} + (1-p_{\text{err}})^{N} - p_{\text{err}}^{N}}, \\[1.5em] 0 & p_{\text{att}} < \dfrac{(1-p_{\text{err}})^{N} - p_{\text{err}}^{N}}{p_{\text{err}}^{N-M} + (1-p_{\text{err}})^{N} - p_{\text{err}}^{N}}, \end{cases}
$$

$$
p_{\text{no-dec}} := \Pi\left(\frac{p_{\text{err}}^{N-M} - \beta}{(1 - p_{\text{err}})^{N-M} + p_{\text{err}}^{N-M}} \right),
$$

$$
\alpha := (1 - p_{\text{att}}) \sum_{k=0}^{\frac{N-1}{2}} \binom{n}{k} p_{\text{err}}^{N-k} (1 - p_{\text{err}})^{k},
$$

$$
\beta := \frac{1 - p_{\text{att}}}{p_{\text{att}}} \left((1 - p_{\text{err}})^{N} - p_{\text{err}}^{N} \right),
$$

$$
\gamma := \sum_{k=0}^{\frac{N-1}{2}} \binom{N-M}{k} p_{\text{err}}^{N-M-k} (1 - p_{\text{err}})^{k},
$$

$$
\eta := \sum_{k=\frac{N+1}{2}-M}^{\frac{N-1}{2}} \binom{N-M}{k} p_{\text{err}}^{N-M-k} (1 - p_{\text{err}})^{k},
$$

and $\Pi : \mathbb{R} \to \mathbb{R}$ denotes the projection function into the interval $[0, 1]$:

$$
\Pi(x) = \begin{cases} 0 & x < 0, \\ x & x \in [0, 1], \\ 1 & x > 1. \end{cases}
$$

$\square$

An important (and convenient) feature of the mixed saddle-point equilibrium in (8.10) is that the defender does not need to know the precise value of the attack probability p_{att} to implement its policy. This is because, to compute $p_{\text{no-cons}}$, the defender only needs to know whether or not

$$
p_{\text{att}} \geq \frac{(1 - p_{\text{err}})^{N} - p_{\text{err}}^{N}}{p_{\text{err}}^{N-M} + (1 - p_{\text{err}})^{N} - p_{\text{err}}^{N}},
\tag{8.11}
$$

which is convenient because, in real applications, the value of p_{att} may be very hard to know. It is also worth noting that, when the error probability p_{err} is small, the right-hand side of (8.11) is close to 1 and therefore $p_{\text{no-cons}}$ is only nonzero for values of p_{att} very close to one. Otherwise, $p_{\text{no-cons}} = 0$ and the majority rule is always used.

While the detector's policy may depend little on p_{att}, that is not the case for the saddle-point probability of an estimation error v^{*}. In fact, one can show that this probability scales with the number of sensors as

$$
p_{\text{err}}^{\frac{N+1-2M}{2}},
$$

which shows that each one of the M potentially compromised sensors effectively decreases the total number of useful sensors from N to $N - 2M$.

Theorem 8.1 is only valid for $p = 1/2$ and parameters M and N that satisfy (8.6)–(8.8). The restriction to the case $p = 1/2$ mostly simplifies the formulas and could be easily lifted. The assumption in (8.6) poses an upper bound on the maximum number of sensors that can be attacked and is not surprising since, even without sensor errors (i.e., $p_{\text{err}} = 0$), the detector can only make use of the measurements if less than half of the sensors have been attacked. The inequality (8.8) is also not restrictive since the left-hand side is typically very close to one. However, the inequalities in (8.7) are restrictive when N is large, because the right-hand side of both inequalities converges to zero as N goes to infinity. This limitation of Theorem 8.1 is the main motivation for the results in the Section 8.2.2.

Before proceedings it is worth noting that, in general, the mixed saddle-point in (8.10) is not unique and there will exist other mixed saddle-point involving different sets of pure policies. However, all other mixed saddle-point will have the value in (8.9), as all mixed saddle-point of a zero-sum game must lead to the same value. Moreover, because of the order interchangeability property of zero-sum games (Hespanha 2017), the mixed defender's policy in (8.10) remains a security policy regardless of whether or not the attackers restricts theirs attention to the deception deception/no-deception policies considered above.

8.2.2 Approximate Saddle-Point Equilibrium

In view of (8.7), the estimation policy provided by Theorem 8.1 is only optimal for a number of sensors N roughly below $2/p_{\text{err}}$. We shall see in this section that when the number of sensors is large and the probability of sensor error is not very small, a threshold estimation policy like the majority rule is almost optimal. However, the proof of this result requires a completely different approach, which is described next.

The random variable

$$\bar{Z} := \sum_{i=1}^{N} Z_i,$$

that the detector uses in the estimation policy (8.3) may have different distributions depending on the value of θ and whether or not there is an attack. Specifically,

$$\bar{Z} = \begin{cases} R & \text{w.p. } 1 - p_{\text{att}}, \\ S + W & \text{w.p. } p_{\text{att}}, \end{cases} \tag{8.12}$$

where the random variable R equals the sum of all the Y_i and therefore its distribution is

$$R \sim \begin{cases} \text{Binom}(N, p_{\text{err}}) & \theta = 0, \\ \text{Binom}(N, 1 - p_{\text{err}}) & \theta = 1; \end{cases} \tag{8.13}$$

the random variable S equals the sum of the $N - M$ sensors that have not been compromised by the attacker and therefore has distribution

$$S \sim \begin{cases} \text{Binom}(N - M, p_{\text{err}}) & \theta = 0, \\ \text{Binom}(N - M, 1 - p_{\text{err}}) & \theta = 1; \end{cases} \tag{8.14}$$

and the random variable W equals the sum of the readings of the M sensors compromised by the attacker. The distribution of W is selected by the attacker and may depend on the value of θ, with the constraint that its support must lie in the set $\{0, 1, \ldots, M\}$.

This perspective motivates the following general problem: Suppose that one wants to estimate the random variable θ with Bernoulli distribution (8.1) based on a measurement $\bar{Z}$ of the form (8.12) where the conditional distributions of R and S given θ are known and the conditional distribution of W given θ is selected by an adversary, but its support is limited to a given subset $I \subset \mathbb{R}$. As before, we formulate this as a zero-sum game where the estimator wants to minimize the probability of an estimation error, whereas the attacker wants to maximize this probability.

We allow the *estimation policy* to be stochastic and represent it by a function $p_{\hat{\theta}=1} : \mathbb{R} \to [0, 1]$, with the understanding that, when the estimator observes a value $\bar{z} \in \mathbb{R}$ for (8.12), the estimate of θ is given by

$$\hat{\theta} = \begin{cases} 1 & \text{w.p. } p_{\hat{\theta}=1}(\bar{z}), \\ 0 & \text{w.p. } 1 - p_{\hat{\theta}=1}(\bar{z}). \end{cases}$$

Denoting by ρ_x, σ_x, and ω_x, respectively, the conditional distributions of R, S, and W given that $\theta = x \in \{0, 1\}$, we can use the law of total probability, to express the probability of an estimation error as follows:

$$J(p_{\hat{\theta}=1}, \omega_0, \omega_1) := P(\hat{\theta} \neq \theta) = p + p_{\text{att}}(1 - p) \iint_{I \mathbb{R}} p_{\hat{\theta}=1}(\bar{y} + \bar{w}) \sigma_0(d\bar{y}) \omega_0(d\bar{w})$$

$$- p_{\text{att}} \, p \iint_{I \mathbb{R}} p_{\hat{\theta}=1}(\bar{y} + \bar{w}) \sigma_1(d\bar{y}) \omega_1(d\bar{w})$$

$$+ (1 - p_{\text{att}})(1 - p) \int_{\mathbb{R}} p_{\hat{\theta}=1}(\bar{y}) \rho_0(d\bar{y}) - (1 - p_{\text{att}}) p \int_{\mathbb{R}} p_{\hat{\theta}=1}(\bar{y}) \rho_1(d\bar{y}). \qquad (8.15)$$

For the above formula to be well defined, we assume that the attacker is only allowed to select distributions (ω_0, ω_1) in a set $\mathcal{A}$ containing all pairs of distributions (ω_0, ω_1) for which the integrals in (8.15) exist.

The result that follows is based on first considering an auxiliary game that replaces the binomial distributions defined by (8.13)–(8.14) by Gaussian distributions and then using the mixed saddle-point for this "relaxed" version of the game to compute ϵ-saddle-point of our original game. We recall that a pair $(u^*, d^*) \in \mathcal{U} \times \mathcal{D}$ is an ϵ-saddle-point with respect to a criterion $J : \mathcal{U} \times \mathcal{D} \to \mathbb{R}$ when

$$J(u^*, d^*) - \epsilon \leq J(u^*, d^*) \leq J(u, d^*) + \epsilon, \quad \forall u \in \mathcal{U}, d \in \mathcal{D}.$$

For $\epsilon = 0$, an ϵ-saddle-point is just a regular saddle-point.

Theorem 8.2 Suppose that

$$M < N(1 - 2p_{\text{err}}), \quad p_{\text{err}} \in (0, 1/2),$$

and that the function

$$g(\bar{z}) := \frac{1 - p_{\text{att}}}{\sqrt{np_{\text{err}}(1 - p_{\text{err}})2\pi}} \left((1 - p)e^{-\frac{(\bar{z} - np_{\text{err}})^2}{2np_{\text{err}}(1 - p_{\text{err}})}} - p \, e^{-\frac{(\bar{z} - n(1 - p_{\text{err}}))^2}{2np_{\text{err}}(1 - p_{\text{err}})}} \right)$$

$$+ \frac{p_{\text{att}}}{\sqrt{(N - M)p_{\text{err}}(1 - p_{\text{err}})2\pi}} \left((1 - p)e^{-\frac{(\bar{z} - M - (N - M)p_{\text{err}})^2}{2(N - M)p_{\text{err}}(1 - p_{\text{err}})}} - p \, e^{-\frac{(\bar{z} - (N - M)(1 - p_{\text{err}}))^2}{2(N - M)p_{\text{err}}(1 - p_{\text{err}})}} \right),$$

has a unique zero $\bar{z} = z^*$. Then the function

$$p_{\hat{\theta}=1}^*(\bar{z}) := \begin{cases} 0 & \bar{z} < \bar{z}^*, \\ 1 & \bar{z} \geq \bar{z}^*, \end{cases} \qquad (8.16)$$

for the defender and the Dirac distributions

$$\omega_0^*(\bar{w}) = \delta(\bar{w} - M), \qquad \omega_1^*(\bar{w}) = \delta(\bar{w}), \qquad (8.17)$$

for the attacker form a 2ϵ-saddle-point with value $J(f^*, \omega_0^*, \omega_1^*)$ satisfying

$$\left| J(f^*, \omega_0^*, \omega_1^*) - p - \int_{z^*}^{\infty} g(\bar{z}) d\bar{z} \right| \leq \epsilon = O\left(\frac{1}{\sqrt{N - M}} \right).$$

$\square$

While the problem formulation enables both players to select mixed policies, the 2ϵ-saddle-point defined by (8.16)–(8.17) actually corresponds to pure policies of the form

$$\hat{\theta} = \begin{cases} 0 & \bar{Z} := \sum_{i=1}^{N} Z_i < z^*, \\ 1 & \bar{Z} := \sum_{i=1}^{N} Z_i \geq z^*, \end{cases} \qquad \delta(\theta) = \delta_{\text{dec}}(\theta) := \begin{cases} 0 & \theta = 0, \\ M & \theta = 1 \end{cases}$$

and when $p = 1/2$ and there are no sensor errors ($p_{\text{err}} = 0$), the threshold value z^* in Theorem 8.2 is precisely equal to $N/2$ for every attack probability and we obtain the majority rule μ_{maj}. As the a-priori probability p deviates from $1/2$ and p_{err} increases, the threshold z^* changes to reflect the a-priori information and the probability of sensor error.

8.3 Sensor-Reveal Games

In sensor-reveal games, we continue to consider a scenario where the defender wants to estimate an unknown variable θ based on a vector $Y := (Y_1, Y_2, \ldots, Y_N)$ of N binary "noisy" sensor measurements that provide recommendations for $\hat{\theta}$. However, now the N sensors are heterogeneous with the model (8.2) replaced by

$$P(Y_i = 1 | \theta = 0; Y_{j \neq i}) = p_{\text{fp}}^i, \qquad P(Y_i = 0 | \theta = 1; Y_{j \neq i}) = p_{\text{fn}}^i,$$

$\forall i \in \{1, 2, \ldots, N\}$, where p_{fp}^i and p_{fn}^i can be regarded as the probabilities of a false positive and a false negative associated with the ith sensor. This model is more general than the one in (8.2) because we now allow the two types of errors (false positives and false negatives) to have different probabilities and also allow different sensors to have different error probabilities (thus the superscript ^{i} in p_{fp}^i and p_{fn}^i).

A more fundamental difference with respect to Section 8.2 is that the defender's decision $\hat{\theta}$ is now based on a single sensor measurements Y_σ revealed by the attacker, where $\sigma \in \{1, 2, \ldots, N\}$ denotes the index of the sensor that the attacker selects to reveal, which the defender can observe and use to inform its decision. Another difference is that we now enable the attacker to select the actual value of the unknown variable θ. This is reasonable when $\theta = 1$ corresponds to some form of cyber attack launched by the same agent that manipulates the sensors.

A sensor-reveal game can also be viewed as a two-player partial information game: The *defender* selects an estimate $\hat{\theta}$ for θ based on the pair (σ, Y_σ) that includes the index σ of the sensor revealed by the attacker and the corresponding sensor recommendation Y_σ. This means that deterministic (pure) policies for the defender are functions μ that map the pair $(\sigma, Y_\sigma) \in \{0, 1\} \times \{1, 2, \ldots, N\}$ into the binary estimate:

$$\hat{\theta} = \mu(\sigma, Y_\sigma). \tag{8.18}$$

The *attacker* makes two decisions: it decides whether or not to launch an attack (i.e., selects the value of $\theta \in \{0, 1\}$) and also the index σ of which sensor to reveal. In the model considered here, the attacker can decide which sensor to reveal (i.e., can select the sensor index σ), but does not know which measurements Y_σ will be reported to the defender. In particular, the attacker does not know a-priori whether or not the sensor recommendation Y_σ is correct in the sense that it is equal to θ.

We also consider a more sophisticated cost structure for the players: The defender seeks to minimize a cost of the form

$$J_{\text{def}} := AP(\hat{\theta} = 1, \theta = 0) + BP(\hat{\theta} = 0, \theta = 1), \tag{8.19}$$

where $A > 0$ and $B > 0$ are parameters that establish the cost of a false detection and of a missed detection, respectively. When $A = B$, the defender's cost (8.19) matches the measurement manipulation game cost in (8.4), but in sensor manipulation games we go beyond zero-sum games and consider a cost for the attacker that includes terms related to the benefit reaped from launching an attack and penalties for revealing different sensors. Specifically, we assume that the attacker minimizes a cost of the form

$$J_{\text{att}} := -R\theta + CP(\hat{\theta} = 1, \theta = 1) - FP(\hat{\theta} = 1, \theta = 0) + S_\sigma, \tag{8.20}$$

where $R \geq 0$ is a reward associated with engaging in the cyber attack (i.e., choosing $\theta = 1$), $C \geq 0$ is the cost of being caught, S_σ the cost of revealing sensor σ, and $F \geq 0$ is a rewards to the attacker for generating a false alarm. In what follows, we assume that the reward $R - C$ associated with setting $\theta = 1$ and being caught is smaller than the reward F of generating a false alarm, i.e.,

$$R - C < F, \tag{8.21}$$

which excludes the trivial solution for the attacker to always select $\theta = 1$.

The problem becomes especially interesting when the different sensors have different levels of "reliability" and it is costly for the attacker to reveal sensors that convey to the defender very little information about the true value of θ (i.e., sensors with large values for p_{fp}^i and p_{fn}^i). Here, "costly" may mean that S_i is large but also that revealing the recommendation of a sensor that is not very informative can be taken as an indication that $\theta = 1$.

Consistent with standard terminology, a *pure Nash equilibrium* for this game is thus a (pure) defender's policy μ^* and a (deterministic) choice (θ^*, σ^*) for the attacker such that

1. when the defender uses the decision rule (8.18) with $\mu = \mu^*$, the attacker's cost J_{att} in (8.20) is minimized for the pair (θ^*, σ^*), over all possible $(\theta, \sigma) \in \{0, 1\} \times \{1, 2, \dots, N\}$; and
2. when the attacker selects (θ^*, σ^*), the defender's cost J_{def} in (8.19) is minimized by using $\mu = \mu^*$ in (8.18).

A *mixed Nash equilibrium* follows a similar definition, but with the deterministic choices replaced by distributions over the sets of all deterministic policies. Specifically, a mix Nash equilibrium consists of a probability distribution over all possible policies $\mu : \{0, 1\} \times \{1, 2, \dots, N\} \rightarrow \{0, 1\}$ for the defender and a probability distribution over all possible $(\theta, \sigma) \in \{0, 1\} \times \{1, 2, \dots, N\}$ for the attacker. We shall see, however, that there is no need to randomize over the sensor selection σ, just over μ and θ.

8.3.1 Nash Equilibria

For the purpose of computing Nash equilibria for this game, it is convenient to consider the extensive form decision tree depicted in Figure 8.1, where the branches represent the players' decisions

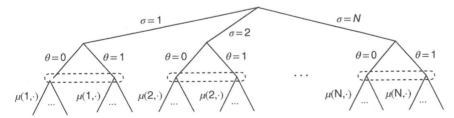

Figure 8.1 Extensive form representation of the sensor-reveal game, with the top two branches corresponding to attacker decisions (which sensor σ to select and whether or not to set $\theta = 1$) and the bottom branch to the defender's selection of the maps $Y_i \mapsto \mu(i, Y_i)$, $i \in \{1, 2, \dots, N\}$ that define the estimate in (8.18).

and the dashed ellipses represent the information sets for the defender, i.e., sets of decision points that are indistinguishable based on the information available to the defender (Hespanha 2017; Kamhoua et al. 2020, Chapter 2). This representation of the game permits the independent analysis of each subtree corresponding to a particular choice for σ by the attacker. To this effect, suppose that the attacker selected a particular sensor $\sigma = i \in \{1, 2, \ldots, N\}$ and consider the pure (i.e., deterministic) choices that each player needs to consider on the subtree corresponding to $\sigma = i$:

1. The attacker must select either $\theta = 0$ or $\theta = 1$.
2. The defender must select the map $Y_i \mapsto \mu(i, Y_i)$ that defines the estimate $\hat{\theta}$ in (8.18) as a function of Y_i. Since each Y_i can only take two values $\{0, 1\}$ and $\mu(i, Y_i)$ also only takes two values, there are three possible choices for the map $Y_i \mapsto \mu(i, Y_i)$:

$$\hat{\theta} = \mu(i, Y_i) := 0, \quad \forall Y_i \in \{0, 1\}, \text{ or}$$
$$\hat{\theta} = \mu(i, Y_i) := 1, \quad \forall Y_i \in \{0, 1\}, \text{ or}$$
$$\hat{\theta} = \mu(i, Y_i) := Y_i, \quad \forall Y_i \in \{0, 1\}.$$

The first two options correspond to ignoring the sensor recommendation Y_i and always setting $\hat{\theta} = 0$ or $\hat{\theta} = 1$, respectively, whereas the third option corresponds to always agreeing with the sensor recommendation and setting $\hat{\theta} = Y_i$. Technically, there is a fourth option of always disagreeing with the sensor recommendation and setting $\hat{\theta} = 1 - Y_i$, but this policy would always be dominated by one of the other three, as long as the sensor is somewhat "informative" in the sense that its error probabilities satisfy

$$p_{fp}^i + p_{fn}^i \leq 1. \tag{8.22}$$

Straightforward computations can be used to show that the problem corresponding to the subtree $\sigma = i$ in Figure 8.1 can be represented by the following 2×3 bimatrix game

$$A_{att}^i := \begin{bmatrix} & \hat{\theta} = 0 & \hat{\theta} = 1 & \hat{\theta} = Y_i \\ \theta = 0 & S_i & S_i - F & S_i - Fp_{fp}^i \\ \theta = 1 & S_i - R & S_i - R + C & S_i - R + C\left(1 - p_{fn}^i\right) \end{bmatrix}, \tag{8.23a}$$

$$B_{def}^i := \begin{bmatrix} & \hat{\theta} = 0 & \hat{\theta} = 1 & \hat{\theta} = Y_i \\ \theta = 0 & 0 & A & Ap_{fp}^i \\ \theta = 1 & B & 0 & Bp_{fn}^i \end{bmatrix}, \tag{8.23b}$$

where A_{att}^i and B_{def}^i should be viewed as cost matrices for the attacker and defender, respectively, and each row and column is labeled with the corresponding policies for the attacker and defender, respectively (as enumerated above).

The following result provides explicit formulas for a mixed Nash equilibrium for the bimatrix game (8.23a) associated with the attacker's decision to reveal sensor $\sigma = i$.

Theorem 8.3 Assuming that (8.21) holds and that the ith sensor is informative in the sense of (8.22), the bimatrix game (8.23a) has a mixed Nash equilibrium of the form

$$y_{att}^i {}^* = \begin{cases} \left[\dfrac{B(1-p_{fn}^i)}{Ap_{fp}^i + B(1-p_{fn}^i)} \quad \dfrac{Ap_{fp}^i}{Ap_{fp}^i + B(1-p_{fn}^i)} \right]', & \bar{C}^i \geq R, \\[3ex] \left[\dfrac{Bp_{fn}^i}{A(1-p_{fp}^i)+Bp_{fn}^i} \quad \dfrac{A(1-p_{fp}^i)}{A(1-p_{fp}^i)+Bp_{fn}^i} \right]', & \bar{C}^i < R, \end{cases} \tag{8.24a}$$

$$z_{def}^i {}^* = \begin{cases} \left[\dfrac{\bar{C}^i - R}{\bar{C}^i} \quad 0 \quad \dfrac{R}{\bar{C}^i} \right]', & \bar{C}^i \geq R, \\[3ex] \left[0 \quad \dfrac{R - \bar{C}^i}{C + F - \bar{C}^i} \quad \dfrac{C + F - R}{C + F - \bar{C}^i} \right]', & \bar{C}^i < R, \end{cases} \tag{8.24b}$$

with values

$$
J_{\text{att}}^{i}{}^{*} = S_i - F \begin{cases} \dfrac{R p_{\text{fp}}^{i}}{\bar{C}^i}, & \bar{C}^i \geq R, \\[2ex] \dfrac{(R-C)(1-p_{\text{fp}}^{i})+C p_{\text{fn}}^{i}}{C+F-\bar{C}^i}, & \bar{C}^i < R, \end{cases} \tag{8.25a}
$$

$$
J_{\text{def}}^{i}{}^{*} = \begin{cases} \dfrac{A B p_{\text{fp}}^{i}}{A p_{\text{fp}}^{i}+B(1-p_{\text{fn}}^{i})}, & \bar{C}^i \geq R, \\[2ex] \dfrac{A B p_{\text{fn}}^{i}}{A(1-p_{\text{fp}}^{i})+B p_{\text{fn}}^{i}}, & \bar{C}^i < R, \end{cases} \tag{8.25b}
$$

where $\bar{C}^i := C(1 - p_{\text{fn}}^{i}) + F p_{\text{fp}}^{i}$. □

We conclude from Theorem 8.3 that the attacker minimizes its cost by revealing the sensor i that leads to the smallest value of its cost $J_{\text{att}}^{i}{}^{*}$ in (8.25a). This provides the last piece of the policy for the attacker:

$$
\sigma = \arg \min_i S_i - F \begin{cases} \dfrac{R p_{\text{fp}}^{i}}{\bar{C}^i} & \bar{C}^i \geq R, \\[2ex] \dfrac{(R-C)(1-p_{\text{fp}}^{i})+C p_{\text{fn}}^{i}}{C+F-\bar{C}^i} & \bar{C}^i < R, \end{cases}
$$

which corresponds to the top branch of the decision tree depicted in Figure 8.1. We thus conclude that the attacker's selection of the sensor index σ can be deterministic (pure), but the selection of θ will typically be mixed and given by the distribution in (8.24a) for $i = \sigma$. The defender's selection of the maps $Y_i \mapsto \mu(i, Y_i)$ will typically also be mixed and given by the distributions in (8.24b), which typically depend on the sensor $\sigma = i$ that has been revealed.

8.4 Conclusions and Future Work

We considered sensor manipulation problems in which an attacker tries to induce a cyber defense system to make the wrong decision. In measurement manipulation games, the cyber defense system is forced to make a decision based on a number of sensor recommendations that may have been altered by an attacker, whereas in sensor-reveal games the attacker selects which sensors the defender is allowed to use in making a decision.

The results for measurement manipulation games in Section 8.2 can be summarized as follows: For scenarios with a relatively small number of sensors (roughly $N \leq 2/p_{\text{err}}$) the game has a saddle-point equilibrium in mixed policies and the corresponding policy for the defender involves randomizing between a majority policy μ_{maj} that agrees with the most sensors and a somewhat unexpected policy $\mu_{\text{no-cons}}$ that most of the times follows a majority-like rule, but decides in opposition to all the sensor recommendations when these are in consensus. For the general case (which includes $N > 2/p_{\text{err}}$), it is shown that there exists an approximate Nash equilibrium corresponding to a deterministic voting-like decision rule that follows the sensor recommendations when more than a certain number of them agrees. While this rule is not optimal, the level of suboptimality decreases to zero as the number of sensors grows.

The results for sensor-reveal games in Section 8.3 provide a closed form solution for mixed Nash equilibrium policies for both players. Under mild assumptions, the defender's policy involves randomizing between three deterministic policies: a policy that ignores Y_σ and always selects $\hat{\theta} = 0$, another one that also ignores Y_σ and always selects $\hat{\theta} = 1$, and a third policy that accepts the sensor revealed and selects $\hat{\theta} = Y_\sigma$. The probabilities associated with each of these policies depend on the index σ of the sensor that has been revealed.

In the interest of clarity, this paper is focused on stylized versions of decision making processes based on data that has been manipulated. For practical application, complexity may need to be added to these models, especially in what regards the nature of the sensor measurements. This is particularly true for the sensor-reveal games, for which we restricted our attention to decisions based on a single binary sensor $Y_\sigma \in \{0, 1\}$. The paper (Hespanha and Garagic 2019) on which Section 8.3 is based considers more general (nonbinary) sensors, but further research is needed to extend this work to sensor models for which the Y_i are not necessarily conditionally independent, given θ. Similarly, for many cyber defense problems a binary decision $\hat{\theta} \in \{0, 1\}$ is too restrictive, as one may want to decide among one of several alternative defense actions, some of which could include deploying additional cyber defense sensors. This chapter touches upon a rich area of research with numerous open problems that are crucial for the design of cyber defense mechanisms robust to information manipulation. Among these it is worth highlighting that cyber defense mechanisms interact with attackers in a repeated fashion over time, which provides opportunities for mutual learning (Kamhoua et al. 2020, Part 3).

References

T. Başar and G. J. Olsder. *Dynamic Noncooperative Game Theory*. Academic Press, London, 1995.

Cuong T. Do, Nguyen H. Tran, Choongseon Hong, Charles A. Kamhoua, Kevin A. Kwiat, Erik Blasch, Shaolei Ren, Niki Pissinou, and Sundaraja Sitharama Iyengar. Game theory for cyber security and privacy. *ACM Computing Surveys (CSUR)*, 50(2):1–37, 2017.

Jairo Giraldo, David Urbina, Alvaro Cardenas, Junia Valente, Mustafa Faisal, Justin Ruths, Nils Ole Tippenhauer, Henrik Sandberg, and Richard Candell. A survey of physics-based attack detection in cyber-physical systems. *ACM Computing Surveys (CSUR)*, 51(4):1–36, 2018.

João Pedro Hespanha. *Noncooperative Game Theory: An Introduction for Engineers and Computer Scientists*. Princeton Press, Princeton, NJ, 2017.

João Pedro Hespanha and Denis Garagic. Sensor-reveal games. In *Proceedings of the 58th Conference on Decision and Control, Nice, France, 11–13 December*, 2019.

Charles A. Kamhoua, Christopher D. Kiekintveld, Fei Fang, and Quanyan Zhu, editors. *Game Theory and Machine Learning for Cyber Security*. Wiley-IEEE Press, 2020.

Kyriakos G. Vamvoudakis, João Pedro Hespanha, Bruno Sinopoli, and Yilin Mo. Detection in adversarial environments. *IEEE Transactions on Automatic Control, Special Issue on the Control of Cyber-Physical Systems*, 59(12): 3209–3223, 2014.

9

Adversarial Gaussian Process Regression in Sensor Networks

Yi Li[1], Xenofon Koutsoukos[1], and Yevgeniy Vorobeychik[2]

[1] *Electrical Engineering and Computer Science, Vanderbilt University, Nashville, TN, USA*
[2] *Computer Science and Engineering, Washington University in St. Louis, St. Louis, MO, USA*

9.1   Introduction

Cyber-physical systems (CPS) are characterized by a control loop mapping sensor measurements of the environment state features to control decisions. Such systems are central in safety critical domains ranging from critical infrastructure, such as electric power and water facilities, to airplane navigation and autonomous automobile driving. Since measurements are typically noisy and sensors can be faulty, it is common to deploy anomaly detection systems to engender robustness into a CPS control loop. Moreover, anomaly detection may serve as the first line of protection against integrity attacks on sensors, limiting the impact of such attacks, with Gaussian process methods, such as Gaussian process regression (GPR) being popular choices (Cheng et al. 2015; Leyva et al. 2017; Amraee et al. 2018). Nevertheless, even armed with anomaly detectors, CPS can be vulnerable to *stealthy attacks* (Ghafouri et al. 2018), whereby an attacker submits measurements of compromised sensors in a way to ensure that they appear normal, while influencing behavior of the control loop. The general principle behind stealthy attacks on CPS by Ghafouri et al. (2018) is that the role of many control loops is to maintain a particular state feature, such as pressure or temperature, in a safe range. Thus, if associated *critical sensors* convey that the state feature is either too high or too low, controller attempts to adjust the state to compensate. For example, if reactor temperature is measured to be too low, the controller attempts to raise the temperature to keep it in the desired range. This introduces a vulnerability: if an attacker can modify the sensor reading to incorrectly report a low temperature, the controller's reaction may well cause it to increase above the safe range, with potentially disastrous consequences. The idea behind the attack is thus to maximize or minimize the reading of a critical sensor, doing so without tripping the anomaly detection alarm.

While Ghafouri et al. (2018) serves as a conceptual precursor, the approach cannot be applied to GPR-based anomaly detection, where the predictions of GPR are random variables, in contrast to the conventional regression approaches, which make deterministic predictions. Moreover, since the attack model of GRP-based anomaly detection is not defined previously, the robustness of anomaly detection as measured by the false anomaly detection alarm has not been addressed.

We consider the problem of vulnerability of CPS with GPR-based anomaly detection to stealthy attacks, as well as the problem of making such systems robust. First, we define the attack model on GPR-based anomaly detection. Due to the nonlinearity and non-convexity in the attack optimization problem, we present a novel approach to find the optimal stealthy attack. Then, we further consider the associated robustness problem via a game theoretical framework, in which the

Game Theory and Machine Learning for Cyber Security, First Edition.
Edited by Charles A. Kamhoua, Christopher D. Kiekintveld, Fei Fang, and Quanyan Zhu.
© 2021 The Institute of Electrical and Electronics Engineers, Inc. Published 2021 by John Wiley & Sons, Inc.

defender considers sensor selection, in addition to the choice of detection thresholds, as a lever for making anomaly detection more robust to attacks. Finally, we evaluate our work using the Tennessee-Eastman process control system as a case study. As our experiments demonstrate, allowing the defender to select sensors significantly enhances our ability to limit the impact of stealthy attacks.

The paper is organized as follows. Section 9.3 presents an overview of GPR in the context of anomaly detection for CPS. Section 9.4 then presents a novel stealthy attack on GPR, followed by a novel approach for robust GPR in Section 9.5. Finally, Section 9.6 presents experimental results, showing both the considerable impact that our attack can have despite conservative anomaly detection thresholds, and the significantly increased robustness to attack achieved by our approach for robust GPR-based anomaly detection.[1]

9.2 Related Work

The existing works on anomaly detection in CPS include several different models, with Gaussian process regression (GPR) (Williams and Rasmussen 2006) a popular choice due to its power and flexibility (Cheng et al. 2015; Leyva et al. 2017; Ghafouri et al. 2017; Amraee et al. 2018). However, other approaches based on machine learning have also been explored (Nader et al. 2014; Junejo and Goh 2016). An important strand in this literature also involves leveraging physical models in anomaly detection (Cárdenas et al. 2011; Urbina et al. 2016). Cárdenas et al. (2011) studied the use of physical models for anomaly detection. However, most of this work does not consider attacks vulnerability of anomaly detection to stealthy attacks. Similarly, there is an extensive literature on the problem of sensor selection in non-adversarial settings (Krause et al. 2008b; Joshi and Boyd 2009; Shamaiah et al. 2010). The extension to consider adversarial sensor selection has received some attention (Krause et al. 2008a; Laszka et al. 2015) but focuses largely on robustness to denial-of-service attacks on sensors, rather than the integrity attacks that we consider. An important precursor to our work is Ghafouri et al. (2018), who consider robust anomaly detection in the context of stealthy integrity attacks. However, this work considers conventional regression which yields deterministic predictions of sensor values, in contrast to GPR, where prediction is a random variable, adding a nontrivial technical challenge to the problem of stealthy attacks. Moreover, Ghafouri et al. do not consider the problem of sensor selection, focusing solely on tuning anomaly detection thresholds. As our experiments demonstrate, the ability to select sensors accounts for *most* of the robustness in our setting. Finally, our work is related to the broader literature on security and game theory (Alpcan and Basar 2003; Tambe 2011), as well as adversarial machine learning (Vorobeychik and Kantarcioglu 2018), with the primary difference stemming from the specific structure of our problem which requires a novel solution approach to address.

9.3 Anomaly Detection with Gaussian Process Regression

Since sensors can be faulty, it is critical to ensure that measurement errors do not impact the control loop. An important class of approaches for accomplishing this is regression-based anomaly detection, where a detected anomaly is either rapidly corrected, or alternative inputs (such as predicted, rather than measured sensor values) are used in the control loop. We start by describing this class of approaches generically, given a collection of L sensors.

---

1 Source code: https://www.dropbox.com/sh/guhnr8a7awtghre/AADxGU4z0isogccPeTUE4Mgfa?dl=0.

Let $\mathbf{y} = (y_1, \ldots, y_L)$ represent a vector of sensor measurements. For each sensor s, a predictor $f_s(\tilde{\mathbf{y}}_{-s})$ is learned from the past data of joint measurements of all sensors (which is assumed to be normal), which maps from observed readings of sensors other than s, $\tilde{\mathbf{y}}_{-s}$, to a predicted reading of sensor s, $\hat{y}_s$. The detection system then compares the difference between the predicted and observed measurements, $|\hat{y}_s - \tilde{y}_s|$, and triggers an alarm when this difference is large.

Next, we describe how the Gaussian process can be used in anomaly detection. The key advantage of the Gaussian process over alternatives is that it directly captures variance of the prediction and, consequently, enables a principled approach to judge anomalies based on confidence intervals.

Consider a zero-mean Gaussian process $y(x) \sim \mathcal{GP}(0, \mathcal{K}(x, x')) \in \mathbb{R}$, where $x, x' \in \mathbb{R}^d$ and $\mathcal{K}$ is a covariance function (for example, the squared exponential kernel, $\exp(-\lambda||x - x'||_2^2)$.) Suppose a training set D has n observations, and $D = \{(x_i, y_i)|i = 1, \ldots, n\}$. Let $\mathbf{x} = \text{col}(x_1, \ldots, x_n) \in \mathbb{R}^{n \times d}$ and $\mathbf{y} = [y_1, \ldots, y_n]^T \in \mathbb{R}^n$. Suppose the observations $\mathbf{y}$ is Gaussian:

$$\mathbf{y} \sim \mathcal{N}(0, K(\mathbf{x}))$$

where $K(\mathbf{x}) \in \mathbb{R}^{n \times n}$ is a covariance matrix, $K_{i,j}(\mathbf{x}) = \mathcal{K}(x_i, x_j)$. Given a new point x_*, we aim to predict the value $y_*|D$. According to the fact that $y(x)$ is a Gaussian process:

$$\begin{bmatrix} \mathbf{y} \\ y_* \end{bmatrix} \sim \mathcal{N}\left(0, \left[\begin{array}{c|c} K(\mathbf{x}) & \mathbf{k}_*(\mathbf{x}, x_*) \\ \hline \mathbf{k}_*(\mathbf{x}, x_*)^T & k_{**}(x_*) \end{array} \right] \right) \tag{9.1}$$

where, $\mathbf{k}_*(\mathbf{x}, x_*) = [\mathcal{K}(x_1, x_*), \ldots, \mathcal{K}(x_n, x_*)]$, and $k_{**}(x_*) = \mathcal{K}(x_*, x_*)$. Then, the prediction can be made by, $y_*|D \sim N(\mu, \sigma^2)$ where

$$\mu = \mathbf{k}_*^T(\mathbf{x}, x_*)K(\mathbf{x})^{-1} \cdot \mathbf{y} \tag{9.2}$$

$$\sigma = -\mathbf{k}_*^T(\mathbf{x}, x_*)K^{-1}(\mathbf{x})\mathbf{k}_*^T(\mathbf{x}, x_*) + k_{**}(x_*) \tag{9.3}$$

Therefore, given a collection of historical observations of the system, the predictor f_s for each sensor s can be model via above process. In the running time, for each sensor s, given the observed reading, $(\tilde{y}_s, \tilde{\mathbf{y}}_{-s})$. The predictions $\hat{y}_s$ can be calculated from $f_s(\tilde{\mathbf{y}}_{-s})$, which are Gaussian and determined by (9.2) and (9.3). Then, we can use the predictions and further check whether the reading $\tilde{y}_s$ is anomalous based on the confidence interval of $\hat{y}_s$. When $\tilde{y}_s$ is not inside the τ_s confidence interval of $\hat{y}_s$, $\tilde{y}_s$ is flagged as anomalous. Formally, $\tilde{y}_s$ is suspicious, if $F(\tilde{y}_s) > \tau_s$, where $F(\cdot)$ is the corresponding conditional Gaussian CDF, or equivalently:

$$|\tilde{y}_s - \mu(\hat{y}_s)| > \sqrt{2}\sigma(\hat{y}_s)\text{erf}^{-1}(\tau_s) \tag{9.4}$$

9.4 Stealthy Attacks on Gaussian Process Anomaly Detection

Clearly, if the anomaly detector is deployed, the attacker can no longer attack with impunity. We now ask whether one can design attacks more carefully, so as to successfully and significantly change the sensor readings of target sensors while remaining stealthy—that is, avoiding detection by the GP-based anomaly detection system.

We formalize the problem of stealthy attacks on GP-based anomaly detection as an optimization problem. Our threat model is an attacker who can compromise up to H sensors, capturing the fact that integrity attacks (such as man-in-the-middle attack) on individual sensors can be technically quite challenging. However, once a sensor is successfully compromised, we allow the attacker to make arbitrary modifications to sensor values, within a basic normal value range (for example, pressure readings cannot be negative) $[y_s^{\min}, y_s^{\max}]$ for each sensor s that is easy for defenders to check (we can think of this range as imposing an additional simplistic anomaly detector independently for each sensor). Let S_c be the set of critical sensors; the attacker's goal is to maximize or minimize observed measurements of these sensors. We focus on minimizing here; maximization can be handled analogously since the objective is linear. The primary motivation for this goal is that sensor measurements, particularly when it comes to critical sensors, impact the control loop. This impact is often of the following form: the controller aims to keep the system state—say, pressure—in a safe range. If the sensor reading the pressure shows that pressure is too low, the controller will cause the pressure to rise to reach the safe range. Consequently, if the attacker can compromise the pressure sensor to read that the pressure is low, it effectively causes the controller to increase pressure, potentially above the safe range. The attacker's decision variables correspond to the change in sensor value for each sensor s, which we denote by $\Delta \tilde{y}_s$. We omit the subscript s when we refer to the *vector* of changes to all sensors and use $\Delta \tilde{\mathbf{y}}_{-s}$ to denote the vector of changes to measured values for all sensors other than s. We also denote by $\tilde{y}_s$ the original measured value of the sensor s, and, as above, $\tilde{\mathbf{y}}_{-s}$ is the vector of measurements of all sensors other than s, while $\tilde{\mathbf{y}}$ is the vector of all sensor measurements. Then the attacker's optimization problem can be represented as the following mathematical program:

$$\min_{s^* \in S_c} \min_{\Delta \tilde{\mathbf{y}}} \Delta \tilde{y}_{s^*} \tag{9.5a}$$

$$\text{s.t. :} \tag{9.5b}$$

$$\forall s, \quad |\tilde{y}_s + \Delta \tilde{y}_s - \mu(\hat{y}_s)| \leq \sqrt{2} \sigma(\hat{y}_s) \text{erf}^{-1}(\tau_s) \tag{9.5c}$$

$$\forall s, \quad \hat{y}_s = f_s(\tilde{\mathbf{y}}_{-s} + \Delta \tilde{\mathbf{y}}_{-s}) \tag{9.5d}$$

$$\forall s, \quad y_s^{\min} \leq \Delta \tilde{y}_s + \tilde{y}_s \leq y_s^{\max} \tag{9.5e}$$

$$||\Delta \tilde{\mathbf{y}}||_0 \leq H \tag{9.5f}$$

Here, Constraint 9.5c ensures that the attack is not flagged as anomalous; Constraint 9.5e requires that measured sensor values are feasible (e.g. physically realizable), and Constraint 9.5f ensures that at most H sensors are actually attacked. Observe that in this formulation, the first minimum over the critical sensors $s^* \in S_c$ can be eliminated as we can simply solve the optimization problem for each critical sensor independently and then choose the solution with the largest impact (i.e. the smallest optimal $\Delta \tilde{y}_{s^*}$). We therefore henceforth focus on the simpler objective $\min_{\Delta \tilde{\mathbf{y}}} \Delta \tilde{y}_{s^*}$ for some critical sensor s^*.

Even with the above simplification, the mathematical program (9.5) is clearly quite challenging. The major complication which qualitatively distinguishes this problem from the prior work on attacking regression-based anomaly detection (Ghafouri et al. 2018) is that the predictor $f_s(\cdot)$ is now a *random variable* (Gaussian Process). In Constraint 9.5d, this random variable is explicitly identified by $\hat{y}_s$, and the stealth constraint (Constraint 9.5c) uses its mean, $\mu(\hat{y}_s)$, as well as its variance $\sigma(\hat{y}_s)$. The most important consequence is that both the mean and the variance functions are potentially non-convex, as they depend on the potentially non-convex kernel function. This is further complicated by the non-convex l_0 constraint which ensures that at most H sensors are attacked.

In order to solve the non-convex optimization problem (9.5) we apply the feasible direction local search method, which is a variant of *feasible direction methods* developed by Zoutendijk (1960). At the high level, in each iteration of this iterative method, we start with a feasible solution computed in the previous iteration and find a descent direction. We then move in the feasible space slightly in the descent direction to obtain a new solution. In this move, we first ignore the l_0 Constraint 9.5f. Then, we find the closest feasible solution for which the l_0 constraint also holds.

Let D be the feasible domain formed by (9.5c)–(9.5e), and D' be the one formed by (9.5c)–(9.5f). We rewrite the feasible domain D as an abstract set of inequalities

$$D = \{g_j(\Delta \tilde{\mathbf{y}}) \geq 0, \quad j = 1, \ldots, J\}$$

where $g_j(\cdot)$ are differentiable functions.

Step 1: For each iteration, $\Delta \tilde{\mathbf{y}}^{(k)}$ is a feasible solution in the kth step. We use following LP to find a locally feasible descent direction of the optimization problem (9.5) in the kth step:

$$\max_{\mathbf{d}^{(k)}} \alpha \tag{9.6a}$$

$$\text{s.t. :} \tag{9.6b}$$

$$\mathbf{e}_{s^*} \cdot \mathbf{d}^{(k)} \leq -\alpha \tag{9.6c}$$

$$\forall j = 1, \ldots, J, g_j(\Delta \tilde{\mathbf{y}}^{(k)}) + \nabla g_j(\Delta \tilde{\mathbf{y}}^{(k)}) \cdot \mathbf{d}^{(k)} \geq \alpha, \tag{9.6d}$$

where $\mathbf{e}_s = (0, \ldots, 1, \ldots, 0)$ is the sth unit vector of the standard Euclidean basis and $\mathbf{d}^{(k)}$ is the $\underbrace{\qquad}_{s}$

direction we aim to find. When $\alpha \geq 0$, Objective 9.6a and Constraint 9.6c guarantee that the optimal solution of this LP yields a descent direction. When $\Delta \tilde{\mathbf{y}}^{(k)}$ reaches one of the edges of the feasible domain D, there is a small positive ϵ such that $\exists j, 0 < g_j(\Delta \tilde{\mathbf{y}}^{(k)}) < \epsilon$. In this case, the term $\nabla g_j(\Delta \tilde{\mathbf{y}}^{(k)}) \cdot \mathbf{d}^{(k)}$ in Constraint 9.6d forces the computed direction to follow the angle into the interior of the feasible domain. If $\alpha \leq 0$, the iteration terminates, since such a direction does not exist.

Step 2: Find the solutions along the feasible descent direction in the feasible domain D. Let $\Delta \tilde{\mathbf{y}}^{(k)} + \beta_{\max} \mathbf{d}^{(k)}$ be the point where some constraints are first activated such that $\beta_{\max}$ is the maximal length that the current iteration can climb along the direction $\mathbf{d}^{(k)}$. More precisely,

$$\beta_{\max} = \min \{\beta | g_j(\Delta \tilde{\mathbf{y}}^{(k)} + \beta \mathbf{d}^{(k)}) = 0, j = 1, \ldots, J \text{ and } \beta \geq 0\}$$

Thus, the solution set is the segment from $\Delta \tilde{\mathbf{y}}^{(k)}$ to $\Delta \tilde{\mathbf{y}}^{(k)} + \beta_{\max} \mathbf{d}^{(k)}$, represented by

$$\{\Delta \tilde{\mathbf{y}}^{(k)} + \beta \mathbf{d}^{(k)} | \beta \in [0, \beta_{\max}]\}$$

We calculate the solution set as follows. Starting from $\Delta \tilde{\mathbf{y}}^{(k)}$, we find $\beta_{\max}$ by incrementally climbing along $\mathbf{d}^{(k)}$ with the step length ϵ. Assume $\beta_{\max}$ is found at $p_{\max}$ step ($\beta_{\max} = p_{\max} \epsilon$), which forms the solution set

$$Z = \{\Delta \tilde{\mathbf{y}}^{(k)} + p\epsilon \mathbf{d}^{(k)} | p = 0, 1, 2, \ldots, p_{\max}\}.$$

Step 3: For the solutions found in Step 2, we project them onto the domain satisfying the l_0 constraint and recheck their feasibility. More precisely, for each solution, we keep its H elements (at most H sensors are attacked) having the maximum sum of their absolute values and set the rest to zero. Consequently, the resulting solution is the closet point to the original one with respect to l_1 distance. Then, if the underlying solution is not in the feasible domain D', it will be discarded. Let $S = \{r_H(\mathbf{z}) = \arg\min_{\mathbf{z}'} ||\mathbf{z} - \mathbf{z}'||_1 : \mathbf{z} \in Z\}$ represent the set of projected solutions obtained in Step 3 corresponding to every solution $\mathbf{z}$ in the solution set obtained in Step 2. Then, the optimal solution in the current iteration, $\Delta \tilde{\mathbf{y}}^{(k+1)}$, is the solution from this set S that most minimizes the objective value of the original problem.

9.5 The Resilient Anomaly Detection System

Next, we consider the issue of hardening the anomaly detection system against attacks of the kind we discussed in the previous section. Suppose a CPS has a collection of critical sensors, the readings of which are considered as the direct inputs of system controller (e.g. the temperature reading for thermostats). To protect the system behavior from being adversely affected by the adversarially compromised readings of critical sensors while maintaining a low false alarm rate, the defender can leverage two types of decisions. First, the defender typically has many choices for sensor placement (indeed, the problem of optimal and resilient sensor placement has received independent attention in the literature (Krause et al. 2008a, b; Laszka et al. 2015), and these choices can be made explicitly trading off resilience and false positive rate. Second, the defender can choose the confidence level for anomaly detectors that also trades off false alarm rate and resilience to attacks. Next, we describe each of these decisions in more detail.

Sensor selection: The defender considers the following sensor selection problem. Let S be the set of N possible sensor locations, and suppose that the defender can place at most L sensors. Let $\theta \in \{0, 1\}^N$ be a binary vector representing sensor placement decision with $\theta_i = 1$ if a sensor is placed at location i and $\theta_i = 0$ otherwise. We further assume that the critical sensors are always selected and $\theta_{i \in S_c} \equiv 1$ (otherwise, the controller cannot work due to the lack of the direct inputs). The budget constraint can then be represented as $\sum_{i=1}^{N} \theta_i \leq L$.

Confidence level: the second set of decisions the defender makes is in choosing the confidence levels τ for the anomaly detectors. This choice directly translates into the trade-off we wish to capture: a narrower confidence interval will lead to more false alarms but will also tighten the space within which the attacker can implement a stealthy attack, thereby limiting attack impact.

9.5.1 Resilient Anomaly Detection as a Stackelberg Game

We model the interaction between the defender, charged with designing a resilient sensor and anomaly detection system, and the attacker who can execute an integrity attack against this system, as a Stackelberg game. In this game, the defender moves first, choosing where sensors are placed among the finite set of locations S, as well as the confidence thresholds τ of the Gaussian process anomaly detection system. The attacker then computes an attack in response to these decisions. We seek a *Strong Stackelberg Equilibrium (SSE)* of this game, where the attacker breaks ties in the defender's favor, and the defender chooses an optimal set of decisions to commit to, accounting for the attacker's optimal response (given this tie-breaking rule). Next, we describe this game formally.

We first define the false alarm rate FA and assume that sensor measurements $\mathbf{y}$ before the attack follow the Gaussian process distribution. The false alarm at sensor s is an event that a normal measured sensor value is misclassified as an anomaly, indicated by the following threshold function:

$$\text{thr}_s(\mathbf{y}, \theta, \tau_s) = \begin{cases} 1 & |y_s - \mu(\hat{y}_s)| > \sqrt{2}\sigma(\hat{y}_s)\text{erf}^{-1}(\tau_s) \\ 0 & \text{otherwise} \end{cases}$$

where $\hat{y}_s = f_s(\mathbf{y}_{-s}, \theta)$. When $\text{thr}_s(\mathbf{y}, \theta, \tau_s) = 1$, the actual reading $\mathbf{y}$ is misclassified as an anomaly and sensor s fires a false alarm. We further define that the system fires a false alarm if *any* of its sensors have a false alarm. To capture this, let $\text{thr}(\mathbf{y}, \theta, \tau) = \bigvee_s \text{thr}_s(\mathbf{y}, \theta, \tau_s)$ where $s \in \{i | \theta_i = 1\}$. Therefore, we define the false alarm rate of the overall system as: $\text{FA}(\theta, \tau) = \mathbb{E}_\mathbf{y} \text{thr}(\mathbf{y}, \theta, \tau)$. In practice, we assume that we sample the actual readings of the system in some time interval between $t = 0$ and $t = T$ and denote by $\{\mathbf{y}^{(t)}\}_{t=0}^{T} = \{\mathbf{y}^{(0)}, \dots, \mathbf{y}^{(T)}\}$. Then, the false alarm rate is the average value of threshold function on the samples, which is

$$\text{FA}(\theta, \tau) = \frac{1}{1+T} \sum_{t=0}^{T} \text{thr}(\mathbf{y}^{(t)}, \theta, \tau). \tag{9.7}$$

Now, we construct the full game. The game $G = (\mathcal{A}, \mathcal{V}, \mathcal{I})$ consists of:

- Players, $\mathcal{I} = \{a, d\}$, where a is the attacker and d the defender.
- Joint action space of the two players, $\mathcal{A} = \mathcal{A}_d \times \mathcal{A}_a$. The defender aims to solve above decision-making problems and we denote defender's actions as (θ, τ). The defender's budget is limited by the upper bound of number of placed sensors, L, and the false alarm rate, $\text{FA}_{\max}$. Thus, $\mathcal{A}_d = \{(\theta, \tau) | \sum_i \theta_i \leq L, \text{FA}(\theta, \tau) \leq \text{FA}_{\max}\}$. On the other hand, we make the worst-case assumption that the attacker knows the actual sensor readings $\mathbf{y}$ at the time of the attack. Then, the attacker needs to decide the attack patterns $\Delta \mathbf{y}$ based on the value of $\mathbf{y}$.
- The utility functions $\mathcal{V}_a, \mathcal{V}_d$. The utility of the attacker is the expected attack impact on the sampled readings $\{\mathbf{y}^{(t)}\}_{t=0}^{T}$, which is $\mathcal{V}_a(\{\Delta \mathbf{y}^{(t)}\}_{t=0}^{T}, \theta, \tau) = \frac{1}{T+1}\sum_{t=0}^{T}\min_{s \in S_c}\Delta y_s^{(t)}$. We assume that our game is zero-sum.

Since our game is zero-sum, its Nash equilibrium and SSE are equivalent and can be found using the following maxmin program:

$$\max_{\theta, \tau} \min_{\{\Delta \mathbf{y}^{(t)}\}_{t=0}^{T}} \mathcal{V}_d(\{\Delta \mathbf{y}^{(t)}\}_{t=0}^{T}, (\theta, \tau)) \tag{9.8a}$$

s.t. :

$$\sum_{t=0}^{N} \theta_t \leq L, \quad \text{FA}(\theta, \tau) \leq \text{FA}_{\max} \tag{9.8b}$$

$$\{\mathbf{y}^{(t)}\}_{t=0}^{T} \quad \text{solve (9.5).} \tag{9.8c}$$

9.5.2 Computing an Approximately Optimal Defense

In this section, we present our approach to approximately solve the game defined above.

Let $\Delta \mathbf{y} = \mathcal{SA}(\mathbf{y}, \theta, \tau)$ be result of the stealthy attack given sensor selection decision θ and confidence levels τ, where $\mathbf{y}$ is the underlying sensor value vector corresponding to θ. The utility of the defender *given the attack* is then $\mathcal{V}_d = \frac{1}{T+1}\sum_{t=0,\ldots T}\min_{s \in S_c}\mathcal{SA}(\mathbf{y}^{(t)}, \theta, \tau)_s$. We call this defender's problem as the *master problem* and term the attacker's problem of computing a stealthy attack in response to the defender's decision, $\mathcal{SA}(\mathbf{y}, \theta, \tau)$, the *slave problem*.

Clearly, exhaustively exploring the defender's options is intractable. Our approach for solving the defender's problem proceeds in two steps: first, an algorithm for finding confidence level thresholds *given* sensor placement decisions θ, and second, an algorithm for sensor placement which computes confidence levels as a subroutine. We begin with the former.

Our approach for finding confidence level thresholds is to start at an initial configuration $\tau^{(0)}$ with some small values $\tau_s^{(0)}, \forall s \in \{i | \theta^{(0)} = 1\}$, which yields high resilience to attacks, but at the cost of a high false alarm rate. Then, we iteratively find the sensor with the highest false alarm rate and increase the associated thresholds τ_s until the false alarm rate drops below the upper bound, taking advantage of the fact that both the defender's utility and the false alarm rate decrease monotonically.

Given the above algorithm for computing thresholds given sensor placement choices θ, we now tackle the problem of resilient sensor selection. For this, we use the Best-First search algorithm. Specifically, given a particular sensor selection decision θ, there are two possible actions:

- *Forward selection*: if the number selected sensors is below the budget, we can add a new unselected sensor, and

- *Backward elimination*: we can eliminate a selected sensor, if such sensor does not belong to the set of critical sensors.

In the algorithm, "best" is defined with respect to the objective value of a subset of selected sensors (which is never above the budget L), and the search proceeds for a specified number of iterations.

9.6 Experiments

We evaluate our approach using the Tennessee-Eastman process control system (TE-PCS).

Tennessee-Eastman Process Control System: TE-PCS is a widely studied industrial process, which consists of five main process components: a reactor, a separator, a stripper, a compressor and a mixer (Downs and Vogel 1993). In this evaluation, we consider five critical sensors corresponding to safety constraints of TE-PCS (e.g. the upper bound of the pressure of the reaction container). We further assume that the system designer can select at most 15 sensors from 22 possibilities.

The model we use is the Simulink model of TE-PCS (Bathelt et al. 2015) with the implementation of the decentralized control law as proposed by Ricker (1996). For the anomaly detector, we first run simulation modeling the system operation for 72 hours and record the sensor measurements and control inputs. We take 225 timesteps periodically between 0 and 72 hours and record the all of sensor readings as the training set and train the collections of Gaussian process regression models under different sensor selection patterns. Then, from 20 to 60 hours, we record the sensor readings for every hour and get the testing set with 40 instances.

Stealthy Attacks: In the rest of the section, two baseline detectors are considered. Both baseline detectors have the same sensor selection pattern (xmeas(1) − xmeas(15) in Table 4; Downs and Vogel 1993) and fixed thresholds of confidence level (one is with 95% confidence level, called B1, and the other is with 99% confidence level, called B2). According to the definition of false alarm rate (9.7), B1 has 45% false alarm rate, while the false alarm rate of B2 is 20%. In this subsection, we perform our stealthy attacks on the baseline detectors and test the attack impact (i.e. the amount by which the attacker changes the measured sensor value).

First, we evaluate the effectiveness of our stealthy attacks on a single critical sensor (the reactor pressure). The results are presented in Figure 9.1a and b. Figure 9.1a shows the attack impact on both B1 and B2 with a fixed attack budget ($H = 9$, that is, the attacker can attack at most nine sensors), varying the confidence thresholds τ (95 vs. 99%). We can readily observe that the more conservative threshold (99%) in terms of limiting false positives is also significantly more susceptible to attacks compared to the more aggressive threshold. This exhibits the trade-off we explicitly consider in designing robust anomaly detectors. Figure 9.1b shows the results on B2 with different attack budgets ($H = 6$ and $H = 9$, i.e. the attacker can attack at most 6 and 9 sensors, respectively). As one would expect, a stronger attacker (one with a larger budget) tends to effect a significantly greater impact on the system.

Next, we consider stealthy attacks on the aforementioned critical sensor set. Figure 9.1c and d present the attack impact on both B1 and B2 among the five critical sensors. Figure 9.1e shows the attack impact on the system for different attack budgets. We again observe similar trends: both the attacker budget and conservativeness of the anomaly detector confidence thresholds have a considerable influence on the impact of the attack. Nevertheless, impact is considerable throughout.

Resilient Detector: To evaluate our resilient detection approach, we fix the attack budget to $H = 9$, the maximum number of iterations of the Best-First search algorithm to 10,

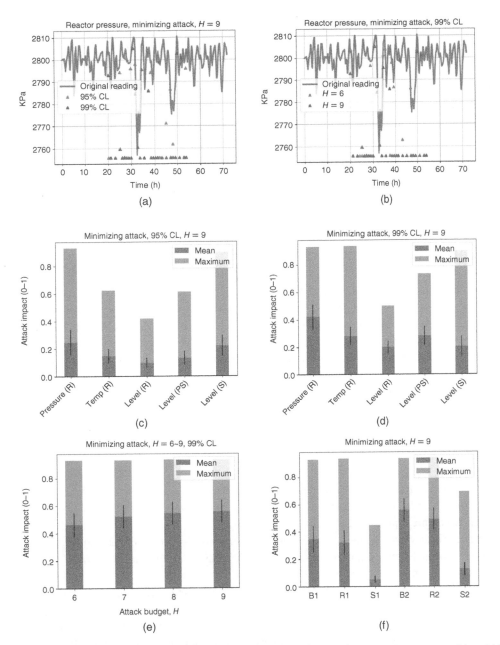

Figure 9.1 (a) and (b) show the effectiveness of our stealthy attacks on reactor pressure. (c) and (d) show the attack outcomes among critical sensors. (e) Illustrates the attack outcomes with different attack budgets. (f) Compares the performance between the resilient detectors and the baseline one. Attack impact is the difference between sensor measurements after and before the attack. Whiskers in the bar graphs indicate the 95% confidence level.

- R1 and R2: R1 and R2 are the resilient detectors with optimized thresholds of confidence level and the original set of selected sensors (i.e. the same set of sensors as used by the baseline detectors).
- S1 and S2: S1 and S2 are the resilient detectors with optimized sensor selection patterns and optimized thresholds of confidence level.

To compare the resilient detectors with the baseline detectors B1 and B2, we let the upper bound of the false alarm rate of R1 and S1 equal to the false alarm rate of B1 (45%). Similarly, both R2 and S2 have the same upper bound of false alarm rate which equals to the false alarm rate of B2 (20%).

Figure 9.1f presents the results, which are quite instructive. While we see the general improvement in robustness (reduction in attack impact) moving fom Bx to Rx to Sx (with $x \in \{1, 2\}$), it is clear that much of the final improvement is due to the careful choice about which sensors are used, with the tuning of thresholds given the original sensor placement providing only limited robustness. This observation serves as important motivation for jointly considering the problem of optimizing sensor and threshold selection in designing robust anomaly detectors, in contrast to prior work which considered each problem in isolation.

9.7 Conclusions

We considered a setting where a cyber-physical system (CPS) is monitored by a Gaussian process regression-based anomaly detection system. First, we presented a stealthy attack on this system that aims to maximize damage while appearing normal to the detector, presenting a novel approach to approximately solve this nontrivial optimization problem. Next, we presented a model of robust anomaly detection for CPS as a Stackelberg game and developed a novel approach for solving this game, in which the defender can decide which sensors to use as well as how to set anomaly thresholds. Our experiments demonstrate that the attack can be quite effective on a baseline system, but our Stackelberg game solution approach is significantly more robust.

References

Tansu Alpcan and Tamer Basar. A game theoretic approach to decision and analysis in network intrusion detection. In *IEEE Conference on Decision and Control, Maui, HI, USA, 9–12 December 2003*, volume 3, pp 2595–2600, 2003.

Somaieh Amraee, Abbas Vafaei, Kamal Jamshidi, and Peyman Adibi. Anomaly detection and localization in crowded scenes using connected component analysis. *Multimedia Tools and Applications*, 77(12):14767–14782, 2018.

Andreas Bathelt, N. Lawrence Ricker, and Mohieddine Jelali. Revision of the tennessee eastman process model. *IFAC-PapersOnLine*, 48(8):309–314, 2015.

Alvaro A. Cárdenas, Saurabh Amin, Zong-Syun Lin, Yu-Lun Huang, Chi-Yen Huang, and Shankar Sastry. Attacks against process control systems: Risk assessment, detection, and response. In *Proceedings of the 6th ACM Symposium on Information, Computer and Communications Security*, pp 355–366. ACM, 2011.

Kai-Wen Cheng, Yie-Tarng Chen, and Wen-Hsien Fang. Gaussian process regression-based video anomaly detection and localization with hierarchical feature representation. *IEEE Transactions on Image Processing*, 24(12): 5288–5301, 2015.

James J. Downs and Ernest F. Vogel. A plant-wide industrial process control problem. *Computers & Chemical Engineering*, 17(3):245–255, 1993.

Amin Ghafouri, Aron Laszka, Abhishek Dubey, and Xenofon Koutsoukos. Optimal detection of faulty traffic sensors used in route planning. In *Proceedings of the 2nd International Workshop on Science of Smart City Operations and Platforms Engineering*, pp 1–6. ACM, 2017.

Amin Ghafouri, Yevgeniy Vorobeychik, and Xenofon Koutsoukos. Adversarial regression for detecting attacks in cyber-physical systems. In *International Joint Conference on Artificial Intelligence, Stockholm, Sweden, 13–19 July 2018*, pp 3769–3775, 2018.

Siddharth Joshi and Stephen Boyd. Sensor selection via convex optimization. *IEEE Transactions on Signal Processing*, 57(2):451–462, 2009.

Khurum Nazir Junejo and Jonathan Goh. Behaviour-based attack detection and classification in cyber physical systems using machine learning. In *Proceedings of the 2nd ACM International Workshop on Cyber-Physical System Security*, pp 34–43. ACM, 2016.

A. Krause, B. McMahan, C. Guestrin, and A. Gupta. Robust submodular observation selection. *Journal of Machine Learning Research*, 9:2761–2801, 2008a.

A. Krause, A. Singh, and C. Guestrin. Near-optimal sensor placements in Gaussian processes: Theory, efficient algorithms and empirical studies. *Journal of Machine Learning Research*, 9:235–284, 2008b.

Aron Laszka, Yevgeniy Vorobeychik, and Xenofon Koutsoukos. Resilient observation selection in adversarial settings. In *IEEE Conference on Decision and Control, Osaka, Japan, 15–18 December 2015*, pp 7416–7421, 2015.

Roberto Leyva, Victor Sanchez, and Chang-Tsun Li. Video anomaly detection with compact feature sets for online performance. *IEEE Transactions on Image Processing*, 26(7):3463–3478, 2017.

Patric Nader, Paul Honeine, and Pierre Beauseroy. lp-norms in one-class classification for intrusion detection in scada systems. *IEEE Transactions on Industrial Informatics*, 10(4):2308–2317, 2014.

N. Lawrence Ricker. Decentralized control of the tennessee eastman challenge process. *Journal of Process Control*, 6(4):205–221, 1996.

Manohar Shamaiah, Siddhartha Banerjee, and Haris Vikalo. Greedy sensor selection: Leveraging submodularity. In *2010 49th IEEE Conference on Decision and Control (CDC)*, pp 2572–2577. IEEE, 2010.

Milind Tambe. *Security and Game Theory: Algorithms, Deployed Systems, Lessons Learned*. Cambridge University Press, 2011.

David I. Urbina, Jairo A. Giraldo, Alvaro A. Cardenas, Nils Ole Tippenhauer, Junia Valente, Mustafa Faisal, Justin Ruths, Richard Candell, and Henrik Sandberg. Limiting the impact of stealthy attacks on industrial control systems. In *Proceedings of the 2016 ACM SIGSAC Conference on Computer and Communications Security*, pp 1092–1105. ACM, 2016.

Yevgeniy Vorobeychik and Murat Kantarcioglu. *Adversarial Machine Learning*. Morgan & Claypool, 2018.

Christopher K. Williams and Carl Edward Rasmussen. Gaussian processes for machine learning. *The MIT Press*, 2(3):4, 2006.

Guus Zoutendijk. *Methods of Feasible Directions: A Study in Linear and Non-linear Programming*. Elsevier, 1960.

10

Moving Target Defense Games for Cyber Security: Theory and Applications

AbdelRahman Eldosouky and Shamik Sengupta

Computer Science and Engineering Department, University of Nevada, Reno, Reno, USA

10.1 Introduction

In the era of the Internet of Things (IoT), cyber systems have emerged as a revolutionary technology that transforms the way humans interact with engineered systems. Cyber systems can be seen as systems that provide cyber services and are connected to the Internet or to local networks. Cyber systems can cut across a variety of domains that range from large-scale sensing to small mobile applications. Examples of cyber systems include wireless networks, sensor networks, smart supply chains (Abdel-Basset et al. 2018), smart homes (Lin and Bergmann 2016), and smart cities (Jin et al. 2014). The applications which utilize cyber systems extend to cyber-physical systems such as unmanned aerial vehicles (French et al. 2019; Eldosouky et al. 2019), smart grids (Tosh and Sengupta 2016; Pazos-Revilla et al. 2019), intelligent transportation systems (Baza et al. 2019; Ferdowsi et al. 2019), and many other critical infrastructure systems (Eldosouky et al. 2015, 2017). Thus, it is needless to mention the importance of protecting such cyber systems given that they are vital to modern day cities and communities.

Owing to their essential role, protecting and securing cyber systems and maintaining their proper operation has recently attracted significant attention (Sanjab et al. 2017; Salama et al. 2019). This interest was supported by the unprecedented growth in the rate of cyber attacks in the recent years. Cyber systems, acting as defenders against cyber attacks, have been continuously developing, implementing, and testing new security tools to monitor, detect, and prevent cyber threats. Attackers, on the other hand, spare no effort in developing new attack techniques to explore and exploit the vulnerabilities within the cyber systems. To find these vulnerabilities, attackers need to collect data about the systems such as network parameters, security tools being used, hardware structure, systems' credentials, and encryption techniques being implemented (Zhu et al. 2011). Once attackers find such vulnerabilities, they can keep using the same vulnerability over and over to affect the system, until the defender is able to detect the attack and thwart it. This gives the attackers the upper hand as it is enough to discover one vulnerability to cause a lot of damage, while the defenders need to fully protect their systems. However, such attack-proof system are almost impossible to build and, thus, attackers can always find new vulnerabilities and perform zero-day attacks (Bilge and Dumitraş 2012).

However, facing such certain threats and attacks, the defenders are not helpless. One critical advantage, for the defenders over the attackers, is that they better know their systems' configurations and parameters. This advantage was the main enabler for a powerful defense technique known as *moving target defense (MTD)* (Okhravi et al. 2014). The premise behind MTD is the continuous randomization of a system's configuration (e.g. IP addresses and cryptographic keys) to

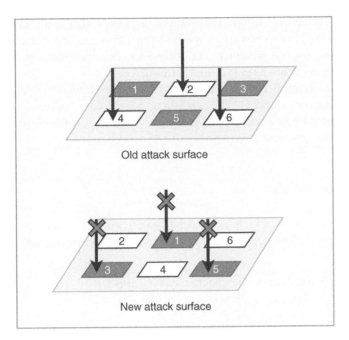

Figure 10.1 Shifting the attack surface. The upper figure shows the vulnerable points which the attacker uses to perform its attacks. The lower figure shows the effect of shuffling the system's configurations such that the old attacks can no longer affect the system.

increase the attacker's uncertainty and cost for performing a successful attack. In the context of MTD, this process is known as *attack surface shifting* (Manadhata and Wing 2011). The attack surface is defined as the subset of the system's resources or the set of ways in which an adversary can exploit to potentially attack the system and cause damage. Thus, the size of the attack surface is an indication to the level of a system's insecurity. The attack surface can be shifted by changing the system's configuration, thus, the system will appear differently to an attacker. The system, in this case, will appear as a *moving target* to the attacker, and, thus, harder to strike.

Figure 10.1 shows the effect of shifting the attack surface. The figure shows a number of points, e.g. ports or IP addresses that can be exploited by an attacker. In particular, it is assumed that an attacker can perform its attacks through the points 2, 4, and 6. Randomizing the locations of these points, the attacker will be faced with the points 1, 3, and 5 instead, which cannot be used to perform the same previous attacks. Hence, the attacker has to spend more effort to find the new points of the vulnerable points or to develop new attacks that can affect the newly faced points. This is similar to the case where the attacker is using a specific IP address to get access to the system, if the IP addresses are randomized, the attacker will need to start over to find a new vulnerable IP address. The goal of the defender, in this and similar examples, is to harden the attacker's mission by changing the attack surface. Note that, shifting the attack surface does not always fully cover the vulnerable points of the system as it can provide a partial protection based on the available configurations or resources.

While MTD is an effective defense mechanism to thwart cyber attacks, applying MTD techniques in a system, e.g. a cyber network, faces several challenges that range from optimizing the randomization to evaluating the costs and the benefits of the randomization (Jafarian et al. 2012; Jajodia et al. 2012; Zhu and Başar 2013; Casola et al. 2013; Xu et al. 2014; McDaniel et al. 2014; Zhuang et al. 2014; Marttinen et al. 2014; Carter et al. 2014). Therefore, there was a dire need to put MTD in theory to analytically study its benefits and side effects, rather than just changing a systems' configurations randomly. To this end, this chapter studies the theoretic models of MTD with the main focus on the game-theoretic models. Then, a promising type of games, that can best model many MTD problems, is studied and a case study is presented. The chapter also discusses the recent applications of MTD in literature and its applications in machine learning.

The rest of this chapter is organized as follows: Section 10.2 presents the main theoretic models of MTD. In Section 10.3, a special type of games to model MTD scenarios is introduced. A case study pertaining to applying MTD in wireless networks is analyzed in Section 10.4. MTD applications are discussed in Section 10.5. Finally, conclusions are drawn in Section 10.6.

10.2 Moving Target Defense Theory

With the continuous uphill battle against cyber attacks, MTD is seen as an essential defense mechanism for configurable systems. However, when applying MTD, defenders are faced by crucial questions such as:

- What are the possible configurations that will make the system *a moving target*?
- How long each configuration should be deployed, i.e. when to apply MTD?
- How to maximize the benefits from applying MTD, e.g. choosing from the available configurations?
- What are the costs associated with applying MTD?

To answer these and similar questions, many works, in literature, have introduced theoretic frameworks that allow the defenders (systems' managers) to effectively apply MTD (Manadhata et al. 2013; Okhravi et al. 2014; Xu et al. 2014; McDaniel et al. 2014; Zhuang et al. 2014; Han et al. 2014; Hong et al. 2018). First, in Okhravi et al. (2014), the authors defined the five dominant domains, in which MTD techniques can be applied, to be networks, platforms, runtime environments, software, and data. These domains capture the essential components of a cyber system which can be found in most systems. All these domains are considered to be dynamic, and, hence,they can be modified or altered to mislead the attacker, in a MTD scenario. In Okhravi et al. (2014), a number of attack phases were considered along with the possible MTD techniques to thwart them, in each of the aforementioned dominant domains. The authors also discussed the drawbacks of applying MTD and identified the possible research directions in all of these domains. Although the work in Okhravi et al. (2014) is not based on a theoretic analysis, it paved the way for a concrete understanding of the advantages and the weaknesses of applying MTD techniques. In Xu et al. (2014), the authors presented different methods to apply MTD techniques along with some methods to evaluate the MTD methodologies that are based on the difficulty to compromise the system and the probability of attack success. The authors, in Xu et al. (2014), proposed a three-layer model to evaluate the effectiveness of MTDs in software. These layers capture low-level contexts in separate programs, model damage propagation between different programs, and provide a user interface to expresses evaluation results.

The work in McDaniel et al. (2014) defined a higher class called system agility in which MTD can be considered as a subclass. The system agility is defined as any reasoned modification to a system or environment in response to a functional, performance, or security need. The authors discussed the parameters that control the timing and a system's ability to employ changes in order to improve its security. These parameters involve the deep understanding of how to respond to attacks, the available maneuvers, and when those maneuvers should be executed to improve the security. The work identified three main challenges to system agility in the form of questions to the system's managers. These questions are how to conceal the maneuvers from the adversary, how to sustain security across layers considering the dependency between the layers, and finally the cost of these maneuvers. In Zhuang et al. (2014), the authors proposed a foundation for defining the theory of MTD. They defined the key problems related to MTD such as selecting the next valid configuration of the system, configuration space (all valid re-configurations), and the timing problem to apply MTD. They also defined the MTD Entropy Hypothesis, which measures the effectiveness of an MTD system as a relation in the system's entropy.

The authors in Han et al. (2014) proposed a framework to facilitate the understanding and to quantify the effectiveness of MTD. The framework uses cyber epidemic dynamics to investigate the defender's characteristics when deploying an MTD technique. The authors addressed the important concept of undesired configurations that are necessary for the security purpose but not for the functionality. The framework covers the case of maximizing the time during which the system can be in an undesired configuration and the case of calculating the minimum time to stay in an underside configuration to limit the costs.

More recently, the authors in Hong et al. (2018) proposed a framework consisting of a set of dynamic security metrics to evaluate and compare the effectiveness of MTD techniques. The framework is based on temporal graphs and it can capture the dynamic changes made by MTD techniques. The graph is used to measure both the attack and the defense efforts defined as two separate metrics to evaluate an MTD technique. Finally, these metrics were used to compare different MTD techniques.

Finally, it is noteworthy that some other works in literature have studied the theoretical aspects of MTD without explicitly defining these techniques as MTD. For instance, in Manadhata and Wing (2011), an attack surface metric was introduced to study the effectiveness of a system's security. This work was extended in Manadhata (2013) to analytically formulate the changes in the attack surface, which correspond to applying MTD in recent works. The authors also discussed the trade-off between security and usability when changing a system's parameters.

While the models in Manadhata and Wing (2011), Manadhata (2013), Okhravi et al. (2014), Xu et al. (2014), McDaniel et al. (2014), Zhuang et al. (2014), Han et al. (2014), Hong et al. (2018) provide some theory to define MTD and they can help answer some of the fundamental questions related to MTD, they cannot fully capture the behavior of the defender and the attacker in MTD scenarios. Hence, more in-depth models were needed to study MTD scenarios under attacker–defender interactions. Such decision-making interactions are best modeled using game theory; therefore, we next cover some game-theoretic models for MTD.

10.2.1 Game Theory for MTD

As applying moving target defense involves an interaction between the defender and the attacker, game theory has recently attracted the attention as a suitable tool for implementing MTD techniques under such interactions. Thus, many works in literature have proposed different game theoretic models for MTD (Zhu and Başar 2013; Manadhata 2013; McDaniel et al. 2014; Carter et al. 2014; Clark et al. 2015; Eldosouky et al. 2016; Maleki et al. 2016).

In McDaniel et al. (2014), the authors proposed that tunable hierarchical games can be used to model MTD scenarios. The hierarchical structure can benefit to determine the level of risk at one game level based on the output ot the previous game level. At each game level, the players can apply the knowledge they learned from the previous level, and thus the game can be progressively refined at each level. However, the paper does not provide a game formulation nor the equilibrium analysis.

The work in Zhu and Başar (2013) proposed a zero-sum stochastic game to a feedback-driven multi-stage MTD. A feedback learning framework was used to implement MTD based on real-time data and the observations made by the system. The purpose of the learning algorithm, for the defender, is to monitor its current state and to update its randomized strategy based on its observation. An attacker has to launch a multi-stage attack starting from network scanning and packet sniffing to illegitimate authentication and service interruption. The game was considered a multi-layer as the attack is been carried out through multiple stages, and so the defense mechanism is developed at each layer of the system. Finally, the paper discussed the trade-off between security and usability, i.e. the utility gain from risk reduction and reconfiguration cost. The paper formulated the game analytically and equilibrium stochastic joint strategy was derived through studying the system's associated continuous-time dynamics.

The authors in Carter et al. (2014) analyzed a system in which the defender has a number of different platforms to run a critical application while the attacker has a set of predefined attacks against some of these platforms. The paper formulated a two-player, incomplete, leader–follower game in which the defender has imperfect information while the attacker has perfect information. The attacker was assumed to have perfect information as it can observe the moves of the defender. The defender, on the other hand, was considered to be unaware of the available moves to the attacker (e.g. which platforms are vulnerable), nor its chosen attack. In this game, the defender acts as the leader, where it can select a specific platform and the attacker is a follower that will respond by selecting an exploit. The paper considered two types of attackers, i.e. static and adaptive, and derived the optimal strategy selection against each type. However, the paper did not study the equilibrium of the defined games.

A two-player stochastic game was introduced In Manadhata (2013) to help the defender determine its optimal defense strategy based on the attacker's profile. In the formulated game, each player seeks to maximize the reward from selecting a specific configuration by considering the expected discounted payoff. The paper proposed to use the concept of stationary equilibrium strategies; however, no equilibrium analysis was discussed.

The authors in Maleki et al. (2016) formulated a framework to model MTD games using a two-player Markov modeling. The Markov property allows each player to calculate its optimal probability, of choosing actions, given the static actions of its opponents. Since the game uses a Markov model, the regular Nash equilibrium of game theory does not apply to this game.

In Clark et al. (2015), the authors formulated two game-theoretic models for the problem of IP address randomization using decoys. The first game is a static zero-sum game in which the defender decides on the states to implement, while the attacker attempts to detect the decoy. The game was analytically formulated and it was proven that the game has a unique Nash equilibrium that can be reached using fictitious play. The second game is a two-player Stackelberg game in which the defender chooses an IP address randomization policy, while the attacker chooses the rate at which it scans for decoys. The game was shown to have an equilibrium, under certain conditions. We note that, while these two games are well formulated, they are very specific to the application and cannot be generalized to other types of MTD scenarios.

The work in Eldosouky et al. (2016) introduced a general framework to model MTD games using a special type of stochastic games, i.e. *single controller stochastic games* (Filar and Raghavan 1984). Single-controller stochastic games allow one player to control the transitions between the game states. In MTD context, this property was mapped to the defender's actions of changing the system's parameters where each configuration represents a state. The paper formulated the game analytically and the stationary equilibrium strategies were derived.

Table 10.1 Comparison of the game-theoretic models for MTD.

	Game type	Game features	Game formulation	Equilibrium analysis
McDaniel et al. (2014)	Not defined	Hierarchical game	No	No
Zhu and Başar (2013)	Stochastic game	Zero-sum Multi stage	Yes	Yes
Carter et al. (2014)	Leader-follower game	Incomplete Imperfect information	Yes	No
Manadhata (2013)	Stochastic game	Stationary strategies	Yes	No
Maleki et al. (2016)	Markov game	—	Yes	No
Clark et al. (2015)	Dynamic game	Zero-sum	Yes	Yes
Eldosouky et al. (2016)	Stochastic game	Single controller Non-zero sum	Yes	Yes

We believe that single-controller stochastic games, used in Eldosouky et al. (2016), are the most suitable type of games for MTD problems for the following reasons:

- The system's possible configurations can each be modeled as a state in the stochastic model.
- The goal of MTD is to change the system's configurations, and, thus, the defender is the single controller of the configurations shift (states transition).
- Single-controller stochastic games were proven to always have an equilibrium solution, for discounted stochastic games.
- Numerical models such as Lemke and Howson (1964) can be used to compute the equilibrium in these games, as shown later.

Defenders in MTD scenarios can then use single-controller stochastic games to predict the outcomes (benefits and costs) of selecting each system's configuration (represented as a state in the game). We will discuss this types of games in details in Section 10.3 and provide its equilibrium analysis.

Finally, we summarize the game-theoretic models, discussed in this section, in Table 10.1, and we give a brief comparison between the different models based on their characteristics.

10.3 Single-Controller Stochastic Games for Moving Target Defense

10.3.1 Stochastic Games

Stochastic games, first introduced by Shapley (1953), are a type of dynamic games in which the game has a finite set of states. The game starts at one state, in which the players can select actions from their sets of finite actions. Based on the combination of the players' selected actions, the game will move to another state with some probability and the players will get a reward based on the selected actions and the current state.

Figure 10.2 shows an example of a simple stochastic game which has three states $S1$, $S2$, and $S3$. Let the game has two players p_1 and p_2. Assume the game starts at state $S1$. Each player will chose an action at $S1$, let a_1 be the action of player p_1 and a_2 be the action of player p_2. Then, at $S1$, each player will get a reward, that is, function $f(S1, a_1, a_2)$. Based on these actions, the game will move to another state with some probability. For instance, the game can move to state $S2$ with a probability q_1, move to $S3$ with a probability q_2, or remain in $S1$ with a probability $1 - q_1 - q_2$. In

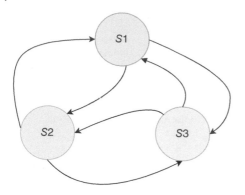

Figure 10.2 A simple stochastic game with three states.

each of these states, the same process is repeated, i.e. the players achieve a reward and the game moves to another state, with some probability.

In such dynamic games, the players need to calculate their accumulated utilities over all states. Summing the utilities over all states is done based on the type of the game. There are usually two types of stochastic (dynamic) games:

- Finite stochastic games: in finite games, the number of maximum transitions is defined, i.e. the number of time steps of the game. In this case, the players simply sum their utilities over all the possible time steps. The players, here, use the notion of expected utilities, which is the utility at each state multiplied by the probability of being at this state.
- Discounted stochastic games: in this type of games, the number of time steps is infinite, or not known in advance to the players. In this case, the expected utilities will sum infinity, otherwise a discount factor is used. A discount factor, β, is a positive value such that $0 < \beta < 1$. This value helps the players to consider future utilities in a discounted manner. Such that, the reward in the immediate time step is weighted more than future rewards. Thus, future time steps will have less impact on the player's decision at a specific time step.

To this end, a stochastic game Ξ can be formulated as a tuple $\langle \mathcal{N}, \mathcal{S}, \mathcal{A}, \mathcal{P}, \mathcal{U} \rangle$, where $\mathcal{N}$ is the set of players, $\mathcal{S}$ is the set of the finite game states, and $\mathcal{A}$ is the set of actions defined for each player at every state. Note that, the players can have different actions at different states. Finally, $\mathcal{P}$ is the set of transition probabilities between the states and $\mathcal{U}$ is the set of utilities each player will get for a given combination of actions and state.

Different approaches were proposed in literature (Dermed and Isbell 2009) to solve stochastic games. The solution, here, means finding the set of actions (strategy) that each player will follow in the game to maximize its utility given the other player's actions (strategy). As there can be multiple solutions available for each player, game theory usually looks to finding the *equilibrium solution*. A game is said to be at equilibrium, when no player can improve its utility by solely changing its actions. Thus, solving the game hereinafter means finding such equilibrium strategies for the players. Players' strategies are defined as the players' chosen actions at each time step and state.

In stochastic games, a special type of such strategies is known as *stationary strategies*, in which the player's actions depend only on the state not the time step (Nowak and Raghavan 1993). Such that, each player chooses the same action whenever the game reaches a specific state regardless the time step. If at each state, the player selects a specific action with probability $p = 1$, then the strategy is called a pure stationary strategy. If the player chooses between actions with some probabilities, then it is called a *mixed stationary strategy*. For discounted stochastic games, the existence of Nash equilibrium points in stationary strategies was proven to always exist (çois Mertens 2002).

10.3.2 Single-Controller Stochastic Games

Single-controller stochastic games (Parthasarathy and Raghavan 1981; Nowak and Raghavan 1993) are a special type of stochastic games in which the state transition is controlled by the actions of only one of the players, the controller. When a game reaches a specific state, both players take actions and they receive rewards based on these actions, like regular stochastic games. However, the game will move to a state, in the following time step, based on the controller's action only.

As discussed earlier, we believe that this type of games is most suitable for MTD problems in which the defender aims at randomizing the system's parameters, as the goal of MTD is to change system parameters in order to harden the attacker's mission. The states of the games can be defined as all the possible configurations of the system. The defender can then take actions to move between these configurations and each action selection will correspond to moving the game to another state. Thus, the defender can accurately compute its stationary strategy based on its certainty about the states transitions.

The problem of finding the Nash equilibrium in single-controller stochastic games was studied in literature. For instance, in Nowak and Raghavan (1993), the authors proposed a scheme that can find a Nash equilibrium point for discounted non-zero sum single-controller stochastic games. The key idea is to form a bimatrix game (one matrix for each player), such that the rows and the columns of each matrix represent the pure stationary strategies for each player. The elements of these matrices are calculated to represent the accumulated discounted utilities over all states (recursion) for every pair of strategies (row and column intersection). Then, any mixed strategy Nash equilibrium of this bimatrix game can be used to obtain a Nash equilibrium of the stochastic game. A numerical example is given next to highlight this concept.

10.3.2.1 Numerical Example

Consider a game that has two states. In an MTD scenario, the defender will have two actions representing selecting between each of these two states. Similarly, the attacker is assumed to have two actions of attacking profiles (both can be applied at either state).

A defender's strategy can then be given as $f = [a \ b]$, where a is the action of moving to state s_1 and b is the action of moving to state s_2.

The defender's and attacker's strategies' permutation will be given as:

$$F = \begin{bmatrix} 1 & 1 \\ 1 & 2 \\ 2 & 1 \\ 2 & 2 \end{bmatrix} G = \begin{bmatrix} 1 & 1 \\ 1 & 2 \\ 2 & 1 \\ 2 & 2 \end{bmatrix}$$

where F and G represent the defender's and the attacker's permutations, respectively.

Since each player has four different permutations, each of the formulated matrices (bimatrix) will be of size $4 \cdot 4$ representing all the possible combinations of the players' permutations. The elements of these matrices will be the accumulated utilities for each player resulting from starting at combination and considering all the future transitions.

Now suppose that the mixed Nash equilibrium for the bimatrix game, calculated from any numerical algorithm such as Lemke and Howson (1964), is given by:

$$x^* = \begin{bmatrix} 0.2 \\ 0.1 \\ 0.3 \\ 0.4 \end{bmatrix} y^* = \begin{bmatrix} 0.0 \\ 0.1 \\ 0.4 \\ 0.5 \end{bmatrix}$$

where each row in x^* and y^* represents the players' probabilities of selecting a strategy in F and G, respectively.

Finally, the stochastic game equilibrium strategies can be calculated by summing the probabilities of choosing each action over the corresponding states. For example, the defender can choose action 1 at state 1 twice in F with probabilities 0.2 and 0.1 (from x^*). When summed, the defender knows that when the game is at state 1, it will choose action 1 with a probability 0.3. Following the same approach, we can compute the full equilibrium matrices as follows:

$$E^* = \begin{bmatrix} 0.3 & 0.5 \\ 0.7 & 0.5 \end{bmatrix} H^* = \begin{bmatrix} 0.1 & 0.4 \\ 0.9 & 0.6 \end{bmatrix}$$

where E^* and H^* are the defender's and the attacker's equilibrium solutions, respectively, and that the rows of the matrices represent players' actions and the columns represent the game states.

10.4 A Case Study for Applying Single-Controller Stochastic Games in MTD[1]

Consider a wireless sensor network that consists of a BS and a number of wireless nodes. The network is deployed for sensing and collecting data about some phenomena in a given geographic area. Sensors will collect data and use multi-hop transmissions to forward this data to a central receiver or BS. The multiple access follows a slotted Aloha protocol. Time is divided into slots and the time slot size equals the time required to process and send one packet. Sensor nodes are synchronized with respect to time slots. We assume that nodes are continuously working and so every time slot there will be data that must be sent to the BS.

All packets sent over the network are assumed to be decrypted using a given encryption technique and a previously shared secret key. All the nodes in the system are pre-programmed with a number of encryption techniques along with a number of encryption keys per technique, as what is typically done in sensor networks (Casola et al. 2013). The BS chooses a specific encryption technique and key by sending a specific control signal over the network including the combination it wants to use. We note that the encryption technique and key sizes should be carefully selected in order not to consume a significant amount of energy when encrypting or decrypting packets. Increasing the key size will increase the amount of consumed energy particularly during the decryption (Lee et al. 2010). Since the BS is mostly receiving data, it spends more time decrypting packets rather than encrypting them and, thus, it will be highly affected by key size selection.

In our model, an eavesdropper is located in the communication field of the BS and it can listen to packets sent or received by the BS. As packets are encrypted, the attacker will seek to decrypt the packets it receives in order to get information. The attacker knows the encryption techniques used in the network and so it can try every possible key on the received packets until getting useful information. This technique is known as brute-force attack.

The idea of using multiple encryption techniques was introduced in Casola et al. (2013). However, in this work, each node individually selects one of these technique to encrypt transmitted packets. The receiving node can know the used technique by a specific field in the packet header. Large encryption keys were used which require a significant amount of power to be decrypted. Nonetheless, these large keys are highly unlikely to be revealed using a brute-force attack in a reasonable time. Here, we propose to use small encryption keys to save energy and, in conjunction with that, we enable the BS to change the encryption method in a way that reduces the chance that

---

1 The case study presented in this section is based on the work in Eldosouky et al. (2016).

the encryption key is revealed by the attacker. This is the main idea behind MTD. In this model, the encryption key represents the attack surface, and by changing the encryption method, the BS will make it harder for the eavesdropper to reveal the key and get the information from the system.

Naturally, the goals of the eavesdropper and the BS are not aligned. On the one hand, the BS wants to protect the data sent over the network by changing encryption method. On the other hand, the attacker wants to reveal the used key in order to get information. To understand the interactions between the defender and the attacker, one can use game theory to study their behavior in this MTD scenario. The problem is modeled as a game in which the attacker and the defender are the players. As the encryption method should be changed over time and depending on the attacker's actions, we must use a dynamic game.

We formulate a stochastic game Ξ described by the tuple $\langle \mathcal{N}, \mathcal{S}, \mathcal{A}, \mathcal{P}, \mathcal{U}, \beta \rangle$ where $\mathcal{N}$ is the set of the two players: the defender p_1, the BS, and the attacker p_2, the eavesdropper. $\mathcal{S}$ is the set of game states and $\mathcal{A}$ is the set of actions defined for each player at every state. $\mathcal{P}$ is the set of transition probabilities between states. $\mathcal{U}$ is the set of utilities each player will get for a given combination of actions and state. Finally, $0 < \beta < 1$ is a discount factor.

The defender can choose to use one of the N available encryption techniques or to use the current encryption technique with one of the M available encryption keys predefined for this technique. Each game state is well defined by the current encryption technique and key combination. Therefore, there will be $K = N \cdot M$ states, i.e. $\mathcal{S} = \{s_1, s_2, \ldots, s_K\}$. In each state $s \in \mathcal{S}$, each player has a set of actions $\mathcal{A}_i$. Let $\mathcal{A}_1 = \{a_1^1, a_2^1, \ldots, a_K^1\}$ be the defender's actions which represent the choice of a specific technique and key combination among the available K combinations. Let $\mathcal{A}_2 = \{a_1^2, \ldots, a_N^2\}$ be the action set of the attacker which represents the set of techniques that the attacker is trying to decrypt.

In each state $s \in \mathcal{S}$ and for each action pair in $\mathcal{A}_1 \times \mathcal{A}_2$, there is an outcome (payoff) for each player. This outcome depends on the current state and actions taken by both players in this state. This outcome is defined by player-specific utility functions in $\mathcal{U}$. For given actions $a^1 \in \mathcal{A}_1$ and $a^2 \in \mathcal{A}_2$, the defender's utility at state s_i is given by:

$$U_1(a^1, a^2, s_i) = R_1(a^2) + T_1(a^1, a^2, s_i) - P_1(s_i), \tag{10.1}$$

where R_1 is the reward gained from protecting a packet. This reward depends on the attacker's action as the defender will obtain a higher reward if the eavesdropper is attacking another encryption technique. P_1 is the power used to decrypt a packet and it depends on the technique (state). T_1 is the transition reward that the defender will gain from applying MTD and choosing a key-technique combination. This reward depends on the current system state, the defender's action taken at this state (which determines the next state), and attacker's action.

Similarly, the attacker's utility at state s_i for given actions $a^1 \in \mathcal{A}_1$ and $a^2 \in \mathcal{A}_2$ will be given by:

$$U_2(a^1, a^2, s_i) = R_2(a^1, a^2, s_i) - P_2(s_i), \tag{10.2}$$

where R_2 is the attacker's reward from examining the encryption keys for a given technique. Here, if the attacker can examine more keys, it will get closer to revealing the actual key. This reward depends on the attacker's action, current encryption technique (state), and defender's action. P_2 is the power used to decrypt a packet that depends also on the current technique such that states using the same encryption technique will require the same amount of power from the attacker.

Based on these rewards, the game is nonzero sum. Thus means, every player will try to maximize its reward and the sum of rewards is not zero. This stochastic game also exhibits an interesting property pertaining to the fact that the transition probabilities in $\mathcal{P}$ depend only on the actions of the defender. Therefore, next we use *single-controller stochastic games* to model and solve the formulated game.

10.4.1 Equilibrium Strategy Determination

In this game, the defender is considered the controller as it can select the actions which move the game from a state to another. The defender will also control when to apply the MTD, i.e. it will determine the duration between the time steps of the stochastic game. Assuming that the attacker has enough power, it can complete the brute-force attack in time t_i for $i = 1, 2, \ldots, N$ for each one of the encryption techniques. Then, the defender should choose the time step t to take the next action as follows:

$$t < \min{(t_i)}, \quad i = 1, 2, \ldots, N. \tag{10.3}$$

By doing this, the defender can make sure that it takes a timely action before the attacker succeeds in revealing one of the keys.

The accumulated utility of player i at state s will be

$$\Phi_i(f, g, s) = \sum_{t=1}^{\infty} \beta^{t-1} \cdot U_i(f(s_t), g(s_t), s_t), \tag{10.4}$$

where f and g are the strategies adopted by the defender and attacker, respectively. The strategy specifies a vector of actions to be chosen at each of the states, e.g. $f = [f(s_1), \ldots, f(s_K)]$ for all the K states. Actions $f(s_t)$ and $g(s_t)$ are the actions chosen at s_t, which is the state of the game at time t, according to strategies f, g. State $s_t \in \mathcal{S}$ is determined by the defender's action at time $t - 1$. The game is assumed to start at a specific state $s = s_1$. Note that the utility in (10.4) is always bounded at infinity due to the fact that $0 < \beta < 1$.

When designing the bimatrix, the defender needs to calculate the accumulated utility when choosing each pure strategy against all of the attacker's pure strategies. The defender, as a controller, can know the next state resulting from its actions, and, thus, it sums the utilities in all states using the discount factor β. Let X be the defender's accumulated utility matrix for all defender's pure strategies' permutations and all attacker's pure strategies' permutations. We let $F_{i\bullet} = [f_1, f_2, \ldots, f_{K^K}]$ be a matrix of all defender's pure strategies' permutation where each row represents actions in this strategy and similarly $G_{i\bullet} = [g_1, g_2, \ldots, g_{N^K}]$ the matrix of all attacker's pure strategies' permutation. Then each element $X_{i,j}$ of X will be given by:

$$X_{i,j} = \sum_{\mathcal{S}} \Phi_1(F_{i\bullet}, G_{j\bullet}, s), \quad \forall i, j, \tag{10.5}$$

where $i = 1, \ldots, K^K$ and $j = 1, \ldots, N^K$. The attacker can only calculate its payoffs at time $t = 1$, as the attacker cannot know in advance the actions taken at each state and hence the reward it will get in future. Similarly, let Y be the attacker's accumulated utility matrix, then each element $Y_{i,j}$ of Y will

$$Y_{i,j} = \sum_{\mathcal{S}} \Phi_2(F_{i\bullet}, G_{j\bullet}, s), \quad \forall i, j, \tag{10.6}$$

where i and j are the same as the defender's case, and $\Phi_2(F_{i\bullet}, G_{j\bullet}, s)$ is only evaluated at time $t = 1$.

The solution of the bimatrix could be obtained by algorithms such as Lemke and Howson (1964), which is proven to always terminate at a solution and hence finds a mixed Nash equilibrium of the bimatrix game. This solution is then used as in Nowak and Raghavan (1993) to find the equilibrium of the stochastic game. Let (x^*, y^*) be any mixed strategy Nash equilibrium point for the bimatrix game (X, Y). Each (x^*, y^*) is a vector of probabilities with which each player can choose each strategy in all the strategies permutations.

As each strategy represents the set of actions per all states, the equilibrium point to the stochastic game, i.e the probability of choosing each strategy, can be calculated as:

$$E^*_{i,j} = \sum_{l=1, i=F_{l,j}}^{K^K} x^*_l, \ i = 1, \ldots, K, j = 1, \ldots, K,$$

$$H^*_{i,j} = \sum_{l=1, i=G_{l,j}}^{K^K} y^*_l, \ i = 1, \ldots, N, j = 1, \ldots, K, \tag{10.7}$$

where $x^*_l \in x^*$ and $y^*_l \in y^*$ are the elements of x^*, y^* that represent strategies' probabilities. Each element $E^*_{i,j}$ of E^* and $H^*_{i,j}$ of H^* is the probability of taking action i in state j for the defender and the attacker, respectively. The summations in (10.7) give the probabilities of one action i which satisfies the condition. This is repeated for all values of i to get a column which is all actions' probabilities in one state. Different values of j give the rest of the states. E^* is a $K \cdot K$ matrix that gives the probability of each of the defender's K actions in each of the K states. Similarly, H^* is an $N \cdot K$ matrix that gives the probability of each of the attacker's N actions in each of the K states. These matrices are the *equilibrium strategies* for both players.

These probabilities specify the behavior of the game. The defender in each state will choose an action (selecting an encryption method) with some probability and so the game will move to another state (encryption method). Then, again in the new state, the defender chooses a new action and so on. Using this process, the defender will keep moving between encryption methods which effectively implements a highly randomized MTD.

Finally, the value (expected utility) of each player at equilibrium can be computed by applying the equilibrium strategies and finding the accumulated payoffs of both players. These expected utilities are calculated by following all the possible transitions due to defender's actions in each state. Let $v^*_i(s)$ be player's i value at state s:

$$v^*_i(s) = \Phi_i(E^*, H^*, s) \ s \in \mathcal{S}, \tag{10.8}$$

As the players get these values at equilibrium, both players will not have an incentive to deviate from these equilibrium strategies. The player who deviates will get a lower value when the other player uses its equilibrium strategy. This can be expressed as:

$$v^*_1(s) \geq \Phi_1(\hat{E}, H^*, s), \quad s \in \mathcal{S},$$
$$v^*_2(s) \geq \Phi_2(E^*, \hat{H}, s), \quad s \in \mathcal{S}, \tag{10.9}$$

for any $\hat{E}$ and $\hat{H}$ other than the equilibrium strategies.

10.4.2 Simulation Results and Analysis

For our simulations, we choose a system that uses two encryption techniques with two different keys per technique. Thus, the number of system states is four and the defender has four actions in each state. For the bimatrix, the attacker has $2^4 = 16$ different strategy permutations and the defender has $4^4 = 256$ different strategy permutations. The power values are set to 1 and 3 to pertain to the ratio between the power consumption in the two different encryption techniques. These values are the same for both players. We set R_1 and R_2 to be 10 and 5 depending on the opponent's actions. We choose these values to be higher than the power values in order for the utilities to be positive. The transition reward is set to 5 and 10 for switching to another state defined by another key or another technique, respectively.

Table 10.2 Attacker's and defender's equilibrium strategies.

	Attacker		Defender			
	a_1	a_2	a_1	a_2	a_3	a_4
s_1	0.7436	0.2564	0.0000	0.6622	0.1681	0.1697
s_2	0.7436	0.2564	0.4441	0.0195	0.1697	0.3667
s_3	0.3482	0.6518	0.4441	0.3667	0.0195	0.1697
s_4	0.3482	0.6518	0.4441	0.3667	0.1697	0.0195

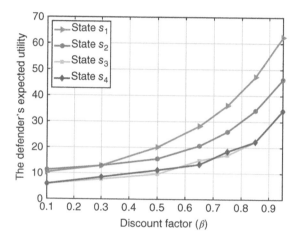

Figure 10.3 The defender's expected utility in each state against discount factor β. Source: From Eldosouky et al. (2016). © 2016, IEEE.

First, we run simulations when there is no transition cost, $q = 0$. The equilibrium strategies for both the attacker and defender are shown in Table 10.2. Note that actions a_1, a_2 represent the selection of two keys for the same encryption technique and actions a_3, a_4 represent two keys for another technique. Table 10.1 shows the probabilities over all actions for each player. These probabilities show how players should select actions in every state. For the defender, if it starts in state s_3, then it should move to state s_1 with the highest probability and move to state s_2 with a very similar probability. This is because the defender will change the technique and so gets a higher transition reward. We can see that the probability of moving to the same state is always very low and can reach 0 as in state s_1. The probability of moving to a state that has a similar encryption key is always less than that of moving to a state with different technique as the transition reward will be lower. For the attacker, the probability of attacking the same technique that is used in the current state is always higher than attacking any other technique.

In Figure 10.3, we show the effect of the discount factor on the defender's utility at equilibrium in every state. First, we can see that all utility values at all states increase as the discount factor increases. This is due to the fact that increasing the discount factor will make the defender care more about future rewards thus choosing the actions that will increase these future rewards. Figure 10.3 also shows that the defender's values at states 1 and 2 are higher than at states 3 and 4. This because states 1 and 2 adopt the first encryption technique which uses less power than the encryption technique used in states 3 and 4. The difference mainly arises in the first state before switching to other states and applying the discount factor. Clearly, changing the discount factor has a big effect on changing the equilibrium strategy, and, thus, the game will move between states with different probabilities resulting in a different accumulated reward.

Figure 10.4 Percentage increase in the defender's expected utility when using the equilibrium strategy and when using equal probabilities over actions. This is shown in each state as function of the discount factor β. Source: From Eldosouky et al. (2016). © 2016, IEEE.

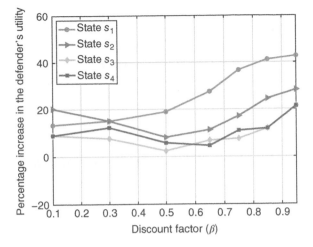

Figure 10.5 The defender's expected utility in each state for different power combinations.

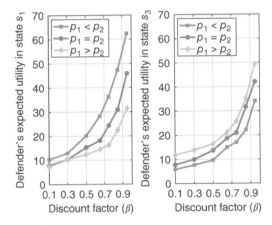

In Figure 10.4, we study the effect of applying the proposed MTD technique against the case when the defender decides to use equal probabilities over its actions in each state, i.e. all entries equal 0.25 as there are four actions per state. Figure 10.4 shows the percentage of increase in the defender's expected utility. We can see that the minimum increase is nonzero which means that the defender will not gain from deviating from equilibrium strategies. Moreover, at high discount factor values, i.e. $\beta > 0.75$, the percentage increase is higher than that at lower β values in all states. The percentage increase ranges from 5% to about 40% at $\beta = 0.75$ depending on the state, and it can reach values between 20% and above 40% at $\beta > 0.95$. This is due to the fact that, at higher β values, future state transitions have higher impact on calculating equilibrium strategies and the defender considers more state changes in the future. This makes equilibrium strategies differ more from equal probabilities. For other β values, the percentage increase depends on how different the equilibrium strategy from the equal allocation scheme.

In Figure 10.5, we study the effect of changing the power on the defender's expected utility at equilibrium. We study three cases, first when the power required for technique 1 is less than the power required for technique 2, similar to the previous experiments. Then, we study the cases in which they are equal and in which technique 1 requires more power than technique 2. Here, we set $\beta = 0.75$.

From Figure 10.5 (left), we can see that when the first technique's power is less than the second technique, the defender gets higher reward in state s_1. This is because in state s_1, the defender begins the game using technique 1 and, hence, it gets higher reward as a result of using less power. The same can be noticed from Figure 10.5 (right) where the defender gets higher reward in state s_3 when the second technique's power is less than the first technique. This happens as the in state s_3 the defender begins the game using technique 2 which uses less power, in this case. This lower power results in a higher reward. Similar interpretations can be made about different states and power combinations in Figure 10.5 which shows the effect of first state parameters on the expected utility.

10.5 Moving Target Defense Applications

Moving target defense techniques have been used in numerous works in literature to thwart cyber attacks. The methods of MTD were applied in different applications such as wireless sensor networks, Android applications, network configurations, and IP addresses randomization (Manadhata and Wing 2011; Jafarian et al. 2012; Casola et al. 2013; McDaniel et al. 2014; Marttinen et al. 2014; Eldosouky et al. 2016). Thus, the goal of this section is to highlight some applications that can utilize MTD without focusing on their theoretic models.

Utilizing MTD for wireless sensor networks security was addressed in Casola et al. (2013). The authors proposed two different reconfigurations for the sensor networks at different architectural layers. The first reconfiguration is applied at a proposed security layer through using a number of cryptographic techniques. Each node, in the network, was allowed to choose its encryption method per packet by adding a special identifier to the packet header identifying the chosen technique. The second reconfiguration is applied at the physical layer through changing the node's firmware. Changing the node's firmware was shown to incur higher costs as it requires the nodes to be off for some time.

Selective jamming attacks were addressed in Marttinen et al. (2014) where the authors used MTD to protect different network parts from being isolated by an attacker. The goal of the attacker in the formulated problem is to isolate a subset of the network by jamming the signals sent from this subnetwork. The work in Marttinen et al. (2014) also provided practical MTD solutions such as address flipping and random address assignment.

The use of software-defined networking in applying MTD was discussed in Jafarian et al. (2012). The authors defined an MTD technique that assigns virtual IPs to the hosts of the network beside their reals IPs. Software-defined networking was used to manage the IP translation between the real and the virtual IPs.

Other MTD applications include the work in McDaniel et al. (2014) in which the authors applied MTD to adaptively quarantine vulnerable code parts in Android applications. Finally, the work in Eldosouky et al. (2016) applied MTD in a wireless sensor network to change the encryption keys and techniques.

10.5.1 Internet of Things (IoT) Applications

The Internet of Things (IoT) is seen as a large-scale ecosystem that integrates a heterogeneous mix of devices, sensors, and wearable devices. Thus, it can be seen as a major enabler for a variety of smart services. However, the spread of IoT systems brought forward many security threats as the devices and services can be attacked remotely. Thus, it is no wonder that many IoT applications have adopted MTD as an effective mechanism to mitigate the effects of such attacks.

MTD, by nature, seeks to randomize the system's configuration. In the IoT, the system's configuration can essentially include encryption keys, network parameters, or IP addresses. For instance, the authors in Sherburne et al. (2014) applied MTD by frequently changing the IP address of the IoT devices to increase the security. Low-powered personal area networks were used to evaluate the MTD implementation in terms of security improvement and network latency. The authors have also proposed to add a new layer of security to the IoT to be responsible for applying the MTD approach.

In Zeitz et al. (2017), the authors proposed a micro-moving target IPv6 defense that utilizes different lightweight hash algorithms to limit the time the attackers may conduct reconnaissance on the systems. In Mahmood and Shila (2016), the authors proposed a framework that changes the code loaded and used on context-aware IoT devices. Each device was allowed to load the minimal piece of code required for operation and to request the download of any new code securely from the cloud. Thus, at any point of time, no node will have the full code which minimizes the risks in case a node is compromised.

Other applications that benefit from the IoT are smart health services (Baza et al. 2020). This integration of health services and the IoT is known as m-Health (Istepanian et al. 2011). M-Health IoT systems include a number of smart devices and sensors that can monitor a patient's medical conditions and send the collected data to remote physicians over the Internet. Since m-health systems carry sensitive information, preserving the privacy of the data in such system is a top priority for the defenders. In Eldosouky and Saad (2018), the authors proposed a framework that allows m-Health systems to dynamically change their cryptographic keys in order to increase the uncertainty on an attacker and secure the data. The smart health devices are allowed to update their keys locally to eliminate the risk of revealing the new keys in case they are being shared with a gateway. The framework also considered the cost associated with applying MTD.

10.5.2 Machine Learning Applications

With the widespread of the applications that use machine learning, MTD techniques found their way to secure such applications. In Sengupta et al. (2018), the authors proposed an MTD technique to increase the robustness of deep neural networks against adversarial attacks. The basic idea of using MTD is to randomize the selection of a neural network, among a group of trained networks, to use its classification output. Thus, the attacker will not be able to determine which trained network is being used in the classification process.

MTD techniques were also shown to help to defend against adversarial machine learning attacks. In Roy et al. (2019), the authors proposed an MTD approach to switch among machine learning algorithms to defend against adversarial attacks. The authors formulated the problem using game theory and studied both the rewards and the costs of applying MTD.

Finally, machine learning can help to maximize the benefits from applying MTD. In Colbaugh and Glass (2012), the authors proposed a framework to predict and counter the adversary attacks, using machine learning. The model also aims at limiting the extent to which attackers can learn about the defense strategies. In Eghtesad et al. (2019), the authors used reinforcement learning to maximize the benefits of applying MTD. The defender acts as an agent that maximizes the reward from applying multiple system's configurations. The paper introduced an adaptive model to apply MTD through finding the policy from a partially observable Markov decision process.

10.5.3 Prospective MTD Applications

The authors are considering applying MTD to extend a number of their recent works. For instance, the work in Das et al. (2020) addresses the problem of testing the integrated circuits against hardware trojans. The problem is formulated using game theory to determine the defender's (the tester's) equilibrium strategy due to the interactions with the attacker (the manufacturer). MTD can be applied in this case by randomly changing the testing patterns in order to overcome any deception from the attacker's side. The payoffs associated with each testing pattern need to be identified in order to effectively use MTD and minimize the reconfiguration costs.

Similarly, the work in Kotra et al. (2020) addresses the problem of privacy preserving in anonymized shared data. The problem is modeled using game theory to optimize the selection of k value in *k-anonymization* technique. MTD is seen a valuable extension to the model by allowing the sharing organizations to randomize over multiple anonymization techniques, instead of one. This approach can better preserve the privacy and harden the attacker's mission of identifying the sensitive attributes of the shared data.

Finally, the trade-off between using MTD and the usability of a system represents a promising research direction. Typically, in MTD, the system can switch between a number of configurations to increase its security. However, some of these configurations can cause the system to fail, if they are targeted by an attacker. We propose to study the resilience of such MTD implementations under certain attacks. The goal is to find the optimum reconfiguration for a system in case of a failure due to an attack. Choosing such a configuration is challenging as the system needs to minimize the reconfiguration cost plus ensuring longer stability on the new configuration in order to improve its resilience.

10.6 Conclusions

In this chapter, we have covered the basics of moving target defense and its applications in cyber security. MTD is a technique that seeks to randomize a system's components to reduce the likelihood of a successful attack. MTD alters the attack surface (points that could be exploited) in order to harden the attacker's mission. Examples of using MTD include changing IP addresses of nodes in a network so that they are not identified by the attacker. Then, we have discussed some theoretical MTD models, from the literature. In particular, we have focused on the game-theoretic models as they can better model the interactions between attackers and defenders in MTD scenarios. A brief comparison between the discussed game-theoretic models was also presented.

Next, we have focused on a type of games knows as single-controller stochastic games which we believe can best model MTD scenarios. The proposed game model, along with its equilibrium analysis, has been covered in details. In a case study, we have examined the use of MTD in a wireless network security problem. In particular, we have formulated a nonzero sum stochastic game between the attacker and the defender. The game follows the framework of single-controller stochastic games in which the states' transition is determined by defender's actions. This feature enables the defender to change the system's parameters before the attacker can reveal them. Then, we have studied the equilibrium of the proposed game and used simulation results to highlight the importance of the proposed game.

Finally, we presented some recent works that utilize MTD in different applications. In particular, we have focused on the use of MTD in IoT applications, especially health-related applications. Machine learning applications have also been covered as they represent a promising research direction for MTD given the widespread of applications that use machine learning.

References

M. Abdel-Basset, G. Manogaran, and M. Mohamed. Internet of things (iot) and its impact on supply chain: A framework for building smart, secure and efficient systems. *Future Generation Computer Systems*, 86:614–628, 2018.

M. Baza, M. Nabil, N. Lasla, K. Fidan, M. Mahmoud, and M. Abdallah. Blockchain-based firmware update scheme tailored for autonomous vehicles. In *IEEE Wireless Communications and Networking Conference (WCNC), Marrakesh, Morocco, April 2019*, pp 1–7, 2019.

M. Baza, A. Salazar, M. Mahmoud, M. Abdallah, and K. Akkaya. On sharing models instead of data using mimic learning for smart health applications. In *IEEE International Conference on Informatics, IoT, and Enabling Technologies (ICIoT), Doha, Qatar, February 2020*, pp 231–236, 2020.

L. Bilge and T. Dumitraş. Before we knew it: An empirical study of zero-day attacks in the real world. In *Proceedings of the ACM Conference on Computer and Communications Security, Raleigh, NC, USA, October 2012*, pp 833–844, 2012.

K. M. Carter, J. F. Riordan, and H. Okhravi. A game theoretic approach to strategy determination for dynamic platform defenses. In *Proceedings of the First ACM Workshop on Moving Target Defense, Scottsdale, AZ*, pp 21–30, November 2014.

V. Casola, A. D. Benedictis, and M. Albanese. A moving target defense approach for protecting resource-constrained distributed devices. In *IEEE 14th International Conference on Information Reuse and Integration (IRI)*, San Francisco, CA, pp 22–29, August 2013.

A. Clark, K. Sun, L. Bushnell, and R. Poovendran. A game-theoretic approach to IP address randomization in decoy-based cyber defense. In *International Conference on Decision and Game Theory for Security*, pp 3–21. Springer, 2015.

J.-F. çois Mertens. Stochastic games. *Handbook of Game Theory with Economic Applications*, 3:1809–1832, 2002.

R. Colbaugh and K. Glass, Predictive moving target defense. Sandia National Lab.(SNL-NM), Albuquerque, NM (United States). Tech. report, 2012.

T. Das, A. Eldosouky, and S. Sengupta. Think smart, play dumb: Analyzing deception in hardware trojan detection using game theory. In *Proceedings of the International Conference on Cyber Security and Protection of Digital Services (Cyber Security)*, Dublin, Ireland, June 2020.

L. M. Dermed and C. L. Isbell. Solving stochastic games. In *Advances in Neural Information Processing Systems, Vancouver and Whistler, BC, Canada, December 2009*, pp 1186–1194, 2009.

T. Eghtesad, Y. Vorobeychik, and A. Laszka. Deep reinforcement learning based adaptive moving target defense. *arXiv preprint arXiv:1911.11972*, 2019.

A. Eldosouky, A. Ferdowsi, and W. Saad. Drones in distress: A game-theoretic countermeasure for protecting uavs against GPS spoofing. *IEEE Internet of Things Journal*, 7(4):2840–2854, 2019.

A. Eldosouky and W. Saad. On the cybersecurity of m-health iot systems with led bitslice implementation. In *IEEE International Conference on Consumer Electronics (ICCE), Las Vegas, NV, USA, January 2018*, pp 1–6, 2018.

A. Eldosouky, W. Saad, C. Kamhoua, and K. Kwiat. Contract-theoretic resource allocation for critical infrastructure protection. In *IEEE Global Communications Conference (GLOBECOM), San Diego, CA, USA, December 2015*, pp 1–6, 2015.

A. Eldosouky, W. Saad, and N. Mandayam. Resilient critical infrastructure: Bayesian network analysis and contract-based optimization. *arXiv preprint arXiv:1709.00303*, 2017.

A. Eldosouky, W. Saad, and D. Niyato. Single controller stochastic games for optimized moving target defense. In *IEEE International Conference on Communications (ICC)*, pp 1–6, 2016.

A. Ferdowsi, U. Challita, and W. Saad. Deep learning for reliable mobile edge analytics in intelligent transportation systems: An overview. *IEEE Vehicular Technology Magazine*, 14(1):62–70, 2019.

J. A. Filar and T. Raghavan. A matrix game solution of the single-controller stochastic game. *Mathematics of Operations Research*, 9(3):356–362, 1984.

A. French, M. Mozaffari, A. Eldosouky, and W. Saad. Environment-aware deployment of wireless drones base stations with google earth simulator. In *IEEE International Conference on Pervasive Computing and Communications Workshops (PerCom Workshops), Kyoto, Japan, March 2019*, pp 868–873, 2019.

Y. Han, W. Lu, and S. Xu. Characterizing the power of moving target defense via cyber epidemic dynamics. In *Proceedings of the Symposium and Bootcamp on the Science of Security, Raleigh, NC, USA, April 2014*, pp. 1–12. ACM, 2014.

J. B. Hong, S. Y. Enoch, D. S. Kim, A. Nhlabatsi, N. Fetais, and K. M. Khan. Dynamic security metrics for measuring the effectiveness of moving target defense techniques. *Computers & Security*, 79:33–52, 2018.

R. S. Istepanian, A. Sungoor, A. Faisal, and N. Philip. Internet of m-health things. In *IET Seminar on Assisted Living 2011, London, UK, April 2011*, pp 1–3, 2011.

J. H. Jafarian, E. Al-Shaer, and Q. Duan. Openflow random host mutation: Transparent moving target defense using software defined networking. In *Proceedings of the First Workshop on Hot Topics in Software Defined Networks*, Helsinki, Finland, pp. 127–132, August 2012.

S. Jajodia, A. K. Ghosh, V. Subrahmanian, V. Swarup, C. Wang, and X. S. Wang. *Moving Target Defense II: Application of Game Theory and Adversarial Modeling*, volume 100. Springer, 2012.

J. Jin, J. Gubbi, S. Marusic, and M. Palaniswami. An information framework for creating a smart city through internet of things. *IEEE Internet of Things journal*, 1(2):112–121, 2014.

A. Kotra, A. Eldosouky, and S. Sengupta. Every anonymization begins with k: A game-theoretic approach for optimized k selection in k-anonymization. In *Proceedings of the Sixth International Conference on Advances in Computing & Communication Engineering ICACCE*, Las Vegas, NV, USA, June 2020.

J. Lee, K. Kapitanova, and S. H. Son. The price of security in wireless sensor networks. *Computer Networks*, 54(17):2967–2978, 2010.

C. E. Lemke and J. T. Howson, Jr. Equilibrium points of bimatrix games. *Journal of the Society for Industrial & Applied Mathematics*, 12(2):413–423, 1964.

H. Lin and N. Bergmann. Iot privacy and security challenges for smart home environments. *Information*, 7(3):44, 2016.

K. Mahmood and D. M. Shila. Moving target defense for internet of things using context aware code partitioning and code diversification. In *IEEE 3rd World Forum on Internet of Things (WF-IoT), Reston, VA, USA, December 2016*, pp 329–330, 2016.

H. Maleki, S. Valizadeh, W. Koch, A. Bestavros, and M. van Dijk. Markov modeling of moving target defense games. In *Proceedings of the ACM Workshop on Moving Target Defense, Vienna, Austria, October 2016*, pp 81–92, 2016.

P. K. Manadhata. Game theoretic approaches to attack surface shifting. In *Moving Target Defense II*, pp 1–13. Springer, 2013.

P. K. Manadhata and J. M. Wing. An attack surface metric. *IEEE Transactions on Software Engineering*, 37(3):371–386, 2011.

A. Marttinen, A. M. Wyglinski, and R. Jantti. Moving-target defense mechanisms against source-selective jamming attacks in tactical cognitive radio manets. In *IEEE Conference on Communications and Network Security (CNS)*, San Francisco, CA, pp 14–20, October 2014.

P. McDaniel, T. Jaeger, T. F. L. Porta, N. Papernot, R. J. Walls, A. Kott, L. Marvel, A. Swami, P. Mohapatra, S. V. Krishnamurthy *et al.* Security and science of agility. In *Proceedings of the First ACM Workshop on Moving Target Defense*, Scottsdale, AZ, pp 13–19, November 2014.

A. Nowak and T. Raghavan. A finite step algorithm via a bimatrix game to a single controller non-zero sum stochastic game. *Mathematical Programming*, 59(1–3):249–259, 1993.

H. Okhravi, T. Hobson, D. Bigelow, and W. Streilein. Finding focus in the blur of moving-target techniques. *IEEE Security & Privacy*, 12(2):16–26, 2014.

T. Parthasarathy and T. Raghavan. An orderfield property for stochastic games when one player controls transition probabilities. *Journal of Optimization Theory and Applications*, 33(3):375–392, 1981.

M. Pazos-Revilla, M. Baza, M. Nabil, A. Sherif, M. Mahmoud, and W. Alasmary, Privacy-preserving and collusion-resistant charging coordination schemes for smart grid. *arXiv preprint arXiv:1905.04666*, 2019.

A. Roy, A. Chhabra, C. A. Kamhoua, and P. Mohapatra. A moving target defense against adversarial machine learning. In *Proceedings of the 4th ACM/IEEE Symposium on Edge Computing, Arlington, VA, USA, November 2019*, pp 383–388, 2019.

A. Salama, Eldosouky Mahmoud Abdelrahman. Security of critical cyber-physical systems: Fundamentals and optimization. Ph.D. dissertation, Virginia Tech, 2019.

A. Sanjab, W. Saad, and T. Başar. Prospect theory for enhanced cyber-physical security of drone delivery systems: A network interdiction game. In *IEEE International Conference on Communications (ICC), Paris, France, May 2017*, pp 1–6, 2017.

S. Sengupta, T. Chakraborti, and S. Kambhampati. Mtdeep: boosting the security of deep neural nets against adversarial attacks with moving target defense. In *Workshops at the Thirty-Second AAAI Conference on Artificial Intelligence, New Orleans, LA, USA, February 2018*, 2018.

L. S. Shapley. Stochastic games. *Proceedings of the National Academy of Sciences*, 39(10):1095–1100, 1953.

M. Sherburne, R. Marchany, and J. Tront. Implementing moving target ipv6 defense to secure 6lowpan in the internet of things and smart grid. In *Proceedings of the 9th Annual Cyber and Information Security Research Conference, Oak Ridge, TN, USA, April 2014*, pp 37–40. ACM, 2014.

D. K. Tosh and S. Sengupta. An adaptive game theoretic framework for self-coexistence among cognitive radio enabled smart grid networks. In J. D. Matyjas, S. Kumar, and F. Hu, editors. *Spectrum Sharing in Wireless Networks: Fairness, Efficiency, and Security*, pp. 349–367. CRC Press, 2016.

J. Xu, P. Guo, M. Zhao, R. F. Erbacher, M. Zhu, and P. Liu. Comparing different moving target defense techniques. In *Proceedings of the First ACM Workshop on Moving Target Defense*, Scottsdale, AZ, pp 97–107, November 2014.

K. Zeitz, M. Cantrell, R. Marchany, and J. Tront. Designing a micro-moving target ipv6 defense for the internet of things. In *IEEE/ACM Second International Conference on Internet-of-Things Design and Implementation (IoTDI), Pittsburgh, PA, USA, April 2017*, pp. 179–184, 2017.

Q. Zhu and T. Başar. Game-theoretic approach to feedback-driven multi-stage moving target defense. In *Decision and Game Theory for Security*, pp 246–263. Springer, 2013.

B. Zhu, A. Joseph, and S. Sastry. A taxonomy of cyber attacks on scada systems. In *International Conference on Internet of Things and 4th International Conference on Cyber, Physical and Social Computing, Dalian, China, October 2011*, pp 380–388. IEEE, 2011.

R. Zhuang, S. A. DeLoach, and X. Ou. Towards a theory of moving target defense. In *Proceedings of the First ACM Workshop on Moving Target Defense*, Scottsdale, AZ, pp 31–40, November 2014.

11

Continuous Authentication Security Games

Serkan Sarıtaş[1], Ezzeldin Shereen[1], Henrik Sandberg[2], and György Dán[1]

[1]*Division of Network and Systems Engineering, KTH Royal Institute of Technology, Stockholm, Sweden*
[2]*Division of Decision and Control Systems, KTH Royal Institute of Technology, Stockholm, Sweden*

11.1 Introduction

Cyber-attacks against online payment systems often involve some form of online identity theft or session hijacking. The incidence of such attacks is expected to rise in the future, on the one hand because tools for performing identity theft and session hijacking attacks are becoming widely available, on the other hand due to the proliferation of bringing your own device (BYOD) policies among enterprises and even critical infrastructure operators. As a consequence, in the near future identity theft and session hijacking could become an important attack vector in compromising not only online transactions but also critical infrastructures. Thus, it is essential to mitigate these attacks in order to mitigate advanced persistent threat (APT).

Continuous authentication schemes, as an extension to traditional identity and access management (IAM) strategies, are emerging as a promising technology for detecting identity theft and session hijacking attacks, as described in the following. IAM strategies are designed to protect data security and privacy starting with user authentication and authorization. Conventional authentication schemes rely on three kinds of factors to authenticate and identify a user,

- something the user knows; e.g. login credentials: a password, PIN code, or other secret fact;
- something the user has; e.g. verified device: a smart card or token;
- something the user is; i.e. biometrics; e.g. physiological biometrics (a face, fingerprint, or retinal pattern) or behavioral biometrics (focuses not only on what the user normally does but also on how she does it).

The first factor is simple to implement; however, it provides limited level of security. The second factor requires the user to physically carry an item, which may be inconvenient and the item could be stolen. The third factor is the strongest one, since the physiological and behavioral characteristics are unique for each individual and unlike passwords and tokens, biometrics cannot easily be lost or stolen.

Nonetheless, even if the initial authentication (utilizing single or multiple criteria listed above) is performed relying on multiple factors, many events can subsequently take place during a session (i.e. anomalies during the session are not considered), and thus, the integrity and authenticity of the

Part of this chapter was presented at the 10th Conference on Decision and Game Theory for Security (GameSec 2019) in Stockholm, Sweden, on 30 October–1 November 2019 (Sarıtaş et al. 2019). This work was partly funded by the Swedish Civil Contingencies Agency (MSB) through the CERCES project and by the SSF CLAS project.

session gradually erode over time. The only way to maintain authenticity is to introduce additional authentication factors repeatedly or continuously, i.e. making authentication a continuous process instead of a one-time event. Such an approach can maintain a high level of security, but at the cost of reduced user experience, i.e. there is a trade-off between the user experience and the level of security. Therefore, there is a need for alternative authentication approaches that can be executed in the background in a way that does not significantly affect the user experience.

Continuous authentication solutions introduced in recent years aim at addressing this issue. They passively, transparently, and continuously monitor the authenticity of users in real-time and are typically based on behavioral or physiological characteristics that are unique to each individual. However, the development and deployment of continuous authentication solutions come with some challenges (Ayeswarya and Norman 2019). Complexity is relatively high when compared with the traditional authentication schemes. Moreover, continuous authentication may be considered as invasive, since people may not be comfortable being passively and continuously monitored. In addition to these, the biometrics may change over time (due to injury, soreness, fatigue, etc.), and they cannot be shared during emergency or critical situations.

Continuous authentication typically relies on a machine learning model trained based on recorded behavioral biometrics of the user, e.g. movement patterns of pointing devices, keystroke patterns,[1] transaction characteristics, which is used for detecting anomalous user input in real-time (Dee et al. 2019; Deutschmann et al. 2013). User input that is classified as anomalous is typically rejected and may result in the need for user re-authentication. Due to the real-time execution of continuous authentication schemes, the authenticity of the user is determined continuously, and thus, the amount of potential damage that might be done to the system by a potential attacker can be limited. Unfortunately, continuous authentication inevitably involves false positives, i.e. classification of legitimate user input as illegitimate, which is detrimental to the usability of the system, and thus, it should be kept low. A lower false-positive rate at the same time implies a higher false-negative rate, i.e. lower probability of detection. Finding the optimal parameters for continuous authentication is thus a challenging problem, especially if continuous authentication is used in combination with other solutions for incident detection, such as intrusion detection services (IDS).

We address this problem in this chapter. We consider a system consisting of a defender (information technology (IT) manager, operator) who manages an IDS and continuous authentication for mitigating APT, a user who interacts with operator's multiple resources, and an attacker who tries to execute a rogue command on the system's resources. We formulate the interaction between the attacker and the defender as a dynamic stochastic leader-follower game. We then characterize the optimal attack strategy, and show that it has a threshold structure. Regarding the optimization of the defender's parameters, we provide a characterization of the impact of the parameters of continuous authentication and of the IDS on the utilities of the attacker and the defender. We provide numerical results to illustrate the attacker strategy and the impact of the defender's strategy on the attacker's and defender's expected utilities.

The rest of the chapter is organized as follows. After presenting the related literature in Section 11.2, the problem formulation is provided in Section 11.3. The optimal attack and defense strategies are discussed in Sections 11.4 and 11.5, respectively. In Section 11.6, we provide numerical examples and comparative analyses. Section 11.7 concludes the chapter.

---

1 The use of keystroke patterns as a method of verification or identification is not new. During World War II, the telegraph operators were able to identify each other by their Morse code typing pattern, which is known as *The Fist of the Sender*.

11.2 Background and Related Work

Continuous authentication is increasingly popular in industry and academia alike. Deutschmann et al. (2013) demonstrated the use of keystroke dynamics, mouse movements, and application usage for continuously authenticating users on workstations. Their results showed that keystroke dynamics proved to be the best indicator of user identity. Continuous authentication for smartphone users and users of other wearable electronic devices was considered recently in Dee et al. (2019), based on behavioral information of touch gestures like pressure, location, and timing. Sitová et al. (2016) demonstrated the potential of using other behavioral information like hand movement, orientation and grasp (HMOG) information for continuously authenticating mobile users. Similarly, Peng et al. (2017) demonstrated continuous authentication for wearable glasses, such as Google glass. Ferro et al. (2009) showed that car owners or office workers could be continuously authenticated by sensors on their seats. Similar ideas have been proposed for military and battlefield applications for continuously authenticating soldiers by their weapons and suits (Castiglione et al. 2017).

There is also a significant body of previous works that used game-theoretic approaches for modeling various problems related to authentication in network security. Cooperative authentication in Mobile Ad-hoc Networks (MANETs) was considered in Yunchuan et al. (2014), where many selfish mobile nodes need to cooperate in authenticating messages from other nodes while not sacrificing their location privacy. In Xiao et al. (2016), a game was used to model the process of physical layer authentication in wireless networks, where the defender adjusts its detection threshold in hypothesis testing while the attacker adjusts how often it attacks. The problem of secret (password) picking and guessing was modeled in Khouzani et al. (2015) as a game between a defender (the picker) and an attacker (the guesser). Slightly similar to our work is Yang et al. (2018), where the authors consider a game between monitoring nodes and monitored nodes in wireless sensor networks, where the monitoring nodes decide the duration of behavioral monitoring, and the monitored nodes decide when to cooperate and when not to cooperate. Nonetheless, to the best of our knowledge, our work is the first to propose a game-theoretic approach for secure risk management considering continuous authentication.

11.3 Problem Formulation

As illustrated in Figure 11.1, we consider a system that consists of an organization that maintains a corporate network (e.g. a critical infrastructure operator), an employee u of the organization that uses resources $\mathcal{R} \triangleq \{1, 2, \ldots, R\}$ on the corporate network, and an attacker denoted by a. For ease of exposition, we consider that time is slotted, and we use t for indexing time slots. Our focus is on the interaction between the organization and the attacker, which we model as a dynamic discrete stochastic game with imperfect information. Following common practice in game-theoretic models of security, we assume that the attacker is aware of the strategy of the defender (information technology (IT) manager, operator), while the defender (information technology (IT) manager, operator) is not aware of the actions taken by the attacker over time, and hence of the attacker's knowledge. Note that in practice, the defender is typically not aware of the existence of an attacker until a system compromise is detected. In this section, we first describe the system model, then we define the actions of the operator and of the attacker.

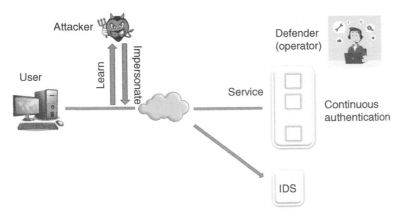

Figure 11.1 Illustration of the considered system including a system operator (defender, IT manager), a legitimate user, and an attacker. Continuous authentication and an IDS are used for detecting and mitigating attacks.

11.3.1 User Behavior

We focus on a user u that interacts with the operator's resources (e.g. servers, control systems, etc.) through generating data traffic. In each time slot t, the user interacts with one of the resources $r \in \mathcal{R}$ chosen at random with probability α_r such that $\sum_{r \in \mathcal{R}} \alpha_r = 1$. We refer to the resource that the user interacts with in time slot t as the active resource. We denote by $\Lambda_r(t)$ the amount of traffic generated by the user in time slot t towards resource $r \in \mathcal{R}$. We consider that $\Lambda_r(t)$ is Poisson distributed with parameter λ_r (this is equivalent to the common assumption that arrivals can be modeled by a Poisson process, with intensity λ_r/ι, where ι is the length of the time slot) whenever r is the active resource, and $\Lambda_r(t) = 0$ otherwise. As a model of productivity we consider that successful interaction of the user with resource r in a time slot generates immediate reward v_r for the operator.

11.3.2 Intrusion Detection System Model

We consider that the operator deployed an intrusion detection service (IDS) in its infrastructure. State-of-the-art IDSs typically look for anomalous behavior in hosts and in the network,[2] detection thus requires attacker activity. A detection by the IDS is followed by an investigation by a security threat analyst, which implies that a potential attacker would be detected and eliminated. We denote by m_r the per time slot operation cost of the IDS regarding resource r, which determines its ability to detect an attacker (e.g. m_r determines the number of security threat analysts that can be hired on resource r), as discussed later.

11.3.3 Model of Continuous Authentication

In order to mitigate identity theft, e.g. through session hijacking and remote access tool-kits, the operator uses continuous authentication (also referred to as behavioral authentication) for verifying that the traffic received from user u was indeed generated by user u. Continuous authentication is typically based on a statistical characterization of the user behavior, e.g. through training a machine learning model. For ease of exposition, let us consider that the user behavior

---

2 The threat of generative adversarial networks (GAN) aided-attacks on ML-based IDSs is considered in Chapter 17.

(e.g. keystroke patterns) can be modeled by a Gaussian distribution[3] $B_u \sim \mathcal{N}(\mu_u, \sigma_u)$ for all resources $r \in \mathcal{R}$, with mean μ_u and variance σ_u; i.e. the user behavior[4] is the same for all resources $r \in \mathcal{R}$. This model is admittedly simplistic, but the analysis we develop in the following does not depend on it; the analysis can be performed for any classifier for which a characterization of the receiver operating characteristic (ROC) is possible.

We consider that continuous authentication is performed in every time slot, i.e. the user behavior during a time slot is analyzed at the end of the time slot by comparing the user behavior model and the actual behavior of the user during the time slot. If the user fails the test because the actual behavior is not close enough to the model then the user is blocked from accessing the resources. Note that the user could be blocked due to a false positive (FP) even if there is no attacker. Let us denote by η_r the false-positive rate for resource r. Using our model of the user behavior this is equivalent to saying that the system applies a detection threshold of $t_r = \Phi_B^{-1}(1 - \eta_r)$ for resource r, where Φ_B is the cumulative distribution function (CDF) of B, and the test result is positive if $B > t_r$, and negative otherwise.

The decision process of continuous authentication can also be modeled with memory (i.e. the decisions from past affect the current decision), instead of the memoryless version described above. One such alternative model is the Trust Model introduced by Bours (2012) and Deutschmann and Lindholm (2013). In this model, the genuineness of the user is adjusted at the end of every time slot. In particular, if the actual behavior of the user during the time slot has a small deviation from the user behavior model, then the system's trust in the genuineness of the user will increase, which is called a reward. Otherwise, i.e. if the deviations are large, then the trust of the system will decrease, which is called a penalty. The penalty or reward decision is based on a comparison between the deviations and thresholds. Typical trust models have two levels, as investigated in Bours (2012) and Deutschmann and Lindholm (2013), i.e. there is only one threshold value. The main difference between these two models is the calculation of the penalty and reward values. The two-level trust model can be extended to a three or a four-level trust model as proposed in Bours and Mondal (2015). In all of these trust models, if the trust of the system drops below a predefined lockout threshold then the user is blocked from accessing the resources and is required to re-authenticate herself. Ideally, a legitimate user would never reach the lockout threshold, while the trust level of an attacker should decrease fast to a value below the lockout threshold.

11.3.4 System States without an Attacker

Observe that if there is no attacker then the system S can be in two different states[5]: the blocking state (BL) and the unblocking state (UB). In state BL access to all resources is blocked, the user cannot interact with any of the resources and hence cannot generate a reward. In state UB the user can interact with the system, it can use one of the resources during a time slot, and thus, it can generate reward v_r if it interacts with resource r successfully.

The system state can switch between UB and BL upon each time slot. It switches from UB to BL if the user fails continuous authentication in a time slot; note that this could happen due to an FP or due to a true positive (TP), i.e. input generated by an attacker as discussed later. Furthermore,

---

3 In practice, for a better representation of the keystroke patterns, a Gaussian Mixture Model (GMM) can be considered (Hosseinzadeh and Krishnan 2008). The experiments conducted in Hosseinzadeh and Krishnan (2008) to verify only short texts were extended for the purpose of continuous user monitoring and transparent authentication in Çeker and Upadhyaya (2015). Furthermore, since human response times have power-law tails, the generalized inverse-Gamma distribution can be used for their description (Ma et al. 2016), and accordingly, inter-keystroke time distribution may be modeled by the Gamma distribution (Cardaioli et al. 2019).
4 Since only a single user case is considered here, the subscript u is omitted for simplicity; i.e. $B \sim \mathcal{N}(\mu, \sigma)$.
5 We will introduce an additional state AD when there is an attacker in Section 11.3.6.

to allow productivity, we consider that if the system is in state BL in time slot t it switches to UB in time slot $t + 1$ with probability q; i.e. $\Pr(S(t + 1) = \text{UB} \mid S(t) = \text{BL}) = q$. Consequently, if there is no input from the user in time slot t and the system was in state UB then it will stay in state UB, as an FP cannot be generated due to lack of user activity. Otherwise, i.e. if the user is active for resource r and provides some input then the system stays in state UB and generates a reward if there is no FP. Thus, given that the user is utilizing resource r, the probability of staying in the unblocked state is $P_r \triangleq e^{-\lambda_r} + (1 - e^{-\lambda_r})(1 - \eta_r)$.

Putting all this together, without an attacker we can model the continuous authentication security system as a discrete time Markov chain with state space $\{\text{UB}, \text{BL}\}$, as shown in Figure 11.2a, and the state transition probabilities are

$$\Pr(S(t + 1) = \text{UB} \mid S(t) = \text{UB}) = \sum_{r=1}^{R} \alpha_r P_r \triangleq \overline{P},$$

$$\Pr(S(t + 1) = \text{BL} \mid S(t) = \text{UB}) = 1 - \overline{P},$$

$$\Pr(S(t + 1) = \text{UB} \mid S(t) = \text{BL}) = q,$$

$$\Pr(S(t + 1) = \text{BL} \mid S(t) = \text{BL}) = 1 - q. \tag{11.1}$$

11.3.5 Attack Model

We consider an attacker that compromises a system component at cost C_a, e.g. the user's computer, which allows it to observe the traffic generated by the user and to craft packets that appear to originate from user u. The compromise could happen through identity theft or through session hijacking, as in recent security incidents. We will use the term *listening* when the attacker is observing the user traffic and the term *attacking* when the attacker is crafting packets.[6] The attacker can also decide not to do anything during a time slot; we refer to this as *waiting*. By listening to user traffic the attacker can essentially perform a model extraction attack against the classifier used for continuous authentication, which then allows it to generate traffic that is likely classified legitimate by the classifier.

The attacker can choose between $2R + 1$ actions in every time slot: wait ($l(t) = 0$, $a(t) = 0$), listen to resource r ($l(t) = r$, $a(t) = 0$), and attack resource r ($l(t) = 0$, $a(t) = r$), where $l(t) = r$ stands for listening to resource r and $a(t) = r$ stands for attacking resource r for some $r \in \mathcal{R}$. The purpose of listening is to collect behavioral information about the user, so as to learn to imitate legitimate user behavior that would pass continuous authentication. The purpose of attacking is to execute a rogue command on the resource, but in order for the attack to be successful, the system has to be in state UB, the user-generated input must not cause a false alarm if the user is active for any resource, and the attacker generated input should pass continuous authentication. If in time slot t the attack is successful, then the attacker obtains immediate reward c_r, which is a penalty for the defender. Motivated by that many attacks have a monetary reward, we consider that the future reward of the attacker is discounted by a discount factor ρ. In what follows, we first define the actions of the attacker at time slot t and then provide expressions for the attacker's expected utility in Section 11.4.2.

---

6 It is possible to make an analogy between the attacker considered in this chapter and the adversary in Chapter 14, which discusses two stealth attack scenarios in real 5G systems. In particular, the adversary learns the 5G transmit behavior by making an exploratory attack; i.e. *listening*, and launches an evasion attack over the air to manipulate inputs to the deep learning-based spectrum sensing classifier; i.e. *attacking*. The goal of the adversary is to decide when to jam 5G communications.

11.3.5.1 Listening ($l(t) = r$, $a(t) = 0$)

While listening, the attacker can observe the behavior of the user so as to learn to imitate the user's behavior for a successful attack. Learning during time slot t is determined by the traffic $\Lambda_r(t)$ generated by the user towards resource r and by the learning rate γ_r for resource r. The total amount of observation of the attacker about the user until time slot t can be expressed as $L(t) = \sum_{\tau=0}^{t-1} \sum_{r=1}^{R} \mathbb{1}_{\{l(\tau)=r\}} \Lambda_r(\tau)$, where $\mathbb{1}_{\{D\}}$ is the indicator function of an event D.

Listening in a time slot requires activity from the attacker, and thus, the IDS could detect the attacker in the time slot. We denote by $\delta_{l_r}(m_r)$ the probability that the IDS detects the attacker in a time slot when the attacker is listening to resource r. We make the reasonable assumption that $\delta_{l_r}(m_r)$ is a concave function of m_r, $\delta_{l_r}(0) = 0$ and $\lim_{m_r \to \infty} \delta_{l_r}(m_r) = 1$ for $r \in \mathcal{R}$, where m_r is the per time slot operation cost of the IDS for resource r, as defined previously.

11.3.5.2 Attacking ($l(t) = 0$, $a(t) = r$)

When attacking, the attacker generates and sends rouge input to the resource, trying to impersonate the legitimate user. Whether or not the attacker can successfully imitate the user depends on the amount of observation $L(t)$ that it has collected about the user. We consider that given $L(t)$ amount of information the attacker can generate input towards resource r following a Gaussian distribution, $\hat{B}_r(L(t)) \sim \mathcal{N}(\hat{\mu}_r(L(t)), \hat{\sigma}_r(L(t)))$, where $\hat{\mu}_r(L(t)) = \mu(1 + e^{-\gamma_r L(t)})$ and $\hat{\sigma}_r(L(t)) = \sigma(1 + e^{-\gamma_r L(t)})$. Since the user behavior $B \sim \mathcal{N}(\mu, \sigma)$ is a Gaussian random variable, and $\hat{B}_r \sim \mathcal{N}(\hat{\mu}_r, \hat{\sigma}_r)$ is the random variable generated by the attacker for resource r, we can use the binormal method (Goncalves et al. 2014) for expressing the Receiver Operating Characteristic (ROC) curve of the continuous authentication security system for resource r as

$$\mathrm{ROC}_r(\eta_r, L(t)) = \Phi(a_r + b_r \Phi^{-1}(\eta_r)), \tag{11.2}$$

where η_r is the FP rate for resource r, $\Phi(\cdot)$ is the CDF of the standard normal distribution, $a_r = \frac{\hat{\mu}_r(L(t)) - \mu}{\hat{\sigma}_r(L(t))}$, and $b_r = \frac{\sigma}{\hat{\sigma}_r(L(t))}$. Note that $\mathrm{ROC}_r(\eta_r, L(t))$ is the TP rate of the detector (i.e. the conditional probability of classifying rogue input as such) for resource r.

By inspecting (11.2), and substituting[7] $\omega = L(t)$, we can observe that

$$\mathrm{ROC}_r(\eta_r, \omega) = \Phi\left(\underbrace{\frac{\mu}{\sigma} - \frac{\mu - \sigma\Phi^{-1}(\eta_r)}{\sigma(1 + e^{-\gamma_r \omega})}}_{\triangleq \xi_{\omega,r}} \right) = \Phi\left(\underbrace{\frac{\mu e^{-\gamma_r \omega} + \sigma\Phi^{-1}(\eta_r)}{\sigma(1 + e^{-\gamma_r \omega})}}_{\triangleq \xi_{\omega,r}} \right) = \Phi(\xi_{\omega,r}).$$

Normally, one would expect that as the number of observations increases, the attacker can successfully imitate the real behavior of the user with a higher probability; i.e. its input is harder to distinguish from real user input. Hence, we can safely assume that $\mathrm{ROC}_r(\eta_r, \omega) = \Phi(\xi_{\omega,r})$ should be a nonincreasing function of ω, or equivalently, it must hold that $\mu \geq \sigma\Phi^{-1}(\eta_r)$ for every resource $r \in \mathcal{R}$.

Just like for listening, the attacker has to perform some activity for attacking, and hence the IDS could detect the attacker. We denote by $\delta_{a_r}(m_r)$ the probability that the IDS detects the attacker in a time slot when it is attacking resource r. Similar to $\delta_{l_r}(m_r)$, we assume that $\delta_{a_r}(m_r)$ is a concave function of m_r, $\delta_{a_r}(0) = 0$ and $\lim_{m_r \to \infty} \delta_{a_r}(m_r) = 1$ for $r \in \mathcal{R}$.

11.3.5.3 Waiting ($l(t) = 0$, $a(t) = 0$)

The attacker can choose to wait, in which case it neither learns nor attacks. Hence it cannot be detected during the time slot.

---

7 For ease of exposition, unless otherwise stated, ω denotes $L(t) = \omega$.

11.3.6 Continuous Authentication Game

We are now ready to formulate the continuous authentication game as a dynamic game,[8] in which the defender (the operator) is the leader and the attacker is the follower. The defender chooses a defense strategy (m_r, η_r) for $\forall r \in \mathcal{R}$, while the attacker decides whether or not to compromise the system at cost C_a, and if it decides to compromise the system, then in every time slot it decides whether to wait, listen, or attack. The game ends when the attacker is detected (AD) by the IDS, i.e. when $S(t) = \text{AD}$. AD is thus an absorbing state. The attacker is assumed to know the system parameters used by the defender, and the defender assumes the existence of such an attacker.[9] The attacker is interested in maximizing its expected reward, while the operator is interested in maximizing its expected utility.

11.4 Optimal Attack Strategy under Asymmetric Information

In what follows we formulate the optimal attack strategy under the assumption that the attacker knows the defense strategy, i.e. $\forall r \in \mathcal{R}$, the false-positive rates η_r, the IDS operation costs m_r, and the detection capabilities $\delta_{l_r}(m_r)$ and $\delta_{a_r}(m_r)$ of the IDS, which are assumed to be constant. On the contrary, the defender cannot observe the actions taken by the attacker. Under these assumptions of information asymmetry the attacker is faced by a Markov decision process (MDP). We thus start with describing the state space and the state transitions of the MDP as a function of the attacker's strategy and of the defender's strategy. We then derive the optimal attack strategy for a given defender strategy.

11.4.1 MDP Formulation

Observe that the state of the system from the perspective of the attacker depends on whether the system is blocked ($S(t) = \text{BL}$) or unblocked ($S(t) = \text{UB}$), and on the amount of observations $L(t)$ it has collected so far. Furthermore, the state transition probabilities and the rewards are affected by the actions of the attacker, hence the optimization problem faced by the attacker can be formulated as an MDP. In the following, we provide the state transition probabilities, depending on the action chosen by the attacker.

11.4.1.1 Waiting ($l(t) = 0$, $a(t) = 0$)

When the attacker is waiting it does not observe the user traffic, neither does it attempt to attack. Thus, the state transition probabilities are determined by the FP rate in state UB, and by the unblocking probability in state BL. As shown in Figure 11.2b, the state transition probabilities are

$$\Pr\left(S(t+1) = \text{UB}, L(t+1) = \omega \mid S(t) = \text{UB}, L(t) = \omega, l(t) = 0, a(t) = 0\right) = \overline{P},$$

$$\Pr\left(S(t+1) = \text{BL}, L(t+1) = \omega \mid S(t) = \text{UB}, L(t) = \omega, l(t) = 0, a(t) = 0\right) = 1 - \overline{P},$$

$$\Pr\left(S(t+1) = \text{UB}, L(t+1) = \omega \mid S(t) = \text{BL}, L(t) = \omega, l(t) = 0, a(t) = 0\right) = q,$$

$$\Pr\left(S(t+1) = \text{BL}, L(t+1) = \omega \mid S(t) = \text{BL}, L(t) = \omega, l(t) = 0, a(t) = 0\right) = 1 - q. \tag{11.3}$$

It is worth to note that whenever the system is in state BL, waiting is preferred by the attacker. To see why, observe that in state BL the user cannot interact with the resources, hence the

---

8 Basics of game theory are presented in Chapter 2.
9 We provide a discussion of this assumption and possible alternatives in the Conclusion.

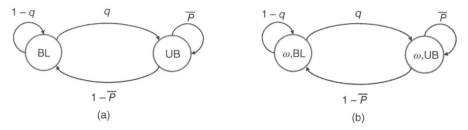

Figure 11.2 State transitions and corresponding probabilities when the attacker (a) does not compromise the system and (b) is waiting.

attacker cannot increase its total number of observation about the user, and thus, listening cannot be an optimal action for the attacker due to the possibility of being detected. Similarly, when the system is in state BL, an attack would be blocked, but the attacker could be detected. Therefore, as shown in Figures 11.2b, 11.3a and 11.3b, there is no state transition to state AD from state BL.

11.4.1.2 Listening ($l(t) = r, a(t) = 0$)

When the attacker is listening to resource r it is detected by the IDS with probability $\delta_{l_r}(m_r)$. While listening to resource r, the attacker's amount of observation about the user does not change (i.e. $N = 0$) if the user does not generate any traffic or if the user is active for another resource. Note that if the user is active for another resource, no false alarm should be generated to stay in the unblocking state. On the contrary, if the user is active for the same resource that the attacker is listening to and the user generates traffic (i.e. $N \geq 1$), then the attacker can observe and learn, as long as the user-generated traffic does not cause an FP. Thus, as depicted in Figure 11.3a, the transition probabilities are

$$\Pr\left(S(t+1) = \text{AD} \mid S(t) = \text{UB}, L(t) = \omega, l(t) = r, a(t) = 0\right) = \delta_{l_r}(m_r),$$

$$\Pr\left(S(t+1) = \text{UB}, L(t+1) = \omega \mid S(t) = \text{UB}, L(t) = \omega, l(t) = r, a(t) = 0\right)$$

$$= (1 - \delta_{l_r}(m_r))\left(\alpha_r e^{-\lambda_r} + (1 - \alpha_r)\left(\frac{\overline{P} - \alpha_r P_r}{1 - \alpha_r}\right)\right)$$

$$= (1 - \delta_{l_r}(m_r))\left(\overline{P} - \alpha_r(1 - e^{-\lambda_r})(1 - \eta_r)\right),$$

$$\Pr\left(S(t+1) = \text{UB}, L(t+1) = \omega + N \mid S(t) = \text{UB}, L(t) = \omega, l(t) = r, a(t) = 0\right)$$

$$= (1 - \delta_{l_r}(m_r))\alpha_r(1 - \eta_r)\frac{e^{-\lambda_r}\lambda_r^N}{N!}, \text{ for } N = 1, 2, \dots,$$

$$\Pr\left(S(t+1) = \text{BL}, L(t+1) = \omega \mid S(t) = \text{UB}, L(t) = \omega, l(t) = r, a(t) = 0\right)$$

$$= (1 - \delta_{l_r}(m_r))(1 - \overline{P}). \tag{11.4}$$

11.4.1.3 Attacking ($l(t) = 0, a(t) = r$)

When the attacker is attacking resource r it is detected by the IDS with probability $\delta_{a_r}(m_r)$. For the attack to be successful, (i) the attacker should not be detected by the IDS, (ii) the attacker generated input must pass continuous authentication (false negative), and (iii) if the user is active for any resource then the user traffic must not cause an FP. If any of these three conditions is not met,

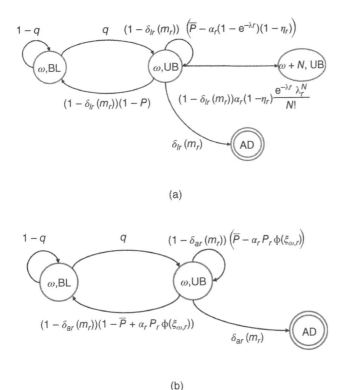

Figure 11.3 State transitions and corresponding probabilities when the attacker is (a) listening to resource r and (b) attacking resource r.

the system switches to state BL (or to AD in case of (i)). As shown in Figure 11.3b the transition probabilities are

$$\Pr(S(t+1) = \text{AD} \mid S(t) = \text{UB}, L(t) = \omega, l(t) = 0, a(t) = r) = \delta_{a_r}(m_r),$$

$$\Pr(S(t+1) = \text{UB}, L(t+1) = \omega \mid S(t) = \text{UB}, L(t) = \omega, l(t) = 0, a(t) = r)$$

$$= (1 - \delta_{a_r}(m_r)) \left(\alpha_r P_r(1 - \Phi(\xi_{\omega,r})) + (1 - \alpha_r) \left(\frac{\overline{P} - \alpha_r P_r}{1 - \alpha_r} \right) \right)$$

$$= (1 - \delta_{a_r}(m_r)) \left(\alpha_r P_r(1 - \Phi(\xi_{\omega,r})) + \overline{P} - \alpha_r P_r \right)$$

$$= (1 - \delta_{a_r}(m_r)) \left(\overline{P} - \alpha_r P_r \Phi(\xi_{\omega,r}) \right),$$

$$\Pr(S(t+1) = \text{BL}, L(t+1) = \omega \mid S(t) = \text{UB}, L(t) = \omega, l(t) = 0, a(t) = r)$$

$$= (1 - \delta_{a_r}(m_r))(1 - \overline{P} + \alpha_r P_r \Phi(\xi_{\omega,r})). \tag{11.5}$$

11.4.2 Optimality of the Threshold Policy

In what follows we prove that the optimal attacker strategy is a threshold policy. We do so by providing a dynamic programming recursion for the attacker reward. In order to formulate the recursion, let us denote by $J_t(L(t) = \omega, S(t) = \text{UB})$ and $J_t(L(t) = \omega, S(t) = \text{BL})$ the expected total reward of the attacker starting from time slot t with observation $L(t) = \omega$ when $S(t) = \text{UB}$ and $S(t) = \text{BL}$, respectively. For notational convenience, we will use $J(\omega, \text{UB})$ and $J(\omega, \text{BL})$ henceforth. Observe that the total expected reward of the attacker corresponds to $J(0, \text{UB})$. Then, based on the states, actions and corresponding state transition probabilities, and accounting for the discount factor ρ, we can formulate the following dynamic programming recursion for the attacker reward

$$
\begin{aligned}
J(\omega, \text{BL}) &= \rho q J(\omega, \text{UB}) + \rho(1-q)J(\omega, \text{BL}) \\
&\Rightarrow J(\omega, \text{BL}) = \frac{\rho q}{1-\rho(1-q)} J(\omega, \text{UB}),
\end{aligned}
\tag{11.6}
$$

$$
J(\omega, \text{UB}) = \begin{cases}
\rho \overline{P} J(\omega, \text{UB}) + \rho(1-\overline{P})J(\omega, \text{BL}) & l = 0, \; a = 0 \\
\rho(1-\delta_{l_r}(m_r))\left(O_r(\omega) + (1-\overline{P})J(\omega, \text{BL})\right) & l = r, \; a = 0 \; , \\
\rho(1-\delta_{a_r}(m_r))A_r(\omega) & l = 0, \; a = r
\end{cases}
\tag{11.7}
$$

where

$$
O_r(\omega) = \left(\overline{P} - \alpha_r(1-e^{-\lambda_r})(1-\eta_r)\right)J(\omega, \text{UB}) + \alpha_r(1-\eta_r)\sum_{n=1}^{\infty}\frac{e^{-\lambda_r}\lambda_r^n}{n!}J(\omega+n, \text{UB}),
$$

$$
A_r(\omega) = \underbrace{\left(\overline{P} - \alpha_r P_r \Phi(\xi_{\omega,r})\right)}_{\triangleq \tilde{q}_{\omega,r}}\left(\frac{c_r}{\rho} + J(\omega, \text{UB})\right) + \left(1 - \underbrace{\left(\overline{P} - \alpha_r P_r \Phi(\xi_{\omega,r})\right)}_{\triangleq \tilde{q}_{\omega,r}}\right)J(\omega, \text{BL})
$$

$$
= \tilde{q}_{\omega,r}\frac{c_r}{\rho} + \tilde{q}_{\omega,r}J(\omega, \text{UB}) + (1-\tilde{q}_{\omega,r})J(\omega, \text{BL}).
$$

Observe that the parameter $\tilde{q}_{\omega,r}$ is a nondecreasing function of ω because $\text{ROC}_r(\eta_r, \omega) = \Phi(\xi_{\omega,r})$ is a nonincreasing function of ω.

To bring in some structural insight and to simplify the expressions, let us substitute (11.6) into (11.7), thus for $l = 0$ and $a = 0$ we obtain

$$
J(\omega, \text{UB}) = \rho\frac{\overline{P}(1-\rho) + \rho q}{1 - \rho + \rho q}J(\omega, \text{UB}).
\tag{11.8}
$$

This leads us to the following proposition.

Proposition 11.1 Let $\rho < 1$. Then waiting cannot be optimal in state *UB*.

Proof: By (11.8), if waiting is to be optimal in state UB then its reward must be $J(\omega, \text{UB}) = 0$, which cannot be optimal.

A consequence of the above result is that if the system is in state UB then the attacker prefers either listening or attacking during the time slot. On the contrary, waiting is the optimal action in state BL, as discussed in Section 11.4.1.

Proposition 11.2 The optimal attacker strategy satisfies the following Bellman optimality equations for the attacker reward:

$$
J_\omega \triangleq J(\omega, \text{UB}) = \begin{cases}
\dfrac{K_{\omega,r}}{1-U_r}, \; l = r, \; a = 0 & \text{if} \quad \mathcal{L}_\omega^* > \mathcal{A}_\omega^* \text{ and } r = \arg\max_{i\in\{1,\ldots,R\}}\mathcal{L}_{\omega,i} \\[3mm]
\dfrac{C_{\omega,r}}{1-T_{\omega,r}}, \; l = 0, \; a = r & \text{if} \quad \mathcal{L}_\omega^* \le \mathcal{A}_\omega^* \text{ and } r = \arg\max_{i\in\{1,\ldots,R\}}\mathcal{A}_{\omega,i}
\end{cases}
\tag{11.9}
$$

Here, the additional parameters are defined as

$$
\mathcal{L}_\omega^* = \max_{i\in\{1,\ldots,R\}}\mathcal{L}_{\omega,i} = \max_{i\in\{1,\ldots,R\}}\frac{K_{\omega,i}}{1-U_i},
$$

$$\mathcal{A}_\omega^* = \max_{i\in\{1,\dots,R\}} \mathcal{A}_{\omega,i} = \max_{i\in\{1,\dots,R\}} \frac{C_{\omega,i}}{1 - T_{\omega,i}},$$

$$K_{\omega,r} = \rho(1 - \delta_{l_r}(m_r))\alpha_r(1 - \eta_r)\sum_{n=1}^{\infty} \frac{e^{-\lambda_r}\lambda_r^n}{n!}J(\omega + n, \mathrm{UB}),$$

$$U_r = \rho(1 - \delta_{l_r}(m_r))\left(\left(\frac{\overline{P}(1 - \rho) + \rho q}{1 - \rho + \rho q} - \alpha_r(1 - e^{-\lambda_r})(1 - \eta_r)\right)\right),$$

$$C_{\omega,r} = (1 - \delta_{a_r}(m_r))\tilde{q}_{\omega,r}c_r,$$

$$T_{\omega,r} = \rho(1 - \delta_{a_r}(m_r))\frac{\tilde{q}_{\omega,r}(1 - \rho) + \rho q}{1 - \rho + \rho q}.$$

Remark 11.1   Notice that $\mathcal{L}_{\omega,i}$ (resp. $\mathcal{A}_{\omega,i}$) is the reward for listening to (resp. attacking) resource r when the attacker's amount of observation about the user is ω, and the maximum listening (resp. attacking) reward corresponding to that amount of observation is denoted by $\mathcal{L}_\omega^*$ (resp. $\mathcal{A}_\omega^*$).

Proof: Consider that the attacker prefers listening to resource r in state UB. We can substitute (11.6) into (11.7), to obtain for $l = r$ and $a = 0$,

$$J(\omega, \mathrm{UB}) = \rho(1 - \delta_{l_r}(m_r))O_r(\omega) + \rho(1 - \delta_{l_r}(m_r))(1 - \overline{P})\frac{\rho q}{1 - \rho(1 - q)}J(\omega, \mathrm{UB})$$

$$= \rho(1 - \delta_{l_r}(m_r))\left(\left(\overline{P} - \alpha_r(1 - e^{-\lambda_r})(1 - \eta_r)\right)J(\omega, \mathrm{UB})\right.$$

$$\left. + \alpha_r(1 - \eta_r)\sum_{n=1}^{\infty}\frac{e^{-\lambda_r}\lambda_r^n}{n!}J(\omega + n, \mathrm{UB})\right)$$

$$+ \rho(1 - \delta_{l_r}(m_r))(1 - \overline{P})\frac{\rho q}{1 - \rho(1 - q)}J(\omega, \mathrm{UB})$$

$$= \underbrace{\rho(1 - \delta_{l_r}(m_r))\left(\left(\frac{\overline{P}(1 - \rho) + \rho q}{1 - \rho + \rho q} - \alpha_r(1 - e^{-\lambda_r})(1 - \eta_r)\right)\right)}_{\triangleq U_r}J(\omega, \mathrm{UB})$$

$$+ \underbrace{\rho(1 - \delta_{l_r}(m_r))\alpha_r(1 - \eta_r)\sum_{n=1}^{\infty}\frac{e^{-\lambda_r}\lambda_r^n}{n!}J(\omega + n, \mathrm{UB})}_{\triangleq K_{\omega,r}}$$

$$= U_r J(\omega, \mathrm{UB}) + K_{\omega,r}. \tag{11.10}$$

Using the same substitution, if the attacker prefers attacking resource r in state UB, i.e. for $l = 0$ and $a = r$, we obtain

$$J(w, \mathrm{UB}) = \rho(1 - \delta_{a_r}(m_r))\left(\tilde{q}_{\omega,r}\frac{c_r}{\rho} + \tilde{q}_{\omega,r}J(\omega, \mathrm{UB}) + (1 - \tilde{q}_{\omega,r})\frac{\rho q}{1 - \rho(1 - q)}J(\omega, \mathrm{UB})\right)$$

$$= \underbrace{(1 - \delta_{a_r}(m_r))\tilde{q}_{\omega,r}c_r}_{\triangleq C_{\omega,r}} + \underbrace{\rho(1 - \delta_{a_r}(m_r))\frac{\tilde{q}_{\omega,r}(1 - \rho) + \rho q}{1 - \rho + \rho q}J(\omega, \mathrm{UB})}_{\triangleq T_{\omega,r}}$$

$$= T_{\omega,r}J(\omega, \mathrm{UB}) + C_{\omega,r}. \tag{11.11}$$

Based on (11.10) and (11.11), the attacker reward in state UB can be expressed as

$$J_\omega \triangleq J(\omega, \mathrm{UB}) = \begin{cases} U_r J_\omega + K_{\omega,r} & l = r, \ a = 0 \\ T_{\omega,r} J_\omega + C_{\omega,r} & l = 0, \ a = r \end{cases} = \begin{cases} \dfrac{K_{\omega,r}}{1 - U_r} & l = r, \ a = 0 \\ \dfrac{C_{\omega,r}}{1 - T_{\omega,r}} & l = 0, \ a = r \end{cases}.$$

Since the attacker aims for the maximum reward $J_\omega = \max\left\{\mathcal{L}_\omega^*, \mathcal{A}_\omega^*\right\}$, we obtain the backward induction in the statement.

11.4.2.1 Optimality of Listening

Given the recursion, we now establish a relationship between the attacking reward and the listening reward and show that there is a threshold $\tilde{\omega}_{r,s}$ such that listening to resource r is preferred over attacking resource s for $\omega < \tilde{\omega}_{r,s}$. Note that $\tilde{q}_{\omega,r}$ is a nondecreasing function of ω, and thus, the reward $\frac{C_{\omega,r}}{1-T_{\omega,r}}$ of attacking resource r is a nondecreasing function of ω; i.e. more observation is always at least as good for the attacker. We thus continue with the analysis of the advantage of attacking with more observation.

Proposition 11.3   Let $\mathcal{R}_{\omega,r} \triangleq \dfrac{\mathcal{A}_{\omega+1,r}}{\mathcal{A}_{\omega,r}} = \dfrac{\frac{C_{\omega+1,r}}{1-T_{\omega+1,r}}}{\frac{C_{\omega,r}}{1-T_{\omega,r}}}$ be the ratio between the attacking rewards of two consecutive amounts of observation regarding resource r. Then, $\mathcal{R}_{\omega,r}$ is a monotonic decreasing function of ω and $\lim_{\omega \to \infty} \mathcal{R}_{\omega,r} = 1$ for $r = 1, \dots, R$.

Proof: $\mathcal{R}_{\omega,r}$ can be expanded as follows:

$$\mathcal{R}_{\omega,r} = \frac{\dfrac{(1-\delta_{a_r}(m_r))\tilde{q}_{\omega+1,r}c_r}{1-\rho(1-\delta_{a_r}(m_r))\frac{\tilde{q}_{\omega+1,r}(1-\rho)+\rho q}{1-\rho+\rho q}}}{\dfrac{(1-\delta_{a_r}(m_r))\tilde{q}_{\omega,r}c_r}{1-\rho(1-\delta_{a_r}(m_r))\frac{\tilde{q}_{\omega,r}(1-\rho)+\rho q}{1-\rho+\rho q}}} = \frac{\tilde{q}_{\omega+1,r}}{\tilde{q}_{\omega,r}} \frac{1-\rho(1-\delta_{a_r}(m_r))\frac{\tilde{q}_{\omega,r}(1-\rho)+\rho q}{1-\rho+\rho q}}{1-\rho(1-\delta_{a_r}(m_r))\frac{\tilde{q}_{\omega+1,r}(1-\rho)+\rho q}{1-\rho+\rho q}}. \qquad (11.12)$$

Since $\tilde{q}_{\omega,r}$ is a nondecreasing function of ω; i.e. $\tilde{q}_{\omega+1,r} \geq \tilde{q}_{\omega,r}$, from (11.12) we obtain $\mathcal{R}_{\omega,r} \geq 1$. Furthermore, since $\lim_{\omega \to \infty} \tilde{q}_{\omega,r} = \alpha_r(1 - \eta_r + \eta_r e^{-\lambda_r})(1 - \eta_r)$, we have that $\lim_{\omega \to \infty} \mathcal{R}_{\omega,r} = 1$. In general, it can be shown that $\frac{d\mathcal{R}_{\omega,r}}{d\omega} < 0$ for a continuous extension of $\mathcal{R}_{\omega,r}$; i.e. assuming $\omega \in [0, \infty)$ rather than $\omega \in \{0, 1, \dots\}$.

Next, in order to compare the rewards for listening to resource r and for attacking resource s, let us define the incremental observation gain

$$\chi_{\omega,(r,s)} \triangleq \frac{\rho(1 - \delta_{l_r}(m_r))\alpha_r(1 - \eta_r)}{1 - U_r} \sum_{n=1}^{\infty} \frac{e^{-\lambda_r}\lambda_r^n}{n!} \left(\prod_{i=0}^{n-1} \mathcal{R}_{\omega+i,s} \right). \qquad (11.13)$$

Lemma 11.1 $\chi_{\omega,(r,s)}$ is a decreasing function of ω for every $r, s \in \mathcal{R}$. Furthermore, $\lim_{\omega \to \infty} \chi_{\omega,(r,s)} < 1$.

Proof: The first part of the lemma follows from that $\mathcal{R}_{\omega,s}$ is a decreasing function of ω. To prove the second part of the lemma, since $\lim_{\omega \to \infty} \mathcal{R}_{\omega,s} = 1$, observe that

$$\lim_{\omega \to \infty} \chi_{\omega,(r,s)} = \frac{\rho(1 - \delta_{l_r}(m_r))\alpha_r(1 - \eta_r)}{1 - \rho(1 - \delta_{l_r}(m_r))\left(\left(\frac{\overline{p}(1-\rho)+\rho q}{1-\rho+\rho q} - \alpha_r(1 - e^{-\lambda_r})(1 - \eta_r)\right)\right)}(1 - e^{-\lambda_r}) < 1.$$

A consequence of the above result is the following.

Corollary 11.1   If $\chi_{\omega=0,(r,s)} > 1$ then there exists a critical value $\tilde{\omega}_{(r,s)}$ such that $\chi_{\omega=\tilde{\omega}_{(r,s)}-1,(r,s)} > 1$ and $\chi_{\omega=\tilde{\omega}_{(r,s)},(r,s)} \leq 1$. Otherwise; i.e. if $\chi_{\omega=0,(r,s)} \leq 1$ then $\tilde{\omega}_{(r,s)} = 0$.

Observe that $\tilde{\omega}_{(r,s)}$ can be calculated a priori (before the game-play) for a given set of parameters.[10] Furthermore, we can observe the following properties of $\tilde{\omega}_{(r,s)}$ with respect to the defender parameters:

Proposition 11.4 $\tilde{\omega}_{(r,s)}$ is a decreasing function of $\delta_{l_r}(m_r)$ and $\delta_{a_s}(m_s)$.

Proof: The result follows from that $\chi_{\omega,(r,s)}$ is a decreasing function of $\delta_{l_r}(m_r)$ and $\delta_{a_s}(m_s)$. Also, $\mathcal{R}_{\omega,s}$ is a decreasing function of $\delta_{a_s}(m_s)$.

We are now ready to prove that the optimal attacker strategy is indeed a threshold strategy.

Theorem 11.1   The attacker prefers listening to resource r over attacking resource s for $\omega < \tilde{\omega}_{(r,s)}$.

Proof: The proof is based on comparing the listening reward $\mathcal{L}_{\omega,r} = \frac{K_{\omega,r}}{1-U_r}$ and the attacking reward $\mathcal{A}_{\omega,s} = \frac{C_{\omega,s}}{1-T_{\omega,s}}$,

$$
\begin{aligned}
\mathcal{L}_{\omega,r} = \frac{K_{\omega,r}}{1-U_r} &= \frac{\rho(1-\delta_{l_r}(m_r))\alpha_r(1-\eta_r)}{1-U_r} \sum_{n=1}^{\infty} \frac{e^{-\lambda_r}\lambda_r^n}{n!} J_{\omega+n} \\
&\geq \frac{\rho(1-\delta_{l_r}(m_r))\alpha_r(1-\eta_r)}{1-U_r} \sum_{n=1}^{\infty} \frac{e^{-\lambda_r}\lambda_r^n}{n!} \mathcal{A}_{\omega+n}^* \\
&\geq \frac{\rho(1-\delta_{l_r}(m_r))\alpha_r(1-\eta_r)}{1-U_r} \sum_{n=1}^{\infty} \frac{e^{-\lambda_r}\lambda_r^n}{n!} \frac{C_{\omega+n,s}}{1-T_{\omega+n,s}} \\
&= \frac{C_{\omega,s}}{1-T_{\omega,s}} \frac{\rho(1-\delta_{l_r}(m_r))\alpha_r(1-\eta_r)}{1-U_r} \sum_{n=1}^{\infty} \frac{e^{-\lambda_r}\lambda_r^n}{n!} \frac{\frac{C_{\omega+n,s}}{1-T_{\omega+n,s}}}{\frac{C_{\omega,s}}{1-T_{\omega,s}}} \\
&= \mathcal{A}_{\omega,s} \underbrace{\frac{\rho(1-\delta_{l_r}(m_r))\alpha_r(1-\eta_r)}{1-U_r} \sum_{n=1}^{\infty} \frac{e^{-\lambda_r}\lambda_r^n}{n!} \left(\prod_{i=0}^{n-1} \mathcal{R}_{\omega+i,s}\right)}_{\chi_{\omega,(r,s)}}.
\end{aligned}
\tag{11.14}
$$

By the above, listening to resource r is preferred over attacking resource s when $\chi_{\omega,(r,s)} > 1$; i.e. $\omega < \tilde{\omega}_{(r,s)}$, which proves the theorem.

Note that we cannot compare the attacking and listening rewards based on (11.14) after the critical value of $\omega \geq \tilde{\omega}_{(r,s)}$, since $\chi_{\omega,(r,s)} \leq 1$.

11.4.2.2 Optimality of Attacking
The next step in proving the optimality of the threshold policy is to show that attacking is optimal above the threshold established in the previous subsection. In what follows, we do so by using value

---

10 The sum in (11.13) can be partitioned into $\sum_{n=1}^{C}$ and $\sum_{n=C+1}^{\infty}$ for any arbitrary C, and $\chi_{\omega,(r,s)}$ can be approximated from below by utilizing $\mathcal{R}_{\omega,s} \geq 1$ in the latter one. Then, the corresponding $\tilde{\omega}_{(r,s)}$ can be calculated accordingly.

iteration. Observe that Theorem 11.1 implies that listening to resource r is preferred over attacking resource s for $\omega < \tilde{\omega}_{(r,s)}$. In what follows, we focus on the optimal strategy, i.e. listening or attacking, for $\omega \geq \max_{r,s \in \mathcal{R}} \tilde{\omega}_{(r,s)}$.

Since the attacker gets an immediate reward c_r only when the attack is successful, the attacker, if listens, only gets a (discounted) reward because of potentially successful attacks in the future. Therefore, for any amount of observation $\hat{\omega} \geq \tilde{\omega}$ there must be some $\overline{\omega} \geq \hat{\omega}$, in which attacking is optimal, as otherwise the attacker gets zero reward if she only listens, which implies that there are infinitely many ω for which attacking is optimal.

Since we already established a backward induction through the Bellman optimality equations for the attacker reward in (11.9), we are ready to apply the value iteration method to obtain the optimal attacker strategy for $\omega \geq \max_{r,s \in \mathcal{R}} \tilde{\omega}_{(r,s)}$ (note that since $\frac{\rho(1-\delta_{l_r}(m_r))\alpha_r(1-\eta_r)}{1-U_r} \sum_{n=1}^{\infty} \frac{e^{-\lambda_r}\lambda_r^n}{n!} < 1$, the Bellman update/operator in (11.9) is a contraction mapping, which guarantees the existence and the uniqueness of an optimal point, that is achievable by the value iteration method).

Theorem 11.2   The attacker prefers attacking resource s over listening to resource r for some $\omega \geq \tilde{\omega}_{(r,s)}$, and after attacking, the attacker always prefers attacking.

Proof: For the initial values of the rewards, we assign zero reward for every ω; i.e. $J_\omega^{(0)} = 0 \; \forall \omega$. Regarding the first iteration of the value updates, since

$$J_{\omega,L,r}^{(1)} = \frac{K_{\omega,r}^{(0)}}{1-U_r} = \frac{\rho(1-\delta_{l_r}(m_r))\alpha_r(1-\eta_r)}{1-U_r} \sum_{n=1}^{\infty} \frac{e^{-\lambda_r}\lambda_r^n}{n!} J_{\omega+n}^{(0)} = 0 \tag{11.15}$$

for $r \in \mathcal{R}$, i.e. all listening rewards are zero, attacking would be the optimal choice for every ω. Then, $J_{\omega,L,r}^{(1)} = 0$ and $J_{\omega,A,s}^{(1)} = \mathcal{A}_{\omega,s} = \frac{c_{\omega,s}}{1-T_{\omega,s}}$ hold $\forall \omega, r, s$, which implies $J_{\omega,*}^{(1)} = \mathcal{A}_\omega^* = \max_{i \in \{1,\dots,R\}} \mathcal{A}_{\omega,i} \; \forall \omega$.

In the second iteration, the attacking rewards do not change; i.e. $J_{\omega,A,s}^{(2)} = \frac{c_{\omega,s}}{1-T_{\omega,s}} \; \forall \omega, s$. Regarding the listening rewards, for every ω, similar to (11.14), we obtain

$$J_{\omega,L,r}^{(2)} = \frac{K_{\omega,r}^{(1)}}{1-U_r} = \frac{\rho(1-\delta_{l_r}(m_r))\alpha_r(1-\eta_r)}{1-U_r} \sum_{n=1}^{\infty} \frac{e^{-\lambda_r}\lambda_r^n}{n!} \mathcal{A}_\omega^*$$

$$= \mathcal{A}_\omega^* \underbrace{\frac{\rho(1-\delta_{l_r}(m_r))\alpha_r(1-\eta_r)}{1-U_r} \sum_{n=1}^{\infty} \frac{e^{-\lambda_r}\lambda_r^n}{n!} \left(\prod_{i=0}^{n-1} \frac{\mathcal{A}_{\omega+i+1}^*}{\mathcal{A}_{\omega+i}^*} \right)}_{\chi_{\omega,(r)}^*}. \tag{11.16}$$

Since $\chi_{\omega,(r)}^*$ is not necessarily a decreasing function of ω, in the second iteration, the monotonic relation between the listening and attacking rewards cannot be directly established as in the single-resource case discussed below:

Remark 11.2   If there is a single resource; i.e. $R = 1$, then $\chi_{\omega,(1)}^*$ is a decreasing function of ω, which results in an optimality of attacking over listening for $\chi_{\omega,(1)}^* \leq 1$, or equivalently $\omega \geq \tilde{\omega}_{(1,1)}$. Thus, in the single-resource case, listening is optimal for $\omega < \tilde{\omega}_{(1,1)}$, and attacking is optimal for $\omega \geq \tilde{\omega}_{(1,1)}$.

Even though (11.16) may not end up with two distinct convex sets of ω which stands for listening-optimal and attacking-optimal ω (as in the single-resource case), at the end of value iteration algorithm, $J_{\omega,*}^{(\infty)}$ converges to the attacker rewards which correspond to the optimality of listening or attacking for every ω. Since there are infinitely many ω for which attacking is optimal, the attacker reaches that amount of observation by listening to the inputs from the user, and then she always prefers attacking over listening.

11.5 Optimal Defense Strategy

The operator maintains multiple resources and manages its security risk. We consider two defender types: risk-averse and risk-seeking. The risk-averse defender aims at finding system parameters so that it is not rational for the attacker to compromise the system, thus a rational adversary would not attack the system. The risk-seeking defender takes a risk of being attacked by choosing system parameters that lead to low operation cost. Intuitively, the risk-averse defender invests in the IDS and sets a high false alarm rate for continuous authentication security, whereas the risk-seeking defender is exposed to penalty due to successful attacks. In this section, first we define the expected defender utility, then we express it for the case of no attacker (which corresponds to a risk-averse defender) and for the case with an attacker (which corresponds to a risk-seeking defender). The effects of system parameters and trade-offs will be illustrated and discussed in Section 11.6.

11.5.1 Expected Defender Utility

Unlike the expected attacker utility, which is expressed as a discounted sum, the expected defender utility is expressed as the average per time slot utility; i.e. the average utility of the defender. Note that while the attacker obtains a reward only until it is detected, i.e. in expectation over a finite horizon, the defender obtains its reward over an infinite horizon, and may interact with multiple attackers over time. Depending on the system parameters η_r and m_r, in every time slot, the defender

- gets a reward v_r if the user interacts with resource r successfully,
- gets a penalty c_r if the attacker successfully attacks resource r,
- pays $\sum_{r \in \mathcal{R}} m_r$ as the operation cost of the IDS.

Let us define the total defender utility over the next k time slots as $\mathcal{V}_k(\text{UB})$ and $\mathcal{V}_k(\text{BL})$ when the current state is BL and UB, respectively. Then, the expected (average) defender utility can be expressed as $\overline{\mathcal{V}} = \lim_{k \to \infty} \frac{\mathcal{V}_k(\text{UB})}{k}$, if the limit exists. For example, when the current state is UB, then the defender gets an immediate reward v_r if the next state is also UB. However, if the current state is BL, then the defender cannot get a reward in a single time slot. Based on this, now we can investigate the expected defender utility when there is no attacker.

11.5.2 Analysis without an Attacker

As discussed previously, in the case of no attacker, the states and transition probabilities are expressed in (11.1) as shown in Figure 11.2a. Let the expected defender reward (i.e. excluding the penalties and operation costs) be denoted by $\overline{\mathcal{V}}^+$. Note that the defender gets an immediate reward v_r only if the system is in the unblocked state and the user input does not cause a false alarm; i.e. with probability $\overline{P}$. Therefore, it holds that $\mathcal{V}_1^+(\text{UB}) = \overline{P}v_r$. Similarly, when the current state is BL, $\mathcal{V}_1^+(\text{BL}) = 0$ holds. This analysis can be extended to 2 and 3 time slots as follows:

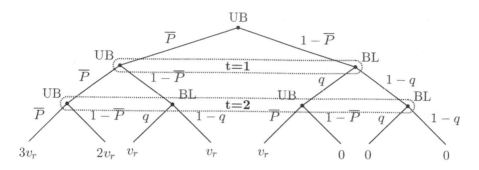

Theorem 11.3   When there is no attacker, the expected defender reward is $\overline{\mathcal{V}}^+ = \lim_{k\to\infty} \frac{\mathcal{V}_k^+(\text{UB})}{k} = \frac{\overline{P}q}{q-\overline{P}+1}v_r$.

Proof: For the general case, observe the following recursion

$$\mathcal{V}_k^+(\text{UB}) = \overline{P}(v_r + \mathcal{V}_{k-1}^+(\text{UB})) + (1 - \overline{P})\mathcal{V}_{k-1}^+(\text{BL})$$
$$\mathcal{V}_k^+(\text{BL}) = q\mathcal{V}_{k-1}^+(\text{UB}) + (1 - q)\mathcal{V}_{k-1}^+(\text{BL})$$

Then, the recursion above can be expressed in vector form as

$$\begin{bmatrix} \mathcal{V}_k^+(\text{UB}) \\ \mathcal{V}_k^+(\text{BL}) \end{bmatrix} = \begin{bmatrix} \overline{P} & 1 - \overline{P} \\ q & 1 - q \end{bmatrix} \begin{bmatrix} \mathcal{V}_{k-1}^+(\text{UB}) \\ \mathcal{V}_{k-1}^+(\text{BL}) \end{bmatrix} + \begin{bmatrix} \overline{P}v_r \\ 0 \end{bmatrix},$$

with an initial condition $\begin{bmatrix} \mathcal{V}_1^+(\text{UB}) \\ \mathcal{V}_1^+(\text{BL}) \end{bmatrix} = \begin{bmatrix} \overline{P}v_r \\ 0 \end{bmatrix}$. Then, letting $\mathcal{A} \triangleq \begin{bmatrix} \overline{P} & 1 - \overline{P} \\ q & 1 - q \end{bmatrix}$, we obtain

$$\begin{bmatrix} \mathcal{V}_k^+(\text{UB}) \\ \mathcal{V}_k^+(\text{BL}) \end{bmatrix} = \sum_{j=0}^{k-1} \mathcal{A}^j \begin{bmatrix} \overline{P}v_r \\ 0 \end{bmatrix}, \tag{11.17}$$

where $\mathcal{A}^0 = \begin{bmatrix} 1 & 0 \\ 0 & 1 \end{bmatrix}$; i.e. 2×2 identity matrix. Since the eigen-decomposition of $\mathcal{A}$ is

$$\mathcal{A} = \begin{bmatrix} \overline{P} & 1 - \overline{P} \\ q & 1 - q \end{bmatrix} = \begin{bmatrix} 1 & \overline{P} - 1 \\ 1 & q \end{bmatrix} \begin{bmatrix} 1 & 0 \\ 0 & \overline{P} - q \end{bmatrix} \left(\begin{bmatrix} 1 & \overline{P} - 1 \\ 1 & q \end{bmatrix} \right)^{-1},$$

(11.17) reduces to

$$\begin{bmatrix} \mathcal{V}_k^+(\text{UB}) \\ \mathcal{V}_k^+(\text{BL}) \end{bmatrix} = \sum_{j=0}^{k-1} \begin{bmatrix} 1 & \overline{P} - 1 \\ 1 & q \end{bmatrix} \begin{bmatrix} 1 & 0 \\ 0 & (\overline{P} - q)^j \end{bmatrix} \left(\begin{bmatrix} 1 & \overline{P} - 1 \\ 1 & q \end{bmatrix} \right)^{-1} \begin{bmatrix} \overline{P}v_r \\ 0 \end{bmatrix}$$

$$= \begin{bmatrix} 1 & \overline{P} - 1 \\ 1 & q \end{bmatrix} \begin{bmatrix} k & 0 \\ 0 & \dfrac{1 - (\overline{P} - q)^k}{1 - (\overline{P} - q)} \end{bmatrix} \left(\dfrac{1}{q - \overline{P} + 1} \begin{bmatrix} q & 1 - \overline{P} \\ -1 & 1 \end{bmatrix} \right) \begin{bmatrix} \overline{P}v_r \\ 0 \end{bmatrix}$$

$$= \dfrac{\overline{P}v_r}{q - \overline{P} + 1} \begin{bmatrix} kq - (\overline{P} - 1)\dfrac{1 - (\overline{P} - q)^k}{1 - (\overline{P} - q)} \\ kq - q\dfrac{1 - (\overline{P} - q)^k}{1 - (\overline{P} - q)} \end{bmatrix}. \tag{11.18}$$

Hence, by (11.18), the expected defender reward is

$$\overline{\mathcal{V}}^+ = \lim_{k \to \infty} \frac{\mathcal{V}_k^+(\text{UB})}{k} = \lim_{k \to \infty} \frac{\overline{P}v_r}{q - \overline{P} + 1} \frac{kq - (\overline{P} - 1)\dfrac{1 - (\overline{P} - q)^k}{1 - (\overline{P} - q)}}{k} = \frac{\overline{P}q}{q - \overline{P} + 1} v_r. \tag{11.19}$$

Note that since there is no attacker, the defender gets no penalty. Thus, the expected defender utility can be expressed as $\overline{\mathcal{V}} = \overline{\mathcal{V}}^+ - \sum_{r \in \mathcal{R}} m_r$, which brings us to the following conclusion.

Corollary 11.2   The defender selects $\eta_r = m_r = 0$ to maximize her expected utility when there is no attacker.

Proof: Since $\overline{\mathcal{V}} = \frac{\overline{P}q}{q - \overline{P} + 1} v_r - \sum_{r \in \mathcal{R}} m_r$, to maximize $\overline{\mathcal{V}}$, $\sum_{r \in \mathcal{R}} m_r$ must be zero. Further, since $\overline{\mathcal{V}}$ is an increasing function of $\overline{P}$, η_r should be selected to maximize $\overline{P}$. Then, $\overline{P} = \sum_{r=1}^{R} \alpha_r P_r = \sum_{r=1}^{R} \alpha_r \left(e^{-\lambda_r} + (1 - e^{-\lambda_r})(1 - \eta_r) \right)$ implies that all false alarm rates η_r must be selected as zero in order to maximize the expected defender utility.

11.5.3   Analysis with an Attacker

To establish an optimal defense strategy, observe that the defender (operator) is the leader, and thus, for any defense strategy (m_r, η_r) it can anticipate the attacker to use the optimal attack strategy derived above. Thus, the defender can make use of the transition probabilities described in Section 11.4.1.

From the perspective of the defender, the system state consists of the triplet $(t, L(t) = \omega, S(t))$ which evolves over time as an MDP, as discussed in Section 11.4.1. Note that the evolution of the triplet $(t, L(t) = \omega, S(t))$ starting from $(0, 0, \text{UB})$ can be represented as an infinite directed graph with countably many vertices (since t and w are discrete, and there are three possible states $S(t)$). Vertices correspond to states and edges to state transitions. For example, if the optimal action of the attacker is to listen to resource r when the amount of observation is ω, the transition from (t, ω, UB) to $(t, \omega + N, \text{UB})$ results in the defender utility of $v_r - \sum_{s \in \mathcal{R}} m_s$ with probability $(1 - \delta_{l_r}(m_r))$ $\alpha_r(1 - \eta_r) \frac{e^{-\lambda_r} \lambda_r^N}{N!}$ for $N = 1, 2, \dots$. In particular, the defender makes use of Table 11.1 depending on the optimal action of the attacker for a given amount of observation ω. The game continues until the attacker is detected, i.e. until $S(t) = \text{AD}$ for some t, and the defender is interested in maximizing its expected (average) utility (i.e. per time slot average utility) through all possible realizations of the events and transitions. However, since the defender does not know $L(t) = \omega$, stochastic averaging is needed to obtain the expected defender utility. In the next section, after presenting the illustrations, we provide a corresponding discussion.

11.6   Numerical Results

In the following we illustrate the optimal strategies and expected utilities of the attacker and defender using results from extensive simulations. For simplicity, we assume a single resource (thus the subscript r is omitted), and we use the default parameter values shown in Table 11.2. For continuous authentication security we consider $\eta \leq 0.01$ to keep the annoyance due to FP low. Furthermore, in accordance with Section 11.3.5, we consider $\delta_l(m) = 1 - e^{\delta_L m}$ and $\delta_a(m) = 1 - e^{\delta_A m}$ with $\delta_A = 7$, $\delta_L = 4.2$ (to make sure that $\delta_a(m) \geq \delta_l(m)$ always holds, which is a

Table 11.1 System state evolution from the perspective of the defender.

Current state	Optimal action	Next state	Probability	Utility
(t, ω, BL)	$l = 0, a = 0$	$(t+1, \omega, \text{UB})$	(11.3)	$-\sum_{s \in R} m_s$
(l, ω, BL)	$l = 0, a = 0$	$(t+1, \omega, \text{BL})$	(11.3)	$-\sum_{s \in R} m_s$
(t, ω, UB)	$l = r, a = 0$	$(t+1, \omega, \text{AD})$	(11.4)	$-\sum_{s \in R} m_s$
(t, ω, UB)	$l = r, a = 0$	$(t+1, \omega, \text{UB})$	(11.4)	$-\sum_{s \in R} m_s$
(t, ω, UB)	$l = r, a = 0$	$(t+1, \omega + N, \text{UB})$	(11.4)	$v_r - \sum_{s \in R} m_s$
(t, ω, UB)	$l = r, a = 0$	$(t+1, \omega, \text{BL})$	(11.4)	$-\sum_{s \in R} m_s$
(t, ω, UB)	$l = 0, a = r$	$(t+1, \omega, \text{AD})$	(11.5)	$-\sum_{s \in R} m_s$
(t, ω, UB)	$l = 0, a = r$	$(t+1, \omega, \text{UB})$	(11.5)	$-c_r - \sum_{s \in R} m_s$
(t, ω, UB)	$l = 0, a = r$	$(t+1, \omega, \text{BL})$	(11.5)	$-\sum_{s \in R} m_s$

Table 11.2 Default parameters.

λ	B	v	c	q	ρ	γ	C_a
10	$\mathcal{N}(100, 0.5)$	0.5	1	0.7	0.98	0.01	5

reasonable assumption), and $m \leq v = 0.5$ (which is due to (11.19); i.e. max. achievable expected defender utility in the case of no attacker).

Figure 11.4 shows the observation/attack threshold $\tilde{\omega}$ (which can only take integer values) to start attacking as a function of detection parameters η and $\delta_l(m)$ (which is an equivalent of m, due to a one-to-one relation between m and $\delta_l(m)$ described above). From Figure 11.4b and c, it can be deduced that the attacking is more desirable when the FP rate is low since the success probability of attack is higher.[11] Furthermore, as proved in Proposition 11.4, it can be observed from Figure 11.4d that the attacker is more willing to attack (instead of listening) as the probability of being detected increases, so that $\tilde{\omega}$ is nonincreasing with $\delta_l(m)$ (and simultaneously with $\delta_a(m)$).

Figure 11.5 illustrates the relation between the expected attacker utility and the defender parameters η and m. In Figure 11.5a, the white area represents the negative utility for the attacker; i.e. the attacker does not prefer to compromise the system, whereas the gray area represents the positive utility for the attacker; i.e. the attacker prefers to compromise the system. Note that the boundary between these two regions; i.e. the attacker is indifferent between compromising and not compromising the system, is clearly observed. Since a higher false-positive rate implies a higher true positive rate, and thus, the system stays in the blocking state longer, the utility of the attacker is a convex decreasing function of η as shown in Figure 11.5c. Similarly, the expected attacker utility is also a convex decreasing function of the detection probabilities, as shown in Figure 11.5d.

Figure 11.5a–d above show the optimal observation/attack threshold and the optimal expected attacker utility as a function of the defender parameters η and m. In the following, we compare different threshold strategies of attacker; i.e. we compare the expected attacker utility with different

---

11 Note that, as illustrated in figure 4a of Sarıtaş et al. (2019), since the system stays in the blocking state longer for the higher FP rates (i.e. as η approaches to 1), attacking would be the preferred action; i.e. the attacker is urged to attack for the lower and higher rates of FP.

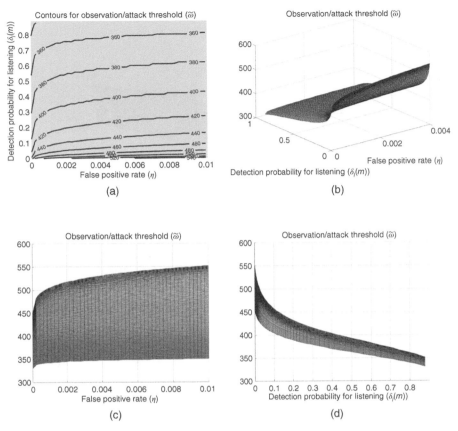

Figure 11.4 Observation/attack threshold ($\tilde{\omega}$) vs. detection parameters. (a) Contour plot of threshold ($\tilde{\omega}$) vs. detection parameters, (b) Threshold ($\tilde{\omega}$) as a function of detection parameters, (c) Threshold ($\tilde{\omega}$) vs. η as a 2D projection of (b), and (d) Threshold ($\tilde{\omega}$) vs. $\delta_l(m)$ as a 2D projection of (b).

observation/attack threshold values. For this part, we assume that the defender parameters are $m = 0.00144$ and $\eta = 0.006$. As it can be observed in Figure 11.6, a threshold strategy with an optimal observation/attack threshold $\tilde{\omega} = 522$ performs better than the other threshold strategies. The intuition behind this is as follows: if the attacker prefers to attack before reaching the optimal threshold, since the attacker does not gather enough amount of observation to attack, the probability of having a successful attack is not high. On the other hand, if the attacker has more observation than the threshold, even though the success probability of the attack would be higher, the expected attacker utility diminishes due to the discounting factor and the probability of being detected while listening.

As discussed in Section 11.5, the optimization of the expected defender utility consists of taking the average of infinitely many states (t, ω, AD) and infinitely many different paths. In order to be able to cover all most-likely states (t, ω, AD) and the corresponding utilities, we implemented Monte Carlo simulations that run the dynamic stochastic game with corresponding states and transition probabilities in Table 11.1 $1 \cdot 10^5$ times for a given parameter set of the defender (η and m), and then we take the average of all results to obtain the expected defender utility. While obtaining the expected average defender utility, we assume that the defender knows the compromising cost C_a of attacker so that the defender can anticipate whether the attacker could compromise the system or not, thus the defender can calculate her corresponding utilities with or without the attacker.

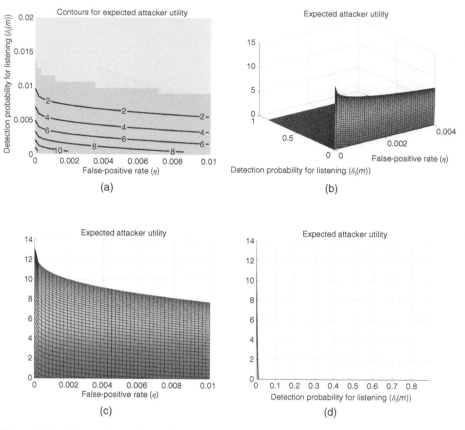

Figure 11.5 Expected attacker utility vs. detection parameters. (a) Contour plot of expected attacker utility vs. detection parameters, (b) Expected attacker utility as a function of detection parameters, (c) Expected attacker utility vs. η as a 2D projection of (b), (d) Expected attacker utility vs. $\delta_l(m)$ as a 2D projection of (b).

Figure 11.7 shows the expected defender utility as a function of detection parameters η and m. The regions with and without the attacker (gray and white, respectively) are clearly visible in Figure 11.7a, and it can be observed from Figure 11.7b that there is a sharp transition on the expected defender utility at the boundary between these two regions. Since the expected defender utility is not dependent on the IDS parameter m as proved in (11.19), in the case of no attacker, the expected defender utility decreases as m increases, which can be observed in Figure 11.7d.

11.7 Conclusion and Discussion

We considered a security risk management in a system combining continuous authentication and intrusion detection as a dynamic discrete stochastic leader-follower game with imperfect information between an attacker and a defender (an operator). We derived a backward recursion for the optimal attacker reward, which allowed us to characterize the optimal attack strategy. Then, based on the optimal strategy of the attacker we expressed the expected utility of the defender. Extensive simulations illustrate the relationship between the optimal strategies/utilities and the defender parameters and show that continuous authentication can be very efficient for security risk reduction, if combined with appropriate incident detection.

Our model and analysis rely on a number of assumptions that are worth discussing. First, we assume that the attacker is aware of the system parameters used by the defender, and is thus able

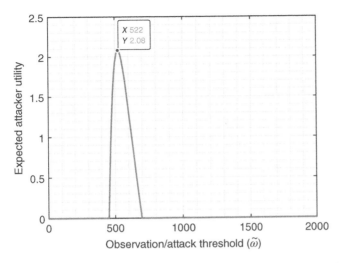

Figure 11.6 Expected attacker utility vs. observation/attack threshold $\tilde{\omega}$ when the defender parameters are $m = 0.00144$ and $\eta = 0.006$.

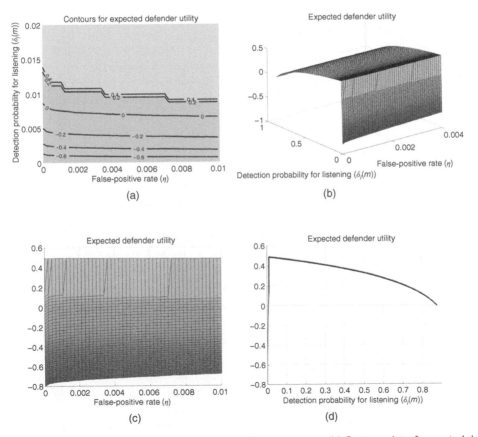

Figure 11.7 Expected defender utility vs. detection parameters. (a) Contour plot of expected defender utility vs. detection parameters, (b) Expected defender utility as a function of detection parameters, (c) Expected defender utility vs. η as a 2D projection of (b), (d) Expected defender utility vs. $\delta_l(m)$ as a 2D projection of (b).

to choose the optimal attack strategy. This may of course not be the case in practice, unless the defender announces the system parameters publicly. Second, we assume that the defender knows this, and can use this knowledge for optimizing the system parameters. Nonetheless, in practice the attacker may not know the system parameters, and-even worse-the defender does not know whether the attacker knows the system parameters. Thus, even a risk-averse defender (one that sets the system parameters so as to make the expected attacker reward negative) may be attacked by an attacker that is not aware of the system parameters and thus counts on a positive expected reward. Our model has many possible interesting extensions. On the one hand, blocking due to failed behavioral authentication may be resource-specific; on the other hand, the behavior of the user may not be the same for all resources. In addition, one could consider multiple users and/or attackers. The model could also be extended to include other detection and mitigation schemes and more sophisticated models of incident response automation. Overall, there is a huge and increasing variety of detection and mitigation schemes, and a comprehensive optimization framework will become a fundamental tool for security risk management.

References

S. Ayeswarya and J. Norman. A survey on different continuous authentication systems. *International Journal of Biometrics*, 11(1):67–99, 2019.

P. Bours. Continuous keystroke dynamics: A different perspective towards biometric evaluation. *Information Security Technical Report*, 17(1):36–43, 2012. Human Factors and Bio-metrics.

P. Bours and S. Mondal. Continuous authentication with keystroke dynamics. In Y. Zhong and Y. Deng, editors, *Recent Advances in User Authentication Using Keystroke Dynamics Biometrics*, volume 2, chapter 3. Science Gate Publishing, Burlington, MA, USA, 2015.

M. Cardaioli, M. Conti, K. S. Balagani, and P. Gasti. Your PIN sounds good! on the feasibility of PIN inference through audio leakage. *arXiv preprint arXiv:1905.08742*, 2019.

A. Castiglione, K. Raymond Choo, M. Nappi, and S. Ricciardi. Context aware ubiquitous biometrics in edge of military things. *IEEE Cloud Computing*, 4 (6):16–20, Nov. 2017.

H. Çeker and S. Upadhyaya. Enhanced recognition of keystroke dynamics using Gaussian mixture models. In *IEEE Military Communications Conference (MILCOM)*, pp. 1305–1310. IEEE, 2015.

T. Dee, I. Richardson, and A. Tyagi. Continuous transparent mobile device touchscreen soft keyboard biometric authentication. In *International Conference on VLSI Design (VLSID)*, pp. 539–540. IEEE, 2019.

I. Deutschmann and J. Lindholm. Behavioral biometrics for DARPA's Active Authentication program. In *2013 International Conference of the BIOSIG Special Interest Group (BIOSIG)*, pp. 1–8, September 2013.

I. Deutschmann, P. Nordström, and L. Nilsson. Continuous authentication using behavioral biometrics. *IT Professional*, 15(4):12–15, 2013.

M. Ferro, G. Pioggia, A. Tognetti, G. D. Mura, and D. De Rossi. Event related biometrics: Towards an unobtrusive sensing seat system for continuous human authentication. In *International Conference on Intelligent Systems Design and Applications*, pp. 679–682, November 2009.

L. Goncalves, A. Subtil, R. M. Oliveira, and P. de Zea Bermudez. ROC curve estimation: An overview. *Revstat—Statistical Journal*, 12:1–20, 2014.

D. Hosseinzadeh and S. Krishnan. Gaussian mixture modeling of keystroke patterns for biometric applications. *IEEE Transactions on Systems, Man, and Cybernetics, Part C (Applications and Reviews)*, 38(6):816–826, 2008.

M. H. R. Khouzani, P. Mardziel, C. Cid, and M. Srivatsa. Picking vs. guessing secrets: A game-theoretic analysis. In *IEEE Computer Security Foundations Symposium*, pp. 243–257. IEEE, 2015.

T. Ma, J. G. Holden, and R. A. Serota. Distribution of human response times. *Complexity*, 21(6):61–69, 2016.

G. Peng, G. Zhou, D. T. Nguyen, X. Qi, Q. Yang, and S. Wang. Continuous authentication with touch behavioral biometrics and voice on wearable glasses. *IEEE Transactions on Human-Machine Systems*, 47(3):404–416, 2017.

S. Sarıtaş, E. Shereen, H. Sandberg, and G. Dán. Adversarial attacks on continuous authentication security: A dynamic game approach. In T. Alpcan, Y. Vorobeychik, J. S. Baras, and G. Dán, editors, *Decision and Game Theory for Security*, pp. 439–458. Springer International Publishing, Cham, 2019. Springer International Publishing.

Z. Sitová, J. Šeděnka, Q. Yang, G. Peng, G. Zhou, P. Gasti, and K. S. Balagani. HMOG: New behavioral biometric features for continuous authentication of smartphone users. *IEEE Transactions on Information Forensics and Security*, 11(5):877–892, 2016.

L. Xiao, Y. Li, G. Han, G. Liu, and W. Zhuang. Phy-layer spoofing detection with reinforcement learning in wireless networks. *IEEE Transactions on Vehicular Technology*, 65(12):10037–10047, 2016.

L. Yang, Y. Lu, S. Liu, T. Guo, and Z. Liang. A dynamic behavior monitoring game-based trust evaluation scheme for clustering in wireless sensor networks. *IEEE Access*, 6:71404–71412, 2018.

G. Yunchuan, Y. Lihua, L. Licai, and F. Binxing. Utility-based cooperative decision in cooperative authentication. In *IEEE INFOCOM*, pp. 1006–1014. IEEE, 2014.

12

Cyber Autonomy in Software Security: Techniques and Tactics

Tiffany Bao and Yan Shoshitaishvili

School of Computing, Informatics, and Decision Systems Engineering, Arizona State University, Tempe, AZ, USA

12.1 Introduction

Our lives have been significantly dependent on software programs-check the applications installed in your computers, tablets, cellphones, and smart devices. Software helps us to be more productive in work and more convenient in life; unfortunately, many software programs have bugs. In particular, security bugs, which we call software vulnerabilities in this chapter, are severe defects that allow attackers to exploit. Attackers, taking advantage of software vulnerabilities, will gain unauthorized access to the vulnerable programs and affect the confidentiality, integrity, and availability of the programs and even the residing computers.

Researchers have been seeking solutions to address software vulnerabilities in the world. Given billions of programs, it is impossible to manually go over the programs and inspect for vulnerabilities. Instead, researchers are looking for approaches to *automatically* discover and mitigate software vulnerabilities, which we call the field of study *Cyber Autonomy in Software Security*.

Note that the notation of cyber autonomy also exists in other realms, such as autonomous vehicles. In the software security domain, specifically, cyber autonomy means to develop *techniques* that automatically discover vulnerabilities, prove the exploitability of a vulnerability, generate the patches, as well as *tactics* that autonomously decide whether and when to patch.

For techniques, we hope to develop techniques to find and fix all software vulnerabilities. However, finding all vulnerabilities is known to be intractable: analogous to the halting machine problem where the goal is to decide whether a program will terminate, deciding whether a program contains a vulnerability may never terminate. The goal of vulnerability discovery techniques is to find more vulnerabilities in a shorter time.

For tactics, the readers may wonder the point of deciding whether or when to release a patch-it seems that we should always release a patch as soon as possible in order to defend against attacks and minimize potential loss. Interestingly, patching right away is not always the best solution in reality, because patch reveals the vulnerability, and that deploying a patch cost time. If a patch is released yet there are vulnerable machines remain unpatched, then attackers may learn the vulnerability from the patch and attack those machines that have not been patched. Therefore, we need to decide the best timing to release a patch.

Under such circumstances, researchers have been developing software security techniques (Bao et al. 2014; Brumley et al. 2008a, b; Shoshitaishvili et al. 2015, 2016) and study optimal strategies based on the status quo of techniques (Bao et al. 2017a; Moore et al. 2010). Techniques and

Game Theory and Machine Learning for Cyber Security, First Edition.
Edited by Charles A. Kamhoua, Christopher D. Kiekintveld, Fei Fang, and Quanyan Zhu.
© 2021 The Institute of Electrical and Electronics Engineers, Inc. Published 2021 by John Wiley & Sons, Inc.

strategy complement each other; techniques execute actions that are determined by strategies, and strategies are calculated from a model which describes a specific cyber security scenario such as what to do with the discovery of a zero-day vulnerability.

In this chapter, we introduce cyber autonomy in software security. We start with a real-world example which helps to understand software vulnerability mitigation process and motivate the techniques and tactics for software vulnerability mitigation. In the next, we will present the model for cyber autonomy, introduce the related techniques, and discuss the algorithm to solve the cyber autonomy model.

12.2 Background

We introduce vulnerability CVE-2017-0144 (2017) as the motivating example for the readers to understand vulnerability life cycle. Comparing to most of the vulnerabilities, CVE-2017-0144 is a complicated vulnerability handling case as it involves multiple parties, and such complexity help us to understand the process comprehensively.

CVE-2017-0144 is a software vulnerability targeting Windows Server Message Block service. Taking advantage of the vulnerability, attackers are able to execute code remotely in compromised computers. CVE-2017-0144 exists in a majority of Windows operating systems, including Microsoft Windows Vista SP2, Windows Server 2008 SP2 and R2 SP1, Windows 7 SP1, Windows 8.1, Windows Server 2012 Gold and R2, Windows RT 8.1, Windows 10 and Windows Server 2016 (CVE-2017-0144 2017). Due to the severity of the vulnerability and the large population of victim users, the attacks of the vulnerability are very serious. The WannaCry attack, one of the attacks using the vulnerability, infected 300 000 machines (Chappell 2017) and caused $4 billion losses (Berr 2017) in the end, which is considered the biggest ransomware attack in history (Raconteur 2017). For more technical details regarding the vulnerability, please refer to the CVE entry (CVE-2017-0144 2017).

The timeline of CVE-2017-0144 is shown in Figure 12.1. We divide the timeline into five phases as follows:

- 2012 (or possibly earlier): The vulnerability is identified by the U.S. National Security Agency. The NSA learned that the vulnerability affects many Windows operating systems, and it decided to withhold the vulnerability and produce its exploit called EternalBlue. The EternalBlue exploit was used as a weapon targeting computers in other countries such as Iran, India and Syria.
- January 2017: The EternalBlue exploit was stolen by the Shadow Broker and NSA noticed the possible leak of the vulnerability.
- February 2017: NSA noticed Microsoft in terms of the vulnerability.
- March 2017: Vulnerability CVE-2017-0144 was published, and the patch was released on March's Patch Tuesday (Microsoft Security Updates 2017).

Figure 12.1 The timeline of vulnerability CVE-2017-0144.

2012	NSA discovered the vulnerability
Jan 2017	NSA learned about a possible leak
Feb 2017	NSA disclosed the vulnerability
Mar 2017	Microsoft released the patch
May 2017	WannaCry attack started

Figure 12.2 Vulnerability lifecycle.

- May 2017: The WannaCry attack started. It was ransomware using vulnerability CVE-2017-0144. The first outbreak lasted for 4 days. Later, this attack was attributed to North Korea, according to the U.S. government (Nakashima and Rucker 2017).

The motivating example demonstrates the lifetime of a vulnerability, shown in Figure 12.2. On the high level, a vulnerability lifetime typically consists of the following events (Bilge and Dumitras 2012):

- Introduction. The vulnerability is introduced into the software and spread through the software users.
- Discovery. The vulnerability is discovered by a party.
- Disclosure. The vulnerability is disclosed by the software vendor or a third party.
- Patch Release. The patch of the vulnerability is released by the software vendor.
- Patch Deployment. The patch is deployed to vulnerable machines, and the vulnerability is fixed.

Although some vulnerabilities' patches are released before the disclosure, for most vulnerabilities, these events happen chronologically. One related work regarding vulnerability disclosure and patch release is vulnerability disclosure policy study. This topic studies the question that, if a third party-for example, Computer Emergency Response Team (CERT) or a bug bounty program-informs a software vendor of a vulnerability, how long of the time window should the third party set ahead of disclosing the vulnerability, in order to minimize the negative effect caused by the vulnerability. In this chapter, we bound patch release and vulnerability disclosure together, considering that a patch is released before or after a vulnerability is disclose in some amount of time.

Exploits of the vulnerability occur any time after the vulnerability is introduced. Specifically, the exploits occurring before vulnerability disclosure are called zero-day exploits, and those occurring after vulnerability disclosure and before patch deployment are called one-day exploits.

Involved parties

From a vulnerability's perspective, we can divide the involved parties into two leagues: the defenders and the attackers. The attackers, including malicious hackers, aim to take advantage of the vulnerability and gain benefits from the exploits. The defenders, including software vendors, the legitimate bug bounty programs, and benign individuals, corporate with each other and mitigate the vulnerability.

In the real world, a party can be both attacker and defender. Consider a country who may want to weaponize a vulnerability for national purpose. It may notify the software vendor and meanwhile attack with the vulnerability. In this case, a party, when making decisions for vulnerabilities, can decide for both attack and defend actions.

12.3 Related Work

Moore et al. (2010) proposed the cyber-hawk model in order to find the optimal strategy for both players. The cyber-hawk model describes a game where each player chooses to either keep vulnerabilities private (and create exploits to attack their opponent) or disclose. They conclude that the first player to discover the vulnerability should always attack, precluding vulnerability disclosure.

This model has three limitations. First, it limits the number of players to 2. Second, it assumes that both players will eventually discover the same zero-day vulnerabilities. Third, it models a one-shot game where players are only allowed to make one choice between attack and disclosure. After this choice, the game is over.

The cyber-hawk model raised new questions that need to be explored, such as if a player determines to use a vulnerability offensively, how soon they should commence the attack. Schramm et al. (2014) proposes a dynamic model to answer this question. The model indicates that waiting reduces a player's chance of winning the game, which implies that if a player determines to attack, he should act as soon as possible.

The Schramm model relies on a key assumption of *full player awareness*, requiring that players know whether, and how long ago, their opponent discovered a vulnerability. This assumption is unlikely to be valid in real-world scenarios because nations keep the retained vulnerabilities (if they have any at all) secret. Additionally, it is still limited to two players and supports only single taken action, after which the game ends.

We observe three things missing in previous models that are vital for choosing players' best strategies in computer security games. First, players in a computer security game often have multiple actions over multiple rounds. As an example of a multiple round game, consider the NSA case. Although NSA claimed to disclose 91% of all zero-day vulnerabilities, it did not mention whether they first exploited before disclosing or not.

Second, players in a computer security game are uncertain as to the other players' state, specifically, whether other players have discovered vulnerabilities. This uncertainty influences players' decision, as a player must account for all the possible states of the other players in order to maximize his expected utility. Previous approaches cannot be extended to handle multiple steps with partial information and dependencies.

Third, both attacking *and* disclosing reveal the information of a vulnerability. Previous work shows that a patch may be utilized by attackers to generate new exploits (Brumley et al. 2008b). However, we show that attacking leaks information, and we introduce the notion of ricochet into the game theory model. In the automated patch-based exploit generation (APEG) (Brumley et al. 2008b) technique, a player infers the vulnerable program point from analyzing a patch and then creates an exploit. Also, in the ricochet attack technique, a player detects an exploit (e.g. through network monitoring or dynamic analysis) and then turns around and uses the exploit against other players. For instance, Costa et al. have proposed monitoring individual programs to detect exploits, and then replaying them as part of their technique for self-certifying filters (Costa et al. 2005), where the filters self-certify by essentially including a version of the original exploit for replay. Both inadvertent disclosures through attack and patching create new game actions, but previous work does not take it into account. Policy makers and other users of previous models (Moore et al. 2010; Schramm et al. 2014) can reach incorrect conclusions and ultimately choose suboptimal strategies.

12.4 Model Setup

The example of vulnerability CVE-2017-0144 motivates the question of how to mitigate a discovered vulnerability, given the consideration that the vulnerability may be discovered and used by adversaries. To answer this question, we need to set up a model that characterizes the scenarios and is able to help us to reason about strategies. We refer the model the Software Vulnerability Mitigation Model. In this section, we will present the high-level design of the model.

Player Setup

A player involved in software security can be an individual, a software vendor, an organization, or a nation. It can be represented as a group a people with various skill sets, or even an autonomous system. Although different players have different technical setup, they own goal is to maximize their utility in terms of a software vulnerability. Therefore, we model the players by dividing the skill set by functionalities. In specific, we consider three components, listed as follows:

- **Vulnerability Awareness Component.** The vulnerability awareness component is responsible for realizing the existence of a vulnerability. When the component acquires a vulnerability, it will trigger the offense and defense components to generate an exploit and a patch for the vulnerability, respectively. In addition, the player will start to decide the strategy for the vulnerability and play the defensive or/and offensive actions.

  There are many ways to learn a vulnerability. For example, the component can get a vulnerability by discovering crashes from programs, detecting attacks through network or studying vulnerability databases such as the Common Vulnerabilities and Exposures (CVE) (CVE-2017-11882 2017) entries. These means can be used simultaneously; one can implement multiple techniques for the vulnerability awareness component, and it will help the autonomous system learn vulnerabilities more effectively and more efficiently.

- **Offense Component.** The offense component aims to weaponize a vulnerability to exploits that carry out the player's intended execution such as to install backdoors, to exfiltrate sensitive information and so on. A player with the generated exploit can choose to attack the other players, and the attack might bring benefits to the attacker. This component can be realized by techniques that generates exploits in general, such as automatic exploit generation (Avgerinos et al. 2011), automatic patch-based generation (Brumley et al. 2008b) and automatic attack ricochet (Bao et al. 2017b).

- **Defense Component.** The defense component is to generate a patch that protects a player from attacks. However, patch releasing takes time, during which a player may still be attacked due to the incomplete process. Patch releasing is also public, in the sense that the released patch will be known by all players in the game. This reflects the reality since patch releasing is usually acted by software vendors, which is a party exclusive from any players and their goal is to patch all users. This component can be realized by automatic patching engine such as Patcherex (Shellphish 2021).

Player States and Player Actions

A player either knows or does not know about a vulnerability. We use player state to represent such awareness. Each player i has a state denoted by θ_i in each round, where $\theta_i \in \Theta_i = \{\neg D, D\}$. $\neg D$ refers to the situation in which a player has not yet learned of a vulnerability, while D refers to the situation in which a player knows the vulnerability.

In each round, players choose one of the following actions: {ATTACK, PATCH, NOP, STOCKPILE}, where the semantics of ATTACK, PATCH, and NOP have their literal meaning, and STOCKPILEmeans holding a zero-day for future use.

Players are limited in their actions by their state. In particular, while a player in state $\neg D$ can only act NOP, a player in state D can choose an action among ATTACK, PATCH and STOCKPILE before the patch is released, and between ATTACK and NOP after the release, depending on their skill at detecting attacks or generating patch-based exploits.

Player Interactions

The players interact with each other as shown in Figure 12.3. There are two kinds of interaction: attacking and patching. For attacking, it has a culprit and a victim, which we call the attacker and

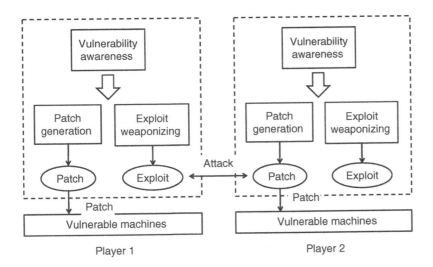

Figure 12.3 The interaction of two players.

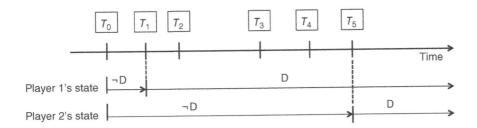

0. A vulnerability is introduced
1. Player 1 realizes the vulnerability
2. Player 1 weaponizes the vulnerability
3. Player 1 launches an attack
4. Player 1 generates a patch
5. Player 1 starts to patch and Player 2 realizes the vulnerability

Figure 12.4 A player interaction example.

the defender. We assume that if the defender detects the attack, he will know the attacker. For patching, we assume that it is public and thus all players know which player releases the patch.

The interactions happen over time. Figure 12.4 shows an instance of two players (player 1 and player 2) with a vulnerability. At the beginning of the interaction, both players are in state of $\neg D$. Player 1 realizes a vulnerability at T_1, after which he weaponizes the vulnerability at T_2 and generates a patch at T_4. After T_2 and T_4, player 1 starts to attack at T_3 and patch at T_5. While player 1 is patching, if player 2 has not yet discovered the vulnerability, he will know the vulnerability from the disclosure of the patch, and he will generate exploit and determine when to attack and when to defend.

Scope

In this chapter, we consider Software Vulnerability Mitigation Modelwithin the scope of the following assumptions:

- We assume that players are monitoring their systems, and may probabilistically detect an attack. We also assume they may be able to then *ricochet* the exploit to other players. We note that the detection may come through monitoring the network (in the case of network attacks), or other measures such as honeypots and dynamic analysis of suspicious inputs. For example, Vigilante (Costa et al. 2005) detects exploits and creates a self-certifying alert (SCA), which is essentially a replayable exploit. We note that such attacks may be detected over a network (e.g. in DEFCON CTFs these are called reflection attacks (Samurai CTF Team 2013)) or via dynamic analysis, as with Vigilante.

- We assume that patching is always public. Once a patch reveals, players can patch their machines, and those who have not yet generated the exploit can take advantage of the patch information. For example, Brumley *et al* shows this can be done in some instances automatically (Brumley et al. 2008b).

- We assume that the defender knows the attacker for the detected attacks. We do not consider the situation that a player eavesdrops the communication of the other players to learn a vulnerability from the attacks between the others.

- We focus on one vulnerability for Software Vulnerability Mitigation Model, assuming that vulnerabilities are used independently. Modeling the game with multiple vulnerabilities are challenging since players are uncertain about what vulnerabilities the other players have. We leave this as a future task.

12.5 Techniques

In this section, we introduce related software security techniques. We divide techniques based on their purpose, and we show the relationships of different technique categories in Figure 12.5. Vulnerability discovery techniques, such as fuzzing, identify vulnerabilities from programs. If the vulnerability is used for attacks, then one can employ automatic exploit generation techniques to weaponize a vulnerability to working exploits. To defend against the vulnerability, one can use automatic patching techniques to generate a patch or directly a patched programs that is robust to the attack of the vulnerability. Moreover, a patch may also reveal the details of a vulnerability; the automatic patch-based automatic exploit generation technique is able to produce an exploit based on the patch information. Also, a weaponized exploit also contains vulnerability information, which attacks ricochet techniques are able to leverage and produce newly customized exploits.

Vulnerability Discovery

Vulnerability discovery is to find and demonstrate that a software bug exists and the bug is a security-related. Vulnerability discovery techniques can be either static or dynamic. Static vulnerability discovery techniques, such as program verification, is to statically prove that a program is

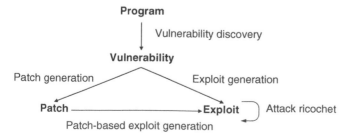

Figure 12.5 Techniques and their relationships.

secure, satisfying all predefined security properties. Dynamic vulnerability discovery, including automatic exploit generation (Avgerinos et al. 2011; Lin et al. 2008; Schwartz et al. 2011), is to run a program with inputs and check if the testing program executes normally. Static approaches are able to cover the entire program, but for binary analysis, resolving the missing information such as indirect jump is still open problem. For source code analysis, static approaches still suffer from false positive issues, and thus applying static approaches to real-world applications remains challenging. Dynamic approaches, on the other hand, does not have the performance issue since it runs the program. However, dynamic approaches cannot cover all possible paths, which makes the discovery results incomplete.

One of the popular vulnerability discovery approaches is fuzzing (AFL 2021; Aschermann et al. 2019; Böhme et al. 2016; Chen et al. 2018; Eddington 2021; Gan et al. 2018; Stephens et al. 2016; Yun et al. 2018). Fuzzing automatically generates different inputs, runs the program with the inputs, and test if the program executes normally. Fuzzing is commonly used for program testing since it is easy to set up. Meanwhile, researchers and developers have been developing advanced tools to improve the coverage of the fuzzed inputs on testing programs, such as Driller (Stephens et al. 2016) and TortoiseFuzz (Wang et al. 2020).

Vulnerability Weaponization

Vulnerability weaponization is to generate an exploit that fulfills the purpose of attacks. For example, if the purpose of an attack is to steal personal credential, then vulnerability weaponization, given a vulnerability will produce an exploit that takes advantage of the vulnerability, compromises the vulnerable computer, and successfully steal the secret.

The challenge of vulnerability weaponization is how to bypass the defense mechanisms in modern operating systems and CPU implementations, such as Address Space Layout Randomization (ASLR) and Supervisor Mode Access Prevention (SMAP). To tackle these challenges, researchers have been looking for approaches such as Return-Oriented Programming (ROP) to bypass extra mitigation (Schwartz et al. 2011), as well as symbolic execution-based approaches to deal with kernel vulnerabilities (Wu et al. 2018).

Some techniques do both vulnerability discovery *and* exploit generation, such as AEG and its derivatives (Avgerinos et al. 2011; Brumley et al. 2008b; Lin et al. 2008). Although they do both jobs, most of the work can be separated into two steps, and they discover vulnerabilities first and then generate corresponding exploits.

Patch Generation

Patch generation is that given a vulnerability to automatically generate a program that is robust to the vulnerability while maintain the original functionality of the vulnerable program. For source code, the key challenge of patch generation is how to maintain the correctness and the availability of the patched program. For binary code, an additional challenge is how to modify the binary code correctly so that the pointers of the binaries will not be affected. Although in most cases source code patching would be sufficient, binary patching is still necessary due to two reasons. First, sometimes the source code of the patched program no longer exists. For example, Microsoft binary-patched the Equation Editor executable in Microsoft Office to fix vulnerability CVE-2017-11882 (2017). Second, the users of a program or a third party security advisor may want to patch without knowing the source code. Therefore, binary patching is necessary in addition to source patching.

Patch-based Exploit Generation

Brumley et al. (2008b) have shown that automated patch-based exploit generation is possible and the technique can be used for generating exploits for real-world vulnerabilities. The existence of such technique highlights that an autonomous system is capable of generating exploits from a released patch, and generating exploits from a released patch is more effective than discovering vulnerabilities from a program. Therefore, we should take into account such technique for the autonomous computer security game.

Attack Ricochet

Attack ricochet is a class of techniques that re-purpose an exploit. It takes an exploit and an attack goal as input, and it produces a new exploit that takes advantage the same vulnerability as the input attack and achieves the specified attack goal. For instance, ShellSwap (Bao et al. 2017b) is an attack ricochet technique that changes the shellcode of original exploits. Attack ricochet techniques enable the victim of an attack to turn the table and become the attacker after he receives the exploit. It implies that a player may be able to learn the vulnerability and generate his own attacks after he gets attacked by his opponents. Also, if a player uses a zero-day vulnerability to attack other players, the player needs to be prepared that he may receive attacks of the same vulnerability due to his own exploits.

12.6 Tactics

In this section, we discuss how to calculate the strategy for Software Vulnerability Mitigation Model. To this end, we will formalize the software security process and investigate approaches that solve the formalized game.

12.6.1 Model Parameters

The Software Vulnerability Mitigation Modelneeds to be general and compatible with known software security events such as the presented motivating example. One requirement is that the model must support players with comprehensive techniques, rather than the simple choices that prior models allow. Figure 12.6 shows the workflow of a player in our model. A player may detect an attack, receive the disclosure from other players, or discover the vulnerability by himself. After a player learns a vulnerability, the strategy generator will compute the strategy for the vulnerability.

Players

All players are participating in a networked scenario. They are capable of finding new vulnerabilities, monitoring their own connections and ricocheting an attack, patching after vulnerability

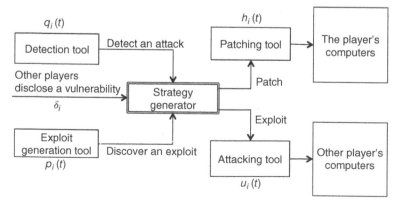

Figure 12.6 A player's parameters in Software Vulnerability Mitigation Model.

Table 12.1 Parameters for player i.

Parameter	Definition
$p_i(t)$	The probability distribution over time that player i discovers a vulnerability at round t.
$q_i(t)$	The probability to launch a ricochet attack with exploits that player i received in the previous round.
$h_i(t)$	The ratio of the amount of patched vulnerable resources over the total amount of vulnerable resources by t rounds *after* the vulnerability is disclosed.
δ_i	The number of rounds required by player i to generate a patch-based exploit after a vulnerability and the corresponding patch are disclosed.
$u_i(t)$	The dynamic utility that player i gains by attacking his opponents at round t.

disclosure, and generating patch-based exploits. Players may have different levels of skills, which are characterized using the parameters listed in Table 12.1.

These parameters are a substantial component of our model. For player i, the vulnerability discovery skill is denoted by $p_i(t)$, which is a function of probability that the player discovers a zero-day vulnerability distributed over rounds. Their level of ricochet ability is characterized by parameter $q_i(t)$, which is the probability of launching a ricochet attack with exploits they recovered from the traffic received in the previous round.

A player's patching skill is represented by function $h_i(t)$. $h_i(t)$ is the ratio of the amount of patched vulnerable resources over the total amount of vulnerable resources by t rounds *after* the vulnerability is disclosed. In the real world, while patching a single computer might take only minutes, patching all vulnerable resources (depending on the organization, containing thousands of instances) might take days to months (Bilge and Dumitras 2012). While one player is patching, other players could possibly attack, and any vulnerable resources that have not been patched will suffer from the attack.

The last parameter, δ_i, describes player i's level of APEG skills, which is the number of rounds required by the player to generate a patch-based exploit after a vulnerability and the corresponding patch are disclosed. Finally, the attacking utility, denoted as $u_i(t)$, encodes the *dynamic* utility that player i gains by attacking his opponents at round t before the patch is released.

Game Parameter Measurements

These parameters can be either measured or estimated with existing techniques. Parameter $p_i(t)$, $h_i(t)$, and $u_i(t)$ can be estimated via current risk management techniques (Dumitras 2019; Sabottke et al. 2015; Xiao et al. 2018). Parameter $q_i(t)$ can be measured by running ShellSwap (Bao et al. 2017b), the automatic attack ricochet technique, and δ_i can be measured by running the APEG tools (Brumley et al. 2008b). Given that these techniques are all public, the above parameters can be accessed or fairly estimated via known tools. Therefore, we assume that the value of model parameters is publicly known by all players.

12.6.2 Formalization

We formalize Software Vulnerability Mitigation Modelas a two-player zero-sum partial observation stochastic game:

$$\text{POSG} = \langle \mathcal{N}^P, \mathcal{A}^P, \Theta^P, \Phi^P, R^P \rangle.$$

In this chapter, we focus on two players ($|\mathcal{N}^P| = 2$) called player 1 and player 2.

Game State

The complete game state Θ^P is defined as $\mathcal{T} \times \mathcal{R} \times \Theta_1 \times \Theta_2$, where $\mathcal{T}$ is the round number, $\mathcal{R}$ is the specific round when the patch is released, and Θ_i is the set of player i's states ($\Theta_i = \{\neg D, D\}$). The round of releasing a patch is $\emptyset$ before a patch is released. The patch release time is needed because it is a public indicator of the discovery of the vulnerability. We use this to bound uncertainty, since after a patch is released every player has the potential (and eventual) understanding of the vulnerability.

State Transition

In each round, the game is in a concrete state, but the players have incomplete information about that state. The players make an observation and then choose an action. The chosen actions transition the game to a new state. The transition function is public for both players. Transitions in a game may be probabilistic. We denote the probability transition function over game states by $\Phi^P : \Theta^P \times \mathcal{A}_1^P \times \mathcal{A}_2^P \to \Delta(\Theta^P)$ and show these transitions in Figure 12.7. This divides the game state into five categories:

a. Neither player discovers a vulnerability (Figure 12.7a, $\langle t, \emptyset, \neg D, \neg D\rangle$). The available action for each player is NOP, and the probability that player i discovers a vulnerability in the current round is $p_i(t)$. Since players discover vulnerabilities independently, the joint probability of player 1 in state θ_1 and player 2 in θ_2 is equal to the product each player is in his respective state.

b. Only one player has discovered a vulnerability (Figure 12.7b, $\langle t, \emptyset, \neg D, D\rangle$ or $\langle t, \emptyset, D, \neg D\rangle$). Suppose player 2 has the exploit, then player 2 has three possible actions while player 1 has one. If player 2 chooses to ATTACK, the probability that player 1 transits to state D is equal to the joint probability of finding the vulnerability by himself, and that detecting player 2's attack. If player 2 chooses to STOCKPILE, the probability that player 1 will be in state D in the next round is equal to the probability that he independently discovers the vulnerability. If player 2 chooses to PATCH, player states remain unchanged and the patch-releasing round will be updated to t.

c. Both players have discovered the vulnerability and they withhold it (Figure 12.7c, $\langle t, \emptyset, D, D\rangle$). If neither player releases a patch, the states and the patch-releasing round remain the same. Otherwise, the game will transition to $\langle t+1, t, D, D\rangle$.

d. One player has disclosed the vulnerability, while the other player has not discovered it (Figure 12.7d, $\langle t, r, D, \neg D\rangle$ or $\langle t, r, \neg D, D\rangle$). Suppose player 1 releases the patch at round r, then player 2 will generate an exploit based on this patch in δ_2 rounds. If player 1 chooses to NOP during those rounds, then player 2 will keep developing the patch-based exploit until the $(r + \delta_2)$th round. Otherwise, if player 1 chooses to ATTACK, then player 2 will detect the attack with probability $q_2(t)$ and transition to state D in the next round after doing so.

e. Both players have discovered the vulnerability, and the vulnerability has been disclosed (Figure 12.7e, $\langle t, r, D, D\rangle$).
In this case, player states and the patch-releasing round remains unchanged.

Utility

Players' utility for each round is calculated according to the reward function for one round $R^P : \mathcal{A}_1^P \times \mathcal{A}_2^P \to \mathbb{R}$. This function is public, but the actual utility per round is secret to players because players do not always know the action of the other players. We assume that the amount of utility that a player gains is equal to the amount that the other player loses, which makes our game zero-sum.

The reward function is calculated using the attacking utility functions $u_{1,2}(t)$ and the patching portion functions $h_{1,2}(t)$. We define the reward function before and after patching separately, which are shown as Tables 12.2 and 12.3. In both tables, player 1 is the row player and player 2 is the column player.

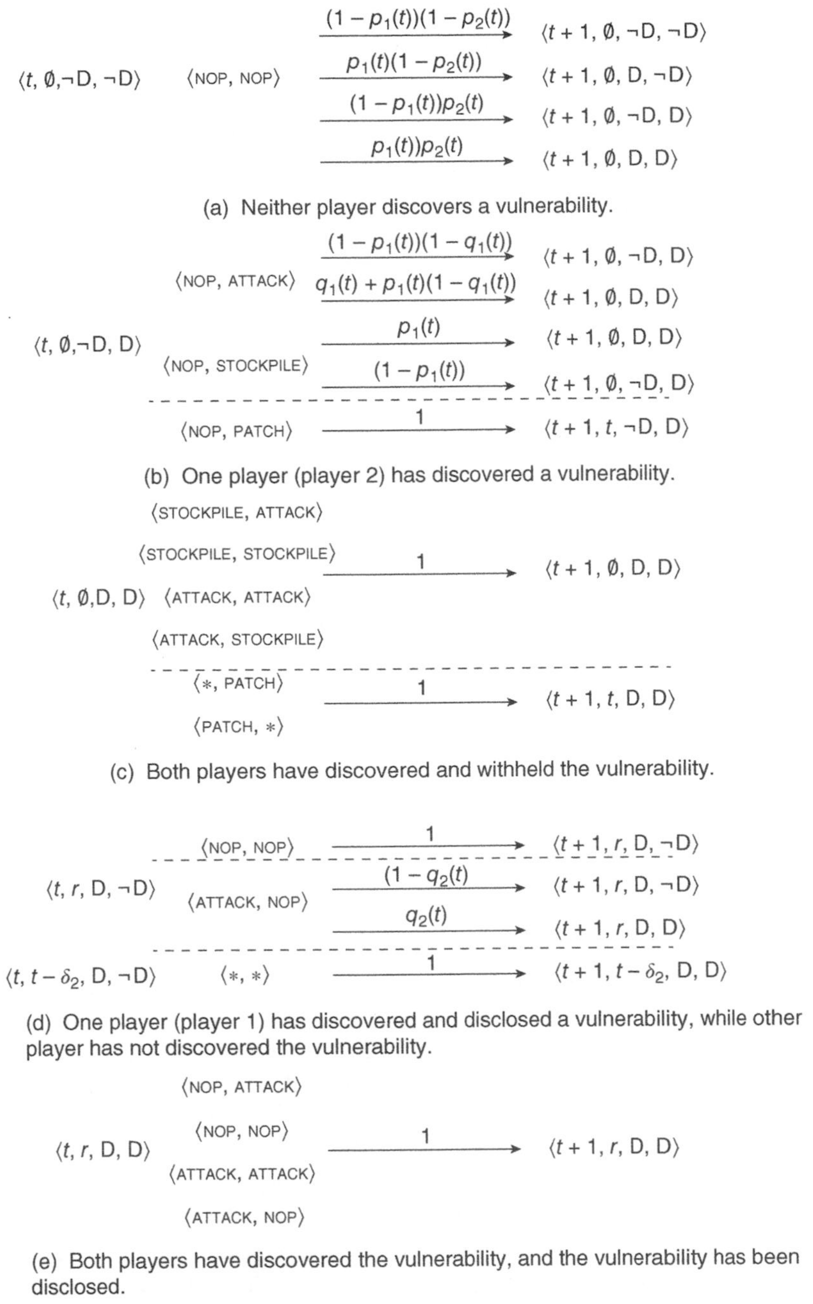

$\langle t, \emptyset, \neg D, \neg D \rangle$ $\langle NOP, NOP \rangle$

$$\xrightarrow{(1-p_1(t))(1-p_2(t))} \langle t+1, \emptyset, \neg D, \neg D \rangle$$
$$\xrightarrow{p_1(t)(1-p_2(t))} \langle t+1, \emptyset, D, \neg D \rangle$$
$$\xrightarrow{(1-p_1(t))p_2(t)} \langle t+1, \emptyset, \neg D, D \rangle$$
$$\xrightarrow{p_1(t))p_2(t)} \langle t+1, \emptyset, D, D \rangle$$

(a) Neither player discovers a vulnerability.

$\langle t, \emptyset, \neg D, D \rangle$

$\langle NOP, ATTACK \rangle$
$$\xrightarrow{(1-p_1(t))(1-q_1(t))} \langle t+1, \emptyset, \neg D, D \rangle$$
$$\xrightarrow{q_1(t)+p_1(t)(1-q_1(t))} \langle t+1, \emptyset, D, D \rangle$$

$\langle NOP, STOCKPILE \rangle$
$$\xrightarrow{p_1(t)} \langle t+1, \emptyset, D, D \rangle$$
$$\xrightarrow{(1-p_1(t))} \langle t+1, \emptyset, \neg D, D \rangle$$

$\langle NOP, PATCH \rangle$
$$\xrightarrow{1} \langle t+1, t, \neg D, D \rangle$$

(b) One player (player 2) has discovered a vulnerability.

$\langle t, \emptyset, D, D \rangle$

$\langle STOCKPILE, ATTACK \rangle$
$\langle STOCKPILE, STOCKPILE \rangle$
$\langle ATTACK, ATTACK \rangle$
$\langle ATTACK, STOCKPILE \rangle$
$$\xrightarrow{1} \langle t+1, \emptyset, D, D \rangle$$

$\langle *, PATCH \rangle$
$\langle PATCH, * \rangle$
$$\xrightarrow{1} \langle t+1, t, D, D \rangle$$

(c) Both players have discovered and withheld the vulnerability.

$\langle t, r, D, \neg D \rangle$

$\langle NOP, NOP \rangle$
$$\xrightarrow{1} \langle t+1, r, D, \neg D \rangle$$

$\langle ATTACK, NOP \rangle$
$$\xrightarrow{(1-q_2(t))} \langle t+1, r, D, \neg D \rangle$$
$$\xrightarrow{q_2(t)} \langle t+1, r, D, D \rangle$$

$\langle t, t-\delta_2, D, \neg D \rangle$ $\langle *, * \rangle$
$$\xrightarrow{1} \langle t+1, t-\delta_2, D, D \rangle$$

(d) One player (player 1) has discovered and disclosed a vulnerability, while other player has not discovered the vulnerability.

$\langle t, r, D, D \rangle$

$\langle NOP, ATTACK \rangle$
$\langle NOP, NOP \rangle$
$\langle ATTACK, ATTACK \rangle$
$\langle ATTACK, NOP \rangle$
$$\xrightarrow{1} \langle t+1, r, D, D \rangle$$

(e) Both players have discovered the vulnerability, and the vulnerability has been disclosed.

Figure 12.7 The transitions of game states. For each sub-figure, the left-hand side is the state of the current round and the right-hand side is the state of the next round.

12.6.3 Finding Equilibriums

Although the POSG model, as an incomplete information game, characterizes the uncertainty inherent in cyber warfare, computing the equilibriums of a general POSG remains open. However, we discover three insights, specific to cyber warfare, that help us reduce the complexity of the game and calculate the Nash equilibrium for our cyber warfare game model.

Table 12.2 The reward matrix at round t before patch is released.

		NOT DISCOVER		DISCOVER	
		NOP	ATTACK	STOCKPILE	PATCH
NOT DISCOVER	NOP	0	$-u_2(t)$	0	0
DISCOVER	ATTACK	$u_1(t)$	$u_1(t) - u_2(t)$	$u_1(t)$	0
	STOCKPILE	0	$-u_2(t)$	0	0
	PATCH	0	0	0	0

The table shows the reward of player 1 (the row player). The reward of player 2 (the column player) is the negative value of that of player 1.

Table 12.3 The reward matrix at round t after a patch is released.

		NOT DISCOVER	DISCOVER	
		NOP	ATTACK	NOP
DISCOVER	NOP	0	$-u_2(t)h_1(t-r)$	0
	ATTACK	$u_1(t)/h_2(t-r)$	$u_1(t)h_2(t-r) - u_2(t)h_1(t-r)$	$u_1(t)h_2(t-r)$
NOT DISCOVER	NOP	0	$-u_2(t)h_1(t-r)$	0

The table shows the reward of player 1 (the row player). The reward of player 2 (the column player) is the negative value of that of player 1.

First, if player i releases a patch, then all players subsequently know player i has found the vulnerability. We use this to split the cyber war game into two phases: before disclosure and after disclosure.

Second, players can probabilistically bound how likely another player is to discover a vulnerability based upon their skill level. This is because the probability is inferred based on players' attributes, such as the discovery probability, ricochet probability, and those attributes are public to all players.

Finally, although players are uncertain about the state of the other players (which they represent as a probability distribution of player states), they know the probability of their opponents being in a state given the public information of the opponents, such as the vulnerability discovery probability (e.g. based upon prior zero-day battles) and the ricochet probability.

Based on the above insights, we convert the POSG model to a stochastic game model by encoding the belief of each player into the game state. In our game, the belief of a player is the probability that the player thinks the other player has found the vulnerability. We can compute the Nash equilibrium for the converted stochastic game by dynamic programming. We will also discuss the observation of players' strategy after vulnerability disclosure.

Stochastic Game

As our assumption that a player's belief about the state of opponent players can be estimated from the globally-known player properties, the POSG model reduces to a much more tractable stochastic model in the predisclosure phase. We define the stochastic game (SG) model

$$\text{SG} = \langle \mathcal{N}^S, \mathcal{A}^S, \Theta^S, \Phi^S, R^S \rangle$$

We retain the definition of players in POSG, $\mathcal{N}^P = \mathcal{N}^S = \{\text{player 1}, \text{player 2}\}$.

Player Actions

The player action in SG is defined as a combination of player actions under different player states. For example, if player i plays ATTACK in state D and NOP in state ¬D, the corresponding action in the SG model is {D : ATTACK, ¬D : NOP}. For each player action a_i, we will use $a_i[D]$ and $a_i[¬D]$ to denote the action in state D and ¬D, respectively.

Game State

The game state Θ^S in the SG model is defined as $\Theta^S = \mathcal{T} \times \mathcal{R} \times \mathbb{R} \times \mathbb{R}$. Besides the current round number $\mathcal{T}$ and the patch-releasing round number $\mathcal{R}$, a game state includes the beliefs of the two players about each other, which is the probability that a player has discovered a vulnerability from the other player's perspective, $b_i \in [0, 1]$. A game state $\theta^S \in \Theta^S$ can be represented as $\theta^S = \langle t, r, b_1, b_2 \rangle$, in which player 2 thinks the probability that player 1 has discovered the vulnerability is b_1, and player 1 thinks the probability that player 2 has discovered the vulnerability is b_2.

Contrary to the POSG model, the game states of the SG model include the uncertainty of a player about the other player's state. In each round of the game, players know their own states; although they do not know the other player's state, they infer the likelihood of the other player's state based on the other player's parameters. In addition, a player also knows the other player's beliefs about the game state because the player also knows the parameters of himself. Therefore, we are able to convert to the SG model under the structure of the game states above.

State Transition

We define the state transition function of the SG model as $\Phi^S : \Theta^S \times \mathcal{A}_1^S \times \mathcal{A}_2^S \to \Delta(\Theta^S)$. We represent the probability that a game transitions to θ^S using $\Phi^S(\cdot)[\theta^S]$. The transition between the game states is shown in Figure 12.8. The game states are divided by the time before and after vulnerability disclosure, because the actions and information available to players are different between the two phases.

Phase I. Before Disclosure (Figure 12.8a)

Suppose the game is in state $\langle t, \emptyset, b_1, b_2 \rangle$. If neither player acts ATTACK, the probability that player i discovers the exploits at the current round is $p_i(t)$ and the probability that player i discovers the exploit by the current round is $1 - (1 - b_i)(1 - p_i(t))$. The game transits to state $\langle t + 1, \emptyset, 1 - (1 - b_1)(1 - p_1(t)), 1 - (1 - b_2)(1 - p_2(t)) \rangle$.

If a player chooses to ATTACK, the probability that their opponent will acquire the exploit in the current round is the joint probability that the opponent discovers the vulnerability by himself and that he detects the exploit. Meanwhile, if the opponent detects the exploit, they will be certain that the attacker has the exploit. For example, if player 1 's action is {D : ATTACK, ¬D : NOP} while player 2 's action is {D : STOCKPILE, ¬D : NOP}, the game will transition to $\langle t + 1, \emptyset, 1 - (1 - b_1)(1 - p_1(t)), 1 - (1 - b_2)(1 - p_2(t))(1 - q_2(t)) \rangle$ with the probability of $1 - q_2(t)$ and $\langle t + 1, \emptyset, 1, 1 - (1 - b_2)(1 - p_2(t))(1 - q_2(t)) \rangle$ with the probability of $q_2(t)$.

Similarly, if both players act ATTACK in state D, the game will transition to one of four possibilities. If neither player detects the exploit, the game state will be $\langle t + 1, \emptyset, 1 - (1 - b_1)(1 - p_1(t))(1 - q_1(t)), 1 - (1 - b_2)(1 - p_2(t))(1 - q_2(t)) \rangle$. If player 1 detects the exploit while player 2 does not, the game state will be $\langle t + 1, \emptyset, 1 - (1 - b_1)(1 - p_1(t))(1 - q_1(t)), 1 \rangle$. If player 2 detects the exploit while player 1 does not, the game state will be $\langle t + 1, \emptyset, 1, 1 - (1 - b_2)(1 - p_2(t))(1 - q_2(t)) \rangle$. Finally, if both players detect the exploit, the game state will be $\langle t + 1, \emptyset, 1, 1 \rangle$.

If one player acts PATCH, both players will patch immediately. If player 1 releases a patch, the game will transition to $\langle t + 1, t, 1, b_2 \rangle$, as everyone is certain that player 1 has the exploit.

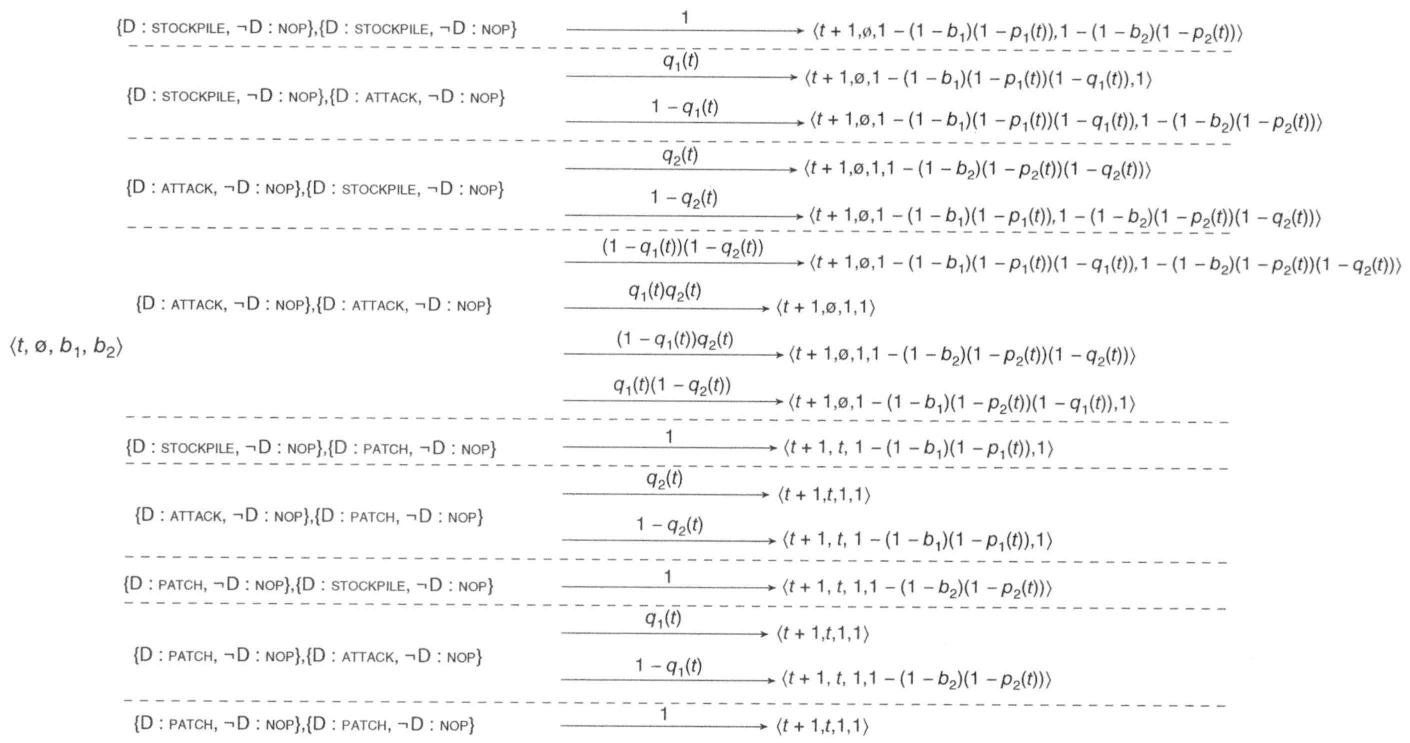

Figure 12.8 transition diagram (a)

$\{D : \text{STOCKPILE}, \neg D : \text{NOP}\}, \{D : \text{STOCKPILE}, \neg D : \text{NOP}\}$
$\xrightarrow{\;1\;} \langle t + 1, \varnothing, 1 - (1 - b_1)(1 - p_1(t)), 1 - (1 - b_2)(1 - p_2(t))\rangle$

$\{D : \text{STOCKPILE}, \neg D : \text{NOP}\}, \{D : \text{ATTACK}, \neg D : \text{NOP}\}$
$\xrightarrow{\;q_1(t)\;} \langle t + 1, \varnothing, 1 - (1 - b_1)(1 - p_1(t))(1 - q_1(t)), 1\rangle$
$\xrightarrow{\;1 - q_1(t)\;} \langle t + 1, \varnothing, 1 - (1 - b_1)(1 - p_1(t))(1 - q_1(t)), 1 - (1 - b_2)(1 - p_2(t))\rangle$

$\{D : \text{ATTACK}, \neg D : \text{NOP}\}, \{D : \text{STOCKPILE}, \neg D : \text{NOP}\}$
$\xrightarrow{\;q_2(t)\;} \langle t + 1, \varnothing, 1, 1 - (1 - b_2)(1 - p_2(t))(1 - q_2(t))\rangle$
$\xrightarrow{\;1 - q_2(t)\;} \langle t + 1, \varnothing, 1 - (1 - b_1)(1 - p_1(t)), 1 - (1 - b_2)(1 - p_2(t))(1 - q_2(t))\rangle$

$\{D : \text{ATTACK}, \neg D : \text{NOP}\}, \{D : \text{ATTACK}, \neg D : \text{NOP}\}$
$\xrightarrow{\;(1 - q_1(t))(1 - q_2(t))\;} \langle t + 1, \varnothing, 1 - (1 - b_1)(1 - p_1(t))(1 - q_1(t)), 1 - (1 - b_2)(1 - p_2(t))(1 - q_2(t))\rangle$
$\xrightarrow{\;q_1(t)q_2(t)\;} \langle t + 1, \varnothing, 1, 1\rangle$
$\xrightarrow{\;(1 - q_1(t))q_2(t)\;} \langle t + 1, \varnothing, 1, 1 - (1 - b_2)(1 - p_2(t))(1 - q_1(t))\rangle$
$\xrightarrow{\;q_1(t)(1 - q_2(t))\;} \langle t + 1, \varnothing, 1 - (1 - b_1)(1 - p_1(t))(1 - p_2(t))(1 - q_1(t)), 1\rangle$

$\langle t, \varnothing, b_1, b_2\rangle$

$\{D : \text{STOCKPILE}, \neg D : \text{NOP}\}, \{D : \text{PATCH}, \neg D : \text{NOP}\}$
$\xrightarrow{\;1\;} \langle t + 1, t, 1 - (1 - b_1)(1 - p_1(t)), 1\rangle$

$\{D : \text{ATTACK}, \neg D : \text{NOP}\}, \{D : \text{PATCH}, \neg D : \text{NOP}\}$
$\xrightarrow{\;q_2(t)\;} \langle t + 1, t, 1, 1\rangle$
$\xrightarrow{\;1 - q_2(t)\;} \langle t + 1, t, 1 - (1 - b_1)(1 - p_1(t)), 1\rangle$

$\{D : \text{PATCH}, \neg D : \text{NOP}\}, \{D : \text{STOCKPILE}, \neg D : \text{NOP}\}$
$\xrightarrow{\;1\;} \langle t + 1, t, 1, 1 - (1 - b_2)(1 - p_2(t))\rangle$

$\{D : \text{PATCH}, \neg D : \text{NOP}\}, \{D : \text{ATTACK}, \neg D : \text{NOP}\}$
$\xrightarrow{\;q_1(t)\;} \langle t + 1, t, 1, 1\rangle$
$\xrightarrow{\;1 - q_1(t)\;} \langle t + 1, t, 1, 1 - (1 - b_2)(1 - p_2(t))\rangle$

$\{D : \text{PATCH}, \neg D : \text{NOP}\}, \{D : \text{PATCH}, \neg D : \text{NOP}\}$
$\xrightarrow{\;1\;} \langle t + 1, t, 1, 1\rangle$

(a)

Figure 12.8 The transition of game states in the stochastic game. The left-hand sides of the arrows are game states with possible strategies. The right-hand sides of the arrows are the possible game states after transition. (a) Before disclosing a vulnerability. (b) After disclosing a vulnerability. Suppose player 1 disclose a vulnerability.

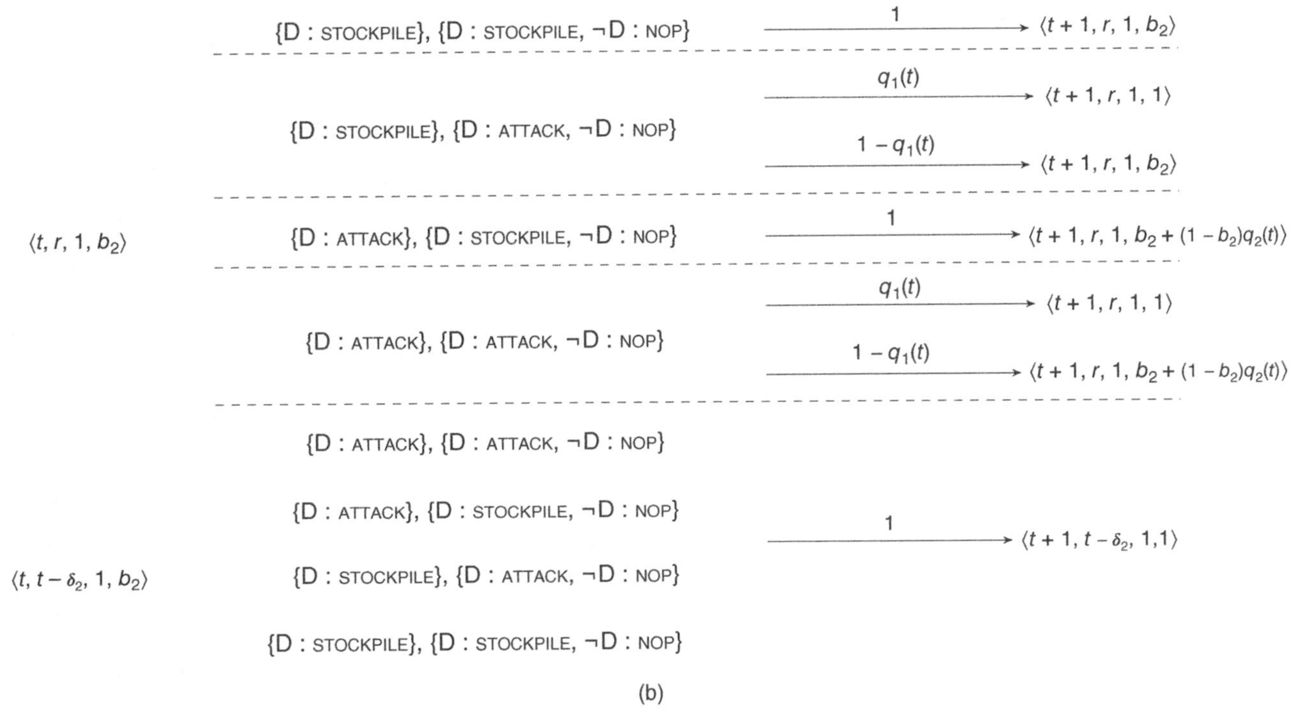

$\langle t, r, 1, b_2 \rangle$

$\{D : \text{STOCKPILE}\}, \{D : \text{STOCKPILE}, \neg D : \text{NOP}\}$ $\xrightarrow{\ 1\ }$ $\langle t+1, r, 1, b_2 \rangle$

$\{D : \text{STOCKPILE}\}, \{D : \text{ATTACK}, \neg D : \text{NOP}\}$ $\xrightarrow{\ q_1(t)\ }$ $\langle t+1, r, 1, 1 \rangle$

$\xrightarrow{\ 1-q_1(t)\ }$ $\langle t+1, r, 1, b_2 \rangle$

$\{D : \text{ATTACK}\}, \{D : \text{STOCKPILE}, \neg D : \text{NOP}\}$ $\xrightarrow{\ 1\ }$ $\langle t+1, r, 1, b_2 + (1-b_2)q_2(t) \rangle$

$\{D : \text{ATTACK}\}, \{D : \text{ATTACK}, \neg D : \text{NOP}\}$ $\xrightarrow{\ q_1(t)\ }$ $\langle t+1, r, 1, 1 \rangle$

$\xrightarrow{\ 1-q_1(t)\ }$ $\langle t+1, r, 1, b_2 + (1-b_2)q_2(t) \rangle$

$\langle t, t-\delta_2, 1, b_2 \rangle$

$\{D : \text{ATTACK}\}, \{D : \text{ATTACK}, \neg D : \text{NOP}\}$

$\{D : \text{ATTACK}\}, \{D : \text{STOCKPILE}, \neg D : \text{NOP}\}$ $\xrightarrow{\ 1\ }$ $\langle t+1, t-\delta_2, 1, 1 \rangle$

$\{D : \text{STOCKPILE}\}, \{D : \text{ATTACK}, \neg D : \text{NOP}\}$

$\{D : \text{STOCKPILE}\}, \{D : \text{STOCKPILE}, \neg D : \text{NOP}\}$

(b)

Figure 12.8 (*Continued*)

If player 2 releases a patch, the game will transit to $\langle t+1, t, b_1, 1 \rangle$ and if both player release a patch, the game will transition to $\langle t+1, t, 1, 1 \rangle$.

Phase II. After Disclosure (Figure 12.8b)

After disclosure, both players will know the vulnerability so they will stop searching for it. Also, the player disclosing a vulnerability is public so both players know that the player is in state D. Suppose player 1 discloses a vulnerability in round r. In response, player 2 starts APEG and will generate the exploit by round $r + \delta_2$. Meanwhile, player 2 still has the chance to ricochet attacks if player 1 attacks. Therefore, the belief of player 2 's possession of the exploit will increase if player 1 attacks in the previous round.

Utility

We calculate players' utility by the single-round reward function $R^S : \Theta^S \times A_1^S \times A_2^S \to \mathbb{R}$. Given a game state and players actions, the single-round reward is equal to the expected reward over player states.

Given player i and a player state θ_i, the probability that the player is in state θ_i when the SG game state is $\theta^S = \langle t, r, b_1, b_2 \rangle$, which is denoted by $P(\theta^S, \theta_i)$, is equal to

$$P(\theta^S, \theta_i) = P(\langle t, r, b_1, b_2 \rangle, \theta_i) = \begin{cases} b_i & \text{if } \theta_i = \neg D \\ 1 - b_i & \text{if } \theta_i = D \end{cases}$$

Recall the reward function for the POSG model $\mathcal{R}^P : A_1^P \times A_2^P \to \mathbb{R}$ takes as input the players' actual actions and produces as output the utility of one player (because the utility of the other is the negative value for zero-sum game). We calculate the reward for the SG model using $\mathcal{R}^P$. In specific, we have

$$R^S(\theta^S, a_1^S, a_2^S) = \sum_{\theta_1} \sum_{\theta_2} P(\theta^S, \theta_1) P(\theta^S, \theta_2) R^P(a_1^S[\theta_1], a_2^S[\theta_2])$$

Computing Nash Equilibriums

A Nash equilibrium is a strategy profile where neither player has more to gain by altering its strategy. It is the stable point of the game when both players are rational and making their best response. Let $\mathrm{NE}^S : \Theta^S \to \mathbb{R}$ denote player 1 's utility when both players play the Nash equilibrium strategy in the SG model. Since the game is a zero-sum game, the utility of player 2 is equal to $-\mathrm{NE}^S$.

We compute the Nash equilibrium inspired by the Shapley method (Shapley 1953), which is a dynamic programming approach for finding players' best responses. For game state $\theta^S = \langle t, r, b_1, b_2 \rangle$, the utility of player 1 is equal to sum of the reward that player 1 gets in the current round and the expected utility that he gets in the future rounds. In the future rounds, players will continue to play with their best strategies, so the utility in the future rounds is equal to the one that corresponds to the Nash equilibrium in the future game states. Therefore, the utility of the Nash equilibrium of a game state is as following:

$$\mathrm{NE}^S(\theta^S) = \max_{a_1^S \in A^S} \min_{a_2^S \in A^S} \left\{ R^S(\theta^S, a_1^S, a_2^S) + \sum_{\theta \in \Theta^S} \Phi^S(\theta^S, a_1^S, a_2^S)[\theta] \cdot \mathrm{NE}^S(\theta) \right\} \qquad (12.1)$$

In theory, the game could go for infinite rounds when neither players discloses a vulnerability. In this case, the corresponding utility will be equal to 0, positive infinity or negative infinity. However, for implementation, we need to set a boundary to guarantee that the recursive calculation of Nash equilibrium will stop. We introduce MAX_t to denote the maximum round of the game, and we assume that

$$\mathrm{NE}^S(\langle t, r, b_1, b_2 \rangle) = 0, \text{ if } t \geq \mathrm{MAX}_t. \qquad (12.2)$$

Optimize the Post-Disclosure Game

Equation 12.1 is only applicable for calculating the Nash equilibrium of the SG model. Nonetheless, we find an optimized way to compute the Nash equilibrium after the vulnerability is disclosed ($r \neq \emptyset$). The optimized approach is based on the finding that if a player discloses a vulnerability, the other player should attack right after he generates the exploits. We call the player who discloses the vulnerability the explorer, and the other player who witness the disclosure of the vulnerability the observer. Intuitively, disclosure implies that the explorer has discovered the vulnerability, and the observer's attack will not reveal to the explorer any new information about the vulnerability. Therefore, there is no collateral damage if the observer attacks, and the observer's best strategy is to constantly attack until his adversary completes patching. We formally prove the finding as follows.

Theorem 12.1 If one player discloses a vulnerability, the best response of the other player is {D : ATTACK, ¬D : NOP}.

Proof: Without loss of generality, we assume that player 1 discloses a vulnerability, and the current game state for the SG model is SG$\langle t, r, 1, b_2 \rangle$. The corresponding game state for the POSG model is either POSG$\langle t, r, D, D \rangle$ or POSG$\langle t, r, D, \neg D \rangle$, shown in Figure 12.9.

If player 2 has not discovered the vulnerability, then the actual game state is POSG$\langle t, r, D, \neg D \rangle$. Player 2 can only play NOP, so their action is NOP when they are in state ¬D.

If player 2 has discovered the vulnerability, then the actual game state is POSG$\langle t, r, D, D \rangle$. Player 2 chooses actions between ATTACK and STOCKPILE, and the game will deterministically transition to state POSG$\langle t + 1, r, D, D \rangle$. Recall that $R^P(a_1, a_2)$ represents the utility of player 1 when player 1 chooses action a_1 and player 2 chooses action a_2. Let NE$^P(\theta^P)$ denote the utility of player 1 in state θ^P when both players play the Nash equilibrium strategy. Thus, we have

$$
\begin{aligned}
\text{NE}^P(\langle t, r, D, D \rangle) \\
= \max_{a_1^P \in \mathcal{A}^P} \min_{a_2^P \in \mathcal{A}^P} \left\{ R^P(a_1^P, a_2^P) + \text{NE}^P(\langle t+1, r, D, D \rangle) \right\} \\
= \max_{a_1^P \in \mathcal{A}^P} \min_{a_2^P \in \mathcal{A}^P} R^P(a_1^P, a_2^P) + \text{NE}^P(\langle t+1, r, D, D \rangle)
\end{aligned}
\tag{12.3}
$$

According to the reward matrix in Table 12.3, ATTACK dominates STOCKPILE, and the best strategy for player 2 in state D is ATTACK. Therefore, the best strategy for player 2 is {D : ATTACK, ¬D : NOP}.

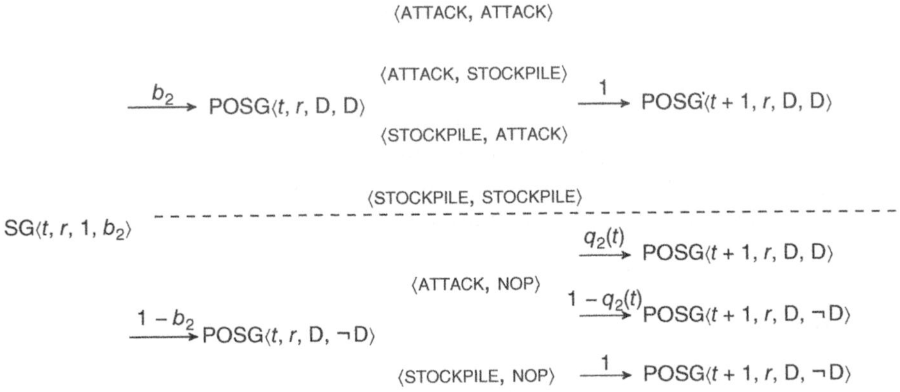

Figure 12.9 The relationship between the SG and the POSG models after disclose a vulnerability. SG$\langle \cdot \rangle$ denotes the game state of the SG model and POSG$\langle \cdot \rangle$ denotes the game state of the POSG model. Suppose player 1 discloses a vulnerability.

A special case is that both players disclose a vulnerability at the same round. Under this situation, both players will attack right after disclosure, since both players are the observers and the explorers. As observers, the players will attack once they can; as explorers, the players know how to exploit.

Given the above theorem, the SG model after disclosure becomes a Markov decision process in which the explorer makes a decision given a state of the stochastic game.

Next, we discuss how to compute the best response for the explorer. Given a game state, the explorer chooses one action between ATTACK and STOCKPILE. Suppose player 1 is the explorer.

First, we discuss the algorithm to compute the best response for the POSG game state. If player 2 is in state D, then the game state is $\langle t, r, D, D \rangle$. According to Table 12.3, player 1 should play ATTACK:

$$NE^P\left(\langle t, r, D, D \rangle\right) = R^P(\text{ATTACK}, \text{ATTACK}) + NE^P\left(\langle t+1, r, D, D \rangle\right) \tag{12.4}$$

Let $\Phi^S(\theta^X)[\theta^Y]$ be the probability that a game transitions from θ^X to θ^Y. If player 2 is in state $\neg D$ and the game is in state $\langle t, r, D, \neg D \rangle$, player 1 should choose the action with greater utility according to the following formula:

$$NE^P\left(\langle t, r, D, \neg D \rangle\right)$$

$$= \max_{a_1^P \in \{\text{ATTACK}, \text{STOCKPILE}\}} \left\{ R^P(a_1^P, \text{NOP}) + \sum_{\theta \in \Theta^P} \Phi^P\left(\langle t, r, D, \neg D \rangle\right)[\theta] NE^P(\theta) \right\} \tag{12.5}$$

Finally, given a game state of the SG model $SG\langle t, r, 1, b_2 \rangle$, the best response for player 1 is the action with greater expected value of the utilities over POSG states.

$$NE^S\left(\langle t, r, 1, b_2 \rangle\right)$$

$$= \max_{a_1^P \in \{\text{ATTACK}, \text{STOCKPILE}\}} \left\{ b_2 \cdot NE^P\left(\langle t, r, D, D \rangle\right) + (1 - b_2) \cdot NE^P\left(\langle t, r, D, \neg D \rangle\right) \right\} \tag{12.6}$$

12.6.4 Algorithm

Algorithm 12.1 shows the algorithm of computing the Nash equilibrium before a vulnerability is disclosed. Given a round index and beliefs of the players, the goal is to compute player utility when both players rationally play their best response. If the round index is equal to or larger than MAX_t, which is the maximum number of rounds argument self-configured for the game, then the calculation will stop. Otherwise, the algorithm finds the Nash equilibrium according to Eq. 12.1 and Figure 12.8a. For each Nash equilibrium candidate, if players do not disclose the vulnerability, the game will continue in the before-disclosure phase, else the game will step to the after-disclosure phase.

Algorithm 12.2 shows the computation for the Nash equilibrium after a vulnerability is disclosed. If the game has equal to or more than MAX_t rounds, then the game is over. Otherwise, we update players' state according their APEG skill. If both players have generated the exploit, then both of them should attack. If not, the player who did not disclose the vulnerability should attack once he has generated the exploit. The other player who disclosed the vulnerability should choose between attack and stockpile depending on the sum of the utilities at the current round and that in the future.

12.7 Case Study

In this section, we apply Software Vulnerability Mitigation Modelto the Cyber Grand Challenge scenario and investigate if Software Vulnerability Mitigation Modelis effective in making decisions on software security offense and defense. The Cyber Grand Challenge (CGC) is an automated cyber security competition designed to mirror "real-world challenges" (DARPA 2016b). This competition

---

Algorithm 12.1 The before-disclosure game algorithm.

Input :

t: The index of the current round

b_1, b_2: The probability that player 1 and player 2 have discovered the vulnerability.

Output:

$NE^S(\langle t, \emptyset, b_1, b_2 \rangle)[i]$: The utility of player i under the Nash equilibrium at round t before disclosure.

1 **if** $t >= MAX_t$ **then**

2   Game is over.

3 **end**

4 $\theta^S \leftarrow \langle t, \emptyset, b_1, b_2 \rangle$;

5 $\Theta^T \leftarrow$ set of possible states transiting from θ^S;

6 max $\leftarrow -\infty$;

7 **foreach** $a_1^S \in A^S$ **do**

8   min $\leftarrow \infty$;

9   **foreach** $a_2^S \in A^S$ **do**

10        $t \leftarrow R(\theta^S, a_1^S, a_2^S) + \sum_{\theta \in \Theta^S} \Phi^S(\theta^S, a_1^S, a_2^S)[\theta] NE^S(\theta)$;

11   **if** $min > t$ **then**

12   min $\leftarrow$ t;

13   **end**

14   **end**

15   **if** $max < min$ **then**

16   max $\leftarrow$ min;

17   **end**

18 **end**

19 $NE^S(\langle t, \emptyset, b_1, b_2 \rangle)[1] \leftarrow$ max;

20 $NE^S(\langle t, \emptyset, b_1, b_2 \rangle)[2] \leftarrow$ -max;

21 **return** $NE^S(\langle t, b_1, b_2 \rangle)$;

---

provides an excellent opportunity to evaluate the strategies suggested by our model against those actually carried out by competitors. The CGC final consists of 95 rounds. In this case study, we experimented on the ranking of the third-place team in the Cyber Grand Challenge, Shellphish. Based on their public discussions regarding their strategy, Shellphish simply attacked and patched right away (DARPA 2016a). This made them an optimal subject of this case study, as, since they would use their exploits as soon as possible (rather than stockpiling them), we can closely estimate their technical acumen for the purposes of testing our model. We call our modified, more strategic, player 'Strategic-Shellphish.'

In our experiment, we adapted our model to the CGC final in the following way. First, we update the reward function on the CGC scoring mechanism. As the CGC final is not a zero-sum game, we compute the Nash equilibrium by focusing on the current round. Second, we separate the game by binaries, and for each binary we model Strategic-Shellphish as one player while all the non-Shellphish team as the other player. Third, we estimated the game parameters according to the data from the earlier rounds, then calculated the optimal strategy and applied the strategy in the later rounds. For example, we get the availability score for the patch by deploying it in the earlier

---

Algorithm 12.2 The after-disclosure game algorithm.

Input :
t: The index of the current round
r: The index of the round at which a vulnerability is disclosed
b_1, b_2: The probability that player 1and player 2 have discovered the vulnerability.

Output:
$NE^S(\langle t, r, b_1, b_2 \rangle)$:The player utility under the Nash equilibrium at round t after the vulnerability is disclosed at round r.

1 **if** $t >= MAX_t$ **then**
2 | Game is over.
3 **end**
4 **foreach** $i \in \{1, 2\}$ **do**
5 | **if** $b_i < 1 \ and \ t > r + \delta_i$ **then**
6 | | $b_i \leftarrow 1$;
7 | **end**
8 **end**
9 **if** $b_1 == 1 \ \&\& \ b_2 == 1$ **then**
10 | $NE^S(\langle t, r, b_1, b_2 \rangle) \leftarrow NE^P(\langle t, r, D, D \rangle)$;
11 **end**
12 **else if** $b_2 < 1$ **then**
13 | $NE^S\left(\langle t, r, b_1, b_2 \rangle\right) \leftarrow \max_{a_1^P \in \{\text{ATTACK},\text{STOCKPILE}\}} \left\{ b_2 \cdot NE^P\left(\langle t, r, D, D \rangle\right) + (1 - b_2) \cdot NE^P\left(\langle t, r, D, \neg D \rangle\right) \right\}$;
14 **end**
15 **else**
16 | $NE^S\left(\langle t, r, b_1, b_2 \rangle\right) \leftarrow \min_{a_2^P \in \{\text{ATTACK},\text{STOCKPILE}\}} \left\{ b_1 \cdot NE^P\left(\langle t, r, D, D \rangle\right) + (1 - b_1) \cdot NE^P\left(\langle t, r, \neg D, D \rangle\right) \right\}$;
17 **end**
18 **return** $NE^S(\langle t, r, b_1, b_2 \rangle)$;

---

rounds. The data is from the public release from DARPA, which includes the player scores for each vulnerable binaries in each round.

In the first CGC experiment, we estimated the game parameters by the information of the first 80 rounds of the game and applied the model on the 80-95 rounds. This range included 11 challenge binaries, and we simulated Shellphish's performance, if they had used our model for strategy determinations, across these programs. The score comparison for each vulnerability is shown in Figure 12.10, with the x axis representing the 11 binaries and the y axis representing the scores. The new scores are either higher or equal to the original score. Among these binaries, our model helps improve 5 cases out of 11. The overall score for the 11 vulnerabilities is shown in Figure 12.11.

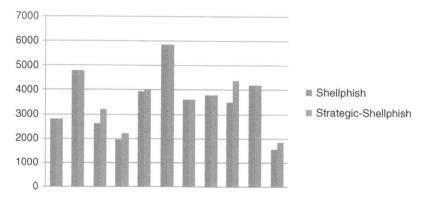

Figure 12.10 The per-vulnerability score comparison between the original Shellphish team and Strategic-Shellphish-the Shellphish team + our model.

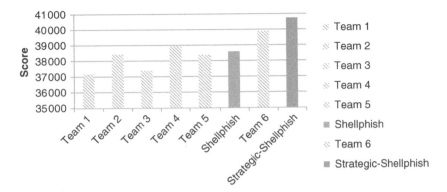

Figure 12.11 The overall score comparison between the original Shellphish team and Strategic-Shellphish for the game round 80–95.

The original Shellphish team got 38 598.7 points, while our model got 40 733.3 points. Moreover, our Strategic-Shellphish team won against all other teams in terms of these 11 vulnerabilities.

We observed that Strategic-Shellphish withdrew the patch of some vulnerabilities after the first round of patching. After the first round of patching, Strategic-Shellphish got the precise score for availability, and this helped it compare the cost of patching to the expected lost in the future rounds.

In the second CGC experiment, we estimated the game parameters by the information of the first 15 rounds. Given the parameters, Strategic-Shellphish calculates the likelihood that the other teams discovers the vulnerability, and it uses our algorithm to determine the best response. Before it is well-aware of the patching cost, we assigned the cost to 0. After the first round of patching, we updated the patching cost and adjusted the future strategy.

The score for the entire game is shown in Figure 12.12. The original Shellphish team got 254 452 points and ranked third in the game. On the other hand, the Strategic-Shellphish got 268 543 points, which is 6000 points higher than the score of the original second-rank team. Our experiment highlights the importance of our model as well as the optional strategy solution. If a team such like Shellphish used our model, it could have achieved a better result compared to its original

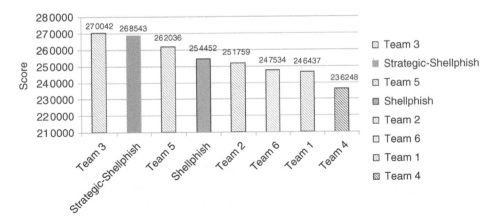

Figure 12.12 The overall score comparison between the original Shellphish team and Strategic-Shellphish for the entire game.

strategy. In fact, in the Cyber Grand Challenge, the difference between third (Shellphish) and second (Strategic-Shellphish) place was $250 000.

12.8 Discussion

Parameter Sensitivity

Our model employs parameters to capture players' different skill levels. These parameters need to be evaluated, and one way is to use a relevant benchmark proposed by Axelrod and Iliev (2014). For example, one can estimate $p_i(t)$ by reasoning about the ratio of the vulnerabilities independently rediscovered in software from the research by Bilge and Dumitras (2012). As the White House stated that the government weighed the likelihood that the other nations re-discover the vulnerability, the evaluation approach should have already existed. In the future, we need to investigate the robustness of the model under different parameter margins.

Deception

Although the game parameters are public, they can be manipulated by players. For example, a player could pretend to be weak in generating exploits by never launching any attack against anyone. In this case, we do not consider player deception, and we leave it for future work.

Inferring the State of Games with Special Parameters

We consider that players infer the states of other players by detecting attacks or learning vulnerability disclosure. However, we do not consider that players could these state by reasoning about game parameters. For example, suppose we have a game with public game parameters denoting that player 1 is able to capture *all* the attacks and player 2 should *always* attack after he generates an exploit. In this case, if player 1 does not detect attacks from player 2, then player 2 has not generated an exploit, and the belief on player 2 should be zero all the time until player 1 detects an attack. To tackle this problem, a possible way is to solve the game separately with different groups of parameter conditions.

The Incentives of Patching

In our model, we consider patching as a defensive mechanism that only prevents players from losing utility. This leads to players not having incentives to disclose a vulnerability. We argue that patching might bring positive benefits to players. For instance, a player would have a better reputation if he chooses to disclose a vulnerability and patch their machines. We leave the consideration of the positive reputation caused by disclosure as future work.

12.9 Conclusion

In this chapter, we introduce the techniques and tactics for cyber autonomy in software security. We set up the Software Vulnerability Mitigation Modelto reason about the defense and offense for software security. We also present the techniques that support to enforce the game strategy, measurements for the value of the game parameters, as well as algorithms that solve the game theoretical model. We use the Cyber Grand Challenge for case study where we show that our model is effective in improving the performance of teams playing as agents in the game. We hope that the reader, by reading this chapter, will get a better understanding of cyber autonomy in software security, learning the related techniques and tactics and realizing how techniques and tactics co-work to achieve security outcomes.

References

AFL. American fuzzy lop, 2021. http://lcamtuf.coredump.cx/afl.

Cornelius Aschermann, Sergej Schumilo, Tim Blazytko, Robert Gawlik, and Thorsten Holz. REDQUEEN: Fuzzing with input-to-state correspondence. In *Proceedings of the Network and Distributed System Security Symposium, San Diego, 24–27 February 2019*, 2019.

Thanassis Avgerinos, Sang Kil Cha, Brent Tze Hao Lim, and David Brumley. AEG: Automatic exploit generation. In *Proceedings of 18th Annual Network and Distributed System Security Symposium*. Internet Society, 2011.

Robert Axelrod and Rumen Iliev. Timing of cyber conflict. *Proceedings of the National Academy of Sciences*, 111(4):1298–1303, 2014.

Tiffany Bao, Jonathan Burket, Maverick Woo, Rafael Turner, and David Brumley. ByteWeight: Learning to recognize functions in binary code. In *Proceedings of the 23rd USENIX Security Symposium*, pp. 845–860. USENIX, 2014.

Tiffany Bao, Yan Shoshitaishvili, Ruoyu Wang, Christopher Kruegel, Giovanni Vigna, and David Brumley. How shall we play a game:a game-theoretical model for cyber-warfare games. In *IEEE 30th Computer Security Foundations Symposium*, pp. 7–21. IEEE, 2017a.

Tiffany Bao, Ruoyu Wang, Yan Shoshitaishvili, and David Brumley. Your exploit is mine: Automatic shellcode transplant for remote exploits. In *Proceedings of the 38th IEEE Symposium on Security and Privacy*. IEEE, 2017b.

Jonathan Berr. WannaCry ransomware attack losses could reach $4 billion, 2017. https://www.cbsnews.com/news/wannacry-ransomware-attacks-wannacry-virus-losses.

Leyla Bilge and Tudor Dumitras. Before we knew it—an empirical study of zero-day attacks in the real world. In *Proceedings of the 2012 ACM Conference on Computer and Communications Security*, pp. 833–844. ACM, 2012.

Marcel Böhme, Van-Thuan Pham, and Abhik Roychoudhury. Coverage-based greybox fuzzing as Markov chain. In *Proceedings of the 2016 ACM SIGSAC Conference on Computer and Communications Security*. ACM, 2016.

David Brumley, James Newsome, Dawn Song, Hao Wang, and Somesh Jha. Theory and techniques for automatic generation of vulnerability-based signatures. *IEEE Transactions on Dependable and Secure Computing*, 5(4): 224–241, 2008a.

David Brumley, Pongsin Poosankam, Dawn Song, and Jiang Zheng. Automatic patch-based exploit generation is possible: Techniques and implications. In *Proceedings of the 2008 IEEE Symposium on Security and Privacy*, pp. 143–157. IEEE, 2008b.

Bill Chappell. WannaCry ransomware: What we know Monday, 2017. https://www.npr.org/sections/thetwo-way/2017/05/15/528451534/wannacry-ransomware-what-we-know-monday.

Chen Chen, Baojiang Cui, Jinxin Ma, Runpu Wu, Jianchao Guo, and Wenqian Liu. A systematic review of fuzzing techniques. *Computers & Security*, 75, 2018. doi:10.1016/j.cose.2018.02.002.

Manuel Costa, Jon Crowcroft, Miguel Castro, Antony Rowstron, Lidong Zhou, Lintao Zhang, and Paul Barham. Vigilante: End-to-end containment of internet worms. In *Proceedings of the 20th ACM Symposium on Operating Systems Principles*, pp. 133–147. ACM, 2005.

CVE-2017-0144. Common Vulnerability and Exposures, 2017. https://cve.mitre.org/cgi-bin/cvename.cgi?name=CVE-2017-0144.

CVE-2017-11882. Common Vulnerability and Exposures, 2017. https://nvd.nist.gov/vuln/detail/CVE-2017-11882.

DARPA. DARPA's Cyber Grand Challenge: Final event program, 2016. https://www.youtube.com/watch?v=n0kn4mDXY6I.

Tudor Dumitras. *How to Predict Which Vulnerabilities Will Be Exploited*. USENIX Association, 2019.

Michael Eddington. Peach fuzzing platform, 2021. https://github.com/MozillaSecurity/peach.

Shuitao Gan, Chao Zhang, Xiaojun Qin, Xuwen Tu, Kang Li, Zhongyu Pei, and Zuoning Chen. Collafl: Path sensitive fuzzing. In *Proceedings of the 2018 IEEE Symposium on Security and Privacy*. IEEE, 2018.

Zhiqiang Lin, Xiangyu Zhang, and Dongyan Xu. Convicting exploitable software vulnerabilities: An efficient input provenance based approach. In *IEEE International Conference on Dependable Systems and Networks With FTCS and DCC (DSN)*, pp. 247–256. IEEE, 2008.

Microsoft Security Updates, 2017. https://technet.microsoft.com/en-us/security/bulletins.aspx.

Tyler Moore, Allan Friedman, and Ariel D. Procaccia. Would a "cyber warrior" protect us? Exploring trade-offs between attack and defense of information systems. In *Proceedings of the Workshop on New Security Paradigms*, pp. 85–94. Association for Computing Machinery, 2010.

Ellen Nakashima and Philip Rucker. U.S. declares North Korea carried out massive WannaCry cyberattack. *The Washington Post*, 19 December 2017. https://www.washingtonpost.com/world/national-security/us-set-to-declare-north-korea-carried-out-massive-wannacry-cyber-attack/2017/12/18/509deb1c-e446-11e7-a65d-1ac0fd7f097e_story.html.

Raconteur. WannaCry: The biggest ransomware attack in history. https://www.raconteur.net/infographics/wannacry-the-biggest-ransomware-attack-in-history, 2017.

Carl Sabottke, Octavian Suciu, Tudor Dumitra?. Vulnerability disclosure in the age of social media: Exploiting twitter for predicting real-world exploits. In *Proceedings of the 24th USENIX Security Symposium*. USENIX, 2015.

Samurai CTF Team. *We Are Samurai CTF and We Won Defcon CTF This Year*. AMA, 2013.

Harrison C. Schramm, David L. Alderson, W. Matthew Carlyle, and Nedialko B. Dimitrov. A game theoretic model of strategic conflict in cyberspace. *Military Operations Research*, 19(1):5–17, 2014.

Edward J. Schwartz, Thanassis Avgerinos, and David Brumley. Q: Exploit hardening made easy. In *Proceedings of the USENIX Security Symposium*, pp. 379–394. USENIX, 2011.

Lloyd Stowell Shapley. Stochastic games. *Proceedings of the National Academy of Sciences*, 39(10):1095, 1953.

Shellphish. Patcherex, 2021. https://github.com/shellphish/patcherex.

Yan Shoshitaishvili, Ruoyu Wang, Christophe Hauser, Christopher Kruegel, and Giovanni Vigna. Firmalice—automatic detection of authentication bypass vulnerabilities in binary firmware. *Proceedings 2015 Network and Distributed System Security Symposium*, (February):8–11, 2015.

Yan Shoshitaishvili, Ruoyu Wang, Christopher Salls, Nick Stephens, Mario Polino, Andrew Dutcher, John Grosen, Siji Feng, Christophe Hauser, Christopher Kruegel, and Giovanni Vigna. (State of) the art of war: Offensive techniques in binary analysis. In *Proceedings of the 2016 IEEE Symposium on Security and Privacy*, pp. 138–157. IEEE, 2016.

Nick Stephens, John Grosen, Christopher Salls, Andrew Dutcher, Ruoyu Wang, Jacopo Corbetta, Yan Shoshitaishvili, Christopher Kruegel, and Giovanni Vigna. Driller: Augmenting fuzzing through selective symbolic execution. In *Proceedings of the Network and Distributed System Security Symposium*. IEEE, 2016.

The MITRE Corporation. CVE: Common Vulnerabilities and Exposures, n.d. https://cve.mitre.org (accessed 18 March 2021).

Yanhao Wang, Xiangkun Jia, Yuwei Liu, Kyle Zeng, Tiffany Bao, Dinghao Wu, and Purui Su. Not all coverage measurements are equal: Fuzzing by coverage accounting for input prioritization. In *Proceedings of the 2020 Network and Distributed System Security Symposium*. NDSS, 2020.

Wei Wu, Yueqi Che, Jun Xu, and Xinyu Xing. FUZE: Towards facilitating exploit generation for kernel use-after-free vulnerabilities. In *Proceedings of the 27th USENIX Security Symposium*. USENIX, 2018.

Chaowei Xiao, Armin Sarabi, Yang Liu, Bo Li, Mingyan Liu, and Tudor Dumitras. From patching delays to infection symptoms: Using risk profiles for an early discovery of vulnerabilities exploited in the wild. In *27th USENIX Security Symposium (USENIX Security 18)*, pp. 903–918. USENIX, 2018.

Insu Yun, Sangho Lee, Meng Xu, Yeongjin Jang, and Taesoo Kim. QSYM: A practical concolic execution engine tailored for hybrid fuzzing. In *27th USENIX Security Symposium*, pp. 745–761, 2018.

Part III

Adversarial Machine Learning for Cyber Security

13

A Game Theoretic Perspective on Adversarial Machine Learning and Related Cybersecurity Applications

Yan Zhou[1], Murat Kantarcioglu[1], and Bowei Xi[2]

[1]*Computer Science Department, University of Texas at Dallas, Richardson, TX, USA*
[2]*Department of Statistics, Purdue University, West Lafayette, IN, USA*

13.1 Introduction to Game Theoretic Adversarial Machine Learning

In the world where data-driven decision making is trending, machine-intelligence has been typically known to exceed human intelligence in both speed and depth. Machine learning in recent years has had a great success in a wide spectrum of modern life. However, machine learning techniques have not been adopted as quickly as many have anticipated in cybersecurity applications. One of the reasons is that in cybersecurity machine learning techniques have to face challenges normal applications do not have to confront, that is, uncertainties caused by adversaries.

Take anti-virus (AV) software as an example. Machine-learning based anti-virus software systems are constantly challenged by obfuscated malware intentionally developed to evade detection by the AV software. The AV defense system and the malicious software are both instrumented with advanced techniques serving opposing interests, with each trying to defeat the other. The adversary attacks the AV system by strategically transforming malicious code to seemingly benign code. The AV system fails to detect the transformed malicious code because of the fundamental assumption its underlying machine learning algorithm makes: malicious data always follows the same distribution as the malicious data used for training. It is a common assumption known as i.i.d, that is, training data and test data are identically independently distributed. When the i.i.d assumption is violated because of adversarial attacks, the contract between the learned model and the data presented for prediction is broken, therefore the learning model can no longer function as designed. Machine learning algorithms become more vulnerable if the adversary has the control of the training data, misguiding the training process from the beginning to produce an erroneous classifier. Another example in cybersecurity is the e-mail spam problem. Spammers have to defeat machine learning-based spam filters for easy profits. As spam filters advance and become more sophisticated, spammers would also become more skilled and adopt more effective strategies to outsmart the spam filters.

The interaction between a machine learning-based system and an adversary in the previous examples is essentially a two-player game. One player's action causes reaction from the other player, to which the player has to respond. It is practically rare in a security domain, that one player has a dominant strategy that guarantees a win regardless of the actions taken by the other player. Instead, to optimize the gain, each player needs to look at the world from the opponent's perspective and respond to the opponent's strategy. The sheer size of all possible strategies each side can adopt is fairly intimidating. On top of that, each player has to conceptualize the motivations and objectives

Game Theory and Machine Learning for Cyber Security, First Edition.
Edited by Charles A. Kamhoua, Christopher D. Kiekintveld, Fei Fang, and Quanyan Zhu.

Table 13.1 Payoff table in prisoner's dilemma.

		Prisoner B	
		Betray	Cooperate
Prisoner A	Betray	$(-2, -2)$	$(0, -5)$
	Cooperate	$(-5, 0)$	$(-1, -1)$

of the opponent based on presumed beliefs and perceptions–whether the opponent would play rationally or with bounded rationality; whether it is a competitive situation in which, in order for someone to win, the other has to lose.

Game theory is a mathematical compass that guides the search for strategies to win games involving two or more players. As in the classic problem of prisoner's dilemma, game theory can provide an equilibrium solution–the best and optimal strategy–for rational players to play when they are facing dilemmas in which the information is incomplete. Each prisoner can choose to betray or cooperate with each other. The highest reward for each prisoner is to betray while the other chooses to cooperate, for example, spending zero years in prison while the other prisoner would spend five years in prison as shown in Table 13.1. Vice versa, the worst case for each prisoner is to cooperate while the other chooses to betray. If both choose to cooperate, each would have the second best reward.

The game is set up in such a way that each rational player has an incentive to make a decision that produces a nonoptimal overall outcome by acting in their self-interests. Game players have specific goals, and make decisions based on their goals and their opponents' responses. Games are characterized by competitiveness among players, roles they play, and the interactions among them.

The main contribution of this book chapter is to provide a holistic view of applying game theory in adversarial machine learning applications, putting together the theory, the analysis, the solution concept, and the case studies in real applications. We provide a bird's eye view of the existing development of game theoretic approaches to the adversarial learning problem, and discuss from a perspective of game theory the practical challenges of mitigating vulnerabilities of machine learning and AI algorithms in the presence of public scrutiny and strategic gaming.

The rest of the chapter is organized as follows. In Section 13.2, we formally define the adversarial learning problem and provide general game theoretic solution frameworks. We review simultaneous games and sequential games in Section 13.3 and provide case studies of a simultaneous game and a sequential game in real applications of spam and web spam classification in Sections 13.4 and 13.5. Finally, in Section 13.6 we discuss practical challenges of applying game theory in cybersecurity applications where players tend not to make rational decisions, and other challenges specific to machine learning algorithms such as high false-positive rate, lack of clean training data, susceptibility to adversarial attacks, and difficulties with evaluation.

13.2 Adversarial Learning Problem Definition

We define an adversarial learning problem over an input space $X \in \mathbb{R}^d$, where d is the number of attributes in the input space. A learning model trained on clean data provides a function f that takes $x \in X$ as input and outputs $y \in \{+1, -1\}$. There exists an adversary, corrupting data at test time by an amount of δ so that a positive (malicious) instance x will be classified as negative (benign), that is, $f(x) \neq f(x + \delta)$. The problem of adversarial learning is to find a robust f so that $P[f(x) \neq f(x + \delta)] < \epsilon$ for an arbitrary $\epsilon > 0$.

The adversarial learning problem is essentially a game between the learning model and the adversary. If we proceed with a pessimistic assumption that one player's gain is the other's loss, the robust solution to the problem will be a minimax strategy: the learner chooses the best action when assuming the adversary is playing optimally to maximize the learner's loss:

$$\min_{w^*} \max_{\delta^*} L(f, x, \delta),$$

where L is the loss function of the learner, w^* is the optimal learning parameter, and δ^* is the optimal distortion the adversary can apply to x to maximize the learner's loss.

If we assume a sequential game in which both players aim to minimize their own costs, the two players will take sequential actions: one player–the leader, acts first; the other player–the follower, observes the leader's action and plays optimally. For example, the learner moves first by setting the learning parameter w and the adversary, after observing w, chooses a distortion δ to minimize the loss incurred, as shown in the following optimization:

$$\min_{w^*} \max_{\delta^*} \quad L_\ell(w, x, \delta)$$
$$s.t. \quad \delta^* \in \arg\min_{\delta} L_f(w, x, \delta)$$

where L_ℓ and L_f are the loss functions of the leader and the follower, respectively.

Note that these game-theoretic frameworks are neither data dependent nor learning algorithm dependent. Both frameworks are applicable to any learning algorithms such as support vector machines and deep neural networks. As long as there are well defined losses for the learner and the adversary, one can always define and solve a game by following this framework regardless of the underlying learning algorithms. In Sections 13.4 and 13.5, we focus on the support vector machine method when we review two types of games in real applications–spam and web spam classification. Similarly, one can apply these frameworks to other machine learning applications such as image classification and hate speech detection. It is important to understand that in order to obtain effective playing strategies, attack models need to be explicitly defined. No robust learning strategies can work universally against any attack models.

13.3 Game Theory in Adversarial Machine Learning

In this section, we review some existing work on applying game theory in the context of adversarial machine learning. First, we review simultaneous games including: (1) zero sum games defined on a pessimistic assumption that one player's gain is the other player's loss; and (2) games in which an equilibrium is reached when all players make rational decisions that they have no incentive to change. Next, we discuss sequential games with a focus on Stackelberg games where one player leads while the other follows. In the sequential games, all players aim to maximize their own gain with the lowest possible cost.

13.3.1 Simultaneous Games

In simultaneous games, all players must choose a strategy to play at the same time, without knowing the strategies taken by other players, for example, the rock-paper-scissors game. Since no players have the knowledge about the actions of other players, they have to act on the perception that other players would play their best strategy.

13.3.1.1 Zero Sum Games

If we model a problem as a strictly competitive game, the players will be considered to have diametrical objectives. The game they play is called a zero sum game in which the total payoff is always zero (or a constant sum game with total payoff as a constant). In such a competitive situation, in order for someone to win, someone else has to lose. Sometimes, game players do not necessarily have completely opposed interests, but instead trying to maximize their own payoffs by playing their best strategies. Problems in the cybersecurity domain are often modeled as zero-sum games. Machine learning models are developed under the assumption that they would always face the worst-case scenario, that is, the adversary's objective is to maximize the loss of machine learning models, and one player's gain is the other player's loss. In zero-sum games, players are most effective by following a minimax strategy–a strategy that minimizes one's own loss while assuming the opponent always plays optimally.

Lanckriet et al. (2002) present a minimax formulation to construct a classifier so that the maximum probability of misclassification of future data is minimized. To ensure robustness, they assume the mean and covariance of class-conditional density are bounded in a convex region, and solve the optimization problem as a second order cone program. Geometrically, the minimax problem can be interpreted as minimizing the maximum of the Mahalanobis distance to each class. El Ghaoui et al. (2003) present a robust classifier for problems where data points are unknown but bounded in given hyper-rectangles. Each data point is represented by the smallest hyper-rectangle that defines all possible choices of data in multidimensional intervals–an interval matrix model for the data. A robust methodology is developed to minimize the worst-case loss over data within given interval bounds, and solutions are found with polynomial-time convex optimization interior-point algorithms. Globerson and Roweis (2006) solve an optimal SVM learning problem to deal with test-time malicious feature deletion. They search for the zero-sum minimax strategy that minimizes the hinge loss of the support vector machine. Dekel and Shamir (2008) consider application-time noise and corruption in data. They assume a worst-case scenario in which the adversary may corrupt any feature subset whose total value is upper-bounded by a predefined parameter. Each feature is assigned an a-priori importance value. A polynomial-size linear program is formulated and a statistical generalization bound for this approach is given. Alternatively, a more efficient online learning problem is formulated by converting a Perceptron into a statistical learning algorithm, and theoretical guarantees are provided by the online-to-batch technique. Recently, Madry et al. (2018) investigate adversarial robustness of neural networks through robust optimization. They define security against adversarial attacks and desired security guarantee using a min–max formulation in a common theoretical framework. The min–max optimization problem captures the essence of adversarial training and existing methods for attacking neural networks against specific constrained optimization models.

13.3.1.2 Nash Equilibrium Games

Actions taken by each player are contingent on the responses of other players. In simultaneous games, we assume all players take their actions without knowing other players' decisions. Suppose there are N players. Let S_i be the set of strategies available to player $i \in \{1, \dots, N\}$. The best response R_i of player i to the rest of players playing $\mathbf{s}_{-i}$ is:

$$R_i(\mathbf{s}_{-i}) = \underset{s_i \in S_i}{\operatorname{argmax}} \ \Pi_i(s_i, \mathbf{s}_{-i})$$

where Π_i is the payoff to player i when other players playing $\mathbf{s}_{-i}$. Given a N-player game, a Nash equilibrium solution is $\mathbf{s} = (s_1, \dots, s_N)$ where $s_i \in R_i(\mathbf{s}_{-i})$ for each player i. Searching for a unique Nash equilibrium strategy is normally the goal. When players play a Nash equilibrium strategy, none could benefit by unilaterally changing to a different strategy. In some cases, a game may have multiple Nash equilibria with different payoffs.

Bruckner and Scheffer (2009) study static prediction games in which both players act simultaneously. Each player has to make a decision without prior information on the opponent's move; therefore the optimal action of either player is not well-defined. If both players are rational and a game has a unique Nash equilibrium, each player can reasonably assume that the opponent would follow the Nash equilibrium strategy to maximize the gain. In other words, if one player follows the equilibrium strategy, the optimal move for the other player is to play the equilibrium strategy as well. When a game has multiple equilibria and each player follows an arbitrary equilibrium strategy, the results may be arbitrarily disadvantageous for either player.

Since each player has no information about the opponent's strategy, players have to make assumptions on how the opponent would act when making their own decisions. Brückner and Scheffer model this type of decision process as a static two-player game with complete information, that is, both players know their opponents' cost function and action space, and both players are rational to seek their own lowest possible costs. Notice that this setting is different from the minimax strategy in which each player's objective is to seek greatest damage on the opponent's cost. Whether a unique Nash equilibrium exists depends on the loss functions, the cost factors, and the regularizers, and Brückner and Scheffer derive sufficient conditions for the existence and uniqueness of a Nash equilibrium.

Dritsoula et al. (2017) provide a game theoretic analysis of the adversarial classification problem. They model the problem as a nonzero sum game in which the defender chooses a classifier to differentiate attacks from normal behaviors. The attacker works on the trade-off between the reward of successful attacks and the cost of being detected, and the defender's trade-off is framed in terms of correct attack detection and the cost of false positives. They propose an efficient algorithm to compute the mixed strategy Nash equilibria.

13.3.2 Sequential Games

Unlike simultaneous games, sequential games assume some players know the moves of other players who have moved first. First movers take the lead and the responding players follow by playing the best strategy against the leaders. In many real applications, the adversarial learning problem is more appropriately modeled as a sequential game between two players. For example, spammers can compute the best strategy against a spam filter if they know the strategy used in the spam filter. One player must commit to its strategy before the other player responds. The advantage the responding player has is partial or complete information of the first player. The responding player can therefore play its optimal strategy against its opponent. However, in some cases, extra information the responding player has can actually lower the player's payoff when the first player anticipates the optimal strategy from the responding player. This type of game is known as a Stackelberg game in which the first player is the leader and its opponent is the follower.

Adversarial learning researched in this area falls into two categories, depending on who plays the role of the leader: the classifier or the adversary. Kantarcioglu et al. (2011) solve for a Stackelberg equilibrium using simulated annealing and the genetic algorithm to discover an optimal set of attributes. Similar work has also been done by Liu and Chawla (2009). The difference between their research lies in that the former assumes both players know each other's payoff function, while the latter relaxed the assumption and only the adversary's payoff function is required. In both cases, the adversary is the leader whose strategies are stochastically sampled while the classifier is the follower that searches for an equilibrium given its knowledge about the adversary. In a more realistic scenario, the classifier is more likely to commit to a strategy before the adversary takes its actions.

The adversary's response is optimal given that it has some knowledge about the classifier's strategy. Example solutions to this type of Stackelberg game are presented by Brückner and Scheffer (2011). They define a Stackelberg prediction game in which the learner minimizes its loss knowing that the adversary is playing its optimal strategy. The adversary's strategy is optimal if it is among the solutions that minimize the adversary's loss given the learner's strategy. They solve bilevel optimization problems for three instances of loss functions.

Alfeld et al. (2017) propose a robust defense against adversarial attacks by modeling the interaction between the predictor and the attacker as a two player, nonzero sum, Stackelberg game. They define a general framework in which the predictor is the leader publishing explicit defense strategies against the adversary. With this framework they compute the optimal defense action for worst-case attacks against linear predictors, which is reduced to computing an induced matrix norm.

Hardt et al. (2016) also model classification as a sequential game between a classifier and an attacker. The attacker makes changes to the input according to a cost function known to both players. For a class of separable cost functions and certain generalizations, they developed computationally efficient learning algorithms that are arbitrarily close to the theoretical optimum. For general cost functions, they show that optimal classification is NP-hard. Their analysis demonstrates that in certain cases keeping classifiers secret is necessary for a good classification performance.

In more complex situations, the player who plays the role of a leader may have to face many opponents of various types, known as the followers. The game is therefore referred to as a *single leader multiple followers* game, often modeled as a Bayesian Stackelberg game that probabilistically chooses the actions that follow the best mixed strategy in which players take an action with a probability. The followers may have many different types and the leader does not know exactly the types of adversaries it may face when solving its optimization problem. However, the distribution of the types of adversaries is known or can be inferred. Paruchuri (2008) present a single-leader–single-follower Bayesian Stackelberg game to model interactions between a security agent and a criminal of unknown type. The security agent has only one type and must commit to its strategy first and stick with it. The criminal plays its best strategy given the knowledge about the security agent's strategy. They solve the Stackelberg game as a mixed integer linear programming problem.

Dasgupta and Collins (2019) provide an excellent survey of game theoretic approaches for adversarial machine learning in cybersecurity applications.

13.4 Simultaneous Zero-sum Games in Real Applications

In this section, we review a case of zero sum games in adversarial learning where the adversary modifies malicious data to evade detection (Zhou et al. 2012). A practical application is e-mail spam filtering. Spammers often disguise spam messages in a seemingly legitimate e-mail to evade detection. A well-known technique used by spammers is referred to as "good word" attack (Lowd 2005). "Good word" attacks change the word distributions in spam messages by inserting "good" words that frequently appear in legitimate e-mail. Such attacks invalidate the distribution that governs the training data, making it more feasible to foil the trained classifier. Transforming data maliciously at test time is also applicable in Web spam.

Suppose we have a set of data $\{(x_i, y_i) \in (\mathcal{X}, \mathcal{Y})\}_{i=1}^n$, where x_i is the ith sample and $y_i \in \{-1, 1\}$ is its label, $\mathcal{X} \subseteq \mathbb{R}^d$ is a d-dimensional feature space, n is the total number of samples. The adversary has an incentive to move the malicious data ($y_i = 1$) in any direction by adding a nonzero displacement vector δ_i to $x_i|_{y_i=1}$. For example, in spam-filtering the adversary may add good words to spam e-mail to defeat spam filters. We assume the adversary is not able to modify the good data ($y_i = -1$), such as legitimate e-mail.

In real world applications, attackers often do not have the freedom to arbitrarily modify data. For example, in credit card fraud detection, a skilled attacker usually chooses the "right" amount to spend using a stolen credit card in order to mimic a legitimate purchase. The attacker needs to balance between risk of exposure and potential profit. A common strategy of the adversary is to move the malicious data close enough to where the good data is frequently observed in the feature space so that the transformed malicious data can evade detection.

13.4.1 Adversarial Attack Models

In many security domains, the number of possible ways to attack a machine learning-based system is astronomical. No general robustness guarantees can be given for an arbitrary attack type. We provide two attack models that are general enough to reflect attack scenarios often encountered in real applications, such as denial-of-service attack and backdoor vulnerabilities. The two attack models are *free-range* and *restrained* attacks, each of which makes a simple and realistic assumption about how much is known to the adversary. The models differ in their implications for (1) the adversary's knowledge of the good data, and (2) the loss of utility as a result of modifying the malicious data. The free-range attack model assumes the adversary has the freedom to move data anywhere in the feature space. The restrained attack model is a more conservative attack model. The model is built under the intuition that the adversary would be reluctant to move a data point far away from its original position in the feature space. The reason is that greater displacement often entails more cost and loss of malicious utility. The concept of malicious utility is introduced here in a common sense manner. Smaller distortions to the original input are always favored. For example, in image classification, if too many pixels are changed, the attack would become perceptible to human eyes.

13.4.1.1 Free-Range Attack

In the free-range attack model, we assume the adversary knows the valid range of each attribute in a given data set. Let $x_j^{\max}$ and $x_j^{\min}$ be the upper bound and lower bound values of the jth feature of a data point x_i. We assume $x_j^{\max}$ and $x_j^{\min}$ are bounded. The attack is defined in the following manner such that the corrupted data are "syntactically" legitimate in the given data domain::

$$C_f(x_j^{\min} - x_{ij}) \le \delta_{ij} \le C_f(x_j^{\max} - x_{ij}), \quad \forall j \in [1, d],$$

where $C_f \in [0, 1]$ controls the aggressiveness of attacks. $C_f = 0$ means no attacks, while $C_f = 1$ permits the widest range of data movement.

It is obvious that this type of attack is most powerful, mimicking scenarios in which the adversary's sole objective is to take down the machine learning system whose reliability is called into questions when the system is under attack. A corresponding real world problem is the denial-of-service attack in which the attackers simply want to disrupt the service. This attack model, when taken into consideration in a learning model, would provide the strongest defense when facing the most severe attacks. The weakness of the attack model, however, is being overly "paranoid" when the attack is mild.

13.4.1.2 Restrained Attack

The restrained attack model is a more realistic model for attacks in which excessive data corruption is penalized. Let x_i be a malicious data point the adversary desires to modify, and x_i^t be a potential target into which the adversary would like to transform x_i. For example, if x_i is an e-mail spam message, x_i^t is the desired "look" of the e-mail spam after the attack. The choice of x_i^t is unidirectional in the sense that it has to be a data point in the good class, such as a legitimate e-mail in which the spam message is embedded. Optimally choosing x_i^t requires a great deal of knowledge about the feature space and sometimes the inner working of a learning algorithm (Dalvi et al. 2004; Lowd and Meek 2005), which may be a great obstacle for the attacker to overcome. For the attacker, more feasible choices of x_i^t include an estimated of the centroid of the good data, a data point sampled from the observed good data, or an artificial data point generated from the estimated good data distribution.

In practical scenarios, the adversary may not be able to change x_i directly to x_i^t as desired since x_i may lose too much of its malicious utility. For example, when too many good words are added to spam e-mail, the spam part may not stand out in the message. Therefore, for each attribute j in the d-dimensional feature space, we assume the adversary adds δ_{ij} to x_{ij} where

$$|\delta_{ij}| \le |x_{ij}^t - x_{ij}|, \quad \forall j \in d,$$

and δ_{ij} is further bounded as follows:

$$0 \le (x_{ij}^t - x_{ij})\delta_{ij} \le \left(1 - C_\delta \frac{|x_{ij}^t - x_{ij}|}{|x_{ij}| + |x_{ij}^t|}\right)(x_{ij}^t - x_{ij})^2,$$

where $C_\delta \in [0, 1]$ is a constant modeling the loss of malicious utility as a result of adding δ_{ij} to x_{ij}. This attack model specifies how much the adversary can force x_{ij} towards x_{ij}^t based on how far apart they are from each other. The term $1 - C_\delta \frac{|x_{ij}^t - x_{ij}|}{|x_{ij}| + |x_{ij}^t|}$ is the percentage of $x_{ij}^t - x_{ij}$ that δ_{ij} is allowed to be at most. When C_δ is fixed, the closer x_{ij} is to x_{ij}^t, the more x_{ij} is allowed to move towards x_{ij}^t percentage wise. The opposite is also true. The farther apart x_{ij} and x_{ij}^t, the smaller $|\delta_{ij}|$ will be. For example, when x_{ij} and x_{ij}^t are located on different sides of the origin, that is, one is positive and the other is negative, then no movement is permitted (that is, $\delta_{ij} = 0$) when $C_\delta = 1$.

This model balances between the needs of disguising maliciousness of data and retaining its malicious utility in the mean time. $(x_{ij}^t - x_{ij})\delta_{ij} \ge 0$ ensures δ_{ij} moves in the same direction as $x_{ij}^t - x_{ij}$. C_δ is related to the loss of malicious utility after the data has been modified. C_δ sets how much malicious utility the adversary is willing to sacrifice for crossing the decision boundary. A larger C_δ means smaller loss of malicious utility, while a smaller C_δ models greater loss of malicious utility. Hence a larger C_δ leads to less aggressive attacks while a smaller C_δ leads to more aggress attacks.

The attack model works great for well-separated data as shown in Figure 13.1a. When data from both classes are near the separation boundary as shown in Figure 13.1b, slightly changing attribute values would be sufficient to push the data across the boundary. In this case, even if C_δ is set to 1, the attack from the above model would still be too aggressive compared with what is needed. We could allow $C_\delta > 1$ to further reduce the aggressiveness of attacks, however, for simplicity and more straightforward control, we instead apply a discount factor C_ξ to $|x_{ij}^t - x_{ij}|$ directly to model the severeness of attacks:

$$0 \le (x_{ij}^t - x_{ij})\delta_{ij} \le C_\xi \left(1 - \frac{|x_{ij}^t - x_{ij}|}{|x_{ij}| + |x_{ij}^t|}\right)(x_{ij}^t - x_{ij})^2,$$

where $C_\xi \in [0, 1]$. A large C_ξ gives rise to a greater amount of data movement, and a small C_ξ sets a narrower limit on data movement. Combining these two cases, the restrained-attack model is given as follows:

$$0 \le (x_{ij}^t - x_{ij})\delta_{ij} \le C_\xi \left(1 - C_\delta \frac{|x_{ij}^t - x_{ij}|}{|x_{ij}| + |x_{ij}^t|}\right)(x_{ij}^t - x_{ij})^2.$$

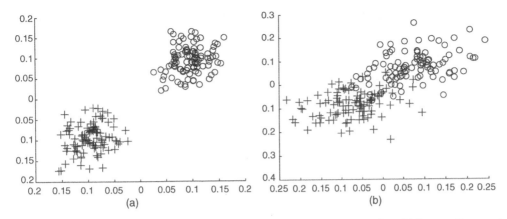

Figure 13.1 Data well separated and data cluttered near separating boundary. (a) Data well separated. (b) Data cluttered near boundary.

13.4.2 Adversarial SVM Learning

We now present an adversarial support vector machine model (AD-SVM) against each of the two attack models discussed in the previous section. We assume the adversary cannot modify data in the good class. Note that this assumption can be relaxed to model cases where good data may also be altered. A minimax strategy is formed to minimize the greatest hinge loss caused by adversarial data corruption. Note that the hinge loss is a loss function defined as $\max(0, 1 - y \cdot f(x))$ given an input x and its target output y while the prediction for x is $f(x)$.

We define two adversarial SVM risk formulations with the free-range attack model and the restrained attack model respectively. We transform each formulation to its dual form to avoid solving a bilinear nonconvex optimization problem.

13.4.2.1 AD-SVM Against Free-range Attack Model

We first consider the free-range attack model. The hinge loss model is given as follows:

$$h(w, b, x_i) = \begin{cases} \max_{\delta_i} \lfloor 1 - (w \cdot (x_i + \delta_i) + b) \rfloor_+ & \text{if } y_i = 1 \\ \lfloor 1 + (w \cdot x_i + b) \rfloor_+ & \text{if } y_i = -1 \end{cases}$$

s.t.

$$\delta_i \preccurlyeq C_f(x^{\max} - x_i)$$
$$\delta_i \succcurlyeq C_f(x^{\min} - x_i)$$

where (w, b) are the parameters (that is, weight and bias) of the hyperplane of SVM, δ_i is the displacement vector for the input x_i, and $\preccurlyeq$ and $\succcurlyeq$ denote component-wise inequality.

Following the standard SVM risk formulation, we have

$$\operatorname*{argmin}_{w,b} \sum_{\{i|y_i=1\}} \max_{\delta_i} \lfloor 1 - (w \cdot (x_i + \delta_i) + b) \rfloor_+$$
$$+ \sum_{\{i|y_i=-1\}} \lfloor 1 + (w \cdot x_i + b) \rfloor_+$$
$$+ \tfrac{1}{2}\|w\|^2$$

Combining cases for positive and negative instances, this is equivalent to:

$$\operatorname*{argmin}_{w,b} \sum_i \max_{\delta_i} \lfloor 1 - y_i(w \cdot x_i + b) - \tfrac{1}{2}(1 + y_i)w \cdot \delta_i \rfloor_+$$
$$+ \tfrac{1}{2}\|w\|^2$$

Note that the worst case hinge loss of x_i is obtained when δ_i is chosen to minimize its contribution to the margin, that is,

$$
\begin{aligned}
f_i \quad &= \quad \min_{\delta_i} \tfrac{1}{2}(1 + y_i)w \cdot \delta_i \\
\text{s.t.} \quad &\quad \delta_i \preceq C_f(x^{\max} - x_i) \\
&\quad \delta_i \succeq C_f(x^{\min} - x_i)
\end{aligned}
$$

This is a disjoint bilinear problem with respect to w and δ_i. Here, we are interested in discovering optimal assignment to δ_i with a given w. We can reduce the bilinear problem to the following asymmetric dual problem over $u_i \in \mathbb{R}^d$, $v_i \in \mathbb{R}^d$ where d is the dimension of the feature space:

$$
g_i \quad = \quad \max \ -\sum_j C_f \left(v_{ij}(x_j^{\max} - x_{ij}) - u_{ij}(x_j^{\min} - x_{ij}) \right)
$$

or

$$
\begin{aligned}
g_i \quad &= \quad \min \ \sum_j C_f \left(v_{ij}(x_j^{\max} - x_{ij}) - u_{ij}(x_j^{\min} - x_{ij}) \right) \\
\text{s.t.} \quad &\quad (u_i - v_i) = \tfrac{1}{2}(1 + y_i)w \\
&\quad u_i \succeq 0 \\
&\quad v_i \succeq 0
\end{aligned}
$$

The SVM risk minimization problem can be rewritten as follows:

$$
\begin{aligned}
\underset{w,b,t_i,u_i,v_i}{\text{argmin}} \quad &\tfrac{1}{2}||w||^2 + C\sum_i \lfloor 1 - y_i \cdot (w \cdot x_i + b) + t_i \rfloor_+ \\
\text{s.t.} \quad &t_i \geq \sum_j C_f \left(v_{ij}(x_j^{\max} - x_{ij}) - u_{ij}(x_j^{\min} - x_{ij}) \right) \\
&u_i - v_i = \tfrac{1}{2}(1 + y_i)w \\
&u_i \succeq 0 \\
&v_i \succeq 0
\end{aligned}
$$

Adding a slack variable and linear constraints to remove the nondifferentiality of the hinge loss, we can rewrite the problem as follows:

$$
\begin{aligned}
\underset{w,b,\xi_i,t_i,u_i,v_i}{\text{argmin}} \quad &\tfrac{1}{2}||w||^2 + C\sum_i \xi_i \\
\text{s.t.} \quad &\xi_i \geq 0 \\
&\xi_i \geq 1 - y_i \cdot (w \cdot x_i + b) + t_i \\
&t_i \geq \sum_j C_f \left(v_{ij}(x_j^{\max} - x_{ij}) - u_{ij}(x_j^{\min} - x_{ij}) \right) \qquad (13.1)\\
&u_i - v_i = \tfrac{1}{2}(1 + y_i)w \\
&u_i \succeq 0 \\
&v_i \succeq 0
\end{aligned}
$$

13.4.2.2 AD-SVM Against Restrained Attack Model

With the restrained attack model, we modify the hinge loss model and solve the problem following the same steps:

$$
h(w, b, x_i) = \begin{cases} \max_{\delta_i} \lfloor 1 - (w \cdot (x_i + \delta_i) + b) \rfloor_+ & \text{if } y_i = 1 \\ \lfloor 1 + (w \cdot x_i + b) \rfloor_+ & \text{if } y_i = -1 \end{cases}
$$

s.t.

$$
(x_i^t - x_i) \circ \delta_i \preceq C_\xi \left(1 - C_\delta \frac{|x_i^t - x_i|}{|x_i| + |x_i^t|} \right) \circ (x_i^t - x_i)^{\circ 2}
$$

$$
(x_i^t - x_i) \circ \delta_i \succeq 0
$$

where δ_i denotes the modification to x_i, $\preceq$ is component-wise inequality, and $\circ$ denotes component-wise operations.

The worst case hinge loss is obtained by solving the following minimization problem:

$$
\begin{aligned}
f_i \quad &= \quad \min_{\delta_i} \tfrac{1}{2}(1+y_i)w \cdot \delta_i \\
\text{s.t.} \quad &(x_i^t - x_i) \circ \delta_i \preceq C_\xi \left(1 - C_\delta \frac{|x_i^t - x_i|}{|x_i| + |x_i^t|}\right) \circ (x_i^t - x_i)^{\circ 2} \\
&(x_i^t - x_i) \circ \delta_i \succeq 0
\end{aligned}
$$

Let

$$
e_{ij} = C_\xi \left(1 - C_\delta \frac{|x_{ij}^t - x_{ij}|}{|x_{ij}| + |x_{ij}^t|}\right)(x_{ij}^t - x_{ij})^2 .
$$

We reduce the bilinear problem to the following asymmetric dual problem over $u_i \in \mathbb{R}^d$, $v_i \in \mathbb{R}^d$ where d is the dimension of the feature space:

$$
\begin{aligned}
g_i \quad &= \quad \max \; -\sum_j e_{ij} u_{ij}, \text{ or} \\
g_i \quad &= \quad \min \; \sum_j e_{ij} u_{ij} \\
\text{s.t.} \quad &(-u_i + v_i) \circ (x_i^t - x_i) = \tfrac{1}{2}(1+y_i)w \\
&u_i \succeq 0 \\
&v_i \succeq 0
\end{aligned}
$$

The SVM risk minimization problem can be rewritten as follows:

$$
\begin{aligned}
\operatorname*{argmin}_{w,b,t_i,u_i,v_i} \quad &\tfrac{1}{2}||w||^2 + C\sum_i \lfloor 1 - y_i \cdot (w \cdot x_i + b) + t_i \rfloor_+ \\
\text{s.t.} \quad &t_i \geq \sum_j e_{ij} u_{ij} \\
&(-u_i + v_i) \circ (x_i^t - x_i) = \tfrac{1}{2}(1+y_i)w \\
&u_i \succeq 0 \\
&v_i \succeq 0
\end{aligned}
$$

After removing the nondifferentiality of the hinge loss, we can rewrite the problem as follows:

$$
\begin{aligned}
\operatorname*{argmin}_{w,b,\xi_i,t_i,u_i,v_i} \quad &\tfrac{1}{2}||w||^2 + C\sum_i \xi_i \\
\text{s.t.} \quad &\xi_i \geq 0 \\
&\xi_i \geq 1 - y_i \cdot (w \cdot x_i + b) + t_i \\
&t_i \geq \sum_j e_{ij} u_{ij} \\
&(-u_i + v_i) \circ (x_i^t - x_i) = \tfrac{1}{2}(1+y_i)w \\
&u_i \succeq 0 \\
&v_i \succeq 0
\end{aligned} \tag{13.2}
$$

The framework of *adversarial SVM* learning is summarized in Algorithm 13.1.

Given training data $\{(x_i, y_i)\}_{i=1}^n$, first choose an attack model anticipated, set the corresponding hyper-parameters, and solve for the risk minimization problem accordingly.

Algorithm 13.1 Adversarial SVM

Input: Training data: $\{(x_i, y_i)\}_{i=1}^n$
Output: Learning parameters: (w, b)
Choose *attack model*;
if *attack model = free range* **then**
| set C_f;
| $(w, b) =$ solve Eq. (13.1);
else
| set (C_δ, C_ξ);
| $(w, b) =$ solve Eq. (13.2);
end
return w, b

13.4.3 Experiment

We test the AD-SVM models on e-mail spam and web spam data sets. In our experiments, we investigate the robustness of the AD-SVM models as we increase the severeness of the attacks. We let x_i^t be the centroid of the good data in our AD-SVM model against restrained attacks. In both applications, to measure malicious utility one can perform saliency analysis to check whether the most critical features indicative of spam or webspam have been modified, which is not covered here.

13.4.3.1 Attack Simulation

All positive (i.e. malicious) data in the test set are attacked and attacks are simulated using the following model in the experiments:

$$\delta_{ij} = f_{\text{attack}}(x_{ij}^- - x_{ij})$$

where x_i^- is a good data point randomly chosen from the test set, and $f_{\text{attack}} > 0$ sets a limit for the adversary to move the test data toward the target good data points. By controlling the value of f_{attack}, we can dictate the severity of attacks in the simulation. The actual attacks on the test data are intentionally designed not to match the attack models in AD-SVM so that the results are not biased. For each parameter C_f, C_δ and C_ξ in the attack models considered in AD-SVM, we tried different values as f_{attack} increases. This allows us to test the robustness of our AD-SVM model in all cases where there are no attacks and attacks that are much more severe than the model has anticipated. We compare our AD-SVM model to the standard SVM and one-class SVM models. We implemented our AD-SVM algorithms in CVX–a package for specifying and solving convex programs (Grant and Boyd 2011).

13.4.3.2 Experimental Results

The two real datasets we used to test our AD-SVM models are: *spam base* taken from the UCI data repository (UCI Machine Learning Repository 2019), and *web spam* taken from the LibSVM website (LIBSVM Data 2019).

The spam base dataset includes advertisements, make money fast scams, chain letters, etc. The spam collection came from the postmaster and individuals who had filed spam. The nonspam

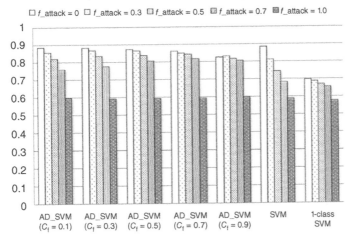

Figure 13.2 Accuracy of AD-SVM, SVM, and 1-class SVM on the *spam base* dataset as attacks intensify. The *free-range* attack is used in the learning model. C_f *increases* as attacks become more aggressive.

e-mail came from filed work and personal e-mails (UCI Machine Learning Repository 2019). The dataset contains 4601 e-mail messages with 39.4% spam. Each message is vectorized as 57 dimensions and one class label. The majority of the attributes represent the frequency of matching words in the e-mail. A few variants of the length of the longest uninterrupted sequence of certain capital letters are also used as the attributes. We divide the data sets into equal halves, with one half T_r for training and the other half T_s for testing. Learning models are built from 10% of random samples selected from T_r. The results are averaged over 10 random runs.

The second data set came from the LibSVM website (LIBSVM Data 2019). Web spam came from the Pascal Large Scale Learning Challenge. Nonspam web data were created by randomly traversing the Internet at well known web-sites. The input to machine learning models is created by treating continuous n bytes as a word and use word count as the feature value and normalize each instance to unit length. We use their unigram data set in which the number of features is 254. The total number of instances is 350,000. We divide the data set into equal halves for training and testing. We use 2% of the training data to build the learning models and report the results averaged over 10 random runs.

Figures 13.2 and 13.3 show the results of the considered learning algorithms, *AD-SVM*, *SVM*, and *one-class SVM*, on the *spam base* dataset as attacks intensify. Detailed accuracy results are also given in Tables 13.2 and 13.3. By assuming free-range and restrained attack models, AD-SVM demonstrates an outstanding performance in terms of predictive accuracy on this dataset. C_δ alone is used in the restrained learning model (as C_ξ has a fixed value of 1). Except for the most pessimistic cases, AD-SVM suffers no performance loss when there are no attacks. On the other hand, it achieved much more superior classification accuracy than SVM and one-class SVM when there are attacks.

Tables 13.4 and 13.5 list the detailed accuracy results on the web spam dataset. AD-SVM with the free-range attack model appears to be too pessimistic when there are no attacks. AD-SVM with the restrained model performs consistently better than SVM and one-class SVM. More importantly, the optimistic settings ($c_\delta > 0.5$) suffer no loss when there are no attacks. The graphical representations of the results are shown in Figures 13.4 and 13.5.

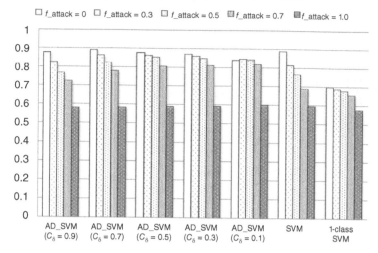

Figure 13.3 Accuracy of AD-SVM, SVM, and 1-class SVM on *spam base* dataset as attacks intensify. The *restrained* attack model is used in the learning model. C_δ *decreases* as attacks become more aggressive.

Table 13.2 Accuracy of AD-SVM, SVM, and one-class SVM on the *spam base* dataset as attacks intensify. The *free-range* attack is used in the learning model. C_f *increases* as attacks become more aggressive. AD-SVM results better than the baseline SVM are bolded.

		$f_{attack} = 0$	$f_{attack} = 0.3$	$f_{attack} = 0.5$	$f_{attack} = 0.7$	$f_{attack} = 1.0$
AD-SVM	$C_f = 0.1$	**0.882**	**0.852**	**0.817**	**0.757**	**0.593**
	$C_f = 0.3$	0.880	**0.864**	**0.833**	**0.772**	**0.588**
	$C_f = 0.5$	0.870	**0.860**	**0.836**	**0.804**	**0.591**
	$C_f = 0.7$	0.859	**0.847**	**0.841**	**0.814**	**0.592**
	$C_f = 0.9$	0.824	**0.829**	**0.815**	**0.802**	**0.598**
SVM		0.881	0.809	0.742	0.680	0.586
One-class SVM		0.695	0.686	0.667	0.653	0.572

Note that we use C_ξ alone in our learning model as C_δ is fixed to 1. Which parameter, C_ξ or C_δ, to use in the restrained attack model can be determined through cross validation on the training data. More discussions on model parameters will be presented in the next section.

13.4.3.3 A Few Words on Setting C_f, C_ξ, and C_δ

Setting parameters properly in the attack models is important. The AD-SVM algorithms assume either a free-range attack model or a restrained attack model. In reality, we might not know the exact attack model or the true utility function of the attackers. However, as Tables 13.2–13.5 demonstrate, although the actual attacks may not match what we have anticipated, our AD-SVM algorithm using the restrained attack model exhibits overall robust performance by setting C_δ or C_ξ values for more aggressive attacks. If we use the restrained attack model, choosing $C_\delta \leq 0.5$ ($C_\xi \geq 0.5$) consistently returns robust results against all f_{attack} values. If we use the free-range attack model, we will have to

Table 13.3 Accuracy of AD-SVM, SVM and one-class SVM on *spam base* dataset as attacks intensify. The *restrained* attack model is used in the learning model. C_δ *decreases* as attacks become more aggressive. AD-SVM results better than the baseline SVM are bolded.

		$f_{attack} = 0$	$f_{attack} = 0.3$	$f_{attack} = 0.5$	$f_{attack} = 0.7$	$f_{attack} = 1.0$
AD-SVM	$C_\delta = 0.9$	0.874	**0.821**	**0.766**	**0.720**	0.579
	$C_\delta = 0.7$	**0.888**	**0.860**	**0.821**	**0.776**	0.581
$(C_\xi = 1)$	$C_\delta = 0.5$	0.874	**0.860**	**0.849**	**0.804**	0.586
	$C_\delta = 0.3$	0.867	**0.855**	**0.845**	**0.809**	0.590
	$C_\delta = 0.1$	0.836	**0.840**	**0.839**	**0.815**	**0.597**
SVM		0.884	0.812	0.761	0.686	0.591
One-class SVM		0.695	0.687	0.676	0.653	0.574

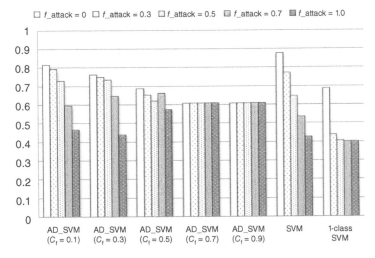

Figure 13.4 Accuracy of AD-SVM, SVM, and 1-class SVM on the *webspam* dataset as attacks intensify. The *free-range* attack is used in the learning model. C_f *increases* as attacks become more aggressive.

set parameter values to avoid the very pessimistic results for mild attacks. Hence choosing $C_f \leq 0.3$ in general returns good classification results against all f_{attack} values.

As a general guideline, the baseline of C_f, C_δ or C_ξ has to be chosen to work well against attack parameters suggested by domain experts. This can be done through cross-validation for various attack scenarios. From there, we gradually increase C_f or C_ξ, or decrease in the case of C_δ. The best value of C_f, C_δ or C_ξ is reached right before performance deteriorates. Also note that it is sufficient to set only one of C_ξ and C_δ while fixing the other to 1. Furthermore, C_f, C_δ and C_ξ do not have to be a scalar parameter. In many applications, it is clear some attributes can be changed while others cannot. A C_f, C_δ/C_ξ parameter vector would help enforce these additional rules.

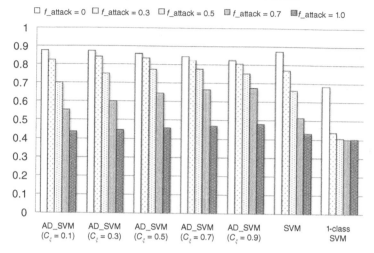

Figure 13.5 Accuracy of AD-SVM, SVM, and 1-class SVM on *webspam* dataset as attacks intensify. The *restrained* attack model is used in the learning model. C_ξ *increases* as attacks become more aggressive.

Table 13.4 Accuracy of AD-SVM, SVM, and one-class SVM on *webspam* dataset as attacks intensify. The *free-range* attack model is used in the learning model. C_f *increases* as attacks become more aggressive. AD-SVM results better than the baseline SVM are bolded.

		$f_{attack} = 0$	$f_{attack} = 0.3$	$f_{attack} = 0.5$	$f_{attack} = 0.7$	$f_{attack} = 1.0$
AD-SVM	$C_f = 0.1$	0.814	**0.790**	**0.727**	**0.591**	**0.463**
	$C_f = 0.3$	0.760	0.746	**0.732**	**0.643**	**0.436**
	$C_f = 0.5$	0.684	0.649	0.617	**0.658**	**0.572**
	$C_f = 0.7$	0.606	0.606	0.606	**0.606**	**0.606**
	$C_f = 0.9$	0.606	0.606	0.606	**0.606**	**0.606**
SVM		0.874	0.769	0.644	0.534	0.427
One-class SVM		0.685	0.438	0.405	0.399	0.399

13.4.4 Remark

Adversarial attacks challenge standard learning models with a different data distribution other than the expected one. By assuming a zero sum game, a minimax strategy can be found for a given attack model. We review two attack models corresponding to practical scenarios in some security domain. We present an adversarial SVM learning model with a minimax formulation to enhance the robustness against each of the two attack models. We demonstrate that the adversarial SVM model is much more resilient to adversarial attacks than standard SVM and one-class SVM models. We also show that optimal learning strategies derived to counter overly pessimistic attack models can produce unsatisfactory results when the real attacks are much weaker. On the other hand, learning models built on restrained attack models perform more consistently as attack parameters vary. One future direction for this work is to add cost-sensitive metrics into the learning models.

Table 13.5 Accuracy of AD-SVM, SVM, and one-class SVM on *webspam* dataset as attacks intensify. The *restrained* attack model is used in the learning model. C_ξ *increases* as attacks become more aggressive. AD-SVM results better than the baseline SVM are bolded.

		$f_{attack} = 0$	$f_{attack} = 0.3$	$f_{attack} = 0.5$	$f_{attack} = 0.7$	$f_{attack} = 1.0$
AD-SVM	$C_\xi = 0.1$	**0.873**	**0.822**	**0.699**	**0.552**	**0.435**
	$C_\xi = 0.3$	0.870	**0.837**	**0.748**	**0.597**	**0.444**
$(C_\delta = 1)$	$C_\xi = 0.5$	0.855	**0.833**	**0.772**	**0.641**	**0.454**
	$C_\xi = 0.7$	0.841	**0.820**	**0.773**	**0.663**	**0.467**
	$C_\xi = 0.9$	0.822	**0.803**	**0.749**	**0.671**	**0.478**
SVM		0.871	0.769	0.659	0.512	0.428
One-class SVM		0.684	0.436	0.406	0.399	0.400

Another direction is to extend the single learning model to an ensemble in which each base learner handles a different set of attacks.

13.5 Nested Bayesian Stackelberg Games

In this section, we review a case where a sequential game is played and the leader has to face multiple adversarial followers (Zhou and Kantarcioglu 2016). We first formulate the adversarial learning problem and review the single leader single follower (SLSF) Stackelberg game between a learner and an adversary that optimally transforms training data. We present strategies to construct component SLSF learning models given adversary types. Finally, we solve the SLMF Stackelberg game with the component SLSF models to counter adversaries of various types.

13.5.1 Adversarial Learning

Assume we are given input samples $x_{i|i=1,...,n} \in \mathcal{X}$ from some input space $\mathcal{X}$ and we want to estimate $y_i \in \mathcal{Y}$ where $\mathcal{Y} = \{+1, -1\}$ in a binary classification problem. In the standard settings of a learning problem, given a learning function $g : \mathcal{X} \to \mathbb{R}$ and a feature vector $\phi(x \in \mathcal{X}) \in \mathbb{R}^N$, we estimate

$$\hat{y}_i = g(w, x_i) = w^T \cdot \phi(x_i).$$

Our estimate $\hat{y}_i = g(w, x_i)_{|w \in \mathbb{R}^N}$ is obtained by optimizing an objective function L. Let $\ell : \mathcal{Y} \times \mathcal{Y} \to \mathbb{R}_+$ be the learner's loss function. Let the objective function L be the learner's loss with L_2 regularization:

$$L = \sum_{i=1}^{n} \ell(\hat{y}_i, y_i) + \lambda ||w||^2 \tag{13.3}$$

where λ is a regularization parameter that weighs the penalty of reduced norms. In cost-sensitive learning where the loss of data in different categories is considered unequal, a cost vector c is used to weigh the loss in Eq. (13.3), and the learner solves the following optimization problem:

$$\underset{w}{\text{argmin}} \, L = \underset{w}{\text{argmin}} \sum_{i=1}^{n} c_i \cdot \ell(\hat{y}_i, y_i) + \lambda ||w||^2 \tag{13.4}$$

The definition of the adversarial learning problem follows naturally from (13.4). Assume there exists an adversary that influences the learning outcome by modifying the data. The classification task becomes an estimate of $\hat{y}_i$ on the transformed data:

$$\hat{y}_i = w^T \cdot \phi(f_t(x_i, w)),$$

where $f_t(x_i, w)$ is a data transformation function used by the adversary to modify the data:

$$f_t(x_i, w) = x_i + \delta_x(x_i, w)$$

where δ_x returns the displacement vector for x_i. Therefore the adversarial learning problem can be defined as the following optimization problem:

$$\operatorname*{argmin}_{w}\operatorname*{argmax}_{\delta_x} L(w, x, \delta_x). \tag{13.5}$$

Note that the above learning problem becomes a disjoint bilinear problem with respect to w and δ_x assuming that the adversary plays its optimal attack $\delta_x^*(w)$.

Our adversarial learning model plays nested Stackelberg games by first fitting a set of learning models on "presumably transformed" training data; and after that learning an optimal distribution of the models. The learner will play the optimal mixed strategy of several learning models with a preassigned probability for each use. A mixed strategy allows the player to randomize over available strategies in a strategy space with a probability distribution.

13.5.2 A Single Leader Single Follower Stackelberg Game

To build a set of learning models, we consider a nonzero sum and nonsimultaneous two-player game between the learner and the adversary. The learner first commits to its strategy that is observable to the adversary and the adversary plays its optimal strategy to minimize the learner's payoff while maximizing its own payoff. This is referred to as a *single leader single follower* (SLSF) game. The setup of the SLSF game is common in practice. For example, e-mail service providers usually have spam filters installed on the server side before providing services to the end user. Sophisticated spammers would obtain firsthand knowledge about the statistical tools implemented in the spam filters before sending out massive amounts of spam. This can be done by probing spam filters with a large number of carefully crafted e-mail messages that are designed to detect the decision boundaries of the spam filters.

The leader's loss is misclassification as defined in Eq. (13.3):

$$L_\ell = \sum_{i=1}^{n} c_{\ell,i} \cdot \ell_\ell(\hat{y}_i, y_i) + \lambda_\ell ||w||^2. \tag{13.6}$$

The follower's loss consists of two parts: the cost of data transformation and its exposure to the leader as a result of misclassification. Therefore the follower's loss is:

$$L_f = \sum_{i=1}^{n} c_{f,i} \cdot \ell_f(\hat{y}_i, y_i) + \lambda_f \sum_{i=1}^{n} ||\phi(x_i) - \phi(f_t(x_i, w))||^2 \tag{13.7}$$

where λ_ℓ, λ_f, c_ℓ, and c_f are the weights of the penalty terms and the costs of data transformation. ℓ_ℓ and ℓ_f are the loss functions of the leader and the follower. The adversarial learning model $\langle w, \delta_x(w) \rangle$ is found by solving the following bilevel optimization problem:

$$\min_{w^*} \max_{\delta_x^*} \quad L_\ell(w, x, \delta_x)$$

$$\text{s.t.} \quad \delta_x^* \in \operatorname*{argmin}_{\delta_x} L_f(w, x, \delta_x) \tag{13.8}$$

It has been shown that when the adversary's loss is convex and continuously differentiable, a Stackelberg equilibrium solution

$$\langle w_s, \delta_x(w_s) \rangle$$

can be found for the optimization problem (Brückner and Scheffer 2011).

13.5.3 Learning Models and Adversary Types

There are two major issues we need to resolve in our adversarial learning framework: defining a set of learning models for the learner to randomly choose from and the types of adversaries that the learner has to face at application time. Another issue we need to resolve is to set the payoff matrices for the single leader multiple-followers game.

13.5.3.1 Learning Models

The Stackelberg equilibrium solution discussed in Section 13.5.2 can be used to generate a set $\mathcal{G} = \{g_s, g_{f_1}, \dots, g_{f_i}, \dots\}$ of learning functions. Learning function g_s is the Stackelberg equilibrium solution

$$g_s(w_s, x) = w_s^T \cdot \phi(x)$$

where w_s is the Stackelberg solution for the leader. The rest of the learning functions are obtained using the follower's solution in the Stackelberg equilibrium. Recall that the follower's solution is the optimal data transformation $\delta_x(w_s)$ given w_s. If we switch roles and let the adversary be the leader and disclose the data transformation δ_x to the learner, we can train a learning function g_f by solving Equation (13.5) as a simple optimization problem. Of course, g_f by itself does not constitute a robust solution to the adversarial learning problem since the adversary could easily defeat g_f with a different data transformation. g_f is better understood as a special case solution under the umbrella of the general Stackelberg solution. However, when the adversary does use δ_x as the data transformation model, g_f typically performs significantly better than the Stackelberg solution in terms of classification error. This is the motivation that we include such learning models in the learner's strategy set in our framework. To determine when and how often to use such learning models during test time, we resort to the solution to the *single leader multiple followers* game discussed in a later section.

We can train a set of g_{f_i} by varying how much impact a data transformation has on the follower's loss function defined in Eq. (13.7). When λ_f is very large, the penalty of data transformation becomes dominant and the adversary's best strategy is not to transform data at all. When λ_f is relatively small, the adversary has more incentives to perform data transformation. However, the diminishing cost of data transformation is counterbalanced by the increase in the cost of misclassification. This prevents the adversary modifying training data arbitrarily. Therefore, we can define a spectrum of learning models specifically characterizing various types of adversaries, from the least aggressive to the most aggressive. Figure 13.6 shows the decision boundaries of two learning

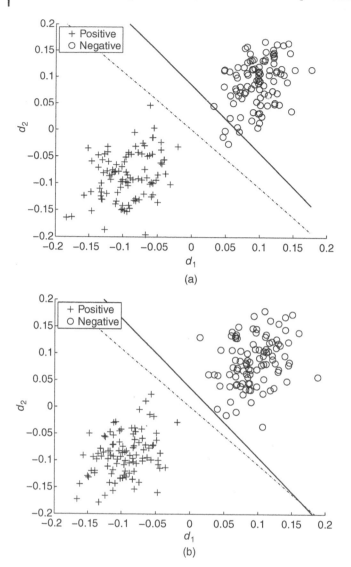

Figure 13.6 Learning models trained on two transformed data distributions in Stackelberg solutions. Dot-dash lines show the Stackelberg learner, solid lines show the linear models trained on the transformed training data. (a) Pessimistic predictor. (b) Optimistic predictor.

models trained on two different Stackelberg equilibrium data transformations. The equilibrium data transformations involve data in both classes. However, how much data can be transformed in either class can be controlled by the cost factors c_f^+ and c_f^- of the adversary. Figure 13.6a shows a pessimistic learning model. Pessimistic models are built to countermeasure aggressive adversaries that make significant data modifications. Figure 13.6b shows a relatively optimistic model, targeting more moderate attacks. Dot-dashed lines in the two plots show the Stackelberg solutions and solid lines show their corresponding linear models trained on the adversary's optimal strategies.

The algorithm that produces the first level SLSF Stackelberg strategies (learning functions) is given in Algorithm 13.2.

Algorithm 13.2 First Level SLSF Stackelberg Strategies

Input: Training data: $\{(x_i, y_i)\}_{i=1}^n$
Output: Learning functions: $\mathcal{G} = \{g_s, g_{f_1}, \dots, g_{f_k}\}$
initialize $\lambda_\ell, \lambda_f, c_\ell, c_f$ in Eq. (13.6) and Eq. (13.7);
$\mathcal{G} = \emptyset$;
$g_s \leftarrow$ solve SLSF Stackelberg game defined in Eq. (13.8);
$\mathcal{G} = \mathcal{G} \cup \{g_s\}$;
set $\{\lambda_{f_1}, \dots, \lambda_{f_k}\}$;
for *each λ in $\{\lambda_{f_1}, \dots, \lambda_{f_k}\}$* **do**
  $\delta_s \leftarrow$ solve SLSF Stackelberg game in Eq. (13.8) with λ;
  $g_f \leftarrow$ solve Eq. (13.5) with δ_s;
  $\mathcal{G} = \mathcal{G} \cup \{g_f\}$;
end
return $\mathcal{G}$

13.5.3.2 Adversary Types

Adversaries may come in many different types. For example, in e-mail spam filtering some spammers' objectives are to successfully send spam contents to the end user, others aim to clog the network by performing denial-of-service attacks; some spammers can modify both spam and legitimate e-mail, others have little privilege and are not entitled to access to legitimate e-mail. It is impossible to implement a spam filter with a single learning model to effectively counter every possible type of adversary. This motivates us to take into account different types of adversaries in our adversarial learning framework.

We model different adversary types by applying different constraints on data transformation. We consider three cases in which the adversary can:

- attack both positive and negative data (e.g. insider attack) such that

  $$x^+ \rightarrow x^- \text{ and } x^- \rightarrow x^+$$

- attack positive data only (e.g. good word attack in spam) such that

  $$x^+ \rightarrow x^-$$

- attack randomly by transforming data freely in the given domain (e.g. malware induced data corruption).

In the first two scenarios, the adversary can transform positive, or negative, or both classes of data with specific goals–transforming data in one class to another; while in the last case, the adversary can transform any data freely in the given domain. Figure 13.7 illustrates an example for each of the three different types of followers.

For each type of adversary, we define three pure attack strategies for the adversary: `mild`, `moderate`, and `aggressive` attacks. Each pure strategy has a different degree of impact on how much the transformed data diverges from the original data distribution.

13.5.3.3 Setting Payoff Matrices for the Single Leader Multiple-followers Game

The solutions to the *single leader and single follower* Stackelberg games produce pure strategies (learning parameters) for the learner: the equilibrium predictor, a pessimistic predictor built on

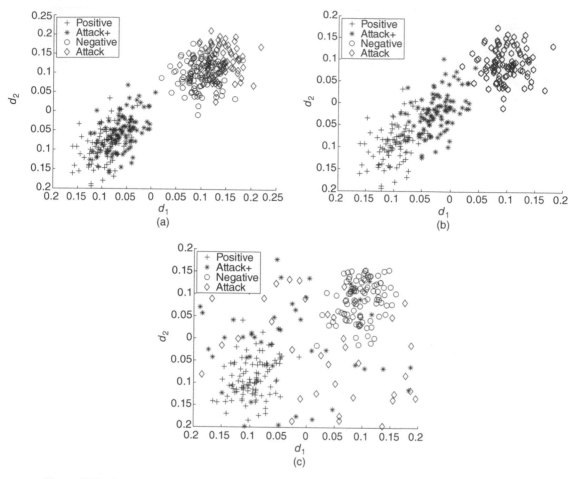

Figure 13.7 Data transformation performed by different types of followers: transforming data in both classes, transforming only positive data, and random transformation. (a) Attack positive + negative. (b) Attack positive only. (c) Random attacks.

the equilibrium data transformation of an aggressive adversary, and an optimistic predictor built on the equilibrium data transformation of a mild adversary. We use this information to compute the payoff matrices of the learner and the adversaries, and pass them to the *single leader multiple followers* game discussed in the next section.

The learner's payoffs are the predictive accuracies of its learning models on the data transformation corresponding to each pure strategy of an adversary. For the three learning models, we refer to the Stackelberg equilibrium predictor as Equi*, the pessimistic SVM model trained on an aggressive adversary's attacks as SVM*^{1} as shown in Figure 13.6a, and the less pessimistic SVM model trained on a less aggressive adversary's attacks as SVM*^{2} as shown in Figure 13.6b. We refer to the three pure strategies of the adversary as `mild`, `moderate`, and `aggressive`, respectively. The `mild` adversary transforms data by adding a small displacement vector. The `moderate` adversary transforms data more aggressively, and the `aggressive` adversary transforms data most aggressively. The payoff matrix for the learner is shown in Table 13.6. R_{ij} denotes the predictive accuracy of predictor i tested on the optimal data transformation of an adversary of type j.

The payoff matrix of the adversary is identical to Table 13.6 except that the rewards of the adversary is the false-negative rate discounted by the amount of data movement that is responsible for

Table 13.6 The payoff matrix of the learner.

	Mild	Moderate	Aggressive
Equi*	R_{11}	R_{12}	R_{13}
SVM*1	R_{21}	R_{22}	R_{23}
SVM*2	R_{31}	R_{32}	R_{33}

the false negatives. The amount of movement is measured as the L_2 distance summed over all data points before and after the data transformation. The payoff matrices of the learner and the followers are taken as input to a *single leader multiple followers* game discussed in the next section.

13.5.4 A Single Leader Multi-followers Stackelberg Game

In a *single leader multiple followers* (SLMF) game (Basar and Olsder 1999), the leader makes its optimal decision prior to the decisions of multiple followers. The Stackelberg game played by the leader is:

$$
\begin{aligned}
\min_{s^\ell, s^*} \quad & F(s^\ell, s^*) \\
\text{s.t.} \quad & G(s^\ell, s^*) \leq 0 \\
& H(s^\ell, s^*) = 0
\end{aligned}
$$

where F is the leader's objective function, constrained by G and H; s^ℓ is the leader's decision and s^* is in the set of the optimal solutions of the lower level problem:

$$
s^* \in \left\{
\begin{aligned}
\operatorname*{argmin}_{s_i} \quad & f_i(s^\ell, s_i) \\
\text{s.t.} \quad & g_i(s^\ell, s_i) \leq 0 \\
& h_i(s^\ell, s_i) = 0
\end{aligned}
\right\} \quad \forall i = 1, \dots, m
$$

where m is the number of followers, f_i is the ith follower's objective function constrained by g_i and h_i.

For the sake of simplicity, we assume the followers are not competing among themselves. This is usually a valid assumption in practice since adversaries rarely affect each other through their actions. In a Bayesian Stackelberg game, the followers may have many different types and the leader does not know exactly the types of adversaries it may face when solving its optimization problem. However, the distribution of the types of adversaries is known or can be inferred from past experience. The followers' strategies and payoffs are determined by the followers' types. The followers play their optimal responses to maximize the payoffs given the leader's strategy. The Stackelberg equilibrium includes an optimal mixed strategy of the learner and corresponding optimal strategies of the followers.

Problem Definition *Given the payoff matrices R^ℓ and R^f of the leader and the m followers of n different types, find the leader's optimal mixed strategy given that all followers know the leader's strategy when optimizing their rewards. The leader's pure strategies consist of a set of generalized linear learning models $\langle \phi(x), w \rangle$ and the followers' pure strategies include a set of vectors performing data transformation $x \to x + \delta x$.*

Given the payoff matrices R^ℓ and R^f discussed in Section 13.5.3.3 (as shown in Table 13.6), let r^f denote the follower's maximum payoff, $\mathcal{L}$ and $\mathcal{F}$ denote the indices of the pure strategies in the leader's policy s^ℓ and the follower's policy s^f. The *single learner multi-followers* Stackelberg game with n possible types can be modeled as follows (Paruchuri 2008):

$$
\begin{aligned}
\max_{s_\ell, s_f, r^{f,t}} \quad & \sum_{i \in \mathcal{L}} \sum_{t \in T} \sum_{j \in \mathcal{F}} p^t R_{ij}^t s_i^\ell s_j^{f,t} \\
\text{s.t.} \quad & \sum_{i \in \mathcal{L}} s_i^\ell = 1 \\
& \sum_{j \in \mathcal{F}} s_j^{f,t} = 1 && \forall t \in T \\
& 0 \le \left(r^{f,t} - \sum_{i \in \mathcal{L}} R_{ij}^{f,t} s_i^\ell \right) \le \left(1 - s_j^{f,t} \right) M && \forall t \in T, \\
& && \forall j \in \mathcal{F} \\
& s_i^\ell \in [0 \dots 1] && \forall i \in \mathcal{L} \\
& s_j^{f,t} \in \{0, 1\} && \forall t \in T, \\
& && \forall j \in \mathcal{F} \\
& r^{f,t} \in \mathbb{R} && \forall t \in T
\end{aligned}
\tag{13.9}
$$

where p^t is the *priori* probabilities of type $t \in T = \{t_1, \dots, t_n\}$, and M is a large constant enforcing the complementary constraint. The above Stackelberg game can be solved as a Mixed-Integer-Quadratic-Programming (MIQP) problem. For a game with a single leader and m followers with n possible types where the m followers are independent of each other and their actions have no impact on each other's decisions, we reduce the problem to solving m instances of the *single leader single follower* game.

The procedure of the nested Bayesian Stackelberg game is given in Algorithm 13.3. We use the efficient Decomposed Optimal Bayesian Stackelberg Solver (DOBSS) (Paruchuri 2008) to solve the MIQP problem. The time complexity of the algorithm is $O(|\mathcal{F}|^{|T|} \cdot |\mathcal{L}| \cdot |\mathcal{F}| \cdot |T|)$.

Algorithm 13.3 Nested Bayesian Stackelberg Game

Input: Training data: $\{(x_i, y_i)\}_{i=1}^n$, $\mathcal{L}$, T, $\mathcal{F}$
Output: Leader's and followers' mixed strategies: s^ℓ, $s^{f,t}$
initialize p^t, $\mathcal{G} = \emptyset$;
$\mathcal{G} \leftarrow$ output of Algo. 13.2;
for i in $\mathcal{L}$ **do**
    for t in T **do**
        for j in $\mathcal{F}$ **do**
            compute payoff $R_{i,j}^t$;
        end
    end
end
s^ℓ, $s^{f,t} \leftarrow$ solve SLMF Stackelberg game defined in Eq. (13.9);
return s^ℓ, $s^{f,t}$

13.5.5 Experiments

In our experiments, all test data are the potential targets of attacks. We use three adversaries, one from each of the three types discussed in Section 13.5.3.2. The first type *Adversary*[*1] can modify both positive and negative data, and the second type *Adversary*[*2] is only allowed to modify positive data as normally seen in spam filtering. The third type of adversary *Adversary*[*3] can transform data freely in the given domain. The prior distribution of the three adversary types is randomly set. Let p be the probability that the adversary modifies negative data. Then for each negative instance x^- in the test set, with probability p, x^- is modified as follows:

$$x^- = x^- + f_a \cdot (x^+ - x^-) + \epsilon$$

where ϵ is local random noise, and x^+ is a random positive data point in the test set. The intensity of attacks is controlled by the attack factor $f_a \in (0, 1)$. The greater f_a is, the more aggressive the attacks are. Similarly, for each positive instance x^+ we modify x^+ as follows:

$$x^+ = x^+ + f_a \cdot (x^- - x^+) + \epsilon$$

where x^- is a random negative data point in the test set. For the third type of attack, x^+ and x^- can be freely transformed in the data domain as follows:

$$x^\pm = \begin{cases} \min(x^{max}, x^\pm + f_a \cdot \delta \cdot (x^{max} - x^{min})) & \delta > 0 \\ \max(x^{min}, x^\pm + f_a \cdot \delta \cdot (x^{max} - x^{min})) & \delta \leq 0 \end{cases}$$

where δ is randomly set and $\delta \in (-1, 1)$, x^{max} and x^{min} is the maximum and minimum values an instance can take. The learner's pure strategy set contains three learning models as discussed in Section 13.5.3.1: (1) Stackelberg equilibrium predictor Equi*; and (2) two SVM models SVM^{*1} and SVM^{*2} trained on equilibrium data transformations. Note that SVM^{*1} and SVM^{*2} are optimal only when the SVM learner knows the adversary's strategy ahead of time. Therefore, SVM*s alone are not robust solutions to the adversarial learning problem. When solving the prediction games, we assume the adversary can modify data in both classes. Note that even though we have two Stackelberg equilibrium predictors Equi^{*1} and Equi^{*2} corresponding to an aggressive and a mild adversary, we use only one in the learner's strategy set and refer to it as Equi* because equilibrium predictors rarely disagree. However, we list the classification errors of both Equi^{*1} and Equi^{*2} in all tables for comparison. SVM^{*1} and SVM^{*2} are trained on the two equilibrium data transformations when λ_f is set to 0.01 and 0.02. The two SVM models are essentially optimal strategies against the adversaries' equilibrium strategies. The learner will choose which learning model to play according to the probability distribution determined in the mixed strategy. The results are displayed as *Mixed* in all tables. We also compare our results to the invariant SVM (Teo et al. 2007) and the standard SVM methods. We test our learning strategy on two artificial datasets and two real datasets. In all of our experiments, we modify the test sets to simulate the three types of adversaries.

13.5.5.1 Artificial Datasets

We generated two artificial datasets to test our learning framework. Samples of the two datasets are shown in Figure 13.8. Each dataset is generated using a bivariate normal distribution with specified means of the two classes, and a covariance matrix. We purposefully generate the datasets so that positive data and negative data are either clearly separated or partially fused. Training and test data are generated separately. Adversarial attacks are simulated in the test data as discussed above.

Detailed results are shown in Tables 13.7. All the results on the artificial data are averaged over 10 random runs. Since the equilibrium solution always chooses the best strategy, the results are the same as the SVM method when there are no attacks. Hence the $f_a = 0$ column is left out for the sake of keeping the size of the table fit. We summarize our observations as follows:

- Equilibrium predictors have similar predictive errors as the invariant SVM and the standard SVM methods.
- SVM^{*1} outperforms SVM^{*2} as expected when attacks are moderate and aggressive; both SVM^{*1} and SVM^{*2} outperform the equilibrium predictors, the invariant and the standard SVMs.

Table 13.7 Error rates of Stackelberg equilibrium*1(Equi*1), SVM trained on transformed data in equilibrium*1 (SVM*1), Stackelberg equilibrium*2(Equi*2), SVM trained on transformed data in equilibrium*2 (SVM*2), Stackelberg mixed strategy (Mixed), invariant SVM, and standard SVM on datasets shown in Figure 13.8a and b. Two single-leader–single-follower equilibriums are obtained by setting $\lambda_i \in \{0.01, 0.02\}$ and $\lambda_\ell = 1$. The probability of modifying negative data is set to $p \in \{0.1, 0.5\}$. The best results are marked with *, and the mixed strategy is bolded for easy comparisons to others.

Dataset I (Figure 13.8a)

$p = 0.1$		$f_a = 0.1$	$f_a = 0.3$	$f_a = 0.5$	$f_a = 0.7$	$f_a = 0.9$
Stackelberg	Equi*1	0.0250 ± 0.0088	0.1560 ± 0.1109	0.3195 ± 0.0962	0.4460 ± 0.0222	0.5190 ± 0.0126
	SVM*1	0.0305 ± 0.0140	$0.0815 \pm 0.0266^*$	$0.2055 \pm 0.1504^*$	$0.3005 \pm 0.0370^*$	$0.4650 \pm 0.0373^*$
	Equi*2	0.0250 ± 0.0088	0.1560 ± 0.1109	0.3195 ± 0.0962	0.4460 ± 0.0222	0.5190 ± 0.0126
	SVM*2	$0.0160 \pm 0.0077^*$	0.1165 ± 0.0884	0.2735 ± 0.1204	0.3865 ± 0.0286	0.5045 ± 0.0117
	Mixed	$\mathbf{0.0273 \pm 0.0149}$	$\mathbf{0.0845 \pm 0.0272}$	$\mathbf{0.2132 \pm 0.1500}$	$\mathbf{0.3563 \pm 0.0655}$	$\mathbf{0.4860 \pm 0.0341}$
Invariant SVM		0.0255 ± 0.0086	0.1560 ± 0.1122	0.3195 ± 0.0962	0.4455 ± 0.0224	0.5190 ± 0.0126
SVM		0.0265 ± 0.0067	0.1595 ± 0.1207	0.3280 ± 0.0952	0.4360 ± 0.0313	0.5205 ± 0.0140
$p = 0.5$		$f_a = 0.1$	$f_a = 0.3$	$f_a = 0.5$	$f_a = 0.7$	$f_a = 0.9$
Stackelberg	Equi*1	$0.0350 \pm 0.0151^*$	0.1520 ± 0.0643	0.3905 ± 0.0480	0.5805 ± 0.0472	0.6905 ± 0.0682
	SVM*1	0.0915 ± 0.0176	0.1515 ± 0.0297	$0.3055 \pm 0.0370^*$	$0.4815 \pm 0.0316^*$	$0.6150 \pm 0.0361^*$
	Equi*2	$0.0350 \pm 0.0151^*$	0.1520 ± 0.0643	0.3905 ± 0.0480	0.5805 ± 0.0472	0.6905 ± 0.0682
	SVM*2	0.0385 ± 0.0147	$0.1365 \pm 0.0385^*$	0.3725 ± 0.0505	0.5520 ± 0.0348	0.6805 ± 0.0652
	Mixed	$\mathbf{0.0631 \pm 0.0328}$	$\mathbf{0.1429 \pm 0.0243}$	$\mathbf{0.3264 \pm 0.0430}$	$\mathbf{0.5042 \pm 0.0478}$	$\mathbf{0.6367 \pm 0.0588}$
Invariant SVM		0.0355 ± 0.0157	0.1530 ± 0.0656	0.3910 ± 0.0482	0.5810 ± 0.0478	0.6900 ± 0.0680
SVM		0.0370 ± 0.0151	0.1545 ± 0.0724	0.3955 ± 0.0474	0.5770 ± 0.0458	0.6845 ± 0.0669

Table 13.7 (Continued)

Dataset II (Figure 13.8b)

$p = 0.1$		$f_a = 0.1$	$f_a = 0.3$	$f_a = 0.5$	$f_a = 0.7$	$f_a = 0.9$
Stackelberg	Equi*1	0.1895 ± 0.0273	0.3050 ± 0.1146	0.3570 ± 0.0747	0.4300 ± 0.0255	0.5020 ± 0.0385
	SVM*1	0.1955 ± 0.0231	0.2920 ± 0.1232	0.3465 ± 0.1232	0.3985 ± 0.0342*	0.5090 ± 0.0731
	Equi*2	0.1895 ± 0.0273	0.3050 ± 0.1146	0.3570 ± 0.0747	0.4300 ± 0.0255	0.5020 ± 0.0385
	SVM*2	0.1810 ± 0.0240*	0.2915 ± 0.1172*	0.3530 ± 0.0985	0.4240 ± 0.0133	0.5045 ± 0.0524
	Mixed	**0.1929 ± 0.0229**	**0.2993 ± 0.1212**	**0.3444 ± 0.0841***	**0.4099 ± 0.0268**	**0.5096 ± 0.0664**
Invariant SVM		0.1835 ± 0.0247	0.3010 ± 0.1167	0.3570 ± 0.0775	0.4350 ± 0.0203	0.4995 ± 0.0403
SVM		0.1840 ± 0.0271	0.2970 ± 0.1169	0.3605 ± 0.0825	0.4330 ± 0.0121	0.4970 ± 0.0429*

$p = 0.5$		$f_a = 0.1$	$f_a = 0.3$	$f_a = 0.5$	$f_a = 0.7$	$f_a = 0.9$
Stackelberg	Equi*1	0.2000 ± 0.0350	0.2985 ± 0.0232	0.4065 ± 0.0207	0.5340 ± 0.0181	0.6190 ± 0.0357
	SVM*1	0.2140 ± 0.0422	0.2955 ± 0.0201	0.3940 ± 0.0094*	0.5240 ± 0.0459*	0.6090 ± 0.0208*
	Equi*2	0.2000 ± 0.0350	0.2985 ± 0.0232	0.4070 ± 0.0211	0.5340 ± 0.0181	0.6195 ± 0.0344
	SVM*2	0.2010 ± 0.0330	0.2875 ± 0.0148*	0.4025 ± 0.0262	0.5365 ± 0.0184	0.6170 ± 0.0309
	Mixed	**0.2074 ± 0.0461**	**0.2900 ± 0.0167**	**0.4021 ± 0.0245**	**0.5284 ± 0.0394**	**0.6150 ± 0.0234**
Invariant SVM		0.1970 ± 0.0342*	0.3010 ± 0.0242	0.4045 ± 0.0211	0.5370 ± 0.0196	0.6210 ± 0.0353
SVM		0.2015 ± 0.0323	0.2960 ± 0.0221	0.4075 ± 0.0204	0.5370 ± 0.0153	0.6220 ± 0.0396

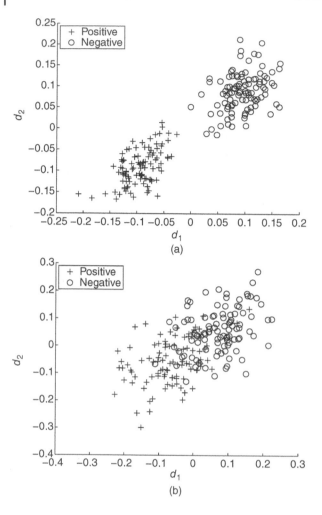

Figure 13.8 Training data of two artificial data sets. (a) Artificial dataset I. (b) Artificial dataset II.

- The mixed strategy lies in between the best and the worst predictors, however, consistently in line with the best results under all circumstances.

In all experiments, when $p = 0.1$, the adversary has 10% chance of modifying the negative data in the test set; when $p = 0.5$, the adversary is allowed to modify the negative data half of the time. Positive data are always modified. Although the classification error rates deteriorate rapidly as more negative data are modified, the best predictors consistently outperform the invariant and the standard SVM methods. The mixed strategy is *either the best predictor or very competitive with the best in all cases*. Note that all three types of adversaries modify all positive data. As we adjust the probability of negative data being attacked, the accuracy drop reflects the impact on the negative data. Therefore, using accuracy (without precision and recall) as the performance measure is sufficient to validate the robustness of the learning models.

In addition to setting the attack factor f_a to a fixed value for each attacked data point in the test set, we also tested the cases where the attack factor $f_a \in (0, 1)$ is completely random under uniform distribution for each attacked sample in the test set. We assume the positive data is always modified by the adversary. In addition, we allow the probability of negative data being attacked to increase gradually from 0.1 to 0.9. Figure 13.9 shows the results on the artificial data set shown

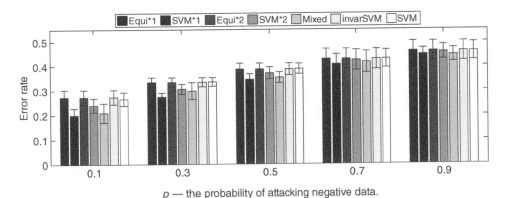

Figure 13.9 Classification error rates (with error bars) of *Equi*[*1], *SVM*[*1], *Equi*[*2], *SVM*[*2], *Mixed*, *invariant SVM*, and *SVM* on the artificial dataset shown in Figure 13.8a. The severity of the attack is completely random for each data point in the test set. The x-axis is the probability that negative data is attacked, ranging from 0.1 to 0.9. The y-axis is the classification error rate.

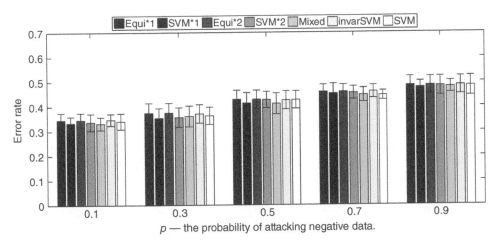

Figure 13.10 Classification error rates (with error bars) of *Equi*[*1], *SVM*[*1], *Equi*[*2], *SVM*[*2], *Mixed*, *invariant SVM*, and *SVM* on the artificial dataset shown in Figure 13.8b. The severity of the attack is completely random for each data point in the test set. The x-axis is the probability that negative data is attacked, ranging from 0.1 to 0.9. The y-axis is the classification error rate.

in Figure 13.8a where data in two classes are well separated. The x-axis is the probability of negative data being modified by the adversary, and the y-axis is the error rate. SVM[*1] and the *Mixed* strategy outperform all the other algorithms. However, their advantage becomes less significant as the amount of negative data being modified increases. Note that in this experimental setting, SVM[*1] appears to be the best consistently. The reason is that SVM[*1] is trained to handle moderate to aggressive attacks. Since f_a is uniformly distributed within $(0, 1)$ in this experiment, SVM[*1] is favored whenever $f_a > 0.3$. SVM[*1] can certainly be disadvantaged as shown in Table 13.7 when the attacks are mild, for example, when $f_a < 0.3$. Figure 13.10 shows the results on the artificial data set shown in Figure 13.8b where data in two classes are fused. In this case, the advantage of SVM[*1] and the *Mixed* strategy is less significant even though they are consistently the winner among all the techniques.

13.5.5.2 Real Datasets

For the two real datasets we again use *spam base* (UCI Machine Learning Repository 2019) and *web spam* (LIBSVM Data 2019). The learning tasks are binary classification problems, differentiating spam or webspam from legitimate e-mail or websites.

Spambase Dataset Recall that in the *spam base* dataset, the task is to differentiate spam from legitimate e-mail. There are 4601 e-mail messages in the dataset including approximately 1800 spam messages. The dataset has 57 attributes and one class label. Training and test datasets are separate. The results are averaged over 10 random runs. The detailed results are shown in Table 13.8. The $f_a = 0$ column is left out for the same reason as explained earlier on the artificial datasets.

This dataset serves as an excellent example to demonstrate the power of the mixed strategy. In the cases where $p = 0.1$, that is, when we assume legitimate e-mail is modified 10% of the time (while spam is always modified), the two equilibrium predictors Equi^{*1} and Equi^{*2} exhibit very stable performance in terms of predictive accuracy. Their error rates fluctuate slightly at 0.37 regardless of how aggressively the test data has been modified. On the other hand, SVM^{*1} and SVM^{*2} significantly outperform the equilibrium predictors Equi^{*1} and Equi^{*2} and the invariant SVM when $f_a \leq 0.5$. However, the performance of SVM^{*1} and SVM^{*2} dropped quickly as the attack gets more intense ($f_a > 0.5$), much poorer than the equilibrium predictors. The mixed strategy, although not the best, demonstrates superb performance by agreeing with the winning models the majority of the time. The standard SVM has similar performance to the equilibrium predictors, behaving poorly as the attack gets intense. When $p = 0.5$, that is, when legitimate e-mail is modified half of the time while all spam is modified, equilibrium predictors still demonstrate very stable performance while the performance of the equilibrium predictors Equi^{*1} and Equi^{*2} deteriorates sharply right after the attack factor increases to 0.3. The mixed strategy, again not the best predictor, demonstrates the most consistent performance among all the predictors given any attack intensity levels.

We also tested the case where the attack factor $f_a \in (0, 1)$ is completely random under uniform distribution for each attacked sample on this dataset. The probability of negative data being attacked increases gradually from 0.1 to 0.9. The results are illustrated in Figure 13.11. Again, we observe similar behavior of all the predictors: stable equilibrium predictors, SVM^{*1}, SVM^{*2}, and SVM progressively deteriorating as p increases. The mixed strategy, although weakened as more negative data is allowed to be modified, *consistently lies in between the equilibrium predictors and the SVM predictors*.

Web Spam Dataset Recall that web spam is a class of web pages that are created to manipulate the search engines. The *web spam* dataset (LIBSVM Data 2019) consists of 350 000 instances and 254 unigram features. We divide the dataset into two equal halves to create non overlapping training and test datasets. We report the results averaged over 10 random runs. Detailed results are shown in Table 13.8.

The equilibrium predictors perform poorly on this data set with error rates around 60%. Their corresponding SVM* predictors are as good as the standard SVM method. When $p = 0.1$–the probability that negative data are modified, the SVM* predictors and the standard SVM predictor significantly outperform the equilibrium predictors. They also outperform the invariant SVM when $f_a < 0.5$. When the attacks intensify and $f_a > 0.5$, the standard SVM predictor and the SVM* predictors were clearly outperformed by the invariant SVM. The mixed strategy consistently agrees with the SVM* predictors. When $p = 0.5$, similar results are observed when $f_a \leq 0.5$. When $f_a > 0.5$, the advantage of SVM* predictors and the standard SVM diminished, and the invariant SVM again exhibits the best performance. The mixed strategy again agreed with the SVM* predictors 100% of the time and *therefore performed better when under less severe attacks*. Note that SVM and SVM*

Table 13.8 Error rates of Equi*[1], SVM*[1], Equi*[2], SVM*[2], Mixed, invariant SVM, and SVM on spam and webspam datasets. Two single-leader–single-follower equilibriums are obtained by setting $\lambda_f \in \{0.1, 0.2\}$ and $\lambda_\ell = 1$. The probability of modifying negative data is set to $\mathbf{p} \in \{0.1, 0.5\}$. Stars(*) mark the best results, and the mixed strategy is bolded.

Spam dataset

$p = 0.1$		$f_a = 0.1$	$f_a = 0.3$	$f_a = 0.5$	$f_a = 0.7$	$f_a = 0.9$
Stackelberg	Equi*[1]	0.3806 ± 0.0142	0.3799 ± 0.0169	0.3740 ± 0.0180	$0.3718 \pm 0.0203^*$	$0.3681 \pm 0.0194^*$
	SVM*[1]	0.2877 ± 0.0468	0.3291 ± 0.0569	0.3365 ± 0.0416	0.3904 ± 0.0273	0.4154 ± 0.0340
	Equi*[2]	0.3806 ± 0.0142	0.3799 ± 0.0169	0.3740 ± 0.0180	0.3718 ± 0.0203	0.3681 ± 0.0194
	SVM*[2]	0.2852 ± 0.0469	0.3218 ± 0.0555	0.3346 ± 0.0418	0.3883 ± 0.0283	0.4137 ± 0.0350
	Mixed	$\mathbf{0.3240 \pm 0.0501}$	$\mathbf{0.3497 \pm 0.0513}$	$\mathbf{0.3553 \pm 0.0461}$	$\mathbf{0.3847 \pm 0.0248}$	$\mathbf{0.3787 \pm 0.0343}$
Invariant SVM		0.3796 ± 0.0141	0.3943 ± 0.0310	0.3886 ± 0.0469	0.3992 ± 0.0699	0.3905 ± 0.0470
SVM		$0.2829 \pm 0.0475^*$	$0.3173 \pm 0.0538^*$	$0.3321 \pm 0.0432^*$	0.3865 ± 0.0289	0.4128 ± 0.0352

$p = 0.5$		$f_a = 0.1$	$f_a = 0.3$	$f_a = 0.5$	$f_a = 0.7$	$f_a = 0.9$
Stackelberg	Equi*[1]	0.3834 ± 0.0154	$0.3807 \pm 0.0186^*$	$0.3810 \pm 0.0198^*$	$0.3879 \pm 0.0192^*$	$0.3867 \pm 0.0198^*$
	SVM*[1]	0.3211 ± 0.0310	0.4088 ± 0.0430	0.4381 ± 0.0379	0.4651 ± 0.0248	0.5156 ± 0.0388
	Equi*[2]	0.3834 ± 0.0154	0.3807 ± 0.0186	0.3810 ± 0.0198	0.3879 ± 0.0192	0.3867 ± 0.0198
	SVM*[2]	$0.3195 \pm 0.0314^*$	0.4058 ± 0.0457	0.4359 ± 0.0376	0.4622 ± 0.0237	0.5148 ± 0.0392
	Mixed	$\mathbf{0.3634 \pm 0.0424}$	$\mathbf{0.3969 \pm 0.0415}$	$\mathbf{0.3989 \pm 0.0316}$	$\mathbf{0.4200 \pm 0.0397}$	$\mathbf{0.4216 \pm 0.0736}$
Invariant SVM		0.3877 ± 0.0422	0.4005 ± 0.0514	0.3916 ± 0.0309	0.4430 ± 0.0879	0.4313 ± 0.0738
SVM		0.3198 ± 0.0342	0.4011 ± 0.0456	0.4348 ± 0.0372	0.4598 ± 0.0232	0.5128 ± 0.0379

Table 13.8 (Continued)

		Webspam dataset				
$p = 0.1$		$f_a = 0.1$	$f_a = 0.3$	$f_a = 0.5$	$f_a = 0.7$	$f_a = 0.9$
Stackelberg	Equi[*1]	0.6060 ± 0.0002	0.6059 ± 0.0003	0.5946 ± 0.0270	0.5950 ± 0.0271	0.5950 ± 0.0271
	SVM[*1]	0.1330 ± 0.0079	0.2004 ± 0.0442	0.3187 ± 0.0282	0.4860 ± 0.0291	0.5544 ± 0.0368
	Equi[*2]	0.6060 ± 0.0002	0.6059 ± 0.0003	0.5946 ± 0.0270	0.5950 ± 0.0271	0.5950 ± 0.0271
	SVM[*2]	0.1320 ± 0.0081	0.1987 ± 0.0427	0.3179 ± 0.0276	0.4853 ± 0.0308	0.5544 ± 0.0370
	Mixed	**0.1323 ± 0.0077**	**0.2000 ± 0.0443**	**0.3188 ± 0.0282**	**0.4854 ± 0.0309**	**0.5544 ± 0.0370**
Invariant SVM		0.3843 ± 0.0217	0.3940 ± 0.0012	0.3793 ± 0.0308	0.4028 ± 0.0284*	0.4055 ± 0.0268*
SVM		0.1311 ± 0.0081*	0.1972 ± 0.0415*	0.3173 ± 0.0272*	0.4848 ± 0.0324	0.5545 ± 0.0373
$p = 0.5$		$f_a = 0.1$	$f_a = 0.3$	$f_a = 0.5$	$f_a = 0.7$	$f_a = 0.9$
morerows="5" Stackelberg	Equi[*1]	0.6059 ± 0.0001	0.6063 ± 0.0002	0.5977 ± 0.0271	0.5752 ± 0.0333	0.5989 ± 0.0264
	SVM[*1]	0.1953 ± 0.0101	0.2608 ± 0.0348	0.4153 ± 0.0248	0.5769 ± 0.0385	0.6035 ± 0.0264
	Equi[*2]	0.6059 ± 0.0001	0.6063 ± 0.0002	0.5977 ± 0.0271	0.5752 ± 0.0333	0.5989 ± 0.0264
	SVM[*2]	0.1947 ± 0.0098	0.2596 ± 0.0343	0.4130 ± 0.0245	0.5762 ± 0.0380	0.6029 ± 0.0256
	Mixed	**0.1950 ± 0.0101**	**0.2598 ± 0.0342**	**0.4147 ± 0.0251**	**0.5770 ± 0.0388**	**0.6034 ± 0.0260**
Invariant SVM		0.3940 ± 0.0012	0.3891 ± 0.0089	0.3954 ± 0.0324*	0.4090 ± 0.0230*	0.4109 ± 0.0266*
SVM		0.1942 ± 0.0094*	0.2586 ± 0.0339*	0.4109 ± 0.0239	0.5754 ± 0.0375	0.6023 ± 0.0249

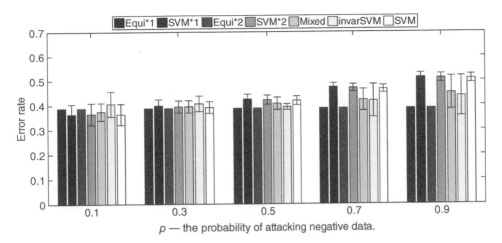

Figure 13.11 Classification error rates (with error bars) of *Equi*[*1], *SVM*[*1], *Equi*[*2], *SVM*[*2], *Mixed*, *invariant SVM*, and *SVM* on the spambase dataset. The severity of the attack is completely random for each data point in the test set. The x-axis is the probability that negative data is attacked, ranging from 0.1 to 0.9. The y-axis is the classification error rate.

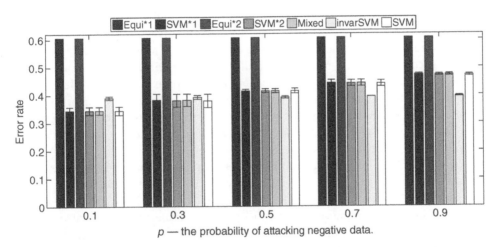

Figure 13.12 Classification error rates (with error bars) of *Equi*[*1], *SVM*[*1], *Equi*[*2], *SVM*[*2], *Mixed*, *invariant SVM*, and *SVM* on the webspam dataset. The severity of the attack is completely random for each data point in the test set. The x-axis is the probability that negative data is attacked, ranging from 0.1 to 0.9. The y-axis is the classification error rate.

performed equally well on this data set. However, the component SLSF equilibrium solutions of the mixed strategy performed rather poorly. The mixed strategy successfully follows the best component in all cases. This experiment attests to the conclusion that the mixed strategy consistently agrees with the best component solutions.

For this dataset, we also tested the case where the attack factor $f_a \in (0, 1)$ is completely random for each attacked sample in the test set. We let the probability of negative data being attacked increase gradually from 0.1 to 0.9. The results are shown in Figure 13.12. Here, we observe similar behavior of all the predictors: disastrous equilibrium predictors, SVM^{*1}, SVM^{*2}, and SVM progressively deteriorating as the amount of negative data being modified increases. The mixed strategy follows the SVM* predictors closely in all cases, and the invariant SVM predictor remains the best when the probability of negative data being modified is greater than 0.5.

Finally, we would like to point out that there does not exist a single technique that works optimally in all possible adversarial scenarios. However, the mixed strategy solution guarantees that it is close to optimal and never fails miserably in an arbitrary adversarial setting.

13.5.6 Remark

We tackle the challenges machine learning applications face when there are adversaries of different types. Existing techniques developed for adversarial learning focus on learning one robust learning model or an ensemble of several of such learning models against a single type of adversary. We present a nested Stackelberg game framework in which the lower level learners provide robust solutions to adversarial data transformations, while at the upper level, the learner plays a *single leader multiple followers* game and searches for an optimal mixed strategy to play. We demonstrate with empirical evidence that the mixed strategy is the best solution in general, without knowing what types of adversaries the machine learning applications are facing in the real world.

13.6 Further Discussions

Game theory attempts to bring mathematical precision to decision-making so that best strategies can be played, even in the arms race between the learner and the adversary. In game theory, we assume the players are rational–they seek to maximize their payoffs or minimize their losses. This assumption is not necessarily valid in real life, especially in cybersecurity domains. What appears irrational to one player may be rational to the opponent player. For example, when we model a game between airport security and terrorists, it may not be a good idea to view the terrorists based on our own experiences. When the opponent is not rational or simply plays poorly, it is important to realize that continuing to play the equilibrium strategy will lead to a losing situation or the loss of opportunity to exploit the opponent's weaknesses.

Behavioral economists have long questioned the strict assumptions about rationality in existing theories in game theory (Aumann 1997). The actual decision-making by individuals is often irrational even in simple decision problems. Individuals typically fail to select the best response from a set of choice. As a matter of fact, optimization is so difficult that individuals are often unable to come up with their best responses to their opponents' moves, unlike what the rational decision theory has always assumed. In response, theories that relax the rationality constraints have been proposed among which include: *Quantal Response Equilibrium* (QRE) (McKelvey and Palfrey 1995), a solution concept that promotes an equilibrium notion with bounded rationality that acknowledges the possibility that players do not always play a best response; *Prospect Theory* (Kahneman and Tversky 1979; Tversky and Kahneman 1992), a solution that introduces payoffs with respect to a reference point at which there is larger slope for losses than for gains and agents tend to overestimate small probabilities and underestimate large probabilities.

The same debate is applicable to mixed strategy games. In the airport security game, how the airport deploys security patrols depends on the response from the terrorists. If the rationality of the terrorists is predictable, it would be best for the airport to deploy its security patrols randomly by playing a mixed strategy. However, when there is a good reason to believe that the terrorists are not going to play the equilibrium strategy, airport security may be better off by playing pure strategies. Generally speaking, unless the odds are strongly in our favor, playing equilibrium strategies would be our best choice. Nevertheless, we should always keep in mind that our calculation of rationality may lead to different behavior, new rules may need to be defined in the game to take this into consideration.

Sometimes, independent rational actions in the interests of each individual player can lead to an undesirable outcome. For example, in the game of the tragedy of the commons, farmers graze their cows in a piece of public land. Each farmer is incentivized to graze as many cows as possible to maximize self-interests, and eventually the land will become useless because of overgrazing. The dilemma is that while each individual is making the maximum profits, the damage is shared by the group, and in the long run the group profits become suboptimal. In this regard, the problem of cybercrime that threatens personal data may not be as serious as the conventional wisdom has convinced us. Cybercrime such as spamming and fishing may be attractive to cybercriminals for easy profits, but for the very same reason there would be too many such cybercriminals to make a lucrative profit easily.

It is important to understand the pitfalls of game theory in addition to the rationality calculation. Game theoretic solutions are not always feasible. In some cases, an equilibrium may not exist; and even when it does exist, playing the equilibrium strategy may involve solving problems that are computationally intractable (Halpern and Pass 2008).

Besides the challenge of adversarial attacks, there are other issues machine learning algorithms need to address before they are ready for practical use. Sommer and Paxson (2010) point out some well known problems of machine learning-based methods in cybersecurity applications, including high false-positive rate, lack of attack-free training data, susceptibility to adversarial attacks, and difficulties with evaluation. The main challenges machine learning-based techniques face include outlier detection, high cost of errors, semantic gap between output and interpretation, and diversity of network traffic. They recommend that machine learning techniques should be used with an understanding of the threat model, data used for training, and the classification results.

Acknowledgements

This work is supported by Air Force Office MURI Grant FA9550-08-1-0265 and FA9550-12-1-0082, NIH Grants 1R01-LM009989 and 1R01HG006844, NSF Grants Career-CNS-0845803, CNS-0964350, CNS-1016343, CNS-1111529, CNS-1228198, Army Research Office Grant W911NF-12-1-0558, SAMSI GDRR 2019-2020 program, ARO award W911NF-17-1-0356, and NSF IIS-1939728.

References

Scott Alfeld, Xiaojin Zhu, and Paul Barford. Explicit defense actions against test-set attacks. *Proceedings of the AAAI Conference on Artificial Intelligence*, 31(1), 2017. https://ojs.aaai.org/index.php/AAAI/article/view/10767

Robert J. Aumann. Rationality and bounded rationality. *Games and Economic Behavior*, 21(1):2–14, 1997. ISSN 0899-8256. doi:doi: 10.1006/game.1997.0585. http://www.sciencedirect.com/science/article/pii/S0899825697905856.

Tamer Basar and Geert Jan Olsder. Dynamic noncooperative game theory. In *Classics in Applied Mathematics*. Society for Industrial and Applied Mathematics, 1999. ISBN 9780898714296. http://books.google.com.tr/books?id=k1oF5AxmJlYC.

M. Bruckner and T. Scheffer. Nash equilibria of static prediction games. In *Advances in Neural Information Processing Systems*. MIT Press, 2009.

Michael Brückner and Tobias Scheffer. Stackelberg games for adversarial prediction problems. In *Proceedings of the 17th ACM SIGKDD International Conference on Knowledge Discovery and Data Mining, 21–24 August*, pp. 547–555. ACM, New York, 2011.

Chih-Chung Chang and Chih-Jen Lin. LIBSVM: A library for support vector machines. *ACM Transactions on Intelligent Systems and Technology*, 2(3):27:1–27:27, 2011. http://www.csie.ntu.edu.tw/~cjlin/libsvm.

Nilesh Dalvi, Pedro Domingos, Mausam, Sumit Sanghai, and Deepak Verma. Adversarial classification. In *Proceedings of the 10th ACM SIGKDD International Conference on Knowledge Discovery and Data Mining, KDD '04, Seattle, WA, USA, 22–25 August*, pp. 99–108. ACM, New York, NY, 2004.

Prithviraj Dasgupta and Joseph B. Collins. A survey of game theoretic approaches for adversarial machine learning in cybersecurity tasks. *AI Magazine*, 40(2):31–43, 2019. ISSN 0738-4602. doi: 10.1609/aimag.v40i2.2847.

O. Dekel and O. Shamir. Learning to classify with missing and corrupted features. In *Proceedings of the International Conference on Machine Learning*, pp. 216–223. ACM, 2008.

L. Dritsoula, P. Loiseau, and J. Musacchio. A game-theoretic analysis of adversarial classification. *IEEE Transactions on Information Forensics and Security*, 12(12):3094–3109, 2017.

Dheeru Dua and Casey Graff. UCI Machine Learning Repository. University of California, Irvine, School of Information and Computer Sciences, 2021. http://archive.ics.uci.edu/ml.

Laurent El Ghaoui, Gert R. G. Lanckriet, and Georges Natsoulis. Robust classification with interval data. Technical Report UCB/CSD-03-1279, EECS Department, University of California, Berkeley, 2003. http://www.eecs.berkeley.edu/Pubs/TechRpts/2003/5772.html.

Amir Globerson and Sam Roweis. Nightmare at test time: Robust learning by feature deletion. In *Proceedings of the 23rd international conference on Machine learning, ICML '06*, pp. 353–360. ACM, 2006.

M. Grant and S. Boyd. CVX: Matlab software for disciplined convex programming, version 1.21. http://cvxr.com/cvx/, April 2011.

Joseph Y. Halpern and Rafael Pass. Game theory with costly computation. *CoRR, abs/0809.0024*, 2008.

Moritz Hardt, Nimrod Megiddo, Christos Papadimitriou, and Mary Wootters. Strategic classification. In *Proceedings of the 2016 ACM Conference on Innovations in Theoretical Computer Science, ITCS '16*, pp. 111–122. Association for Computing Machinery, New York, NY, 2016. ISBN 9781450340571. doi: 10.1145/2840728.2840730. https://doi.org/10.1145/2840728.2840730.

Daniel Kahneman and Amos Tversky. Prospect theory: An analysis of decision under risk. *Econometrica*,47(2):263–291, 1979. https://ideas.repec.org/a/ecm/emetrp/v47y1979i2p263-91.html.

Murat Kantarcioglu, Bowei Xi, and Chris Clifton. Classifier evaluation and attribute selection against active adversaries. *Data Mining and Knowledge Discovery*, 22:291–335, 2011.

G. R. G. Lanckriet, L. E. Ghaoui, C. Bhattacharyya, and M. I. Jordan. A robust minimax approach to classification. *Journal of Machine Learning Research*, 3:555–582, 2002.

LIBSVM Data: Classification, Regression, and Multi-label, 2019. http://www.csie.ntu.edu.tw/~cjlin/libsvmtools/datasets/.

Wei Liu and Sanjay Chawla. A game theoretical model for adversarial learning. In *Proceedings of the 2009 IEEE International Conference on Data Mining Workshops, ICDMW '09*, pp. 25–30. IEEE Computer Society, Washington, DC, 2009.

Daniel Lowd. Good word attacks on statistical spam filters. In *In Proceedings of the Second Conference on Email and Anti-Spam (CEAS), 21–22 July*. Stanford University, CA, USA, 2005.

Daniel Lowd and Christopher Meek. Adversarial learning. In *Proceedings of the 11th ACM SIGKDD International Conference on Knowledge Discovery in Data Mining, KDD '05, Chicago, IL, USA, 21–24 August*, pp. 641–647, 2005. ISBN 1-59593-135-X.

Aleksander Madry, Aleksandar Makelov, Ludwig Schmidt, Dimitris Tsipras, and Adrian Vladu. Towards deep learning models resistant to adversarial attacks. In *International Conference on Learning Representations, Vancouver, BC, Canada, 30 April–3 May*, 2018. https://openreview.net/forum?id=rJzIBfZAb.

Richard D. McKelvey and Thomas Palfrey. Quantal response equilibria for normal form games. *Games and Economic Behavior*, 10(1):6–38, 1995. https://EconPapers.repec.org/RePEc:eee:gamebe:v:10:y:1995:i:1:p:6-38.

Praveen Paruchuri, Jonathan P. Pearce, Janusz Marecki, Milind Tambe, Fernando Ordonez, and Sarit Kraus. Playing games for security: An efficient exact algorithm for solving Bayesian Stackelberg games. In *Proceedings of the 7th International Joint Conference on Autonomous Agents and Multiagent Systems – Volume 2 (AAMAS '08), Estoril, Portugal, 12–16 May*, pp. 895–902. International Foundation for Autonomous Agents and Multiagent Systems, 2008.

Robin Sommer and Vern Paxson. Outside the closed world: On using machine learning for network intrusion detection. In *Proceedings of the 2010 IEEE Symposium on Security and Privacy, SP '10*, pp. 305–316. IEEE Computer Society, Washington, DC, 2010. ISBN 978-0-7695-4035-1. doi: 10.1109/SP.2010.25.

C. H. Teo, A. Globerson, S. T. Roweis, and A. J. Smola. Convex learning with invariances. In *Advances in Neural Information Processing Systems, Vancouver, British Columbia, Canada, 3–6 December*, 2007.

A. Tversky and D. Kahneman. Advances in prospect theory: Cumulative representation of uncertainty. *Journal of Risk and Uncertainty*, 5:297–323, 1992.

UCI Machine Learning Repository, 2019. http://archive.ics.uci.edu/ml/.

Yan Zhou and Murat Kantarcioglu. Modeling adversarial learning as nested Stackelberg games. In James Bailey, Latifur Khan, Takashi Washio, Gillian Dobbie, Joshua Zhexue Huang, and Ruili Wang, editors. *Advances in Knowledge Discovery and Data Mining – 20th Pacific-Asia Conference, PAKDD 2016, Auckland, New Zealand, April 19–22, 2016, Proceedings, Part II. Lecture Notes in Computer Science*, volume 9652, pp. 350–362. Springer, 2016. ISBN 978-3-319-31749-6. doi: http://dx.doi.org/10.1007/978-3-319-31750-2_28.

Yan Zhou, Murat Kantarcioglu, Bhavani Thuraisingham, and Bowei Xi. Adversarial support vector machine learning. In *Proceedings of the 18th ACM SIGKDD international conference on Knowledge discovery and data mining, KDD '12*, pp. 1059–1067. ACM, New York, NY, 2012. ISBN 978-1-4503-1462-6. doi: 10.1145/2339530.2339697. http://doi.acm.org/10.1145/2339530.2339697.

14

Adversarial Machine Learning for 5G Communications Security

Yalin E. Sagduyu[1], Tugba Erpek[1, 2], and Yi Shi[1, 3]

[1] *Intelligent Automation, Inc., Rockville, MD, USA*
[2] *Hume Center, Virginia Tech, Arlington, VA, USA*
[3] *ECE Department, Virginia Tech, Blacksburg, VA, USA*

14.1 Introduction

As the fifth-generation mobile communications technology, 5G supports emerging applications such as smart warehouses, vehicular networks, virtual reality (VR), and augmented reality (AR) with unprecedented rates enabled by recent advances in massive MIMO, mmWave communications, network slicing, small cells, and internet of things (IoT). Complex structures of waveforms, channels, and resources in 5G cannot be reliably captured by simplified analytical models driven by expert knowledge. As a data-driven approach, *machine learning* has emerged as a viable alternative to support 5G communications by learning from and adapting to the underlying spectrum dynamics (Jiang et al. 2016). Empowered by recent advances in algorithmic design and computational hardware resources, *deep learning* shows strong potential to learn the high-dimensional data characteristics of wireless communications beyond conventional machine learning techniques (Erpek et al. 2019) and offers novel solutions to critical tasks of detection, classification, and prediction in 5G systems.

As machine learning becomes a core part of the next-generation communication systems, there is an increasing concern about the vulnerability of machine learning to adversarial effects. To that end, smart adversaries may leverage emerging machine learning techniques to infer vulnerabilities in 5G systems and tamper with the learning process embedded in 5G communications. The problem of learning in the presence of adversaries is the subject to the study of *adversarial machine learning* that has received increasing attention in computer vision and natural language processing (NLP) domains (Kurakin et al. 2016a; Vorobeychik and Kantarcioglu 2018; Shi et al. 2018b). Due to the shared and open nature of wireless medium, wireless applications are highly susceptible to adversaries such as eavesdroppers and jammers that can observe and manipulate the training and testing processes of machine learning over the air. While there is a growing interest in designing attacks on machine learning-driven data and control planes of wireless communications (Erpek et al. 2018; Sagduyu et al. 2019b), adversarial machine learning has not been considered yet for sophisticated communication systems such as 5G.

5G systems are designed to operate in frequency bands from 450 MHz to 6 GHz, and 24.250 to 52.600 GHz (millimeter-wave bands) including the unlicensed spectrum. While some of these

This effort is supported by the US Army Research Office under contract W911NF-17-C-0090. The content of the information does not necessarily reflect the position or the policy of the US Government, and no official endorsement should be inferred.

bands are dedicated to commercial use of 5G, some other ones are opened for *spectrum co-existence* of 5G with other legacy wireless systems. In particular, the US Federal Communications Commission (FCC) has adopted rules for the *Citizens Broadband Radio Service (CBRS) band* to allow the use of commercial communications systems in the 3550–3700 MHz band in an opportunistic manner by treating CBRS users as incumbent. These communications systems including 5G systems are required to vacate the band once the ranging (radar) signal is detected by the Environmental Sensing Capability (ESC) system as an incumbent CBRS user (Citizens Broadband Radio Service 2015). Radar signal detection and classification is a complex problem considering the unpredictable utilization patterns, channel effects and interference from other commercial systems operating in the same band and interference leakage from radar systems operating in different bands. Deep neural networks can capture complex spectrum effects in these bands and perform superior performance compared to conventional signal detection and classification techniques (Souryal and Nguyen 2019; Lees et al. 2019). On the other hand, the use of deep neural networks may expose the system to adversarial attacks, as adversaries can tamper with both data and control plane communications with smart jamming and consequently prevent efficient spectrum sharing of 5G with incumbent users.

Another aspect of 5G that is vulnerable to adversarial machine learning is *network slicing*. As an emerging concept that enables 5G to serve diverse applications (such as IoT and autonomous driving) with different performance requirements (such as throughput and latency) on heterogeneous platforms, network slicing multiplexes virtualized and independent logical networks on a common physical network infrastructure (Zhang 2019). In this setting, both the network slice manager and the user equipment (UE) cannot be trusted in general (NGMN Alliance 2016) and smart adversaries may impersonate their roles. As part of physical layer security, deep learning can be used by 5G for RF fingerprinting to classify and authenticate signals received at network slice manager or host machines. However, by exploring the vulnerabilities of *physical layer signal authentication*, adversarial machine learning provides novel techniques to spoof wireless signals that cannot be reliably distinguished from intended users even when deep neural networks are utilized for wireless signal classification.

Compared to other data domains, there are several unique challenges when we apply adversarial attacks in wireless communications.

1. In computer vision, NLP and other data domains, the input to the machine learning algorithm (e.g. a target classifier) can be directly collected by an adversary, e.g. by querying an online application programming interface (API). However, an adversary in the wireless domain cannot directly collect the same input data (e.g. spectrum sensing data) as the target node in a communication system due to the different channel and interference conditions perceived by the target node and the adversary. As a result, the features (input) to the machine learning algorithm are different for the same instance.
2. Input features of the machine learning algorithm to be used by the adversary may differ from those used by the target communication system depending on the differences in the underlying waveform and receiver hardware characteristics at the adversary and the communication system.
3. The adversary in a wireless domain cannot directly obtain the output (label) of the target machine learning algorithm, since the output is used by the target node only for 5G communications and thus is not available to any other node inside or outside the network. As a result, the adversary needs to observe the spectrum to make sense of the outputs of the target machine learning algorithm but cannot necessarily collect exactly the same outputs (e.g. classification labels).

4. The adversary in a wireless domain cannot directly manipulate the input data to a machine learning algorithm. Instead, it can only add its own transmissions on top of existing transmissions (if any) over the air (i.e. through channel effects) to change the input data (such as spectrum sensing data) indirectly.

By accounting for these challenges, the goal of this chapter is to describe how to apply adversarial machine learning to the 5G communication setting. In particular, we will discuss the *vulnerabilities of machine learning applications in 5G* with two motivating examples and show how adversarial machine learning provides a new attack surface in 5G communication systems:

1. *Attack on spectrum sharing of 5G with incumbent users such as in CBRS*: The ESC system senses the spectrum and uses a deep learning model to detect the radar signals as an incumbent user. If the incumbent user is not detected in a channel of interest, the 5G transmitter, in our case the 5G base station, namely the 5G gNodeB, is informed and starts with communications to the 5G UE. Otherwise, the 5G gNodeB cannot use this channel and the Spectrum Access System (SAS) reconfigures the 5G system's spectrum access (such as vacating this particular channel) to avoid interference with the incumbent signals. By monitoring spectrum dynamics of channel access, the adversary builds a surrogate model of the deep neural network architecture based on its sensing results, predicts when a successful 5G transmission will occur, and jams the communication signal accordingly in these predicted time instances. We consider jamming both data transmission and spectrum sensing periods. The latter case is stealthier (i.e. more difficult to detect) and more energy efficient as it only involves jamming of the shorter period of spectrum sensing before data transmission starts and forces the 5G gNodeB into making wrong transmit decisions. We show that this attack reduces the 5G communication throughput significantly.
2. *Attack on network slicing*: The adversary transmits spoofing signals that mimic the signal characteristics of the 5G UE when requesting a network slice from the host machine. This attack potentially allows the adversary to starve resources in network slices after infiltrating through the authentication system built at the 5G gNodeB. A generative adversarial network (GAN) (Goodfellow et al. 2014) is deployed at an adversary pair of a transmitter and a receiver that implement the generator and discriminator, respectively, to generate synthetic wireless signals that match the characteristics of a legitimate UE's 5G signals that the 5G gNodeB would expect to receive. We show that this attack allows an adversary to infiltrate the physical layer authentication of 5G with high success rate.

These novel attacks leave a smaller footprint and are more energy-efficient compared to conventional attacks such as jamming of data transmissions. As a countermeasure, we present a *proactive defense* approach to reduce the performance of the inference attack that is launched as the initial step to build an adversarial (surrogate) model that captures wireless communication characteristics. Other attacks are built upon this model. The 5G system as the defender makes the adversarial model less accurate by deliberately making a small number of wrong decisions (such as in spectrum sensing, data transmission, or signal authentication). The defense carefully selects which decisions to flip with the goal of maximizing the uncertainty of the adversary while balancing the impact of these controlled decision errors on its own performance.

This chapter is organized as follows. We introduce adversarial machine learning and describe the corresponding attacks in Section 14.2. Then, we extend the use of adversarial machine learning to the wireless communications and discuss the domain-specific challenges in Section 14.3. After identifying key vulnerabilities of machine learning-empowered 5G solutions, Section 14.4 introduces two attacks built upon adversarial machine learning against 5G communications and presents a defense mechanism. Section 14.5 concludes the chapter.

14.2   Adversarial Machine Learning

While there is a growing interest in applying machine learning to different data domains and deploying machine learning algorithms in real systems, it has become imperative to understand vulnerabilities of machine learning in the presence of adversaries. To that end, adversarial machine learning (Kurakin et al. 2016a; Vorobeychik and Kantarcioglu 2018; Shi et al. 2018b) has emerged as a critical field to enable safe adoption of machine learning subject to adversarial effects. One example that has attracted recent attention involves machine learning applications offered to public or paid subscribers via APIs; e.g. Google Cloud Vision (2020) provides cloud-based machine learning tools to build machine learning models. This online service paradigm creates security concerns of adversarial inputs to different machine learning algorithms ranging from computer vision to NLP (Shi et al. 2018c,d). As another application domain, automatic speech recognition and voice controllable systems were studied in terms of the vulnerabilities of their underlying machine learning algorithms (Vaidya et al. 2016; Zhang et al. 2017). As an effort to identify vulnerabilities in autonomous driving, attacks on self-driving vehicles were demonstrated in Kurakin et al. (2016), where the adversary manipulated traffic signs to confuse the learning model.

The manipulation in adversarial machine learning may happen during the training or inference (test) time, or both. During the training time, the goal of the adversary is to provide wrong inputs (features and/or labels) to the training data such that the machine learning algorithm is not properly trained. During the test time, the goal of the adversary is to provide wrong inputs (features) to the machine algorithm such that it returns wrong outputs. As illustrated in Figure 14.1, attacks built upon adversarial machine learning can be categorized as follows.

1. *Attack during the test time.*
   a. *Inference (exploratory) attack*: The adversary aims to infer the machine learning architecture of the target system to build a shadow or surrogate model that has the same functionality as the original machine learning architecture (Barreno et al. 2006; Tramer et al. 2016; Wu et al. 2016; Papernot et al. 2017; Shi et al. 2017; Shi et al. 2018b). This corresponds to a white-box or black-box attack depending on whether the machine learning model such as the deep neural network structure is available to the adversary, or not. For a black-box attack, the adversary queries the target classifier with a number of samples and records the labels. Then, it uses this labeled data as its own training data to train a functionally equivalent (i.e. statistically similar) deep learning classifier, namely a surrogate model. Once the machine learning functionality is learned, the adversary can use the inference results obtained from the surrogate model for subsequent attacks such as confidence reduction or targeted misclassification.
   b. *Membership inference attack*: The adversary aims to determine if a given data sample is a member of the training data, i.e. if a given data sample has been used to train the machine learning algorithm of interest (Nasr et al. 2018; Song et al. 2018; Jia et al. 2019; Leino and Fredrikson 2020). Membership inference attack is based on the analysis of overfitting to check whether a machine learning algorithm is trained for a particular data type, e.g. a particular type of images. By knowing which type of data the machine learning algorithm is trained to classify, the adversary can then design a subsequent attack more successfully.
   c. *Evasion attack*: The adversary manipulates test data of a machine learning algorithm by adding carefully crafted adversarial perturbations (Moosavi-Dezfooli et al. 2015; Papernot et al. 2016; Kurakin et al. 2016b; Shi and Sagduyu 2017). Then, the machine learning model runs on erroneous test data and makes errors in test time, e.g. a classifier is fooled into accepting an adversary as legitimate. The samples with output labels that are closer to the decision

region can be used by the adversary to increase the probability of error at the target machine learning algorithm.

d. *Spoofing attack*: The adversary generates synthetic data samples from scratch rather than adding perturbations to the real ones. The GAN can be used for data spoofing by generating a synthetic dataset that is statistically similar to the original dataset (Davaslioglu and Sagduyu 2018; Shi et al. 2019a). The GAN consists of two deep neural networks, one acting as a generator and the other one acting as a discriminator. The generator generates spoofing signals and the discriminator aims to detect whether the received signal is spoofed, or not. Then, the generator and the discriminator play a mini-max game to optimize their individual performance iteratively in response to each other's actions. After they converge, the generator's deep neural network is used to generate spoofing signals.

2. *Attack during the training time.*

   Poisoning (causative) attack: The adversary manipulates the training process by either directly providing wrong training data or injecting perturbations to the training data such that the machine learning model is trained with erroneous features and thus it makes errors later in test time (Alfeld et al. 2016; Pi et al. 2016; Shi and Sagduyu 2017). This attack is stealthier than the evasion attack (as the training period is typically shorter than the test period). To select which training data samples to tamper with, the adversary first runs samples through the inferred surrogate model and then changes their labels and sends these mislabeled samples as training data to the target classifier provided that their deep learning scores are far away from the decision region of the surrogate model.

3. *Attack during both training and test times.*

   Trojan (backdoor or trapdoor) attack: The adversary slightly manipulates the training data by inserting Trojans, i.e. triggers, to only few training data samples by modifying some data characteristics (e.g. putting stickers on traffic signs) and changing the labels of these samples to a target label (e.g. from the stop sign to the speed limit sign). This poisoned training data may be used to train the machine learning model. In test time, the adversary feeds the target classifier with input samples embedded with the same characteristics that were added as triggers by the adversary during training. The goal of the adversary is to cause errors when machine learning is run on data samples poisoned with triggers. In the meantime, clean (unpoisoned) samples without triggers should be processed correctly. Since only few samples of Trojans are inserted, this attack is harder to detect than both evasion and causative attacks. The disadvantage of the Trojan attack is that it needs to be launched during both training and test times, i.e. the adversary needs to have access to and manipulate both training and test data samples.

Various defense mechanisms have been developed in the literature against adversarial machine learning attacks in computer vision, NLP, and other data domains. The core of the defense is to make the machine learning algorithm robust to the anticipated attacks. One approach against evasion attacks is the certified defense that adds perturbations to training data of the machine learning algorithm and thereby generates a certificate to bound the expected error due to perturbations added later in test time (Cohen et al. 2019). Note that this defense assumes that the erroneous input is generated with perturbations added. We consider a proactive defense mechanism in Section 14.3.3 against adversarial machine learning in wireless communications. The goal of this defense is to provide a small number of carefully crafted wrong inputs to the adversary, as it builds (trains) its attack scheme, and thus prevents the adversary from building a high-fidelity surrogate model.

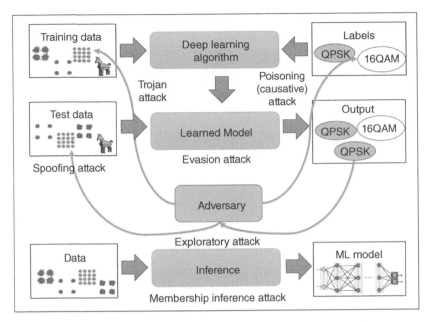

Figure 14.1 Taxonomy of attacks built upon adversarial machine learning.

14.3 Adversarial Machine Learning in Wireless Communications

Machine learning finds rich applications in wireless communications, including spectrum access (Shi et al. 2019b), signal classification (Soltani et al. 2019), beamforming (Mismar et al. 2019), beam selection (Cousik et al. 2020), channel estimation (Dong et al. 2019), channel decoding (Gruber et al. 2017), physical layer authentication (Davaslioglu et al. 2019), and transmitter-receiver scheduling (Abu Zainab et al. 2019). In the meantime, there is a growing interest in bridging machine learning and wireless security in the context of adversarial machine learning (Erpek et al. 2018; Sagduyu et al. 2019). In Section 14.3.1, we discuss how different attacks in adversarial machine learning presented in Section 14.2 can be adapted to the requirements of wireless communications. Then, we identify the domain-specific challenges on applying adversarial machine learning to wireless communications in Section 14.3.2. Finally, we discuss the state-of-the-art defense techniques against the adversarial attacks in wireless communications in Section 14.3.3.

14.3.1 Wireless Attacks Built Upon Adversarial Machine Learning

Figure 14.2 illustrates target tasks, attack types, and attack points of adversarial machine learning when applied to the wireless communications. As the motivating scenario to describe different attacks, we consider a canonical wireless communication system with one transmitter T, one receiver R, and one adversary A. This setting is instrumental in studying conventional jamming (Sagduyu and Ephremides 2009; Sagduyu et al. 2010, 2011) and defense strategies in wireless access, and can be easily extended to a network scenario with multiple transmitters and receivers (Erpek et al. 2018). The communication system shares the spectrum with a background traffic source B, whose transmission behavior is not known by T and A.

In wireless domains, the following attacks have been considered against a machine learning-based classifier.

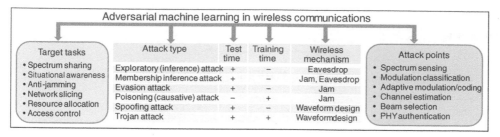

Figure 14.2 Key points of adversarial machine learning in wireless communications.

- *Exploratory (inference) attack.* Transmitter T uses a machine learning-based classifier C_T for a specific task such as spectrum sensing and adversary A aims to build a classifier C_A (namely, the surrogate model) that is functionally the same as (or similar to) the target classifier C_T (Shi et al. 2018e; Erpek et al. 2019). The observations and labels at A will differ compared to the ones at C_T, since T and A are not co-located and they experience different channels.
- *Membership inference attack.* Adversary A aims to determine whether a given data sample is in the training data of C_T. One example of this attack in the wireless domain is to identify whether a target classifier is trained against signals from a particular transmitter, or not (Shi et al. 2020).
- *Evasion attack.* Adversary A aims to determine or craft input samples that the target classifier C_T cannot reliably classify. Wireless examples of evasion attacks include manipulation of inputs to the spectrum sensing (Shi et al. 2018a; Sagduyu et al. 2019a, b), modulation classification (Sadeghi and Larsson 2018; Flowers et al. 2019a, b; Hameed et al. 2019a, b; Kokalj-Filipovic and Miller 2019; Kokalj-Filipovic et al. 2019; Kim et al. 2020a, b), channel access (Wang et al. 2020; Zhong et al. 2020), autoencoder-based physical layer design, (Sadeghi and Larsson 2019a), and eavesdroppers for private communications (Kim et al. 2020).
- *Poisoning (Causative) attack.* Adversary A provides falsified samples to train or retrain the target classifier C_T such that C_T is not properly trained and makes significantly more errors than usual in test time. Poisoning attacks were considered against spectrum sensing at an individual wireless receiver in Sagduyu et al. (2019a, b) and against cooperative spectrum sensing at multiple wireless receivers in Luo et al. (2019, 2020). Note that these poisoning attacks against sensing decisions are extensions of the conventional spectrum sensing data falsification (SSDF) attacks (Sagduyu 2014) to the adversarial machine learning domain, but they are more effective as they manipulate the training process more systematically for the case of machine learning-based spectrum sensors.
- *Spoofing attack*: Adversary A generates spoofing signals that impersonate transmissions originated from T. The GAN can be used to generate synthetic forms of wireless signals that can be used not only to augment training data (e.g. for a spectrum sensing classifier; Davaslioglu and Sagduyu 2018) but also to fool a target classifier (Shi et al. 2019a). To spoof wireless signals in a distributed setting, two adversaries (one transmitter and one receiver) assume roles of a generator and a discriminator of the GAN to spoof and discriminate signals, respectively, with two separate deep neural networks.
- *Trojan attack.* Adversary A provides falsified samples to train the target classifier C_T such that C_T works well in general but provides incorrect results if a certain trigger is activated. For example, small phase shifts can be added as triggers to wireless signals (without changing the signal amplitude) to launch a stealthy Trojan attack against a wireless signals classifier (Davaslioglu and Sagduyu 2019).

14.3.2   Domain-specific Challenges for Adversarial Machine Learning in Wireless Communications

Wireless applications of adversarial machine learning are different from other data domains such as computer vision in four main aspects.

1. The adversary and the defender may not share the same raw data (such as received signals) as channels and interference effects observed by them are different.
2. The type of data observed by the adversary depends on the underlying waveform and receiver hardware of the adversary. While adversarial machine learning may run on raw data samples such as pixels in computer vision applications, the adversary in a wireless domain may have to use different types of available radio-specific features such as I/Q data and received signal strength indicator (RSSI) that may differ from the features used by the target machine learning system.
3. The adversary and the defender may not share the same labels (i.e. machine learning outputs). For example, the defender may aim to classify channel as busy or not during spectrum sensing, whereas the adversary may need to decide on whether the defender will have a successful transmission or not. These two objectives may differ due to different channel and interference effects observed by the adversary and the defender.
4. The adversary may not directly manipulate the input data to the machine learning algorithm, as wireless users are typically separated in location and receive their input from wireless signals transmitted over the air. Therefore, it is essential to account for channel effects when designing wireless attacks and quantifying their impact.

With the consideration of the above challenges, we can apply adversarial machine learning to design attacks on wireless communications. In an example of *exploratory* attack, A collects spectrum sensing data and obtains features for its own classifier C_A. Unlike the traditional exploratory attack, where C_A provides the same set of prediction results as the target classifier C_T, the label that A aims to predict is whether there is a successful transmission from T or not, which can be collected by sensing the acknowledgement message (ACK). For that purpose, A collects both input (features) and output (label) for C_A. If for an instance, A can successfully predict whether there will be an ACK, C_A can predict the correct result. Deep learning is successful in building the necessary classifier C_A for A in exploratory attacks (Erpek et al. 2018).

We can further design an example of *evasion attack* as follows. If A predicts a successful transmission, it can either transmit in the sensing phase (to change features to C_T) (Sagduyu et al. 2019b; Shi et al. 2018a) or in the data transmission phase (to jam data) (Erpek et al. 2018). For the first case, most idle channels are detected as busy by C_T and thus throughput is reduced almost to zero (Sagduyu et al. 2019b). For the second case, many successful transmissions (if no jamming) are jammed and thus throughput is reduced significantly (Sagduyu et al. 2019a).

In a *causative attack*, C_T is updated by additional training data collected over time and A attempts to manipulate this retraining process. For example, if T transmits but does not receive an ACK, the prediction by C_T is incorrect and additional training data are collected. C_T is expected to improve with additional training data. However, A can again either transmit in the sensing phase or in the data transmission phase to manipulate the training process. For the first case, features are changed (Sagduyu et al. 2019b). For the second case, many transmissions that would be otherwise successful are jammed and consequently their labels are changed (Sagduyu et al. 2019a). For both cases, throughput (using updated C_T) is reduced significantly.

Membership inference attack is based on the analysis of overfitting. Note that features can be either useful (used to predict the channel status) or biased (due to the different distributions of training data and general test data) information. C_T is optimized to fit on useful and biased information (F_u and F_b). Fitting on F_b corresponds to overfitting, which provides correct classification on the given training data but wrong classification on general test data. In a white-box attack, A studies parameters in C_T based on local linear approximation for each layer and the combination of all layers. This approach builds a classifier for membership inference that can leak private information of a transmitter such as waveform, channel, and hardware characteristics by observing spectrum decisions based on the output of a wireless signal classifier (Shi et al. 2020).

In the *Trojan attack*, A slightly manipulates training data by inserting triggers to only few training data samples by modifying their phases and changing the labels of these samples to a target label. This poisoned training data is used to train C_T. In test time, A transmits signals with the same phase shift that was added as a trigger during training time. R accurately classifies clean (unpoisoned) signals without triggers but misclassifies signals poisoned with triggers (Davaslioglu and Sagduyu 2019).

14.3.3 Defense Schemes Against Adversarial Machine Learning

The basis of many attacks such as evasion and poisoning attacks discussed in Section 14.3.1 is the exploratory attack that trains a functionally equivalent classifier C_A as a surrogate model for target classifier C_T. Once C_A is built, the adversary can analyze this model to understand the behavior of C_T, which paves the way for identifying the weaknesses of C_T and then designing subsequent attacks. Therefore, a defense mechanism is needed to mitigate the exploratory attack. One *proactive defense* mechanism is to add controlled randomness to the target classifier C_T, such that it is not easy to launch an exploratory attack. For that purpose, transmitter T can transmit when channel is detected as busy or can remain idle when channel is detected as idle. However, such incorrect decisions will decrease the system performance even without attacks. Thus, the key problem is how to maximize the effect of defense while minimizing the impact on system performance. In particular, our approach is to exploit the likelihood score, namely the confidence level, returned by the machine learning classifier such that T performs defense operations only when the confidence is high, thereby maximizing the utility of each defense operation (Shi et al. 2018e). This way, with a few defense operations, the error probability of exploratory attack can be significantly increased (Sagduyu et al. 2019b) and subsequent attacks such as the evasion attack in the sensing phase can be further mitigated. The number of defense operations can be adapted in a dynamic way by monitoring the performance over time.

Possible approaches against the membership inference attack include the following two. The first approach aims to make the distribution in training data similar to the general test data. When we apply C_T on some samples, if a sample is classified with high confidence, it is likely that this sample contributes to overfitting in the training data and it is removed to make the training data similar to general test data. The second approach aims to remove the biased information in F_b. We can analyze the input to any layer in the deep neural network and identify features that play an important role in F_b. Then, we can rebuild C_T on features other than identified ones to remove the impact of overfitting.

Trojan attacks can be detected by identifying potential triggers inserted into training data. Since all malicious samples have a particular trigger, we can apply outlier detection methods such as the one based on median absolute deviation (MAD) or clustering to detect this trigger in the Trojan attack. Once the trigger is detected, any sample with this trigger is discarded or its label is switched to mitigate the attack (Davaslioglu and Sagduyu 2019).

14.4 Adversarial Machine Learning in 5G Communications

We present two scenarios to demonstrate adversarial machine learning-based attacks on 5G systems. The details of the first and second scenario and the performance results are presented in Sections 14.4.1 and Section 14.4.2, respectively.

14.4.1 Scenario 1—Adversarial Attack on 5G Spectrum Sharing

14.4.1.1 Attack Setting

The operation of 5G communications systems is expected to cover the CBRS band, where 5G users need to share the spectrum with the radar signal. The radar is the incumbent (primary) user of the band and the 5G communications system is the secondary user. 5G transmitter (T), namely 5G gNodeB, and receiver (R), namely 5G UE, need to communicate when no radar signal is detected in the band as the background traffic source B. The ESC system senses the spectrum, decides whether the channel is idle or busy by a machine learning-based classifier C_T, and informs its decisions to T. If the channel is idle, T transmits data. R sends an ACK once it receives data from T. An adversary A also senses the spectrum and decides when to perform certain attack actions as shown in Figure 14.3.

We consider the following attack actions built upon adversarial machine learning. First, the adversary A trains the surrogate model C_A in the form of an exploratory attack. Then, A aims to either jam data transmissions such that R cannot receive data transmission from T or jam the spectrum sensing period such that an idle channel is considered as busy. The first part is a conventional jamming attack and the second part is fooling T into wasting transmit opportunities. The second part corresponds to an evasion attack as it manipulates the sensing inputs into the machine learning algorithm used by T for spectrum access decisions.

Since the adversary A only needs to attack when there will be a successful transmission (if there was no attack), it aims to decide whether there is an ACK returned by the receiver R and uses the presence or absence of ACKs as labels to train its machine learning-based classifier C_A. The process of building C_A is not exactly the same as the conventional exploratory attack (such as the one considered in computer vision applications), where C_A should be the same (or as similar as possible) as C_T. Due to different spectrum sensing results at different locations, the inputs to C_A and C_T are different. Further, the output of C_A (i.e. "ACK or "no ACK) is different than the output of C_T (i.e. "idle or "busy). Figure 14.4 illustrates the second step where the adversary trains its adversarial deep learning classifier. Once its classifier is trained, A launches the attack as shown in Figure 14.5.

14.4.1.2 Simulation Setup and Performance Results

For this attack, we set up a scenario, where the distance from T to B is 1000 m and the distance from A to B is 1010 m. T builds C_T based on the sensed signal powers, namely the RSSIs. Each data

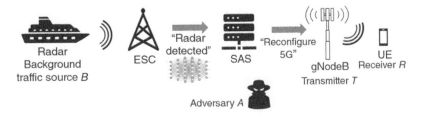

Figure 14.3 Scenario 1—Step 1: the adversary collects training data.

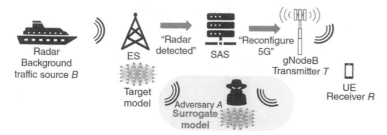

Figure 14.4   Scenario 1—Step 2: the adversary trains adversarial deep learning classifier.

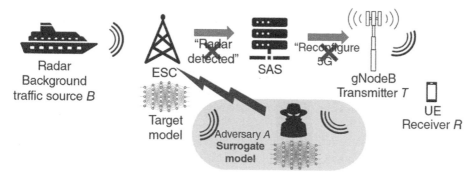

Figure 14.5   Scenario 1—Step 3: the adversary launches its attack.

sample consists of 200 sensing results and T collects 1000 of those samples. Then, one half of these samples are used for training and the other half are used for testing. We train and test classifiers of this chapter in TensorFlow and consider the following classifier characteristics:

- A feedforward neural network is trained for each classifier with backpropagation algorithm using the cross-entropy loss function.
- Rectified linear unit (ReLU) activation function is used at the hidden layers.
- Softmax activation function is used at the output layer.
- Batch size is 100.
- Number of training steps is 1000.

The deep neural network structure of classifier C_T is given as follows.

- The input layer has 200 neurons.
- The first hidden layer is a dense layer of 512 neurons.
- The second hidden layer is a dropout layer with dropout ratio of 0.2.
- The third hidden layer is a dense layer of 512 neurons.
- The fourth hidden layer is a dropout layer with dropout ratio of 0.2.
- The output layer has two neurons.

The monostatic radar signal is simulated in MATLAB as the background signal. Free space model is used to calculate the propagation loss between B and T. The classifier C_T has very good performance in the absence of adversaries. It can correctly detect all idle channel instances and most busy channel instances. The error on busy channel detection is 5.6%. That is, the 5G system can successfully protect 94.4% of radar transmissions while achieving 100% throughput (normalized by the best throughput that would be achieved by an ideal algorithm that detects every idle channel correctly).

Figure 14.6 Generating 5G NR signal at the UE.

The 5G communications system in this scenario uses 5G New Radio (NR) signal. The 5G NR signal is generated using MATLAB 5G Toolbox. The steps used to generate the 5G NR signal are shown in Figure 14.6. The signal includes the transport (uplink shared channel, UL-SCH) and physical channels. The transport block is segmented after the cyclic redundancy check (CRC) addition and low-density parity-check (LDPC) coding is used as forward error correction. The output codewords are 16-QAM modulated. Both data and control plane information are loaded to the time-frequency grid of the 5G signal. Orthogonal frequency-division multiplexing (OFDM) modulation is used with inverse Fast Fourier Transform (iFTT) and Cyclic Prefix (CP) addition operations. The transmit frequency is set to 4 GHz. The subcarrier spacing is 15 kHz. The number of resource blocks used in the simulations is 52. The transmitted waveform is passed through a tapped delay line (TDL) propagation channel model. The delay spread is set to 0.3 μs. Additive white Gaussian noise (AWGN) is added to the signal in order to simulate the signal received at the receiver.

The adversary A collects 1000 signal samples. Each sample includes sensed signal powers as features and "ACK" or "no ACK" as labels. We split data samples by half for training and testing of classifier C_A. The hyperparameters of C_A are the same as C_T except that the input layer has 400 neurons. The classifier C_A can correctly detect all ACK instances and most no-ACK instances. The error probability when there is no ACK is 4.8%. Once C_A is built, the adversary A can successfully jam all data transmissions or all pilot signals from T. Moreover, the unnecessary jamming is minimized, i.e. among all jamming transmissions only 4.48% are performed when there is no ACK.

As mentioned earlier, A can jam either T's data transmission or T's pilot signal. In general, T's data transmission period is much longer than T's pilot signal. Thus, jamming of the pilot signal is more energy-efficient. For example, if the length of data transmission is nine times of the length of pilot signal and the adversary has the energy to jam data transmissions for 20% of time slots, the adversary cannot jam all data transmissions and can only reduce the throughput by 19.08%. With the same amount of energy consumption, the adversary can jam all pilot signals and thus reduce the throughput by 100%.

14.4.2 Scenario 2 — Adversarial Attack on Signal Authentication in Network Slicing

14.4.2.1 Attack Setting

The second scenario considers a spoofing attack on the network slicing application. A classifier is trained at the 5G gNodeB to detect 5G UEs by their signals and then provide certain services after authenticating them. The 5G gNodeB trains the classifier C_S based on the I/Q data that includes both signal power and phase to distinguish signals from a target 5G UE and random noise signals. For that purpose, we use the same deep neural network structure as in scenario 1 except that classifier C_S has the input layer of 400 neurons.

An adversary A aims to transmit similar signals to gain access to 5G-enabled services. For this purpose, it can sense the spectrum to (i) collect signal samples (I/Q data) and (ii) identify whether such a signal is classified as a target user's signal by either monitoring feedback from the 5G gNodeB (regarding which 5G UE is authenticated) or observing which 5G UE starts communicating to the 5G gNodeB as an authenticated user. Once sufficient 5G signal samples are collected, adversary A can apply the GAN to generate synthetic 5G data as spoofing signals and then transmit them to gain access to 5G-enabled services. In this attack, A consists of a pair of transmitter and receiver.

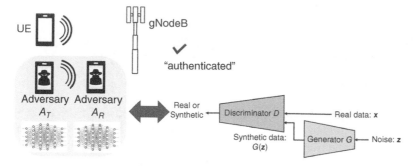

Figure 14.7 Scenario 2–Step 1: the adversary training with the GAN.

Adversary transmitter A_T trains a neural network to build the generator of the GAN and adversary receiver A_R trains another deep neural network to build the discriminator of the GAN. The only feedback from A_R to A_T during the training is whether the signal is transmitted from the 5G UE or A_T. For that purpose, A_T sends a flag along with its signal to A_R to indicate its transmissions and this flag is used to label samples. Note that this spoofing attack serves the same purpose as an evasion attack, namely fooling the 5G gNodeB into making wrong classification decisions. The only difference is that instead of adding perturbations on top of real transmissions (by jamming the channel as in Scenario 1), the adversary A generates new synthetic signals and transmits them directly to the 5G gNodeB. The training process for the adversary A is illustrated in Figure 14.7.

14.4.2.2 Simulation Setup and Performance Results

The deep neural network structure for the generator at A_T is given as follows.

- The input layer has 400 neurons.
- There are three hidden layers, each a dense layer of 128 neurons.
- The output layer has 400 neurons.

The deep neural network structure for the discriminator at A_R is the same as the generator except that its output layer has two neurons.

For this attack, we set up a scenario, where we vary the noise power with respect to the minimum received 5G signal power at the 5G gNodeB and the corresponding signal-to-noise ratio (SNR) is denoted by γ (measured in dB). The 5G gNodeB builds its classifier C_S using 1000 samples, half for training and half for testing. We vary γ as −3, 0, and 3 dB. No matter which γ value is used, C_S can always be built perfectly, i.e. there is no error in distinguishing 5G signals from other (randomly generated) signals. Adversary A collects 1000 samples (5G signals and other signals), applies the GAN to generate synthetic data samples, and transmits them to gain access to services as shown in Figure 14.8. The success probability is shown in Table 14.1. When $\gamma = 3$ dB, the success probability reaches 90%. Note that this approach matches all waveform, channel and radio hardware

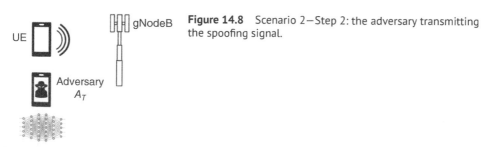

Figure 14.8 Scenario 2–Step 2: the adversary transmitting the spoofing signal.

Table 14.1 Spoofing attack performance.

5G signal SNR, γ (in dB)	Attack success probability
-3	60.6%
0	66.6%
3	90.0%

effects of 5G UE's transmissions as expected to be received at the 5G gNodeB. Therefore, this attack performance cannot be achieved by replay attacks that amplify and forward the received signals.

14.4.3 Defense Against Adversarial Machine Learning in 5G Communications

We use the attack scenario 2 (adversarial attack on 5G signal authentication) as an example to discuss the defense. One proactive defense approach at the 5G gNodeB is to introduce selective errors in denying access to a very small number of requests from intended 5G UEs (i.e. false alarm probability is slightly increased). Note that no errors are made in authenticating non-intended users (i.e. misdetection error is not increased). This is not a deterministic approach such that an intended 5G UE that is denied request in one instance can be authenticated in its next access attempt. The controlled errors made by the 5G gNodeB are inputs to the adversary and thus it cannot train a proper GAN to generate synthetic 5G signals, i.e. spoofed signals can be reliably detected and denied access. Figure 14.9 shows the scenario for defense. Given that such defense actions (selective errors) would also decrease the system performance (as very few intended 5G UEs may be denied access over time), the probability of defense actions should be minimized. Thus, the 5G gNodeB selects signal samples that are assigned high score as intended user by its classifier and denies access to their corresponding transmitter. This approach misleads the adversary most, as the uncertainty in determining the decision boundary in its classifier is maximized for a given number of defense actions.

Let P_{d} denote the ratio of defense actions to all authentication instances. Table 14.2 shows that even a small P_{d}, e.g. $P_{\mathrm{d}} = 0.01$, can significantly decrease the attack success probability, where γ is fixed as -3 dB. On the other hand, there is no need to use a large P_{d}, e.g. larger than 0.05, since the attack success probability converges to roughly 60% quickly.

A similar defense can be applied against scenario 1, where the 5G gNodeB deliberately makes a small number of wrong transmit decisions when accessing the spectrum. Then, the adversary cannot train a proper surrogate model to launch successful attacks on data transmission or spectrum sensing.

Figure 14.9 Scenario 2–Step 3: controlled errors introduced at 5G gNodeB as part of defense.

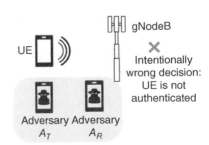

Table 14.2 Defense against spoofing attack

Ratio of defense actions, P_d	Attack success probability
0	90.0%
0.01	68.2%
0.02	61.8%
0.05	59.4%
0.1	62.0%
0.2	61.0%

14.5 Conclusion

The security aspects of machine learning have gained more prominence with the increasing use of machine learning algorithms in various critical applications including wireless communications. In this chapter, we first explained different attack types in adversarial machine learning and corresponding defense methods. Then, we focused on how adversarial machine learning can be used in wireless communications to launch stealthy attacks. We described the challenges associated with designing attacks in wireless domains by accounting for differences from other data domains and unique challenges. Next, we focused on the vulnerabilities of the 5G communication systems due to adversarial machine learning. We considered two 5G scenarios. In the first scenario, the adversary learns 5G's deep learning-driven pattern of spectrum sharing with incumbent user such as in the CBRS band and jams the data and control signals to disrupt 5G communications. Results show that the adversary can significantly reduce the throughput of the 5G communications while leaving only a small footprint. In the second scenario, a spoofing attack is performed by the adversary to pass through the deep learning-based physical layer authentication system at the 5G gNodeB. A GAN is trained to generate the spoofing signal by matching waveform, channel, and radio hardware effects at the receiver. Results show that the attack is successful for a range of SNRs of the 5G signal used during the training. Then, a defense technique is proposed such that controlled errors are made by the 5G system deliberately to fool the adversary into training inaccurate models while minimizing negative effects on its own performance. Novel attacks presented in this chapter highlight the impact of adversarial machine learning on wireless communications in the context of 5G and raise the urgent need for defense mechanisms.

References

N. Abu Zainab, T. Erpek, K. Davaslioglu, Y. E. Sagduyu, Y. Shi, S. Mackey, M. Patel, F. Panettieri, M. Qureshi, V. Isler, and A. Yener. QoS and jamming-aware wireless networking using deep reinforcement learning. In *IEEE Military Communications Conference (MILCOM)*, 2019.

S. Alfeld, X. Zhu, and P. Barford. Data poisoning attacks against autoregressive models. In *AAAI Conference on Artificial Intelligence, Phoenix, AZ, USA, February 2016*, 2016.

M. Barreno, B. Nelson, R. Sears, A. Joseph, and J. Tygar. "Can machine learning be secure?. In *ACM Symposium on Information, Computer and Communications Security, Taipei, Taiwan, March 2006*, 2006.

Citizens Broadband Radio Service. Code of Federal Regulations. Title 47, Part 96. Federal Communications Commission, June 2015.

J. Cohen, E. Rosenfeld, and Z. Kolter. Certified adversarial robustness via randomized smoothing. In *International Conference on Machine Learning (ICML), Long Beach, CA, USA, June 2019*, 2019.

T. S. Cousik, V. K. Shah, J. H. Reed, T. Erpek, Y. E. Sagduyu. Fast initial access with deep learning for beam prediction in 5G mmWave networks. *arXiv preprint, arXiv:2006.12653*, 2020.

K. Davaslioglu and Y. E. Sagduyu. Generative adversarial learning for spectrum sensing. In *IEEE International Conference on Communications (ICC), Kansas City, MO, USA, May 2018*, 2018.

K. Davaslioglu and Y. E. Sagduyu. Trojan attacks on wireless signal classification with adversarial machine learning. In *IEEE DySPAN Workshop on Data-Driven Dynamic Spectrum Sharing (DD-DSS), Newark, NJ, USA, November 2019*, 2019.

K. Davaslioglu, S. Soltani, T. Erpek, and Y. E. Sagduyu. DeepWiFi: Cognitive WiFi with deep learning. *IEEE Transactions on Mobile Computing*, 20(2): 429–444, 2019.

P. Dong, H. Zhang, G. Y. Li, I. S. Gaspar, and N NaderiAlizadeh. Deep CNN-based channel estimation for mmWave massive MIMO systems. *IEEE Journal of Selected Topics in Signal Processing*, 13(5):989–1000, 2019.

T. Erpek, T. O'Shea, Y. Sagdeuyu, Y. Shi, and T. C. Clancy. Deep learning for wireless communications. In *Development and Analysis of Deep Learning Architectures*. Springer, 2019.

T. Erpek, Y. E. Sagduyu, and Y. Shi. Deep learning for launching and mitigating wireless jamming attacks. *IEEE Transactions on Cognitive Communications and Networking*, 5(1):2–14, 2018.

B. Flowers, R. M. Buehrer, and W. C. Headley. Evaluating adversarial evasion attacks in the context of wireless communications. *arXiv preprint, arXiv:1903.01563*, 2019a.

B. Flowers, R. M. Buehrer, and W. C. Headley. Communications aware adversarial residual networks. In *IEEE Military Communications Conference (MILCOM), Norfolk, VA, USA, November 2019*, 2019b.

I. J. Goodfellow, J. P. Abadie, M. Mirza, B. Xu, D. W. Farley, S. Ozair, A. C. Courville, and Y. Bengio. Generative adversarial networks. *International Conference on Neural Information Processing Systems, Montreal, Canada, December 2014*, 2:2672–2680, 2014.

Google Cloud Vision API. https://cloud.google.com/vision (accessed 27 March 2021).

M. Z. Hameed, A. Gyorgy, and D. Gunduz. Communication without interception: Defense against deep-learning-based modulation detection. In *IEEE Global Conference on Signal and Information Processing (GlobalSIP), ON, Canada, November 2019*, 2019a.

M. Z. Hameed, A. Gyorgy, and D. Gunduz. The best defense is a good offense: Adversarial attacks to avoid modulation detection. *arXiv preprint, arXiv:1902.10674*, 2019b.

J. Jia, A. Salem, M. Backes, Y. Zhang, and N. Z. Gong. MemGuard: Defending against black-box membership inference attacks via adversarial examples. In *ACM Conference on Computer and Communications Security (CCS), London, UK, November 2019*, 2019.

C. Jiang, H. Zhang, Y. Ren, Z. Han, K. C. Chen, and L. Hanzo. Machine learning paradigms for next-generation wireless networks. *IEEE Wireless Communications*, 24(2):98–105, 2016.

A. Kurakin, I. Goodfellow, and S. Bengio. Adversarial machine learning at scale. *arXiv preprint, arXiv:1611.01236*, 2016a.

A. Kurakin, I. Goodfellow, and S. Bengio. Adversarial examples in the physical world. *arXiv preprint, arXiv:1607.02533*, 2016b.

S. Kokalj-Filipovic and R. Miller. Adversarial examples in RF deep learning: Detection of the attack and its physical robustness. *arXiv preprint, arXiv:1902.06044*, 2019.

S. Kokalj-Filipovic, R. Miller, N. Chang, and C. L. Lau. Mitigation of adversarial examples in RF deep classifiers utilizing autoencoder pre-training. *arXiv preprint, arXiv:1902.08034*, 2019.

B. Kim, Y. E. Sagduyu, K. Davaslioglu, T. Erpek, and S. Ulukus. Over-the-air adversarial attacks on deep learning based modulation classifier over wireless channels. In *Conference on Information Sciences and Systems (CISS), Princeton, NJ, USA, March 2020*, 2020a.

B. Kim, Y. E. Sagduyu, K. Davaslioglu, T. Erpek, and S. Ulukus. Channel-aware adversarial attacks against deep learning-based wireless signal classifiers. *arXiv preprint arXiv:2005.05321*, 2020b.

B. Kim, Y. E. Sagduyu, K. Davaslioglu, T. Erpek, and S. Ulukus, How to make 5G communications "invisible": Adversarial machine learning for wireless privacy. Asilomar Conference on Signals, Systems, and Computers, 2020.

W. M. Lees, A. Wunderlich, P. J. Jeavons, P. D. Hale, and M. R. Souryal. Deep learning classification of 3.5-GHz band spectrograms with applications to spectrum sensing. *IEEE Transactions on Cognitive Communications and Networking (TCCN)*, 5(2): 224–236, 2019.

K. Leino and M. Fredrikson. Stolen memories: Leveraging model memorization for calibrated white-box membership Inference. In *USENIX Security Symposium, August 2020. Virtual Conference*, 2020.

Z. Luo, S. Zhao, Z. Lu, Y. E. Sagduyu, and J. Xu. Adversarial machine learning based partial-model attack in IoT. In *ACM WiSec Workshop on Wireless Security and Machine Learning (WiseML), July 2020. Virtual Conference*, 2020.

Z. Luo, S. Zhao, Z. Lu, J. Xu, and Y. E. Sagduyu. When attackers meet AI: Learning-empowered attacks in cooperative spectrum sensing. *arXiv preprint, arXiv:1905.014*, 2019.

F. B. Mismar, B. L. Evans, and A. Alkhateeb. Deep reinforcement learning for 5G networks: Joint beamforming, power control, and interference coordination. *IEEE Transactions on Communications*, 68(3): 1581–1592, 2019.

S. M. Moosavi-Dezfooli, A. Fawzi, and P. Frossard. DeepFool: A simple and accurate method to fool deep neural networks. In *IEEE Conference on Computer Vision and Pattern Recognition (CVPR)*, 2015.

M. Nasr, R. Shokri, and A. Houmansadr. Machine learning with membership privacy using adversarial regularization. In *ACM Conference on Computer and Communications Security (CCS), Toronto, Canada, October 2018*, 2018.

NGMN Alliance. 5G security recommendations package 2: network slicing, 2016.

N. Papernot, P. McDaniel, I. Goodfellow, S. Jha, Z. Celik, and A. Swami. Practical black-box attacks against deep learning systems using adversarial examples. In *ACM on Asia Conference on Computer and Communications Security (CCS), Abu Dhabi, United Arab Emirates, April 2017*, 2017.

N. Papernot, P. McDaniel, S. Jha, M. Fredrikson, Z. Celik, and A. Swami. The limitations of deep learning in adversarial settings. In *IEEE European Symposium on Security and Privacy, Saarbruecken, Germany, March 2016*, 2016.

L. Pi, Z. Lu, Y. Sagduyu, and S. Chen. "Defending active learning against adversarial inputs in automated document classification. In *IEEE Global Conference on Signal and Information Processing (GlobalSIP), Washington, DC, USA, December 2016*, 2016.

T. Gruber, S. Cammerer, J. Hoydis, and S. ten Brink. On deep learning-based channel decoding. In *Conference on Information Sciences and Systems (CISS), Baltimore, MD, USA, March 2017*, 2017.

M. Sadeghi and E. G. Larsson. Adversarial attacks on deep-learning based radio signal classification. *IEEE Wireless Communications Letters*, 8(1): 213–216, 2018.

M. Sadeghi and E. G. Larsson. Physical adversarial attacks against end-to-end autoencoder communication systems. *IEEE Communications Letters*, 23(5): 847–850, 2019.

Y. E. Sagduyu. Securing cognitive radio networks with dynamic trust against spectrum sensing data falsification. In *IEEE Military Communications Conference (MILCOM), Baltimore, MD, USA, October 2014*, 2014.

Y. E. Sagduyu, R. Berry, and A. Ephremides. Wireless jamming attacks under dynamic traffic uncertainty. In *IEEE International Symposium on Modeling and Optimization in Mobile, Ad Hoc, and Wireless Networks (WIOPT), Avignon, France, May 2010*, 2010.

Y. E. Sagduyu, R. Berry, and A. Ephremides. Jamming games in wireless networks with incomplete information. *IEEE Communications Magazine*, 49(8): 112–118, 2011.

Y. E. Sagduyu and A. Ephremides. A game-theoretic analysis of denial of service attacks in wireless random access. *Wireless Networks*, 15(5): 651–666, 2009.

Y. E. Sagduyu, Y. Shi, and T. Erpek. IoT network security from the perspective of adversarial deep learning. *IEEE SECON Workshop on Machine Learning for Communication and Networking in IoT (MLCN-IoT), Boston, MA, USA, June 2019*, 2019a.

Y. Sagduyu, Y. Shi, and T. Erpek. Adversarial deep learning for over-the-air spectrum poisoning attacks. *IEEE Transactions on Mobile Computing*, 20 (2):306–319, 2019b.

Y. Shi, K. Davaslioglu, and Y. E. Sagduyu. Generative adversarial network for wireless signal spoofing. In *ACM WiSec Workshop on Wireless Security and Machine Learning (WiseML), Miami, FL, USA, May 2019*, 2019a.

Y. Shi, K. Davaslioglu, and Y. E. Sagduyu. Over-the-air membership inference attacks as privacy threats for deep learning-based wireless signal classifiers. In *ACM WiSec Workshop on Wireless Security and Machine Learning (WiseML), Linz, Austria, July 2020*, 2020.

Y. Shi, K. Davaslioglu, Y. E. Sagduyu, W. C. Headley, M. Fowler, and G. Green. Deep learning for signal classification in unknown and dynamic spectrum environments. In *IEEE International Symposium on Dynamic Spectrum Access Networks (DySPAN), Newark, NJ, USA, November 2019*, 2019b.

Y. Shi, T. Erpek, Y. E. Sagduyu, and J. Li. Spectrum data poisoning with adversarial deep learning. In *IEEE Military Communications Conference (MILCOM), Los Angeles, CA, USA, October 2018*, 2018a.

Y. Shi and Y. E. Sagduyu. Evasion and causative attacks with adversarial deep learning. In *IEEE Military Communications Conference (MILCOM), Baltimore, MD, USA, October 2017*, 2017.

Y. Shi, Y. E. Sagduyu, K. Davaslioglu, and R. Levy. Vulnerability detection and analysis in adversarial deep learning. *Guide to Vulnerability Analysis for Computer Networks and Systems – An Artificial Intelligence Approach*. Springer, Cham, 2018b.

Y. Shi, Y. E. Sagduyu, K. Davaslioglu, and J. H. Li. Generative adversarial networks for black-box API attacks with limited training data. In *IEEE International Symposium on Signal Processing and Information Technology (ISSPIT), Louisville, KY, USA, December 2018*, 2018c.

Y. Shi, Y. E. Sagduyu, K. Davaslioglu, and J. Li. Active deep learning attacks under strict rate limitations for online API calls. In *IEEE Symposium on Technologies for Homeland Security (HST), Woburn, MA, USA, October 2018*, 2018d.

Y. Shi, Y. E. Sagduyu, T. Erpek, K. Davaslioglu, Z. Lu, and J. Li. Adversarial deep learning for cognitive radio security: Jamming attack and defense strategies. In *IEEE ICC Workshop on Promises and Challenges of Machine Learning in Communication Networks (ML4COM), Kansas City, MO, USA, May 2018*, 2018(e).

Y. Shi, Y. E. Sagduyu, and A. Grushin. "How to steal a machine learning classifier with deep learning. In *IEEE Symposium on Technologies for Homeland Security (HST), Waltham, MA, USA, April 2017*, 2017.

L. Song, R. Shokri, and P. Mittal. Privacy risks of securing machine learning models against adversarial examples. In *ACM Conference on Computer and Communications Security (CCS), London, UK, November 2019*, 2018.

S. Soltani, Y. E. Sagduyu, R. Hasan, K. Davaslioglu, H. Deng, and T. Erpek. Real-time and embedded deep learning on FPGA for RF signal classification. In *IEEE Military Communications Conference (MILCOM), Norfolk, VA, USA, November 2019*, 2019.

M. R. Souryal and T. Nguyen. Effect of federal incumbent activity on CBRS commercial service. *IEEE International Symposium on Dynamic Spectrum Access Networks (DySPAN), Newark, NJ, USA, November 2019*, 2019.

F. Tramer, F. Zhang, A. Juels, M. Reiter, and T. Ristenpart. Stealing machine learning models via prediction APIs. In *USENIX Security Symposium, Austin, TX, USA, August 2016*, 2016.

T. Vaidya, Y. Zhang, M. Sherr, C. Shields, D. Wagner, N. Carlini, P. Mishra, and W. Zhou. Hidden voice commands. In *USENIX Security Symposium, Austin, TX, USA, August 2016*, 2016.

Y. Vorobeychik and M. Kantarcioglu. *Adversarial Machine Learning*. Morgan & Claypool, 2018.

F. Wang, C. Zhong, M. C. Gursoy, and S. Velipasalar. Defense strategies against adversarial jamming attacks via deep reinforcement learning. In *Conference on Information Sciences and Systems (CISS), Princeton, NJ, USA, March 2020*, 2020.

X. Wu, M. Fredrikson, S. Jha, and J. F. Naughton. A methodology for formalizing model-inversion attacks. *IEEE Computer Security Foundations Symposium (CSF), Lisbon, Portugal, June 2016*, 2016.

S. Zhang. An overview of network slicing for 5G. *IEEE Wireless Communications (TWC)*. 26(3): 111–117, 2019.

G. Zhang, C. Yan, X. Ji, T. Zhang, T. Zhang, and W. Xu. Dolphinatack: Inaudible voice commands. *arXiv preprint, arXiv:1708.09537*, 2017.

C. Zhong, F. Wang, M. C. Gursoy, and S. Velipasalar. Adversarial jamming attacks on deep reinforcement learning based dynamic multichannel access. In *IEEE Wireless Communications and Networking Conference (WCNC)*, 2020.

15

Machine Learning in the Hands of a Malicious Adversary: A Near Future If Not Reality[1]

Keywhan Chung[1], Xiao Li[2], Peicheng Tang[3], Zeran Zhu[1], Zbigniew T. Kalbarczyk[1], Thenkurussi Kesavadas[2], and Ravishankar K. Iyer[1]

[1] *Coordinated Science Laboratory, University of Illinois at Urbana-Champaign, Urbana, IL, USA*
[2] *Healthcare Engineering Systems Center, University of Illinois at Urbana-Champaign, Urbana, IL, USA*
[3] *Department of Electrical and Computer Engineering, Rose-Hulman Institute of Technology, Terra Haute, IN, USA*

15.1   Introduction

In recent years, researchers and practitioners in the cyber security domain have been investigating and demonstrating the use of machine learning (ML) methods to detect and mitigate the consequences of malicious attacks. For example, ML models have been used to dynamically infer defense strategies (e.g. rules or policies that define suspicious actions and response strategies against detected events) based on data obtained from security monitoring tools/infrastructure (Cao et al. 2015). However, the possibility that adversaries might utilize the same ML technology to advance malicious intentions and conceal suspicious activities has been largely ignored.

This chapter considers a new type of attacks, *AI-driven malware*, where a malicious logic embeds machine learning methods and models to automatically: (i) probe the target system for inferring actionable intelligence (e.g. system configuration or operational patterns) and (ii) customize the attack payload accordingly (e.g. determine the most opportune time to execute the payload so to maximize the impact). Such an AI-driven malware acts as a human attacker, and hence, allows to encode the malicious intention into the ML model and generalize the attack to an extended range of targets. Conceptually, an AI-driven malware can be considered as a skeleton with a structure of the attack strategy/payload with the details to be configured. Upon deployment inside the target system, the skeleton is fleshed out with the details on what, when, and how to attack as inferred from the operational data from the target system.

For instance, current attacks with intelligence encoded into advanced malware (e.g. Stuxnet; Langner (2011)) were tailored to the details of a specific target system. Consequently, such malware is not directly applicable to other systems (e.g. a critical component or parameter that the malware is programmed to tackle might not be present in another system). In contrast, for AI-driven malware, the details of the malware are configured after the malware is deployed in the target system. As a result, the generic payload (as programmed in the malware) can be applied to an extended set of targets.

Machine learning in security attacks. Recent studies demonstrated that sophisticated ML-based techniques, devised to protect systems, can be disarmed with carefully crafted malicious

1 This material is based upon work supported by the National Science Foundation under Grant Nos. 18-16673 and 15-45069. Any opinions, findings, and conclusions or recommendations expressed in this material are those of the authors and do not necessarily reflect the views of the National Science Foundation.

input. Specifically, the use of adversarial machine learning allows attackers to bypass ML-driven security monitors (e.g. spam filtering) or authentication methods (e.g. face recognition) (Huang et al. 2011; Sharif et al. 2016; Papernot et al. 2017). While adversarial learning can become a real obstruction in securing computing infrastructure, we are about to face more advanced threats that are, in fact, *driven by* machine learning techniques. While there has been no public record of a real threat driven by machine learning models, such an advanced malware might already exist and simply remains undetected. A number of researchers have demonstrated methods that utilize machine learning to accelerate the exploitation or bypassing of detectors (Palka and McCoy 2015; Petro and Morris 2017), but attacks driven by artificial intelligence, particularly in targeted attacks, step up the challenge of promptly detecting malicious intentions.

Advanced targeted attacks (ATAs). As reported in Widup (2020), a shared security event recording database is supported by Verizon, targeted attacks are on the rise, with an increasing focus on strategic infrastructure (Symantec Corporation 2019). Unlike the usual threats that impact a large number of random victims (e.g. ransomware or spyware), sophisticated threats are customized for specific targets (e.g. the Stuxnet attack or data breach attacks) to maximize efficiency (e.g. they minimize irrelevant code to reduce traces and customize activities to reflect the configuration of the target system).

In this chapter, we use ATAs to depict a broad spectrum of malicious attacks, including advanced persistent threats (APTs) from Lockheed Martin (2019) and Fire Eye Mandiant Services (2019b), which were introduced to describe state-sponsored attack campaigns delivered by a group of malicious actors (e.g. Stuxnet; Langner (2011) or the attack against Ukraine's power grid; Lee et al. (2016)). However, regardless of whether an attack is conducted by an APT group, a criminal gang, a lone hacker, or a malicious insider, ATAs (in customizing threats to specific targets) go through several stages, which can be represented as a life cycle (discussed in detail in Section 15.2.1).

AI-powered malware for targeted attacks. Often, ATAs are detected after the final stage (i.e. the *actions on objectives* phase). The attacker's intermediate activities are hard to differentiate from benign events, such that no sign of intrusion is detected, or alarms generated from such activities are misidentified as false alerts.

As we discuss in this chapter, targeted attacks powered by ML techniques can increase the efficiency of attacks, reduce the time to success, and compromise the protection of computer infrastructure. As attackers' capabilities and attack's sophistication grow, cyber defenders must understand the mechanisms and implications of the malicious use of AI in order (i) to stay ahead of the threats and (ii) to devise and deploy defenses to prevent theft of data, system damage, or major disruption of critical infrastructure (e.g. the electric power grid).

15.2 AI-driven Advanced Targeted Attacks

15.2.1 Advanced Targeted Attacks

Targeted attacks are on the rise, with an increasing focus on strategic infrastructure (Symantec Corporation 2019; Widup 2020). Unlike the usual threats that impact a large number of random victims (e.g. ransomware or spyware), sophisticated threats are customized for specific targets (e.g. the Stuxnet attack or data breach attacks) to maximize efficiency (e.g. they minimize irrelevant code to reduce traces and customize activities to reflect the configuration of the target system). In our discussion, we use ATAs (as defined in (SentinelOne 2019)) to depict a broad spectrum of malicious attacks, including advanced persistent threats (APTs) from Lockheed Martin (2019) and Fire Eye Mandiant Services (2019b), which were introduced to describe state-sponsored attack campaigns

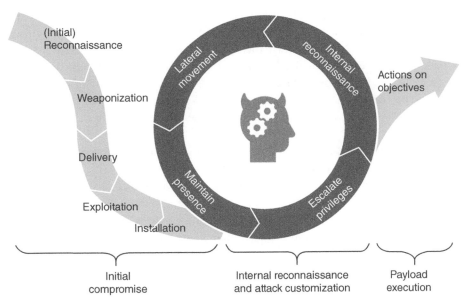

Figure 15.1 The life cycle of advanced targeted attacks (ATAs). The cyclic sequence of procedures—escalate privileges, internal reconnaissance, lateral movement, maintaining presence—can be performed by AI-driven malware.

delivered by a group of malicious actors (e.g. Stuxnet; Langner (2011) or the attack against Ukraine's power grid; Lee et al. (2016)). However, regardless of whether an attack is conducted by an APT group, a criminal gang, a lone hacker, or a malicious insider, ATAs go through several stages, which can be represented as a life cycle.

Life-cycle of ATAs. Figure 15.1 (inspired by Lockheed Martin (2019) and Fire Eye Mandiant Services (2019b)) depicts the sequence of stages taken in advanced targeted attacks. ATAs start with the attacker's collection of publicly available information about the target infrastructure (*initial reconnaissance*). Utilizing the collected information, attackers *weaponize* the payload to initiate a foothold in the target system. Once the payload is ready, the attacker *delivers* the payload to the target system. In this phase, email, web applications, or media devices (e.g. thumb drives or portable hard drives) can be used to deliver the payload. Once executed, the weaponized payload *exploits* vulnerabilities that exist in the target system and *installs* malware on the system. A successful installation often establishes a connection with a remote command and control server, allowing the attacker to communicate with the target system. With control over the system, the attacker collects additional information regarding the system (*internal reconnaissance*), *escalates privileges*, and discovers additional systems internally connected to the compromised system (*lateral movement*). The attacker can take additional steps (e.g. install new backdoors) to sustain the foothold in the system (*maintain presence*). Depending on the final objective, the malicious actors may remain in the system for extended periods of time until they are ready to take *action on the objectives*.

15.2.2 Motivation for Adapting ML method in Malware

Although existing ATAs are already sophisticated, certain restrictions limit the attackers from generalizing the attack or maximizing the impact. In fact, attackers might have numerous reasons to adopt the state-of-the-art learning techniques to achieve goals that would otherwise be distant.

Malicious intentions encoded into the ML model to prevent reverse-engineering. In keeping a system secure, reverse-engineering has played a critical role in revealing attackers' intentions. However, as stated in Kirat et al. (2018), machine learning algorithms are hard to reverse-engineer. For generic malware, a thorough analysis of the binary instructions may reveal the intent of the attacker who designed the malware. But for AI-driven malware, the intent gets encoded into the learning model. Translating a machine-inferred ML model into a human-interpretable logic is a hard problem by itself. The use of machine learning in malware and attacks can increase the chances that the attackers will keep their intentions concealed in the malware, challenge the system administrators studying the attack, and, hence, delay timely response to the attack.

Human intelligence packed into a software package to overcome the air gap barrier and dynamically configure the attack payload. A common defense method for critical computing infrastructure is to isolate the system/network from public access. That protection method can filter out malware that spreads over the network or threats that attempt to establish remote access. Without access to the target system, attackers cannot investigate the target system, so, traditional targeted attacks are not feasible. However, through a human being the weakest link, attackers can overcome the air gap. For instance, in the Stuxnet attack (Langner 2011), a thumb-driven device was used to overcome the air gap and deliver the malware to an isolated system and execute the customized payload, hard-coded into the malware upfront. Although Stuxnet showed that air gaps could be circumvented, the payload delivered to the isolated system remained static with limited capabilities. However, we envisage AI-driven malware that can evade the air gap, substitute human-driven analysis, and automatically collect the intelligence needed to customize the attack payload.

Should such malware become a reality, air gap barriers would not be as effective in restricting attacker capabilities.

Access to rich data that contain target-specific actionable intelligence. The success of ML algorithms relies heavily on having good and sufficient data. The low cost of storage, an increase in communication capacity, and deployment of multiple sensors have led to archiving of a large volume of data on system operations and maintenance. The archived data, combined with real-time information on system operations (e.g. sensor data in a cyber-physical system), provide a rich data stream that an attacker can use to learn about the system configuration/topology or identify communication patterns. In that scenario, a properly designed ML model can be a convenient way to derive the intelligence necessary to devise attack strategies.

Generalizing an ATA to a broader range of targets and scale the attack. When attackers have to manually customize their attack strategies, based on their reconnaissance of a victim, the payload is restricted to a specific system. However, the use of ML models allows flexibility in the attack payload design. Instead of the manual process of reconnaissance and customization, attacker would design the structure (*skeleton*) of the attack strategy/payload and leave the details to be configured. Upon deployment inside the target system, the skeleton is fleshed out by the data (as observed in a given target system). With the payload kept abstract in the implementation phase and driven by observations upon deployment, the malware can take full advantage of automation (in infecting the system, customizing the payload, and delivering the impact), and target an extended range of systems.

Extended presence in the system vs. the risk of detection. In designing a targeted attack, reconnaissance (e.g. learning the system configuration, software/hardware version, and/or I/O access patterns) is critical to success. Hence, once the attacker establishes a foothold in the victim system/network, he/she can collect actionable intelligence, which is the baseline for designing the customized payload. To gather useful intelligence, the attacker needs to maintain remote access to the target system for an extended period of time. For instance, the authors of Fire Eye Mandiant

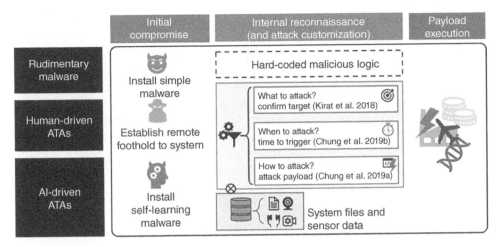

Figure 15.2 Comparison of attack sophistication levels. Detailed steps of the three phases (initial compromise, internal reconnaissance, and payload execution) of ATAs can be found in Figure 15.1.

Services (2019a) reported a global median dwell time[2] of 78 days for targeted attacks investigated by a security firm from October 2017 to September 2018. Security administrators often consider extended connection time as a suspicious anomaly that triggers a thorough investigation, if not an immediate disconnection. Hence, from the attacker's perspective, the longer the foothold, the higher the risk of detection. Such risk, for an attacker, can be eliminated if the reconnaissance and the customization can be automated using machine learning methods. In this case, as the information gathering, data analysis, and attack strategy customization occur in the victim system, the attacker does not need to maintain a foothold, i.e. remote access to the target system for an extended period.

15.2.3 AI to Flesh Out the Details of What, When, and How to Attack from Internal Reconnaissance

In designing attack strategies, malicious actors need to determine what to attack, when to attack, and how to attack. In rudimentary malware or human-driven (traditional) ATAs, such details of the strategies are either hard-coded into the malware or manually inferred by the actor driving the attack (internal reconnaissance). Instead, we envision that, given a well-defined skeleton that structures the attack, AI (acting as a human attacker) can flesh out the details on what, when, and how based on the target-specific details inferred from local data of the target system.

Rudimentary malware. As depicted in Figure 15.2, rudimentary malware gets installed with all logic (e.g. a trigger condition and an attack strategy) hard-coded into the software. Strategies are often generic, such that the success (or efficiency) of the attack remains unknown until its execution (i.e. there is a high chance that the encoded attack strategy is benign in the context of the target system).

Targeted attacks. Common targeted attacks, labeled *human-driven ATAs* in Figure 15.2, establish a remote session between the attacker and the victim system. Utilizing the remote connection and the data available within the system (e.g. system files and sensor data), attackers gather information specific to the victim system in order to verify the target (*what to attack*), design a

---

2 The number of days an attacker is present on a victim network from the first evidence of compromise (Fire Eye Mandiant Services (2019a)).

custom attack payload (*how to attack*), and trigger the attack at the most opportune time (*when to attack*). Iterating through the escalation, reconnaissance, and lateral movements (i.e. the attack stages introduced in Figure 15.1), attackers customize the attack to take into account the specific characteristics of the targeted system.

AI-driven ATAs. In AI-driven ATAs, machine learning models and methods act as human attackers in determining the three instances of attack payload (i.e. what to attack, when to attack, and how to attack). As a result, an attacker can advance threats with more flexibility without needing to keep a foothold in the system. More specifically, upon access to the relevant data, self-learning malware can (i) confirm the attack target, (ii) determine the most opportune time/instance at which to trigger the attack payload, and (iii) infer an attack strategy that is hard to distinguish from accidental (or random) failure.

15.2.3.1 What to Attack: Confirming a Target System

Simple malware either automatically propagates across random targets or is manually installed on the target by the attacker. In targeted attacks, attackers analyze the system to verify the target (upon establishing a foothold) or embed logic that searches for signatures that are specific to the target. For instance, by verifying its signature Langner (2011), the Stuxnet attack limited the execution of the payload to the controllers at the Natanz uranium enrichment plant.

A recent study has demonstrated an approach that embeds a neural network model into malware to verify the target (Kirat et al. 2018). The malware is trained offline (i.e. before delivery to and installation on the target system), with data representing a (group of) target(s) with visual, audio, geo-location, and system-level features. Once installed on the victim system, malware evaluates the embedded neural network model using sensor data available in the system (e.g. a face capture from the web camera). If the model predicts that the victim is likely to be the intended target, malware triggers the execution of the payload (i.e. ransomware, which encrypts the files on the victim machine and keeps them unavailable until the user pays the ransom). In addition to its usage for detecting the target, the machine learning model also acts as an encoder of the trigger condition, which makes it hard for defenders to reverse-engineer the malware to understand the intent of the attacker, and delays the response.

15.2.3.2 When to Attack: Determining the Time to Trigger an Attack Payload

The majority of targeted attacks are detected when the attack payload gets executed (i.e. when the actual damage occurs in the system). As a result, it is critical for the malicious logic to trigger the execution only at the most opportune time, so as to maximize the negative impact on the target system. Simple malware used to rely on conditions (e.g. counters or clocks) to specify an instance to trigger the payload. For sophisticated threats, in which attackers gain control over the target system, the opportune time to trigger the attack is determined by analyzing the data available in the target system (Antonakakis et al. 2017).

Instead of relying on remote control, attackers can embed ML logic in malware to mimic the reconnaissance that might have been performed by the attackers. In Chung et al. (2019b), we demonstrated a learning malware-driven attack on a surgical robot (Raven-II). By exploiting vulnerabilities of the underlying software framework (i.e. ROS,[3]), the malware intercepts the data packets (i.e. reads sensor data exchanged using the publish-subscribe protocol), overwrites the control packets with faulty data, and eventually corrupts the operation of the robotic application.

In Chung et al. (2019b), we have demonstrated the feasibility of AI-driven malware that exploits real vulnerabilities to infer the opportune time to trigger an attack. Despite having limited prior

---

3 The Robot Operating System (ROS) is an open-source framework for programming robotic applications. It is noted for its support for various device drivers, libraries, and communication interfaces.

knowledge on the target robotic application, the learning algorithm was shown to be intuitive, and therefore effective, in reducing the chances of exposing malicious intentions. While that study was conducted in the context of the Raven-II surgical robot, the threat model can be generalized to an extended range of robotic applications or computing infrastructure.

15.2.3.3 How to Attack: Devising the Attack Payload

The success of malware depends on the attack strategy. With more monitors and detectors deployed to secure systems, attackers are challenged by increased risk of exposure/detection.

To maximize the impact and minimize the risk of detection, malicious actors spend a significant amount of time in the target system. While keeping a foothold, the attackers put significant effort into customizing the attack payload so as to gradually achieve their goals (e.g. physical damage to the system; Lee et al. (2016) or exfiltration of massive amounts of data).

Attackers in these targeted attacks manually collect internal information regarding the target system and devise their attack strategies accordingly. However, as demonstrated in Chung et al. (2019a), attackers can take advantage of machine learning algorithms to replace the manual process of designing the attack payload. The attack model in Chung et al. (2019a) is designed to compromise the availability of a large computing infrastructure (i.e. the Blue Waters supercomputer) by corrupting the cyber-physical system that regulates the chilled water that removes heat in the computing infrastructure. As in the case of Stuxnet, the threat model conceals the significance of the attack by disguising the attack as an accidental failure. Furthermore, instead of requiring an attacker to design an attack payload from scratch and hard-code severe events into malware, the threat model infers failure-causing abnormal events from operational data and dynamically injects such events when the attack is triggered.

The learning malware in Chung et al. (2019a) reported seven attack strategies found through analysis of operational data collected for eight months, and three strategies turned out to be effective (i.e. corrupted the chilled water delivery) when executed in a simulated environment. Unlike the traditional approach in which the attacker had to manually investigate the system and tune the attack, the AI-driven malware predicts anomalies in the CPS that are likely to be related to failures in the computing infrastructure. As the method uses unsupervised learning algorithms, it requires minimal knowledge regarding the system, hence, removing the need for a foothold to the system for reconnaissance. Lastly, as the attack strategies are dynamically derived from the environment that the malware observes, the malware can be used in other systems as long as they use a CPS from the same vendor.

15.2.4 Assumptions

In demonstrating AI-driven malware, we have made several assumptions that define the scope of this work. We do not demonstrate the full sequence of actions taken for the initial compromise into the system, but we assume that (a) the target system has vulnerabilities that allow intrusion (unauthorized access to the system). In addition, we assume that (b) the attacker has access to the resources to weaponize the malware. For example, user manuals and software APIs can be helpful in crafting malware. In terms of attack payloads, (c) we assume that fault models (to be used as attack payload) can be derived by exploring the operational/internal data of the target system or analyzing the target application. Upon intrusion into the target system, we assume that (d) the attacker can access the data that contain actionable intelligence for the malware to probe the system. Also, our assumption is that (e) the system administrator cannot detect the malware before impact (e.g. service unavailable or loss of data). Lastly, we assume that (f) accidental failures are less likely to trigger a postmortem investigation that will result in exposing the malware (or attacker).

Our set of assumptions might seem strong. However, as we compare our assumptions with successful attacks (i.e. Stuxnet that targeted the Iranian nuclear program (*Stuxnet*), the Target data breach (*Target*), and the cyberattack on the Ukraine power grid (*Ukraine*)), these assumptions prove to be quite realistic.

15.3 Inference of When to Attack: The Case of Attacking a Surgical Robot

For threats that use self-learning malware, robotic applications turn out to be a fascinating target. Like other cyber-physical systems (CPSes), robotic applications incorporate sensors and actuators that are connected through a network that passes around data. Their (i) *relatively weak security* (Bonaci et al. 2015; Morante et al. 2015; DeMarinis et al. 2018), (ii) *abundance of data that can be used to infer actionable intelligence* (Cheng et al. 2015; Atat et al. 2018; Voas et al. 2018), and (iii) *close proximity to and direct interactions with humans* (such that a successful attack could have a life-threatening impact) (Department of Homeland Security 2019; Giles 2019) make robotic applications a tempting target for advanced threats.

In this section, we demonstrate the feasibility of a self-learning malware that can infer the *opportune time* (i.e. *when* in Section 15.2.3.2) to trigger the attack payload (or take control over the robot) from leaked data available to malicious actors, so that it can maximize the impact. While our attack model applies to any robotic system, we discuss our attack model in the context of the Raven-II surgical robot (Applied Dexterity Website) and its haptic feedback rendering algorithm as a target application. We refer the interested reader to our earlier publication (Chung et al. 2019b) for more details.

15.3.1 Target System

Robots have been adopted across different application domains. For example, in manufacturing, robot manipulators assist human workers; drones are deployed in agriculture, entertainment, and military operations; and surgical robots support surgeons in performing medical procedures. For such applications, robots play a critical role. A robot's failure to make a correct and timely movement can lead to catastrophic consequences, such as injuring people near the robots in factories or risking a patient's life during surgery. This study focuses on the resiliency of a surgical robot against malicious attacks. We use the Raven-II surgical robot (Applied Dexterity Website) and its haptic rendering algorithm as an application to demonstrate the security threat and suggest methods to cope with the risk.

Robot Operating System (ROS). The Robot Operating System (ROS) is an open-source framework for programming robots (Quigley et al. 2009) and is commonly used by various robotic applications. According to its official website, ROS is widely deployed across more than 125 different robots, including mobile robots, drones, manipulators, and humanoids. The framework is being developed to support collaborative development by experts from different domains (e.g. computer vision or motion planning) and provides hardware abstraction, device drivers, libraries, and a communication interface (Quigley et al. 2009). As ROS provides the core underlying runtime environment, the security of ROS is critical in ensuring the correct operation of the robot.

Raven-II and haptic force feedback rendering engine. In this chapter, we study the resiliency of a ROS application in the context of a surgical robot (i.e. Raven-II) and its haptic feedback rendering engine. Leveraging the open-architecture surgical robot, the authors of Li and Kesavadas (2018) present a hardware-in-the-loop simulator for training surgeons in telerobotic surgery. The simulator, in addition to having all the features of the Raven surgical robot, introduces a novel algorithm to provide haptic feedback to the operator and, hence, offer a touch sensation to surgeons. The traditional approach for haptic feedback uses physical force sensors to determine the force applied to the robot. Since the instruments (on which the force sensors are installed) are disposable, that approach turns out to be costly. Instead, the authors of Li and Kesavadas (2018) proposed an indirect haptic feedback rendering approach that does not rely on force sensor measurement but instead uses image sensor data to derive the force feedback.

In our study, the haptic feedback rendering algorithm, as implemented in the augmented Raven-II simulator, utilizes information from a depth map (a matrix of distances from the image sensor to each pixel of a hard surface) to derive the distance from the object (e.g. a patient's tissues) to the robot arm. The algorithm, using the current position of the arm and the measured distance to the object, returns an interactive force value that generates resistance in the haptic device.

Figure 15.3a provides an overview of the software architecture of Raven-II and its haptic feedback rendering algorithm. The haptic algorithm takes input from the Kinect image sensor and the OMNI haptic device to control the Raven-II robot (or its virtual representation in *RVIZ*). The `kinect-Sensor` node parses the image data (as BGR and depth) from the Kinect image sensor, packages the data into ROS messages, and publishes the messages to the ROS core as topics (*kinect/BGR* and *kinect/Depth*, represented as an envelope in Figure 15.3a). The `omni_client` node is connected to the haptic device for user input. The `omni_client` node shares the processed operator input as a topic. A set of nodes, dedicated to running the algorithm, subscribe to the topics from the ROS core and derive the force feedback, which the `omni_client` sends to the haptic device.

The *kinect/Depth* topic from `kinectSensor` is used to derive the distance from the robot arm to the object. However, in deriving the distance, the algorithm needs a reference frame. It leverages the ArUco library (Romero-Ramirez et al. 2018), an OpenSource library commonly used for camera pose estimation. With the ArUco marker (i.e. a squared marker) location fixed and used as a reference point, we can derive the location of the robot arm(s) relative to the marker. The algorithm, using that information, can derive the distance from the robot arm to the object using (i) the transformation from the marker to the robot, (ii) the transformation from the image sensor to the marker, and (iii) the transformation from the image sensor to the object. Because the transformation from the robot arm to the object is evaluated in near-real-time, the algorithm can provide timely haptic force feedback to the OMNI device.

15.3.2 Attack Model

15.3.2.1 Attack Preparation

To deploy the attack, the first step is to identify machines that are running ROS as a master (core) node. Using a network scanning tool, we scan for the default port for ROS masters (i.e. TCP port `11311`) (DeMarinis et al. 2018). Once the master and its IP address are known, we set up ROS on our machine (which mimics a remote attacker) and update the ROS master's Uniform Resource Identifier (URI) variable to that of the identified master. Using the ROS APIs, we search for the topics of interest (i.e. the topics registered to the ROS master are used as a signature for identifying the ROS application).

15.3.2.2 Attack Strategy: ROS-specific MITM

In corrupting a ROS application, we take advantage of the vulnerabilities in ROS and execute a ROS-specific man-in-the-middle attack. As described in Section 15.3.1, ROS provides a set of interfaces (*publish/subscribe*) that ROS nodes can use to communicate; the ROS core serves as the

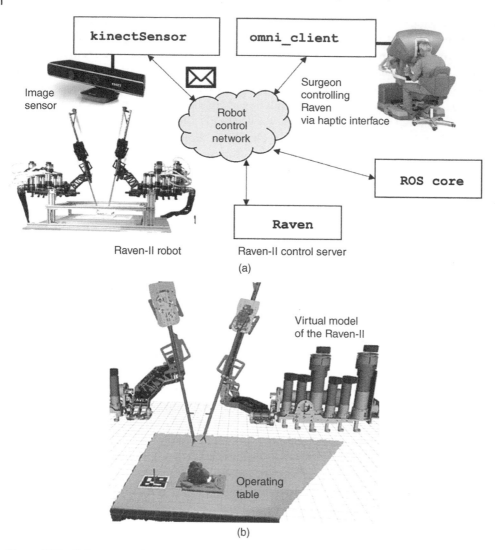

Figure 15.3 Software architecture of Raven-II (a) and its virtual representation in RVIZ (b).

arbitrator. After configuring the ROS setup to connect to the victim ROS master (attack preparation; see Section 15.3.2.1), our malware can initiate a subscriber that eavesdrops on the network communications. To take control of the robot, it kicks out a genuine node and publishes malicious data while masquerading as the original publisher. Without noticing the change in the publisher, the robotic application takes the malicious data as an input (or command) and updates the state of the robot accordingly.

15.3.2.3 Trigger: Inference of Critical Time to Initiate the Malicious Payload

Most security attacks are detected when the attack payload is executed (Sharma et al. 2011). Once detected, the attacker (or the malware that the attacker had installed) is removed from the system. Consequently, in many cases, the attacker may have one chance to execute the payload before being detected. As a result, it is realistic to consider the case in which an attacker tries to identify the ideal time to execute the attack payload (in our case, to inject a fault) in order to maximize the chances of success. A common approach is to embed a trigger function into the malware, which checks for a condition and executes the payload only when the condition is satisfied.

In this study, we present an approach that leverages a well-studied learning technique to infer the *critical time* to trigger the attack payload, so as to maximize the impact.

Inference of object location. During a surgical operation, the robot usually moves within a limited range defined by the nature of the surgical procedure. Hence, the precision in identifying "the time when the robot is touching (or maneuvering close to) the target object" can help in triggering the attack at the most opportune time so as to maximize the impact. For instance, when the robot is moving from its idle location to the patient on the operating table, the robot is operating in an open space without obstacles. On the other hand, when the robot is inside the abdomen of the patient, it is operating in limited space packed with obstacles (e.g. organs) and with blind spots that the image sensor cannot monitor. In that situation, correct operation of the rendering algorithm is critical. Also, the shorter the distance from the robot (at the point of the trigger) to the target object, the less time it takes the surgeon to respond[4] upon discovering the failure of the rendering algorithm (which can be determined only by noticing the lack of force feedback when a surface is touched). In this section, *we analyze the spatial density of the robot end-effector position* throughout the operation *to infer a time* when the robot (i.e. the surgical instrument) is near the object.

Algorithm. We use unsupervised machine learning to determine the location of the target object with respect to the position of the robot's end effector(s). Specifically, we adopted the density-based spatial-clustering algorithm with noise (DBSCAN) (Ester et al. 1996) to accomplish this task.

The DBSCAN algorithm takes two parameters, ϵ and *numMinPonts*. The maximum distance parameter (ϵ) defines the maximum distance between neighboring data points. Iterating over all data points, the algorithm checks for neighbors whose distance from a data point is less than ϵ. If the number of neighbors is less than *numMinPoints*, the data point is considered noise (i.e. the data point is not part of a cluster). Otherwise, the algorithm checks whether the neighbors form a cluster. Any clusters already formed by a neighbor are merged into the current cluster. We consider points (corresponding to the robot's end effector position) within 1 cm of each other to be "close," and define $\epsilon = 10$.

15.3.2.4 Attack Payload: Fault Injection

While the attack strategy in Section 15.3.2.2 is generic to the ROS framework, the payload is specific to the ROS application under study (i.e. Raven-II and its haptic feedback rendering algorithm). As part of assessing the resiliency of the haptic feedback rendering engine, we designed a set of faults that can be injected on-the-fly. The fault models are representative in mimicking realistic cases of (i) loss of information during transmission of data, (ii) data corruption, (iii) a glitch in sensors, and/or (iv) a bug in the software algorithm. None of the faults are specific to the environment (i.e. the faults are not affected by custom settings of the robot in a certain environment). Hence, understanding of the application (without needing to understand certain custom configurations) is sufficient for designing effective faults. In this chapter, we share our results in the context of a sample fault model. The description on the full set of faults and their results can be found in (Chung et al. 2019a).

Sample fault: Shifted depth map. This fault model considers a case in which an entity with malicious intent manipulates a ROS message to obfuscate the visual data provided to the operator. Just as we dropped the depth information from the image sensor in Fault 1, we can overwrite the depth map message with shifted values, causing the rendering algorithm to provide incorrect haptic feedback that the operator will rely on. In our experiment, we shifted the depth map of the object

---

4 Similar to the concept of braking distance when driving a car.

under operation by 50 pixels to the right. As the ROS message that contained the BGR information remained untouched, the 3D rendered image of the object was incomplete. Similarly, an attacker can shift the data in the BGR message and deliver the malicious visual image to the Raven operator.

15.3.3 Results

In this section, we present our results from inferring the time to trigger the attack payload and injecting realistic faults.

Determining attack triggers. We evaluate our accuracy in determining the robot's end effector position with respect to the target object. In Figure 15.4, we present the results of the clustering algorithm (based on DBSCAN) when a surgeon operates on three regions of the target object. Note that in Figure 15.4, an "x" indicates that the point is considered noise, and a circle indicates that the point belongs to a cluster. (Different colors are used to differentiate clusters.)

Impact of attacks on the Raven-II haptic feedback rendering algorithm. This section presents the impacts of executing the attack payloads. Executing the attack models (described in Section 15.3.2) in the experimental setup, we find that the attack payload (Section 15.3.2.4) effectively corrupts the rendering algorithm when injected through the MITM attack models. The shutdown of the (kinectSensor) node did not impact the operation of Raven-II, as the subscriber nodes were still receiving the messages (without acknowledging the malicious publisher).

In Figure 15.5, we show the impact of shifting depth map information during transmission of the information from the publisher to the subscriber. The figure shows that because of the shifted distance measure, the 3D rendering of the left half of the box is flattened (and indeed would be hard to differentiate from the surface, were it not for the colors), and the original surface on the right side of the box has gained volume. As shown in the figure, this fault model can lead to penetration of the object by the robot that would have been prevented by the non-corrupted image.

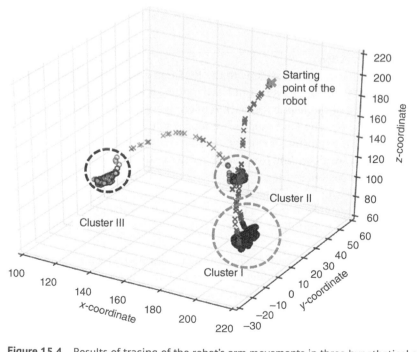

Figure 15.4 Results of tracing of the robot's arm movements in three hypothetical surgical procedures.

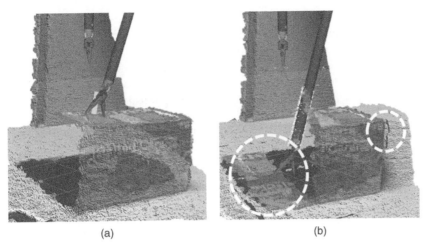

(a)   (b)

Figure 15.5 Simulated Raven operation with (a) uncorrupted depth map, and (b) shifted depth map. Note the problems inside the contents of dotted circles.

15.3.4 Attack timeline

Figure 15.6 depicts the software architecture of the Raven-II robot and how the malware operates. By scanning for ROS applications, the attacker identifies the target application connected to the public network (①). As shown in Figure 15.6, the malware does not necessarily need to be installed on a machine that is running the target application but can be on any machine that has access to the robot control network. If the robot control network is properly isolated from the public network, attackers would need to devise a method to cross the air gap (e.g. as was done in the case of Stuxnet). With access to the control network, malware identifies the *roscore*, which is the central unit that manages the overall operation of the ROS application (②). By registering itself to the roscore, the malware can subscribe to topics (data structures defined to deliver a specific context type) of interest (③). In ④, the learning malware analyzes the subscribed topic (i.e. the x, y, z coordinates of the

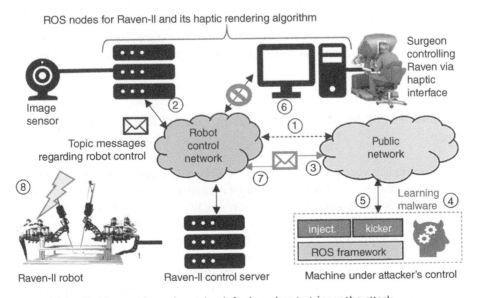

Figure 15.6 AI-driven malware in a robot inferring when to trigger the attack.

robot end point) to determine the most opportune time for an attack, which is a time when the robot is likely to be operating on the object (e.g. patient). The ML-based approach resulted in a reduction of the maximum distance (i.e. the maximum distance from the robot arm to the object at the time of an attack's triggering) by roughly 83%, hence increasing the chance of impact on the patient under operation. When the learning malware triggers the payload execution, a malicious ROS node registers itself to the roscore with the name of a genuine publisher (⑤). Because of the conflict in the namespace, in ⑥, roscore shuts down the genuine publisher (*Haptic node N* in the figure) and forwards the faulty topic messages generated by the malicious node (⑦). In ⑧, the operation of Raven-II gets corrupted by operating with faulty input (from the malicious node) and eventually puts the patient at risk.

15.3.5 Summary

In this section, we demonstrated that our prototype smart malware can utilize the leaked data to trigger an attack at the most opportune time so as to maximize the attack's impact. The inference of the opportune time range for the execution of the attack payload reduces the chances of exposure, which helps the malware disguise the attack as an accidental failure. (Recall that the faults were designed to represent accidental failures.) Without the smart triggering, frequent and likely unsuccessful injections of the fault could make the system administrator aware of the malicious intent behind the sequence of failures. To make things worse, our DBSCAN-based approach does not require extensive prior knowledge of the target robotic application. However, despite the smartness of the malware, our attack is limited in its payload. Unlike the fault in (Alemzadeh et al. 2016), which tackles the time gap between a safety check of the input and its execution (i.e. faults are injected after the safety check), our faults (or faulty scenarios) are injected through the corruption of the raw data, and that corruption might be detected by safety checks (if such checks were implemented as part of the robotic application).

While we demonstrated the feasibility of a smart malware attack in the context of the Raven-II surgical robot and its haptic feedback rendering algorithm, our threat model exploits vulnerabilities in the underlying framework (ROS). Hence, a robotic application running on the most common version of ROS, ROS 1, is vulnerable to MITM attacks and to smart malware that exploits the leaked data. While we leverage vulnerabilities in the ROS, as discussed in Alemzadeh (2016), various entry points exist through which malicious entities could intrude into robotic applications. By leveraging such vulnerabilities, our design for smart malware can be revised to target robotic applications in general.

15.4 Inference of How to Attack: The Case of Attacking a Building Control System

The growing sophistication of cyber-physical systems (CPSes), while enabling new breakthroughs in different application domains, also brings new vulnerabilities that can be exploited by malicious actors. More specifically, a weakness in the cyber security of CPSes and computing infrastructure (CI) can introduce an unforeseen threat. By utilizing the often underestimated dependency of computing infrastructure on surrounding CPSes, an adversary who targets a high-value computing infrastructure (e.g. a data center or supercomputer) can deploy an *indirect attack* by exploiting the relatively weak security of CPSes that are critical to the operation of the computing infrastructure. In this scenario, the vulnerability of the CPS acts as a weak point that lowers the security barrier of the otherwise well-protected system. As demonstrated in Chung et al. (2016), an indirect attack can further thwart detection by replaying known failure scenarios in the CPS such that their impact cascades to the target CI.

In this section, we discuss advances of cyber-attacks in the context of indirectness and automation (driven by self-learning abilities). More concisely, we present a type of smart (self-learning) malware that exploits the weaknesses of a CPS to induce attack strategies (i.e. *how* in Section 15.2.3.2) that (i) masquerades (to monitoring entities of the CPS) as an accidental failure, and (ii) stealthily propagates to the computing infrastructure, causing a major failure that includes a system-wide outage. To be specific, in the context of a real computing facility (i.e. Blue Waters), we present a learning malware logic that mimics failures or abnormal events in the cooling facility (i.e. the National Petascale Computing Facility) that eventually lead to an outage of the HPC housed in the facility. Building on the hypothetical attack scenario in Chung et al. (2016), we implement and evaluate variations of attack scenarios (leveraged by the degree of automation and learning). The most naive approach randomly pollutes the data segments of random packets, whereas the most advanced approach carefully crafts an attack strategy inferred from operational data. We refer the interested reader to our earlier publication (Chung et al. 2019a) for more details.

15.4.1 Target system

As shown in Figure 15.7, we consider a typical configuration of a large-scale computing facility instantiated in the context of the Blue Waters supercomputer. In this configuration, the operational environment (e.g. room temperature, delivery of chilled water to cool the computing nodes) of the computing infrastructure is managed by a building automation system (and associated CPS) that optimizes the control over the actuators, given a set of measurements representing the parameters of the operational environment.

15.4.1.1 Computing Infrastructure: Blue Waters
In our attack model, we set a computing infrastructure, or, to be more specific, its availability, as the final target. In this section, we study the case of Blue Waters, a 13.3-petaflop supercomputer managed by the National Center for Supercomputing Applications (NCSA) at the University of Illinois. Blue Waters is composed of around 300 computing cabinets, each of which houses around 90 nodes (i.e. 3000 16-core processors per cabinet, or 27 000 in total) (Di Martino et al. 2014). To prevent damage to the physical parts from the massive heat generated during the operation of the supercomputer, a liquid-based cooling system has been deployed. For Blue Waters, 72 cooling cabinets are placed alongside the compute nodes, and each cabinet prepares and delivers Freon to the three adjacent compute nodes. The liquid form of Freon, traveling through the rack of compute

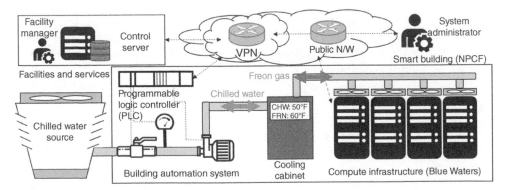

Figure 15.7 Chilled water system overview.

nodes, then absorbs the heat from the compute nodes and returns to the cooling cabinet in the form of a gas.

15.4.1.2 Cyber-physical System: NPCF Building Automation System

The Blue Waters supercomputer is housed in a dedicated building, the National Petascale Computing Facility (NPCF). The 88 000-square-foot building uses the state-of-the-art building automation system (BAS) that is in charge of regulating the environmental parameters (i.e. pressure, flow, and source of the cooling system) of the building, including the server room. A detailed configuration of the system can be found in Chung et al. (2016). The building automation system (which consists of a control server, a set of programmable logic controllers – PLCs, sensors, and actuators) utilizes a set of measurements collected from the chilled water loop to regulate the chilled water delivered to the cooling cabinet, under three modes of operation: campus mode, mix mode, and economic (econ) mode. *Campus mode* is the used mode in most data centers that use chilled water bought from external providers. While the chilled water from such providers is well controlled (i.e. the temperature, flows, and pressure are kept within an agreed range), its usage results in an increase in the cost of operation. To reduce the cost of operation, NPCF has a set of dedicated cooling towers (which use cold temperatures to naturally chill the water) by means of which it can deploy an additional mode of operation: *economic* mode. By taking advantage of cold external temperatures throughout 2/3 of the year to prepare the chilled water, NPCF was able to significantly reduce its cost of operation, which compensated for the construction costs of the water towers after one year of operation. The *mix* mode is an intermediate mode that was introduced to enable a smooth transition between economic and campus modes.

15.4.1.3 Data

The CPS operational data set is an archive of all measurements and control command values within the chilled water system of the building facility that was collected from September 2016 to May 2017. The data set contains 47 distinct parameters collected every five minutes. Sample parameters monitored and collected within the chilled water system include differential pressure, flow, and temperature of the campus input ("CAMP.CHW.DP, FLOW, TEMP"), control valve setting and measurements at the high loop ("CHW.HI.CV, TEMP, FLOW"). This data set is essential for analyzing the operation of the CPS and inferring critical information related to failures of the computing infrastructure.

   The incident reports log incidents related to the computing infrastructure since the deployment of the system (i.e. December 2012). The incidents recorded in the reports include hardware part failures, cooling-system-related problems, and system-wide outages (SWOs) during which all 28,164 compute nodes were shut down. The incident reports are used for validating our approach by cross-validating the attack strategies derived by our smart malware (i.e. subset of the data set that the malware predicted as "related to an SWO of BWs") with the ground truth in a given report (i.e. the status of BWs at the during the timeline of the parsed data set and its cause). In Table 15.1, we present a sample of the incident report.

15.4.2 Attack Model

In this section, in terms of a set of steps (as depicted in Figure 15.8), we describe our approach to developing self-learning malware that is capable of inducing malicious attacks against a CPS that controls the environment in which the computing infrastructure operates. Each procedure is described in terms of its goal, approach, and outcome. Note that the objective of the statistical analysis performed in a procedure is not to perform accurate labeling and identification of key

Table 15.1 Sample Blue Waters incident reports.

Failure cause	Description	Datetime	Node hours down	...
Facility water	Facility water valve replacement...	mm/dd/yyyy	$n^{a)} \times x^{b)}$	...
Water valve actuator	XDP x-y Motor actuator failure	...	...	...

a) Number of nodes affected by the incident.
b) Hours that the system was under impact.

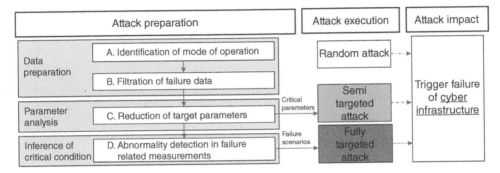

Figure 15.8 Approach and attack modes.

parameters, but to parse information without prior knowledge about the system. Hence, most of our discussions on the results of each procedure do not present the usual performance analysis (e.g. accuracy, true/false positive rate) but instead evaluate the ability of each procedure to achieve its primary goal.

Identification of mode of operation. A BAS relies on multiple input sources to deploy different modes of operation. The control logic and the measurements differ depending on the mode of operation, and hence, it is critical to accurately classify the data by the mode of operation to infer characteristics specific to each mode. We apply (i.e. our malware runs) k-means clustering (Hartigan and Wong 1979) to classify the data by the mode of operation. With the goal of interrupting the cooling capacity, we choose the two parameters that define the cooling capacity delivered to the server room: chilled water flow (*chw_flow*) and temperature difference (*tempDiff*) between the supply and return water. The clustering algorithm, with a k value of two, effectively captures the data clusters (campus and econ mode, colored black and white, respectively, in Figure 15.9). When compared to the ground truth, we achieved a 97.96% true-positive rate (0.04% false-positive rate) for *campus mode*, a 99.76% true-positive rate for *econ mode*, and 98.62% overall accuracy.

Filtration of failure data. To design attacks that masquerade as accidental failures, the primary step is to parse the CPS data that are related to system failures. However, the identification of failures in the computing infrastructure becomes a nontrivial problem, as the information available to the attacker is limited. A feasible approach is to infer the system status from the *returning chilled water temperature*, the only parameter within the measurements that reflects the status of the computing infrastructure of interest. In Figure 15.10, we plot the supply water temperature to the data center (*"CHW.R.HI.CHRS.TEMP"*) and the return temperature from the data center (*"CHW.R.HI.CHWR.TEMP"*) during a month of operation. What we focus on is the occasional nadirs (drops) among the relatively constant (compared to the supply temperature) return temperature. The nadirs indicate events in which the computing infrastructure did not generate heat, which happens only when the computing infrastructure is down. Analysis of CPS operational data

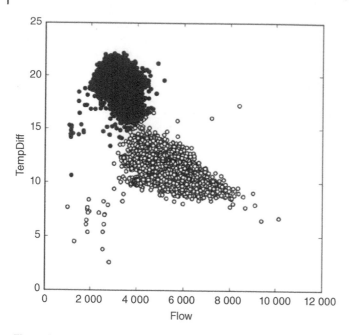

Figure 15.9 K-means clustering for two clusters.

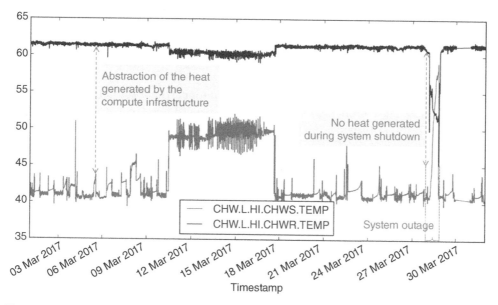

Figure 15.10 SWO observed through chilled water supply and return temperature measurements.

over nine months shows that the nadir detection approach reduces the data of interest from 72.5 K entries (i.e. measurement points) to 2938 entries (a 96% reduction). The 2938 entries can be coalesced into 23 distinct events. By cross-validating those data with the incident reports (which are not available to external entities or attackers) that manually log the failure incidents of Blue Waters, we found that our prediction model captured three out of five system-wide outages that occurred during the period of study. The two SWOs that were not captured with our approach occurred when an outage did not last long enough for its impact to be visible at the chilled water system level.

Reduction of target parameters. Within the control system under study, measurements of 47 parameters are monitored and shared among the controllers. While each and every parameter plays a role in the control logic, not all parameters are closely related to the goal of the attacker, i.e. disrupting the delivery of chilled water to the computing infrastructure. While malware can be designed in a naive manner to pollute random parameters, its likelihood of success would be extremely low despite the high likelihood of exposure (or detection). Instead, we apply a statistical method to carefully choose the target parameters to be altered by our malware. The attacker's goal is to reduce the cooling capacity of the chilled water delivered to the server room, and hence the attacker needs to monitor the final target parameters: the temperature and volumetric flow of the supplied chilled water. Given a series of measurements of the 47 parameters, the malware computes the correlation between each pair of parameters, where a value closer to 1 (or −1) represents the significance of a linear relationship between the two parameters. For each mode of operation, we choose the top 10 parameters with the highest correlation. Example parameters include control valve setting, pump speed, and input water temperature/flow and pressure. In addition to reducing the number of target parameters, our correlation-based approach captures the relationships among the critical parameters to minimize contextual inconsistency during false-data injection. The reduced data set consists of 23 distinct events with 10 critical parameters each. In the next step, we filter out events with no abnormal values of critical parameters.

Filtration of events with abnormal measurements. In this step, we reduce the number of events and **its** critical parameters by selecting those events for which the critical parameters show abnormal values. (We consider those events to be the root cause of the CPS-induced CI outages.) As part of defining an abnormal event for a parameter, we assume that the parameter values can be represented as a Gaussian distribution. Given a Gaussian distribution, measurements with z-scores greater than 3 are considered as outliers. For a stricter bound (to reduce false-positives), we consider a measurement to be abnormal if its z-score exceeds 4. (In a perfect Gaussian distribution, 99.9936% of the measurements have z-scores smaller than 4.) To be specific, for each parameter in the 23 events, we check if there is an abnormal measurement that exceeds a threshold (whose $z = 4$). We remove the parameter from the list of critical parameters if no abnormal measurement/value is detected for this parameter. Also, if all the critical parameters are removed for an event, then we consider that event as "inappropriate for an attack strategy." After running the abnormality analysis over each critical parameter for all events, we are left with a set of events to be implemented as part of an attack strategy.

When we apply (or the malware applies) the abnormality detection method to each candidate parameter of the 23 events, we end up with seven events that each include at least one abnormal measurement (summarized in Table 15.2)

15.4.3 Results

After the malware infers the strategies to be implemented as attacks, the malware identifies the controller (PLC) that uses the critical parameters as inputs or derives the values for the critical parameters as outputs. For the identified PLC, the malware modifies the PLC program (i) to check for attack-triggering conditions and (ii) to overwrite the normal values of the critical parameters with the sequence of abnormal values defined by the attack strategy. The malware completes the preparation of the attack by uploading the modified program to the PLC.

In the remainder of this section, we discuss a sample attack strategy and its impact when executed in a simulated environment in the context of (i) how the actual incident occurred (i.e. the ground truth) and (ii) what we expect to happen when the scenario is implemented as an attack.

Table 15.2 Attack strategies inferred by the fully targeted attack

Anomaly	Recorded incident	Impact on CI
Valve 1B flow	Reduced cooling capacity for emergency outage	Scheduled outage
Return temp.	(Change in mode of operation)[a]	
Return temp.	(Less heat absorbed from CI due to scheduled maintenance)[a]	
Return temp.	(Change in mode of operation)[a]	
Return temp.	(Change in mode of operation)[a]	
Supply temp.	Supply water anomaly due to power interrupt (Figure 15.11)	SWO for 5 hrs
Pump 1 speed	Chilled water loop maintenance	Scheduled outage

a) Inferred strategy that was not related to any incident in the compute infrastructure.

Incident. As shown in Figure 15.11, an abnormality occurred in the temperature of the input chilled water from the campus chilled water plant. Specifically, the input temperature increased by 10°F above the usual value (from 40°F to 50°F) and remained high for 30 minutes. By relating this observation to the corresponding incident report, we found that a power interrupt at the campus chilled water plant had impacted the computing facility. For this incident, the impact did not cascade into the computing infrastructure, because the incident occurred when computing infrastructure was not being heavily used (i.e. the computing infrastructure generated less heat and did not require significant cooling), but the consequences would have been different if the incident had happened during peak hours.

Attack. When this strategy is implemented and triggered as an attack, the malicious PLC logic on the controller overwrites the water temperature measurements as if the temperature had increased and forces the CPS to respond to the apparent abnormality. The CPS, in response to the "abnormality", increases the water flow to compensate for the (fake) loss of cooling capacity, which leads to excess cooling capacity to the CI cooling cabinets in the data center (i.e. a violation of the service-level agreement between the facility and the computing infrastructure). To avoid reaching the dew point, the CI cooling cabinets close the valve to slow down the flow. The attack becomes effective when the parameter values go back to normal. Once the (fake) temperature returns to normal, the CPS starts to close the valves, which reduces the flow of the chilled water to the CI. However, since the CI cooling cabinets do not react in a timely manner (Chung et al. 2016), the CI encounters a shortage in cooling capacity that increases the risk of an EPO.

15.4.4 Attack Timeline

Figure 15.12 shows how the malware progresses. Following the usual sequence of states of ATAs shown in Figure 15.1, attackers perform initial reconnaissance to identify vulnerabilities to exploit so as to gain access to the system and deliver the weaponized malware (①). In ②, a successful preliminary attack leaks credentials, which grant access to the control system/network of the target CPS and install the learning malware (③). Once installed in the system, the learning malware accesses the data archive that keeps the CPS operational data and executes the sequence of learning algorithms (④). The malware classifies measurements that are likely to be related to outages of the target computing infrastructure, clusters data that have similar operational characteristics, and identifies parameters that are critical in regulating the chilled water. The learning malware outputs a set of attack strategies that mimic anomalies in the CPS that could have caused outages in Blue

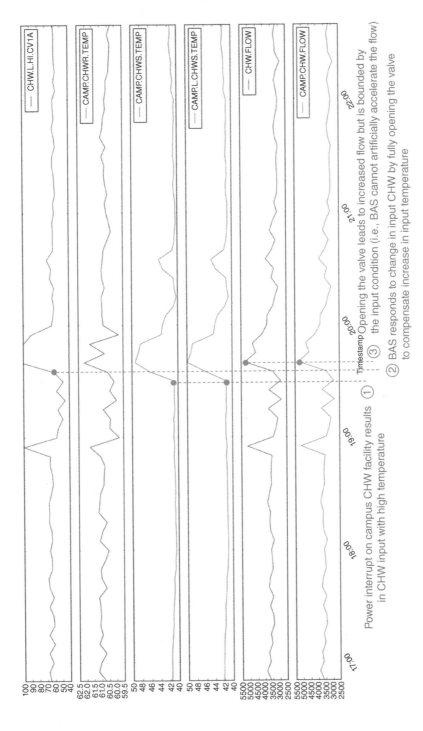

Figure 15.11 Sample attack strategy: supply water temperature abnormality due to power interruption.

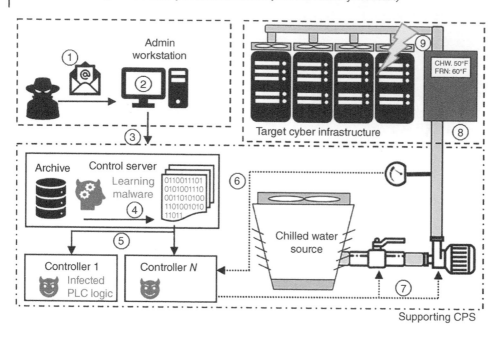

Figure 15.12 AI-driven malware in a CPS inferring how to attack the target.

Waters. The strategies are embedded into the controller logic with a triggering condition, which overwrites the genuine controller logic (⑤). The infected controller logic monitors the critical parameters and checks the triggering condition (⑥). When the condition is satisfied, the infected controller logic overwrites the critical parameter values with the strategic values (from ④) to initiate a failure-triggering abnormal event (⑦). The CPS tries to regulate the chilled water in response to the abnormal event, but, as in the failure incidents, the response is likely to be unsatisfactory (⑧), such that the impact cascades to the CI, triggering an outage (⑨).

15.4.5 Summary

We discussed our malware in the context of the CPS that manages the cooling system of Blue Waters. However, we believe that the same malware (with little or no modification) can be used in other computing facilities (HPCs, data centers, or the cloud) with dedicated cooling systems, as long as they use a CPS from the same vendor. Such a generalization is feasible because the attack strategies are not hard-coded into the malware, but dynamically derived from the environment that the malware observes. If the CPS were from a different vendor, the malware would need to be modified/adapted to account for different ways of programming and uploading a control logic to the controller(s). The rest of the procedures that constitute the presented self-learning malware (i.e. steps A through D, as depicted in Figure 15.8) should remain the same.

Despite the success in inferring actionable intelligence from data, a number of limitations in self-learning malware hint on possible mitigation methods. For instance, the success of the self-learning malware depends on the existence of CPS failures that lead to outages of the computing infrastructure. The events that would benefit the attacker (or malware) are rare (at NPCF, 2.73% of the 5419 incidents from 2013 to 2016 were cooling-system-related (Chung et al. 2016)), and there is no guarantee that the malware can capture all such events. Further, although we focused on indirect (via CPS) attacks against the computing infrastructure, the ultimate success of an attack depends on the workload on the computing infrastructure, which is beyond the control of an attacker. For instance, in the case of cooling system failure, only compute cabinets with high computational load (during the incident) can trigger an EPO.

15.5 Protection from Rising Threats

Examples discussed in this paper demonstrate that AI-driven malware is no longer a remote possibility. As attacker capabilities and attack's sophistication grow, cyber defenders must understand mechanisms and implications of the malicious use of AI to: (i) stay ahead of these threats and (ii) devise and deploy defenses to prevent: data being stolen, systems damaged, or major disruption of critical infrastructure (e.g. electric power grid). AI-driven malware is a relatively new concept, and not everyone considers it practical. The authors of G DATA Software (2019) acknowledge the contribution of Kirat et al. (2018) but indicates that they do not consider the model a real threat. According to the article, such an attack model can be dealt with using existing behavior-based detection methods and signatures that detect usage of certain libraries (i.e. ML libraries) or access to specific files. However, the behavior-based (blacklist) model only applies for known malicious payloads (e.g. the ransomware in Kirat et al. (2018)), so ML-driven threats with non-obvious (hard to differentiate from benign) payloads might not be detectable. In addition, adversarial methods have been reported that bypass signature-based detection methods. For instance, self-learning malware can utilize custom software packages and deploy obfuscation methods to encumber code analysis. In addition to existing cyber security defense methodologies, we find a need to consider additional methods that tackle the new threat (i.e. self-learning malware) and a need for computing infrastructure administrators to develop more comprehensive security awareness.

Prevention. Risk of AI-driven threats can be reduced through fine control and management of data access. Without proper data, only limited intelligence can be inferred, regardless of the effectiveness of the learning model. However, proper management of data access is not a trivial task. New vulnerabilities for open-source or off-the-shelf software packages are reported every day, and defining a breach-prone policy requires modeling of varying use cases and system specifics. In response to vulnerabilities, security patches and version updates become available to users. Nonetheless, there are situations in which such updates are not easily applicable. For instance, although the vulnerabilities in ROS have been known for years, they cannot be removed from the framework without a face-lifting upgrade of ROS, which would require a complete reprogramming of the robotic applications (Chung et al. 2019b). Instead, the authors in (Chung et al. 2019b) (as a mitigation method against AI-driven malware) identify a unique signature that indicates malicious attempts that exploit the vulnerabilities and propose proactive responses to the threat through blocking of the source of the attempt and returning of the robot to a predefined safe state. Deployment of similar monitors/systems can prevent ML-driven advanced threats from making unauthorized data access.

Encryption is a common method deployed in production systems to prevent unauthorized access to the system and its data. However, there are some limitations that keep encryption from becoming the golden key solution to security. In particular, the latency introduced by the encryption/decryption scheme can be crucial for real-time critical systems (such as the Raven-II surgical robot). For instance, referring to a recent measurement with the ROS2 framework (Kim et al. 2018), the deployment of the cryptographic algorithm added a non-constant delay that becomes significant as the size of the data packet increases. Furthermore, due to the number of sequential message passing for a single robot iteration, the latency builds up. In fact, not all systems are time-sensitive and allow the deployment of cryptographic algorithms. However, there are instances within the system where the data are processed in plain text. Hence, with the additional constraints, self-learning malware (instead of intercepting packets in the network layer) only needs to target that particular instance while keeping the current procedure.

Detection. The rising threat of ML-driven malware motivates the deployment of ML technology for mitigation. As discussed in Chung et al. (2019a, b), attackers (and self-learning malware) lack ground truth regarding the data (e.g. system status) and, hence, can only deploy unsupervised learning algorithms. On the other hand, taking advantage of their knowledge of computing systems to be protected (e.g. system configuration or differences between abnormal and benign events), system administrators can deploy monitors running supervised learning algorithms that are trained with the ground truth. Furthermore, system administrators can utilize data from various sources (e.g. different computing nodes or monitors auditing components of the system) and deploy multilayered security monitoring (Cao et al. 2015) that combines intelligence collected from multiple sources. Note that an attacker (if possible) would need to make lateral movements to access different data sources, and that is not a trivial task. Also, data analysis over a larger volume of data increases the computation workload in the victim system, which may expose potential malicious activities.

Aftermath. In addition to adding measures to secure computing infrastructure, the authors of Kirat et al. (2018) perceive a need for methods to reverse-engineer ML models. As presented in Kirat et al. (2018), supervised ML models (e.g. neural networks) can be trained with external data before being embedded in the model. Models trained with data can be considered as encoded logic whose interpretation requires significant effort. Hence, if attackers conceal their intentions in a learning model, current reverse-engineering methods would not be able to decode a network of nodes connected through weighted edges. To fully understand an attacker's intentions, AI-specific reverse-engineering techniques should be considered.

15.6 Related Work

Well-known advanced targeted attacks (ATAs). The discovery of Stuxnet in 2010 is often referred to as a turning point in the history of cybersecurity (Langner 2011). Stuxnet is a highly complicated malware and represents the first cyber-warfare weapon reported. The intelligence of the Stuxnet malware is on its attempt to disguise an attack as accidental failure. Instead of delivering an instant payload to the target (e.g. shutdown the system), Stuxnet resided in the system for a longer period, frequently changing the operational frequency of the hardware actuator to increase the chance of hardware failure. In 2015, a cyberattack on the Ukraine power grid successfully took down the power system (Lee et al. 2016). By carefully orchestrating the attack, the attackers kept foothold over the control system for six months to gather internal information regarding the system and design the attack. Only after ensuring significance consequences, the attackers sent a trigger command to the malware.

Attack models with support from AI. No public record reports the existence of AI-driven threats. However, a few researchers and engineers have investigated the possibility of attacks that use machine learning and AI against cyber security. In Petro and Morris (2017), the authors presented DeepHack, an open-source hacking bot powered with AI technology. This bot, with minimal prior knowledge of applications and databases, learns attack strategies that exploit vulnerabilities specific to the target system. DeepLocker (Kirat et al. 2018), on the other hand, takes advantage of a deep neural network (DNN) for target detection/verification. The authors demonstrate the case where a Ransomware, spread across random systems through various exploits, is executed only on a specific set of targets. A detailed survey of AI-based cyber threats can be found in Kaloudi and Li (2020).

Adversarial learning. While more applications adopt machine learning technologies, it introduces new vulnerabilities that did not exist. The decisions derived by machine learning algorithms rely on probability and the algorithm itself cannot reason on the results. A number of researchers have studied the case where attackers can exploit this unique vulnerability by providing adversarial examples into the learning module for training or evaluation (Huang et al. 2011; Papernot et al. 2017). In Sharif et al. (2016), the authors presented an attack model on facial biometric systems (widely used in surveillance and access control) using a systematic method to generate a pattern, which allows the attacker to evade recognition or impersonate another individual. A number of previous work have proposed ways to improve the robustness of learning models (Gu and Rigazio 2014; Metzen et al. 2017; Lu et al. 2017a). Furthermore, in some applications (as demonstrated in Lu et al. 2017b), the physical constraints of the problem allow the system to overcome adversarial examples, preventing the temporal misclassification from cascading into critical decisions made by the application software.

15.7 The Future

In this chapter, we discussed the state-of-the-art of the potential advanced cyber threats that are empowered by machine learning algorithms. No official records report the real-world existence of such a threat, but, in fact, it might be only a matter of time before one emerges, if it is not already active and simply remaining undetected. Our example applications demonstrate that AI-driven malware represents a real challenge, but with its own limitation. While the structure of the malware can be generalized in terms of functional components, the implementation must adapt to the target specifics. For example, a malware designed to target a system from a specific vendor may not be applicable to another system with different protocols or function APIs. As a result, the attacker must change the interface in implementing the malware while the core of the attack (which is the most time-consuming part in designing malware) remains the same.

References

Homa Alemzadeh. *Data-driven resiliency assessment of medical cyber-physical systems*. PhD thesis, University of Illinois at Urbana-Champaign, 2016.

Homa Alemzadeh, Daniel Chen, Xiao Li, Thenkurussi Kesavadas, Zbigniew T. Kalbarczyk, and Ravishankar K. Iyer. Targeted attacks on teleoperated surgical robots: Dynamic model-based detection and mitigation. In *Proc. of the IEEE Intl. Conf. on Dependable Systems and Networks, Toulouse, France, 28 June–1 July 2016*, pp. 395–406, 2016.

Manos Antonakakis, Tim April, Michael Bailey, Matt Bernhard, Elie Bursztein, Jaime Cochran, Zakir Durumeric, J. Alex Halderman, Luca Invernizzi, Michalis Kallitsis, Deepak Kumar, Chaz Lever, Zane Ma, Joshua Mason, Damian Menscher, Chad Seaman, Nick Sullivan, Kurt Thomas, and Yi Zhou. Understanding the Mirai Botnet. In *Proc. of the USENIX Security Symp., Vancouver, BC, Canada, 16–18 August 2017*, pp. 1093–1110. USENIX Association, 2017.

Applied Dexterity Website. Applied dexterity—driving innovation in surgical robotics, 2019. http://applieddexterity.com (accessed 1 May 2019).

Rachad Atat, Lingjia Liu, Jinsong Wu, Guangyu Li, Chunxuan Ye, and Yang Yi. Big data meet cyber-physical systems: A panoramic survey. *CoRR, abs/1810.12399*, 2018. http://arxiv.org/abs/1810.12399.

Tamara Bonaci, Jeffrey Herron, Tariq Yusuf, Junjie Yan, Tadayoshi Kohno, and Howard Jay Chizeck. To make a robot secure: An experimental analysis of cyber security threats against teleoperated surgical robots. *arXiv preprint arXiv:1504.04339*, 2015. https://arxiv.org/abs/1504.04339.

Phuong Cao, Eric Badger, Zbigniew Kalbarczyk, Ravishankar Iyer, and Adam Slagell. Preemptive intrusion detection: Theoretical framework and real-world measurements. In *Proc. of the Symp. and Bootcamp on the Science of Security, Urbana, IL, USA, 20–22 April 2015*, pp. 5, 2015.

Xiuzhen Cheng, Yunchuan Sun, Antonio Jara, Houbing Song, and Yingjie Tian. Big data and knowledge extraction for cyber-physical systems. *International Journal of Distributed Sensor Networks*, 11(9):1–4, 2015.

Keywhan Chung, Valerio Formicola, Zbigniew T. Kalbarczyk, Ravishankar K. Iyer, Alexander Withers, and Adam J. Slagell. Attacking supercomputers through targeted alteration of environmental control: A data driven case study. In *Intl. Conf. on Communications and Network Security, Philadelphia, PA, USA, 17–19 October 2016*, pp. 406–410, 2016.

Keywhan Chung, Zbigniew T. Kalbarczyk, and Ravishankar K. Iyer. Availability attacks on computing systems through alteration of environmental control: Smart malware approach. In *Proc. of the Intl. Conf. on Cyber-Physical Systems, Montreal, QC, Canada, 15–18 April 2019*, pp. 1–12, 2019a.

Keywhan Chung, Xiao Li, Peicheng Tang, Zeran Zhu, Zbigniew T. Kalbarczyk, Ravishankar K. Iyer, and Thenkurussi Kesavadas. Smart malware that uses leaked control data of robotic applications: The case of Raven-II surgical robots. In *Intl. Symp. on Research in Attacks, Intrusions and Defenses, Beijing, China, 23–25 September 2019*, pp. 337–351, 2019b.

Nicholas DeMarinis, Stefanie Tellex, Vasileios Kemerlis, George Konidaris, and Rodrigo Fonseca. Scanning the internet for ROS: A view of security in robotics research. *arXiv preprint arXiv:1808.03322*, 2018. https://arxiv.org/abs/1808.03322.

Department of Homeland Security. Cyber physical systems security, 2019. https://www.dhs.gov/science-and-technology/csd-cpssec.

Catello Di Martino, Zbigniew Kalbarczyk, Ravishankar K. Iyer, Fabio Baccanico, Joseph Fullop, and William Kramer. Lessons learned from the analysis of system failures at petascale: The case of blue waters. In *Proc. of the Annual Intl. Conf. on Dependable Systems and Networks, Atlanta, GA, USA, 23–26 June 2014*, pp. 610–621, 2014.

Martin Ester, Hans-Peter Kriegel, Jörg Sander, and Xiaowei Xu. A density-based algorithm for discovering clusters in large spatial databases with noise. In *Proc. of the Intl. Conf. on Knowledge Discovery and Data Mining, Portland, OR, USA, 2–4 August 1996*, pp. 226–231, 1996.

Fire Eye Mandiant Services. M-TRENDS 2019—special report, 2019a. https://content.fireeye.com/m-trends/rpt-m-trends-2019.

Fire Eye Mandiant Services. APT1: Exposing one of China's cyber espionage units, 2019b. https://www.fireeye.com/content/dam/fireeye-www/services/pdfs/mandiant-apt1-report.pdf.

G DATA Software. Deeplocker: Interesting, but not yet a threat, 2019. https://www.gdatasoftware.com/blog/2018/08/31012-deeplocker-ai-malware.

Martin Giles. Triton is the World's most murderous malware, and it's spreading. *MIT Technology Review*, 2019. https://www.technologyreview.com/s/613054/cybersecurity-critical-infrastructure-triton-malware/.

Shixiang Gu and Luca Rigazio. Towards deep neural network architectures robust to adversarial examples, 2014 (last updated 9 April 2015). https://arxiv.org/abs/1412.5068.

John A. Hartigan and Manchek A. Wong. Algorithm as 136: A k-means clustering algorithm. *Journal of the Royal Statistical Society. Series C (Applied Statistics)*, 28(1):100–108, 1979.

Ling Huang, Anthony D. Joseph, Blaine Nelson, Benjamin I. P. Rubinstein, and J. D. Tygar. Adversarial machine learning. In *Proc. of the ACM Workshop on Security and Artificial Intelligence, Chicago, IL, USA, 21 October 2011*, pp. 43–58, 2011.

Nektaria Kaloudi and Jingyue Li. The AI-based cyber threat landscape: A survey. *ACM Computing Surveys*, 53(1): Article No.: 20, pp. 1–34, 2020.

Jongkil Kim, Jonathon M. Smereka, Calvin Cheung, Surya Nepal, and Marthie Grobler. Security and performance considerations in ROS 2: A balancing act, 2018. https://arxiv.org/abs/1809.09566.

Dhilung Kirat, Juyoung Jang, and Marc Stoechlin. Deeplocker – concealing targeted attacks with AI locksmithing. *Blackhat USA*, 2018. https://i.blackhat.com/us-18/Thu-August-9/us-18-Kirat-DeepLocker-Concealing-Targeted-Attacks-with-AI-Locksmithing.pdf.

Ralph Langner. Stuxnet: Dissecting a cyberwarfare weapon. *IEEE Security & Privacy*, 9(3):49–51, 2011.

Robert M. Lee, Michael J. Assante, and Tim Conway. Analysis of the cyber attack on the Ukrainian power grid. Technical report, Electricity Information Sharing and Analysis Center (E-ISAC), 2016.

Xiao Li and Thenkurussi Kesavadas. Surgical robot with environment reconstruction and force feedback. In *Proc. of the Annual Intl. Conf. of the IEEE Medicine and Biology Society, Honolulu, HI, USA, 18–21 July 2018*, pp. 1861–1866, 2018.

Lockheed Martin. The cyber kill chain, 2019. https://www.lockheedmartin.com/en-us/capabilities/cyber/cyber-kill-chain.html.

Jiajun Lu, Theerasit Issaranon, and David Forsyth. Safetynet: Detecting and rejecting adversarial examples robustly. In *Proc. of the Intl. Conf. on Computer Vision, Venice, Italy, 22–29 October 2017*, 2017a.

Jiajun Lu, Hussein Sibai, Evan Fabry, and David A. Forsyth. No need to worry about adversarial examples in object detection in autonomous vehicles. *ArXiv, abs/1707.03501*, 2017b.

Jan Hendrik Metzen, Tim Genewein, Volker Fischer, and Bastian Bischoff. On detecting adversarial perturbations. In *Proc. of the Intl. Conf. on Learning Representations, Toulon, France, 24–26 April 2017*, 2017.

Santiago Morante, Juan G. Victores, and Carlos Balaguer. Cryptobotics: Why robots need cyber safety. *Frontiers in Robotics and AI*, 2:23, 2015.

Sean Palka and Damon McCoy. Fuzzing E-mail filters with generative grammars and N-gram analysis. In *Proc. of the USENIX Workshop on Offensive Technologies, Washington, DC, USA, 10–11 August 2015*, 2015.

Nicolas Papernot, Patrick McDaniel, Ian Goodfellow, Somesh Jha, Z. Berkay Celik, and Ananthram Swami. Practical black-box attacks against machine learning. In *Proc. of the ACM on Asia Conf. on Computer and Communications Security, Abu Dhabi, United Arab Emirates, 2–6 April 2017*, pp. 506–519, 2017.

Dan Petro and Ben Morris. Weaponizing machine learning: Humanity was overrated anyway. *DEF CON 25*, 2017. https://www.defcon.org/html/defcon-25/dc-25-speakers.html\LY1\textbackslash#Petro.

Morgan Quigley, Ken Conley, Brian Gerkey, Josh Faust, Tully Foote, Jeremy Leibs, Rob Wheeler, and Andrew Y. Ng. ROS: An open-source robot operating system. In *Proc. of the ICRA Workshop on Open Source Software, Kobe, Japan, 12–17 May 2009*. IEEE, 2009.

Francisco J. Romero-Ramirez, Rafael Muñoz-Salinas, and Rafael Medina-Carnicer. Speeded up detection of squared fiducial markers. *Image and Vision Computing*, 76:38–47, 2018.

SentinelOne. What are advanced targeted attacks? Can you defend against them?, 2019. https://www.sentinelone.com/blog/what-are-advanced-targeted-attacks.

Mahmood Sharif, Sruti Bhagavatula, Lujo Bauer, and Michael K. Reiter. Accessorize to a crime: Real and stealthy attacks on state-of-the-art face recognition. In *Proc. of the ACM SIGSAC Conf. on Computer and Communications Security, Vienna, Austria, 24–28 October 2016*, pp. 1528–1540, 2016.

Aashish Sharma, Zbigniew Kalbarczyk, James Barlow, and Ravishankar Iyer. Analysis of security data from a large computing organization. In *Proc. of the Intl. Conf. on Dependable Systems & Networks, Hong Kong, China, 27–30 June 2011*, pp. 506–517, 2011.

Symantec Corporation. Internet security threat report. Technical report 24, 2019. `https://www`
`.symantec.com/content/dam/symantec/docs/reports/istr-24-2019-en.pdf`.

Jeffrey Voas, Rick Kuhn, Constantinos Kolias, Angelos Stavrou, and Georgios Kambourakis. Cybertrust in the IoT age. *Computer*, 51(7):12–15, 2018.

Suzzane Widup. The VERIS community database. `https://github.com/vz-risk/VCDB`, 2020.

16

Trinity: Trust, Resilience and Interpretability of Machine Learning Models

Susmit Jha[1], Brian Jalaian[2], Anirban Roy[1], and Gunjan Verma[2]

[1] Computer Science Laboratory, SRI International, Menlo Park, CA, USA
[2] US Army Futures Command, US Army Research Laboratory, Adelphi, MD, USA

16.1 Introduction

Deep learning methods have demonstrated impressive performance in many applications such as image processing, speech processing, and natural language processing. This has fueled their rapid adoption in many civilian and military applications. But the fragility, noninterpretability, and consequential lack of trustworthiness have impeded their adoption in security-critical applications such as anomaly detection, authentication, malware recognition, network intrusion detection, and target acquisition. In this chapter, we study the connections between these related challenges of trust, resilience, and interpretability of deep learning models which together hinder the adoption of deep learning models in high-assurance safety-critical applications. We present a new Trinity framework that has a suite of algorithms to establish trust over the safe behavior of systems with machine learning models, resilience against adversarial attacks, and simultaneously improve their interpretability. The neurosymbolic approach in Trinity combines symbolic methods developed for automated reasoning and formal analysis with connectionist deep learning methods that use deep neural networks.

In 2013–2014, machine learning started to reach human-level performance on several benchmark tasks. In image recognition, deep learning powered by CNNs (Krizhevsky et al. 2017) can achieve accuracy close to human levels on ILSVRC data set that contains 14 million images that have been hand-annotated with more than 20 000 categories. For face detection benchmarks, human accuracy is around 97.4% and ensemble deep learning methods can achieve 97.35% (Taigman et al. 2014). Similar success was achieved in applications such as speech recognition and transcription. Microsoft announced in 2017 (Xiong et al. 2018) that they can achieve the word error rate (WER) of 5%. Human error on this benchmark is also 5%.

Machine learning, in particular, deep learning, techniques have made tremendous progress in many applications. Consequently, its use has become ubiquitous particularly in data-rich domains where the power of these methods to exploit large amount of data to automatically infer discriminating features and learn their composition outperforms manual design of features or the curation of expert rules or explicit programming. The impact of deep learning has now spread beyond computer science to physical sciences and even traditional bastions of human creativity such as music and art. In addition to discriminative models, deep learning approaches have shown the potentially to learn generative models that enable automated synthesis.

Game Theory and Machine Learning for Cyber Security, First Edition.
Edited by Charles A. Kamhoua, Christopher D. Kiekintveld, Fei Fang, and Quanyan Zhu.

This rapid progress in the accuracy and scalability of deep learning models has also fueled attempts to integrate these techniques with our social and military infrastructure. The main strength of these methods is the ease of training the deep neural network models with a large amount of data using scalable optimization methods such as stochastic gradient descent. But as recent research has revealed, the same strength also turns out to be a weakness for these models (see Figure 16.1); fairly simple gradient-based techniques applied to the deep neural network enable discovery of adversarial examples imperceptibly similar to natural examples.

While well-trained models provide correct answers on the natural examples, the accuracy of the model drops to almost zero on the adversarial examples. Until recently, machine learning was difficult, even in a nonadversarial setting. And adversarial examples were not interesting to most researchers because mistakes were the rule, not the exception. But the human-level performance of deep learning models has fueled significant investigation into adversarial attack methods. Further, machine learning models exhibit confounding failure modes with difficult to explain failures on some inputs. This has created challenges in trusting machine learning models to make fair and reliable decisions.

In addition to the fragility of deep learning models to adversarial attacks, the good performance of the model on the test data from the training distribution does not transfer to out of distribution data. A number of digital and physically realizable adversarial attack methods have been developed over the last few years.

These limitations make it challenging to adopt deep learning techniques in high-assurance safety-critical applications. In this chapter, we present the Trinity framework that presents an integrated approach to address the three challenges of trust, resilience and interpretability of deep learning models. In the rest of this chapter, we first present methods for increasing trustworthiness of deep learning models. Our approach to trust comprises formal verification techniques, and a scalable novel approach based on top-down analysis by synthesis inspired by human cognition. We then present two approaches to resilience. The first approach is based on learning data

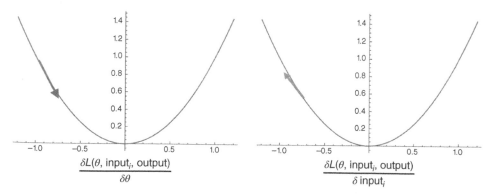

Figure 16.1 While the ease of optimization makes it convenient and scalable to use stochastic gradient descent to find model parameters θ that minimize the loss function for given dataset, it also makes the model vulnerable where finding an adversarial input requires maximizing the same loss function with respect to the input features instead of model parameters. A typical loss function is high dimensional and more irregular than the representative curve above. But the use of gradient for minimizing the loss function during training by finding optimal model parameters is also equally effective in maximizing the loss function by finding the adversarial examples.

manifold, and the second uses attributions to compute a confidence metric over the model's predictions.

16.2 Trust and Interpretability

Deep learning methods use highly expressive and large scale deep neural networks to approximate complex functions. This choice is motivated by the "universal" approximation capability of deep learning models, and the relative ease of training these models with large amounts of data. But unlike other traditional system models, there are few known and scalable methods to verify these large deep neural networks. Trinity adopts a combination of formal verification Dutta et al. (2018a, b) and a novel top-down analysis by synthesis approach.

16.2.1 Formal Methods and Verification

Trinity includes formal methods for verifying deep neural networks. An initial prototype for improving trust of deep learning models is publicly available in an open-source tool, Trinity.[1] At an abstract level, a deep neural network computes a function from a set of inputs to some set of outputs. The formal verification addresses the following: Given a neural network (NN), and constraints (assumptions) which define a set of inputs to the network, provide a *tight* over-approximation (guarantee) of the output set. This serves as one of the main primitives in verification of neural networks. Sherlock uses mixed-integer linear programming (MILP) solver to address this verification problem but it does not merely compile the verification into an MILP problem. Sherlock first uses sound piecewise linearization of the nonlinear activation function to define an encoding of the neural network semantics into mixed-integer constraints involving real-valued variables and binary variables that arise from the (piecewise) linearized activation functions. Such an encoding into MILP is a standard approach to handling piecewise linear functions. As such, the input constraints $\phi(\mathbf{x})$ are added to the MILP and next, the output variable is separately maximized and minimized to infer the corresponding guarantee that holds on the output. This enables us to infer an assume-guarantee contract on the overall deep neural network. Sherlock augments this simple use of MILP solving with a local search that exploits the local continuity and differentiability properties of the function represented by the network. These properties are not exploited by MILP solvers which typically use a branch-and-cut approach. On the other hand, local search alone may get "stuck" in local minima. Sherlock handles local minima by using the MILP solver to search for a solution that is "better" than the current local minimum or conclude that no such solution exists. Thus, by alternating between inexpensive local search iterations and relatively expensive MILP solver calls, Sherlock can exploit local properties of the neural network function but at the same time avoid the problem of local minima, and thus, solve the verification of deep neural networks more efficiently.

Deep neural networks are very common in applications such as image classification and autonomous control. In image classification networks, since each image is a point in the high dimensional pixel space, a polyhedral set may be used to represent all possible bounded perturbations to a given image. If, for such a set, we can guarantee that the output of the

---

1 https://github.com/SRI-CSL/Trinity

classification remains unaltered, then we have proved that the neural network is *robust* to bounded pixel noise. Besides image classification, neural networks are increasingly used in the control of autonomous systems, such as self-driving cars, unmanned aerial vehicles, and other robotic systems. A typical approach to verify these systems involves a *reachability computation* to estimate the forward reachable set as time evolves. Using this, it is possible to prove that, no matter what the initial condition of a system is, it always reaches a target region in finite time. For instance, we wish to prove that, an autonomous car whose inputs are provided by a neural network controller's feedback, will remain within a fixed lateral distance from the center of the road (desired trajectory), while remaining under the speed limit.

Let N be a neural network with n input vectors x, a single output y, and weights $[(W_0, b_0), \ldots, (W_k, b_k)]$. Let F_N be the function defined by such a network. The general problem of verifying neural network and establishing an assume-guarantee contract on its inputs and outputs can be simplified to range estimation problem by suitably transforming the inputs and outputs such that the assumption constraints are described by a polyhedron and the guarantee constraints to be derived over the outputs can be represented as intervals. The universal approximation property of neural networks can be used to approximately encode such transformation as a part of the network itself. Thus, we focus on range estimation problem and rely on reducing other verification problems to it.

The *range estimation* problem is defined as follows:

- INPUTS: Neural Network N, input constraints $P : Ax \leq b$ and a tolerance parameter $\delta > 0$.
- OUTPUT: An interval $[\ell, u]$ such that $(\forall\, x \in P)\, F_N(x) \in [\ell, u]$. i.e, $[\ell, u]$ contains the range of F_N over inputs $x \in P$. Furthermore, we require the interval to be *tight*:

$$(\max_{x \in P} F_N(x) \geq u - \delta), \ (\min_{x \in P} F_N(x) \leq \ell + \delta) .$$

We will assume that the input polyhedron P is compact: i.e, it is closed and has a bounded volume. It was shown in Katz et al. (2017) that even proving simple properties in this setting is NP complete, e.g. proving that there exists an assignment from input set to an output set, which respects the constraints imposed by the neural network. So, one of the fundamental challenges in this problem is to tackle the exponential nature.

Without loss of generality, we will focus on estimating the upper bound u. The case for the lower bound will be entirely analogous. First, we note that a single Mixed-integer linear programming (MILP) problem can be formulated, and then query a solver to directly compute u. Unfortunately, that can be quite expensive in practice. Therefore, our approach will combine a series of MILP feasibility problems alternating with local search steps.

Figure 16.2 shows a pictorial representation of the overall approach. The approach incrementally finds a series of approximations to the upper bound $u_1 < u_2 < \cdots < u^*$, culminating in the final bound $u = u^*$.

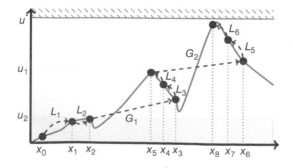

Figure 16.2 A schematic figure showing our approach showing alternating series of local search $L_1, \ldots, L_6$ and "global search" G_1, G_2 iterations. The points x_2, x_5, x_8 represent local minima wherein our approach transitions from local search iterations to global search iterations.

1. The first level u_1 is found by choosing a randomly sampled point $x_0 \in P$.
2. Next, we perform a series of local iteration steps resulting in samples $x_1, \dots, x_i$ that perform gradient ascent in the input space until these steps cannot obtain any further improvement. We take $u_2 = F_N(x_i)$.
3. A "global search" step is now performed to check if there is any point $x \in P$ such that $F_N(x) \geq u_2 + \delta$. This is obtained by solving a MILP feasibility problem.
4. If the global search fails to find a solution, then we declare $u^* = u_2 + \delta$.
5. Otherwise, we obtain a new *witness* point x_{i+1} such that $F_N(x_{i+1}) \geq u_2 + \delta$.
6. We now go back to the local search step.

For neural networks with multiple outputs, we can individually find the bounds for each of the network outputs, and then combine them to form a hyper-rectangle in the output dimensional space. This can be extended to using a template polyhedron to obtain tighter bounds, in the output dimension, described in the next section. In general, we can obtain guarantees on the output from a given class defined by the constraint template used in the minimization/maximization step of the presented approach. Our current implementation in Sherlock built on top of MILP solvers requires the template to be linear.

We did an empirical comparison with a recent solver for deep neural network verification called Reluplex (Katz et al. 2017). The results of the comparison are presented in Figure 16.3. Even though Reluplex is an SMT solver, it can be used to perform set propagation using a binary search wrapper. The comparison shows that Sherlock is much faster than Reluplex used with a binary search wrapper (Dutta et al. 2018). Another set of comparisons was using Sherlock, against a monolithic mixed-integer linear programming (MILP) solver. We have also used Sherlock for verifying properties of various closed-loop cyberphysical systems that have neural networks as controllers (Dutta et al. 2018). We are currently working on scaling our approach to reason over state of the art deep learning models. We are also working on using our approach to complement black-box explaining-mining methods (Jha et al. 2019) to generate interpretable approximations of the neural network models.

16.2.2 Top-down Analysis by Synthesis

Machine learning models are typically bottom-up inferring intent from the effect which makes the learned models fragile and nonresilient. Consider the component of a car's vision system concerned with detecting traffic lanes. A basic method will look for more-or-less straight lines painted on the road, and a bottom-up approach will perform this process afresh as each image frame is processed. But this is inefficientthe traffic lanes in the current image frame are likely to be similar to those in the previous few frames and we should surely use this to seed the searchand it is fragilemissing or scuffed lane markers might cause lanes to go undetected where they could have been extrapolated from previous images. A better approach builds a model of the road and its traffic lanes and uses this to seed the search for lanes in the current image by predicting their location.

There will be some uncertainty in the model and its projection of lanes, so what is sent to the vision system will be a best guess, or perhaps a probability distribution over several such estimates. The vision system will use this to seed its search for lanes in the current image and will send back the difference or "error" between the prediction and its current observation. The error signal is used to refine the model in a way that aims to minimize future prediction errors and thereby bring it closer to reality.

This is an example of "analysis by synthesis" meaning that we formulate hypotheses (i.e., candidate world models) and favor those whose predictions match the input data. In practical applications, we need to consider the level of the "predictions" concerned: do we use the world model

ID	x	k	N	23 cores		Monolithic		Single core		Monolithic		Reluplex
				T	Nc	T	Nc	T	Nc	T	Nc	T
N_0	2	1	100	1 s	94	2.3 s	24	0.4 s	44	0.3 s	25	9.0
N_1	2	1	200	2.2 s	166	3.6 s	29	0.9 s	71	0.8 s	38	1 m 50 s
N_2	2	1	500	7.8 s	961	12.6 s	236	2 s	138	2.9 s	257	15 m 59 s
N_3	2	1	500	1.5 s	189	0.5 s	43	0.6 s	95	0.2 s	53	12 m 25 s
N_4	2	1	1000	3 m 52 s	32E3	3 m 52 s	3E3	1m 20 s	4.8E3	35.6 s	5.3E3	1 h 06 m
N_5	3	7	425	4 s	6	6.1 s	2	1.7 s	2	0.9 s	2	DNC
N_6	3	4	762	3 m 47 s	3.3E3	4 m 41 s	3.6E3	37.8 s	685	56.4 s	2.2E3	DNC
N_7	4	7	731	3.7 s	1	7.7 s	2	3.9 s	1	3.1 s	2	1 h 35 m
N_8	3	8	478	6.5 s	3	40.8 s	2	3.6 s	3	3.3 s	2	DNC
N_9	3	8	778	18.3 s	114	1 m 11 s	2	12.5 s	12	4.3 s	73	DNC
N_{10}	3	26	2340	50 m 18 s	4.6E4	1 h 26 m	6E4	17 m 12 s	2.4E4	18 m 58 s	1.9E4	DNC
N_{11}	3	9	1527	5 m 44 s	450	55 m 12 s	6.4E3	56.4 s	483	130.7 s	560	DNC
N_{12}	3	14	2292	24 m 17 s	1.8E3	3 h 46 m	2.4E4	8 m 11 s	2.3E3	1 h 01 m	1.6E4	DNC
N_{13}	3	19	3057	4 h 10 m	2.2E4	61 h 08 m	6.6E4	1 h 7 m	1.5E4	15 h 1 m	1.5E5	DNC
N_{14}	3	24	3822	72 h 39 m	8.4E4	111 h 35 m	1.1E5	5 h 57 m	3E4	Timeout	–	DNC
N_{15}	3	127	6845	2 m 51 s	1	timeout	–	3 m 27 s	1	Timeout	–	DNC

Figure 16.3 Performance results on networks trained on functions with known maxima and minima. x, number of inputs; k, number of layers; N, total number of neurons; T, CPU time taken; Nc, number of nodes explored. All the tests were run on a Linux server running Ubuntu 17.04 with 24 cores, and 64 GB RAM (DNC: did not complete)

to synthesize the raw data (e.g., pixels) that we predict the sensor will detect, or do we target some higher level of its local processing (e.g., objects)? This is a topic for experimental research, but we would expect a level of representation comparable to that produced by an object detector to be suitable.

The significant attribute of this top-down approach is that it focuses on construction and exploitation of the world model (or models: a common arrangement has a hierarchy of models), in contrast to the sensor-driven character of bottom-up approaches. It is interesting, and perhaps reassuring, to know that it is widely believed to be the way perception works in human (and other) brains, as first proposed by Helmholtz in the 1860s (von Helmholtz 1867). *Predictive Processing* (PP) (Wiese and Metzinger, 2017), also known as predictive *coding* (Clark 2013) and predictive *error minimization* (Hohwy 2013), posits that the brain builds models of its environment and uses these to predict its sensory input, so that much of its activity can be seen as (an approximation to) iterative Bayesian optimization of its models to minimize prediction error. PP has prior "predictions" flowing from cognitive models down to sense organs and Bayesian "corrections" flowing back up that cause the posterior models to track reality. ("Free Energy"; Friston 2010 is a more all-encompassing theory that includes actions: the brain "predicts" that the hand, say, is in a certain place and to minimize prediction error the hand actually moves there.) This is consistent with the fact that the brain has more neural pathways going from upper to lower levels than vice versa: models and predictions are flowing down, and only corrections are flowing up.

We recently proposed an attribution-based approach to perform this analysis by synthesis (Jha et al. 2019) illustrated in Figure 16.4. The overall architecture of the proposed approach is motivated by Dual Process Theory (Groves and Thompson 1970; Evans and Frankish 2009) and Kahneman's decomposition (Kahneman 2011) of cognition into System 1, or the intuitive system, and System 2, or the deliberate system. Typical machine learning models for classification perform anti-causal learning to determine the label (*cause*) from the input instance (*effect*). Such effect-to-cause, i.e. anti-causal reasoning, lacks the natural continuity of causal mechanisms and is often not robust. But we view this model as System 1 and use attribution methods to obtain features with positive and negative attributions. In this example with the MNIST dataset, we see that the adversarial perturbation that causes misclassification of nine into four also significantly changes the attributions. For example, the top part of the perturbed nine (misclassified as four)

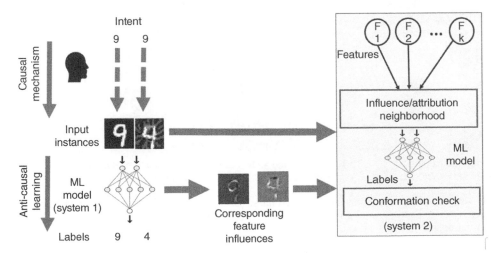

Figure 16.4 Attribution-based approach for analysis by synthesis.

has negative attribution. In deliberative System 2, we perform reasoning in the causal direction and mask the high attribution features (pixels in this case) to obtain a number of input instances in the causal neighborhood of the original image. The model's output does not change (conforms) in the neighborhood of the original image but it is not robust in the neighborhood of the adversarial example. Thus, measuring the degree of conformance in the neighborhood of an input allows us to quantitatively estimate the confidence of the model.

We evaluate out analysis by synthesis approach by evaluating the conformance/confidence of the model on out-of-distribution data. We used MNIST (LeCun and Cortes 2010) with rotation and background for our evaluation. Our approach can track the decrease in the model accuracy on the rotated MNIST dataset with a background image (U. of Montreal 1998). This dataset has MNIST images randomly rotated by 0 to 2π, and with a randomly selected black/white background image. The accuracy and confidence of the model drops with increase in rotation angle (from 0 to 50°) as illustrated in Figure 16.5. The former is expected while the latter is desirable but typically uncommon with modern ML methods. The three examples show how the attribution-based approach can compute quantitative metric that reflects confusability of an input. We later describe how this approach can also be used to detect adversarial examples.

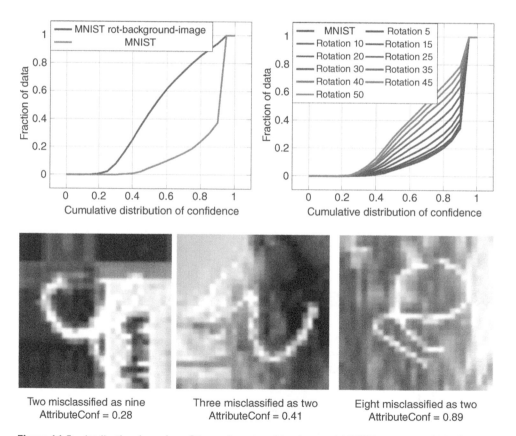

Two misclassified as nine
AttributeConf = 0.28

Three misclassified as two
AttributeConf = 0.41

Eight misclassified as two
AttributeConf = 0.89

Figure 16.5 Attribution-based confidence for rotated-background-MNIST and rotated-MNIST at different angles. Selected examples from rotated-background-MNIST with confidence showing quantitative analysis of attribution-based analysis by synthesis.

16.3   Resilience and Interpretability

Multiple methods have been proposed in the literature to generate as well as defend against adversarial examples. Adversarial example generation methods include both whitebox and black-box attacks on neural networks (Szegedy et al. 2013; Goodfellow et al. 2014; Papernot et al. 2015, 2017), targeting feedforward classification networks (Carlini and Wagner 2016), generative networks (Kos et al. 2017), and recurrent neural networks (Papernot et al. 2016). These methods leverage gradient-based optimization for normal examples to discover perturbations that lead to mispredictionthe techniques differ in defining the neighborhood in which perturbation is permitted and the loss function used to guide the search. For example, one of the earliest attacks (Goodfellow et al. 2014) used a fast sign gradient method (FGSM) that looks for a similar image x' in a "small" L^∞ neighborhood of x. Given a loss function $\text{Loss}(x, l)$ specifying the cost of classifying the point x as label l, the adversarial example x' is calculated as

$$x' = x + \epsilon \cdot \text{sign}(\nabla_x \text{Loss}(x, l_x))$$

FGSM was improved to iterative gradient sign approach (IGSM) in Kurakin et al. (2016) by using a finer iterative optimization strategy where the attack performs FGSM with a smaller step-width α, and clips the updated result so that the image stays within the ϵ boundary of x. In this approach, the i-th iteration computes the following:

$$x'_{i+1} = \text{clip}_{\epsilon,x}(x'_i + \alpha \cdot \text{sign}(\nabla_x \text{Loss}(x, l_x)))$$

In contrast to FGSM and IGSM, DeepFool (Moosavi-Dezfooli et al. 2016) attempts to find a perturbed image x' from a normal image x by finding the closest decision boundary and crossing it. In practice, DeepFool relies on local linearized approximation of the decision boundary. Another attack method that has received a lot of attention is Carlini attack that relies on finding a perturbation that minimizes change as well as the hinge loss on the logits (presoftmax classification result vector). The attack is generated by solving the following optimization problem:

$$\min_\delta [||\delta||_2 + c \cdot \max (Z(x')_{l_x} - \max Z(x')_i : i \neq l_x, -\kappa)]$$

where Z denotes the logits, l_x is the ground truth label, κ is the confidence (raising this will force the search towards larger perturbations), and c is a hyperparameter that balances the perturbation and the hinge loss. Another attack method is projected gradient descent method (PGD) proposed in Madry et al. (2017). PGD attempts to solve this constrained optimization problem:

$$\max_{||x^{adv}-x||_\infty \leq \epsilon} \text{Loss}(x^{adv}, l_x)$$

where ϵ is the constraint on the allowed perturbation norm, and l_x is the ground truth label of x. Projected gradient descent is used to solve this constrained optimization problem by restarting PGD from several points in the l_∞ ball around the data point x. This gradient descent increases the loss function *Loss* in a fairly consistent way before reaching a plateau with a fairly well-concentrated distribution and the achieved maximum value is considerably higher than that of a random point in the data set. In our evaluation, we focus on this PGD attack because it is shown to be a universal first-order adversary (Madry et al. 2017), that is, developing detection capability or resilience against PGD also implies defense against many other first-order attacks.

Defense of neural networks against adversarial examples is far more difficult compared to generating attacks. Madry et al. (2017) propose a generic saddle point formulation where $\mathscr{D}$ is the underlying training data distribution, $\mathrm{Loss}(\theta, x, l_x)$ is a loss function at data point x with ground truth label l_x for a model with parameter θ:

$$\min_{\theta} E_{(x,y) \sim \mathscr{D}} \left[\max_{||x^{\mathrm{adv}} - x||_{\infty} \leq \epsilon} \mathrm{Loss}(\theta, x^{\mathrm{adv}}, l_x) \right]$$

This formulation uses robust optimization over the expected loss for worst-case adversarial perturbation for training data. The internal maximization corresponds to finding adversarial examples and can be approximated using IGSM (Kurakin et al. 2016). This approach falls into a category of defenses that use *adversarial training* (Shaham et al. 2015). Instead of training with only adversarial examples, using a mixture of normal and adversarial examples in the training set has been found to be more effective (Szegedy et al. 2013; Moosavi-Dezfooli et al. 2015). Another alternative is to augment the learning objective with a regularizer term corresponding to the adversarial inputs (Goodfellow et al. 2014). More recently, logit pairing has been shown to be an effective approximation of adversarial regularization (Kannan et al. 2018).

Another category of defense against adversarial attacks on neural networks is defensive distillation methods (Papernot et al. 2015). These methods modify the training process of neural networks to make it difficult to launch gradient-based attacks directly on the network. The key idea is to use distillation training technique (Hinton et al. 2015) and hide the gradient between the presoftmax layer and the softmax outputs. Carlini and Wagner (2016) found methods to break this defense by changing the loss function, calculating gradient directly from presoftmax layer and transferring attack from easy-to-attack network to distilled network. More recently, Athalye et al. (2018) showed that it is possible to bypass several defenses proposed for the whitebox setting. Techniques based on manually identified statistical features (Grosse et al. 2017) or a dedicated learning model (Metzen et al. 2017) trained separately to identify adversarial examples have been previously proposed in literature. These explicit classification methods do not generalize well across different adversarial example generation techniques.

The approach to resilience in trinity falls into the category of techniques that focus on only detecting adversarial examples. In contrast to other defensive methods, our approach does not require any augmentation of training data, modification of the training process or change in the learned model. The design and training of the neural network are independent to the manifold-based filtering described here. Thus, our approach to detection is orthogonal to learning robust machine learning models and can benefit from these methods. Further, we do not require access to the adversarial example generation method, and thus this defense is likely to generalize well across different attack methods. We next present the two approaches to detecting adversarial examples: the first based on identifying data manifold and the second based on attributions and top-down analysis by synthesis.

16.3.1 Manifold-based Defense

Our approach (Jha et al. 2018) relies on just identifying the manifold of typical data which need not be even labeled and hence, this method is more practical in contexts where labeled training data is very difficult to obtain. Manifold-based explanations for adversarial attacks have also motivated defense mechanisms that try to exploit the manifold property of training data. Bhagoji et al. (2018) propose a defense comprising of transforming data to eliminate perturbations in nonprincipal components. Meng and Chen (2017) consider using autoencoders to project an input image to low dimensional encoding and then decode it back to remove perturbations. Xu et al. (2017) propose feature squeezing for images in the input space that effectively forces the data to lie in a low dimensional manifold. Song et al. (2017) also note that the distribution of log-likelihoods show considerable difference between perturbed adversarial images and the training data set which can be used to detect adversarial attacks.

Our approach relies on computing the distance of the new sample point from the manifold of training data. The kernel density estimation can be used to measure the distance $d(x)$ of x from the data manifold of training set. Specifically, $d(x) = \frac{1}{|X|} \sum_{x_i \in X} k(x_i, x)$, where X is the full data set and $k(\cdot, \cdot)$ is a kernel function. In case of using Gaussian kernel, the bandwidth σ needs to be carefully selected to avoid spiky density estimate or an overly smooth density estimate. A typical good choice for bandwidth is a value that maximizes the log-likelihood of the training data (Jones et al. 1996). Further, we can restrict the set of training points to be considered from the full set X to a set of immediate neighbors of x. The neighborhood can be defined using the maximum distance or bound on the number of neighbors. In our experiments, we use L_∞ norm with bound on the number of neighbors which yielded good results.

It has been hypothesized in literature (Bengio et al. 2013; Gardner et al. 2015) that the deeper layers of a deep neural network provide more linear and unwrapped manifolds in comparison to the input space. Thus, the task of identifying the manifold becomes easier as we progress from the input space to the more abstract feature spaces all the way to the logit space. But the adversarial perturbations are harder to detect at higher levels and might get hidden by the lower layers of the neural network. In our experiments, we learned manifolds in input space as well as the logit space. We evaluated our approach on MNIST dataset (LeCun 1998) and CIFAR10 dataset (Krizhevsky et al. 2014).

As the norm bound in the PGD method for generating adversarial examples is increased, the distance of adversarial examples from the manifold increases. While the success of attack on the neural network increases with high norm bound, it also becomes easier to detect these adversarial examples. We observed this behavior to be common across MNIST and CIFAR10 data set as illustrated in Figures 16.6 and 16.7. The distance from manifold monotonically increases in the input space but in the logit space, higher norm bound beyond a threshold allows the attack method to find examples that decrease the distance from logit manifold even though they are farther from the input manifold.

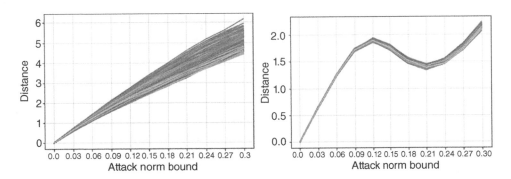

Figure 16.6 Increase in adversarial distance from manifold for MNIST in input space (left) and logit space (right). Each line of different color shows the increase in distance with attack norm for one sample of a 1000 images. The distance monotonically increased in each of the 100 experiments in the input space. The logit space shows increase in distance with norm up to a threshold after which the distance decreases before again increasing. This is because of high norm bound allowing occasional discovery of "clever" adversarial examples that are closer to the logit manifold though farther from the input manifold.

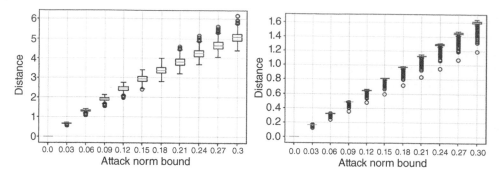

Figure 16.7 Increase in adversarial example's distance from input manifold with increase in attack norm: left: MNIST, right: CIFAR. The boxes in the box plot denote the first and third quartile of the distance at a given attack norm.

16.3.2 Attribution-based Confidence Using Shapley Values

While improving the accuracy and robustness of DNNs has received significant attention, there is an urgent need to quantitatively characterize the limitations of these models and improve the transparency of their failure modes. We need to define an easy to compute confidence metric on predictions of a DNN that closely reflects its accuracy. This can be used to detect adversarial examples as low confidence predictions of the model. Our approach for computing attribution-based confidence used Shapley values for computing importance of input features. Shapley values often employ the notion of a baseline input $\mathbf{x}^b$; for example, the all dark image can be the baseline for images. The baseline can also be a set of random inputs where attribution is computed as an expected value. Let the attribution for the j-th feature and output label i be $\mathcal{A}_j^i(\mathbf{x})$. The attribution for the j-th input feature depends on the complete input x and not just x_j. The treatment for each logit is similar, and so, we drop the logit/class and denote the network output simply as $\mathcal{F}(\cdot)$ and attribution as $\mathcal{A}_j(\mathbf{x})$. For simplicity, we use the baseline input $\mathbf{x}^b = 0$ for computing attributions. We make the following two assumptions on the DNN model and the attributions, which reflect the fact that the model is well-trained and the attribution method is well-founded:

- The attribution is dominated by the linear term. This is also an assumption made by attribution methods based on Shapley values such as Integrated Gradient (Sundararajan et al. 2017) which define attribution as the path integral of the gradients of the DNN output with respect to that feature along the path from the baseline $\mathbf{x}^b$ to the input $\mathbf{x}$, that is,

$$\mathcal{A}_j^i(\mathbf{x}) = (\mathbf{x}_j - \mathbf{x}_j^b) \times \int_{\alpha=0}^{1} \partial_j \mathcal{F}^i(\mathbf{x}^b + \alpha(\mathbf{x} - \mathbf{x}^b)) d\alpha \qquad (16.1)$$

where the gradient of i-th logit output of the model along the j-th feature is denoted by $\partial_j \mathcal{F}^i(\cdot)$.

- Attributions are complete i.e. the following is true for any input $\mathbf{x}$ and the baseline input $\mathbf{x}^b$:

$$\mathcal{F}(\mathbf{x}) - \mathcal{F}(\mathbf{x}^b) = \sum_{k=1}^{n} \mathcal{A}_k(\mathbf{x}) \text{ where } \mathbf{x} \text{ has } n \text{ features.} \qquad (16.2)$$

Shapley value methods such as Integrated Gradient and DeepShap (Sundararajan et al. 2017; Lundberg and Lee 2017) satisfy this axiom too.

The details of the attribution-based approach to compute confidence is presented in Jha et al. (2019).

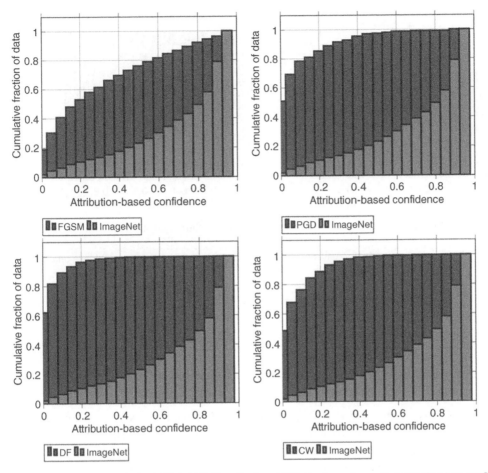

Figure 16.8 ABC metric for FGSM, PGD, DeepFool and CW attacks on ImageNet showing low confidence on attacked images.

We briefly outline the key empirical results here. Figure 16.8 illustrates how the ABC metric reflects the decrease in accuracy under FGSM, PGD, CW and DeepFool adversarial attack of a machine learning model trained on ImageNet (Russakovsky et al. 2015). We also applied physically realizable adversarial patch (Brown et al. 2017) and LaVAN (Karmon et al. 2018) attacks on 1000 images from ImageNet. For adversarial patch attack, we used a patch size of 25% for two patch types: banana and toaster. For LaVAN, we used baseball patch of size 50×50 pixels. Figure 16.9 illustrates how the ABC metric is low for most of the adversarial examples reflecting the decrease in the accuracy of the model and enabling resilient inference.

16.4 Conclusion

The impressive performance of deep learning techniques in applications such as image processing, speech processing, and natural language processing has been limited to recommendation and other low-assurance settings. The major hurdle towards their application in safety-critical social and military infrastructure is their brittleness and fragility to out-of-distribution and adversarial inputs. In this chapter, we describe Trinitya—framework to address the challenges of trust,

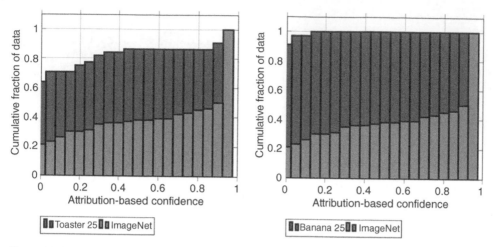

Figure 16.9 ABC metric for physically realizable attacks on ImageNet showing low confidence on attacked images.

resilience and interpretability by exploiting the relationship between them. We show how the techniques developed for interpretability such as attribution methods can be used to also improve trust and resilience of machine learning models.

References

Anish Athalye, Nicholas Carlini, and David Wagner. Obfuscated gradients give a false sense of security: Circumventing defenses to adversarial examples. *arXiv preprint arXiv:1802.00420*, 2018.

Yoshua Bengio, Grégoire Mesnil, Yann Dauphin, and Salah Rifai. Better mixing via deep representations. In *International Conference on Machine Learning, Atlanta, 16 June*, pp. 552–560, 2013. http://proceedings.mlr.press/v28/bengio13.html and https://icml.cc/2013 (accessed 21 March 2021).

Arjun Nitin Bhagoji, Daniel Cullina, Chawin Sitawarin, and Prateek Mittal. Enhancing robustness of machine learning systems via data transformations. In *2018 52nd Annual Conference on Information Sciences and Systems (CISS)*, pp. 1–5. IEEE, 2018.

Tom B Brown, Dandelion Mané, Aurko Roy, Martín Abadi, and Justin Gilmer. Adversarial patch. *arXiv preprint arXiv:1712.09665*, 2017.

Nicholas Carlini and David Wagner. Towards evaluating the robustness of neural networks. *arXiv preprint arXiv:1608.04644*, 2016.

Andy Clark. Whatever next? Predictive brains, situated agents, and the future of cognitive science. *Behavioral and Brain Sciences*, 36(3):181–204, 2013.

Souradeep Dutta, Susmit Jha, Sriram Sankaranarayanan, and Ashish Tiwari. Learning and verification of feedback control systems using feedforward neural networks. *IFAC-PapersOnLine*, 51(16):151–156, 2018.

Souradeep Dutta, Susmit Jha, Sriram Sankaranarayanan, and Ashish Tiwari. Output range analysis for deep feedforward neural networks. In *NASA Formal Methods Symposium*, pp. 121–138. Springer, 2018.

Jonathan St B. T. Evans and Keith Frankish. *In two minds: Dual processes and beyond*, volume 10. Oxford University Press Oxford, 2009.

Karl Friston. The free-energy principle: A unified brain theory? *Nature Reviews Neuroscience*, 11(2):127, 2010.

Jacob R Gardner, Paul Upchurch, Matt J. Kusner, Yixuan Li, Kilian Q. Weinberger, Kavita Bala, and John E. Hopcroft. Deep manifold traversal: Changing labels with convolutional features. *arXiv preprint arXiv:1511.06421*, 2015.

Ian J. Goodfellow, Jonathon Shlens, and Christian Szegedy. Explaining and harnessing adversarial examples. *arXiv preprint arXiv:1412.6572*, 2014.

Kathrin Grosse, Praveen Manoharan, Nicolas Papernot, Michael Backes, and Patrick McDaniel. On the (statistical) detection of adversarial examples. *arXiv preprint arXiv:1702.06280*, 2017.

Philip M. Groves and Richard F. Thompson. Habituation: A dual-process theory. *Psychological Review*, 77(5):419, 1970.

Geoffrey Hinton, Oriol Vinyals, and Jeff Dean. Distilling the knowledge in a neural network. *arXiv preprint arXiv:1503.02531*, 2015.

Jakob Hohwy. *The Predictive Mind*. Oxford University Press, 2013.

Susmit Jha, Uyeong Jang, Somesh Jha, and Brian Jalaian. Detecting adversarial examples using data manifolds. In *MILCOM 2018-2018 IEEE Military Communications Conference (MILCOM)*, pp. 547–552. IEEE, 2018.

Susmit Jha, Sunny Raj, Steven Fernandes, Sumit K. Jha, Somesh Jha, Brian Jalaian, Gunjan Verma, and Ananthram Swami. Attribution-based confidence metric for deep neural networks. In *Advances in Neural Information Processing Systems*, pp. 11826–11837, 2019.

Susmit Jha, Tuhin Sahai, Vasumathi Raman, Alessandro Pinto, and Michael Francis. Explaining ai decisions using efficient methods for learning sparse boolean formulae. *Journal of Automated Reasoning*, 63(4):1055–1075, 2019.

M. Chris Jones, James S. Marron, and Simon J. Sheather. A brief survey of bandwidth selection for density estimation. *Journal of the American Statistical Association*, 91(433): 401–407, 1996.

Daniel Kahneman. *Thinking, Fast and Slow*. Macmillan, 2011.

Harini Kannan, Alexey Kurakin, and Ian Goodfellow. Adversarial logit pairing. *arXiv preprint arXiv:1803.06373*, 2018.

Danny Karmon, Daniel Zoran, and Yoav Goldberg. LaVAN: Localized and visible adversarial noise. *arXiv preprint arXiv:1801.02608*, 2018.

Guy Katz, Clark Barrett, David L. Dill, Kyle Julian, and Mykel J. Kochenderfer. *Reluplex: An Efficient SMT Solver for Verifying Deep Neural Networks*, pp. 97–117. Springer International Publishing, Cham, 2017. ISBN 978-3-319-63387-9. doi: 10.1007/978-3-319-63387-9_5. https://doi.org/10.1007/978-3-319-63387-9_5.

Jernej Kos, Ian Fischer, and Dawn Song. Adversarial examples for generative models. *arXiv preprint arXiv:1702.06832*, 2017.

Alex Krizhevsky, Vinod Nair, and Geoffrey Hinton. The cifar-10 dataset. *Online*: http://www.cs.toronto.edu/kriz/cifar.html, 2014.

Alex Krizhevsky, Ilya Sutskever, and Geoffrey E. Hinton. Imagenet classification with deep convolutional neural networks. *Communications of the ACM*, 6(6):84–90, 2017. doi: 10.1145/3065386. http://doi.acm.org/10.1145/3065386.

Alexey Kurakin, Ian Goodfellow, and Samy Bengio. Adversarial machine learning at scale. *arXiv preprint arXiv:1611.01236*, 2016.

Yann LeCun. The mnist database of handwritten digits. http://yann.lecun.com/exdb/mnist/, 1998.

Yann LeCun and Corinna Cortes. MNIST handwritten digit database. 2010. http://yann.lecun.com/exdb/mnist/.

Scott M. Lundberg and Su-In Lee. A unified approach to interpreting model predictions. In *Advances in Neural Information Processing Systems*, pp. 4765–4774, 2017.

Aleksander Madry, Aleksandar Makelov, Ludwig Schmidt, Dimitris Tsipras, and Adrian Vladu. Towards deep learning models resistant to adversarial attacks. *arXiv preprint arXiv:1706.06083*, 2017.

Dongyu Meng and Hao Chen. Magnet: A two-pronged defense against adversarial examples. In *Proceedings of the 2017 ACM SIGSAC Conference on Computer and Communications Security*, pp. 135–147. ACM, 2017.

Jan Hendrik Metzen, Tim Genewein, Volker Fischer, and Bastian Bischoff. On detecting adversarial perturbations. *arXiv preprint arXiv:1702.04267*, 2017.

Seyed-Mohsen Moosavi-Dezfooli, Alhussein Fawzi, and Pascal Frossard. Deepfool: A simple and accurate method to fool deep neural networks. In *CVPR*, pp. 2574–2582, 2016.

University of Montreal. The rotated MNIST with background image. https://sites.google.com/a/lisa.iro.umontreal.ca/public_static_twiki/variations-on-the-mnist-digits, 1998.

Nicolas Papernot, Patrick McDaniel, Somesh Jha, Matt Fredrikson, Z. Berkay Celik, and Ananthram Swami. The limitations of deep learning in adversarial settings. In *2016 IEEE European Symposium on Security and Privacy (EuroS&P)*, pp. 372–387. IEEE, 2016a.

Nicolas Papernot, Patrick McDaniel, Ananthram Swami, and Richard Harang. Crafting adversarial input sequences for recurrent neural networks. In *Military Communications Conference, MILCOM 2016-2016 IEEE*, pp. 49–54. IEEE, 2016b.

Nicolas Papernot, Patrick McDaniel, Ian Goodfellow, Somesh Jha, Z. Berkay Celik, and Ananthram Swami. Practical black-box attacks against machine learning. In *Proceedings of the 2017 ACM on Asia Conference on Computer and Communications Security*, pp. 506–519. ACM, 2017.

Olga Russakovsky, Jia Deng, Hao Su, Jonathan Krause, Sanjeev Satheesh, Sean Ma, Zhiheng Huang, Andrej Karpathy, Aditya Khosla, Michael Bernstein, Alexander C. Berg, and Li Fei-Fei. ImageNet Large Scale Visual Recognition Challenge. *International Journal of Computer Vision (IJCV)*, 115(3):211–252, 2015. doi: 10.1007/s11263-015-0816-y.

Uri Shaham, Yutaro Yamada, and Sahand Negahban. Understanding adversarial training: Increasing local stability of neural nets through robust optimization. *arXiv preprint arXiv:1511.05432*, 2015.

Yang Song, Taesup Kim, Sebastian Nowozin, Stefano Ermon, and Nate Kushman. Pixeldefend: Leveraging generative models to understand and defend against adversarial examples. *arXiv preprint arXiv:1710.10766*, 2017.

Mukund Sundararajan, Ankur Taly, and Qiqi Yan. Axiomatic attribution for deep networks. In *ICML*, pp. 3319–3328. JMLR.org, 2017.

Christian Szegedy, Wojciech Zaremba, Ilya Sutskever, Joan Bruna, Dumitru Erhan, Ian Goodfellow, and Rob Fergus. Intriguing properties of neural networks. *arXiv preprint arXiv:1312.6199*, 2013.

Yaniv Taigman, Ming Yang, Marc'Aurelio Ranzato, and Lior Wolf. Deep-face: Closing the gap to human-level performance in face verification. In *Proceedings of the IEEE Conference on Computer Vision and Pattern Recognition*, pp. 1701–1708. IEEE, 2014.

Hermann von Helmholtz. *Handbuch der Physiologischen Optik III*, volume 9. Verlag von Leopold Voss, Leipzig, 1867.

Wanja Wiese and Thomas K. Metzinger. Vanilla PP for philosophers: A primer on predictive processing. In Thomas K. Metzinger and Wanja Wiese, editors. *Philosophy and Predictive Processing*, chapter 1. MIND Group, Frankfurt am Main, 2017.

Wayne Xiong, Lingfeng Wu, Fil Alleva, Jasha Droppo, Xuedong Huang, and Andreas Stolcke. The microsoft 2017 conversational speech recognition system. In *IEEE International Conference on Acoustics, Speech and Signal Processing (ICASSP)*, pp. 5934–5938. IEEE, 2018.

Weilin Xu, David Evans, and Yanjun Qi. Feature squeezing: Detecting adversarial examples in deep neural networks. *arXiv preprint arXiv:1704.01155*, 2017.

Part IV

Generative Models for Cyber Security

17

Evading Machine Learning Based Network Intrusion Detection Systems with GANs

Bolor-Erdene Zolbayar[1], Ryan Sheatsley[1], Patrick McDaniel[1], and Mike Weisman[2]

[1]*Pennsylvania State University, Computer Science and Engineering, State College, PA, USA*
[2]*Combat Capabilities Development Command, US Army Research Laboratory, Adelphi, MD, USA*

17.1   Introduction

Network intrusion detection systems (NIDS) are the de facto standard for defending networks against malicious activities. They are able to prevent existing malware such as backdoors, trojans, and rootkits (Cui et al. 2018; Saeed et al. 2013) and detect social engineering attacks such as phishing and man in the middle (Mukherjee et al. 2017; Smadi et al. 2018). Depending on their functionality, they also can block or prevent application attacks such as remote file inclusions and SQL injections (Luong 2010; Rietta 2006). They can provide information that helps to determine the nature of attack such as its source and propagation properties. In general, NIDS can be divided into two main types: misuse-based and anomaly-based. In this work, we specifically focus on ML-based misuse NIDS with multiple-class detection.

ML techniques are broadly used in network intrusion detection systems due to their superior nature of finding hidden patterns and abstract representation of features of network traffic (Lee et al. 2017; Vinayakumar et al. 2019; Zamani and Movahedi 2013). ML-based NIDS are trained with a database of normal and attack traffic observed in the network and then used to detect the observed attacks, their mutations (in misuse-based detection systems), or abnormal behaviors that deviate from normal traffic learned from the database (in anomaly-based detection systems). In the misuse-based NIDS type, ML-based techniques are known to detect attack traffic that are not detectable signature-based techniques. As the size of the database grows larger, the ML-based techniques could get much more effective at detecting attacks than these of signature-based techniques (Lee et al. 2017; Vinayakumar et al. 2019; Zamani and Movahedi 2013).

However, ML models are vulnerable to adversarial examples (carefully crafted inputs by an adversary)(Biggio et al. 2013; Carlini and Wagner 2017; Papernot et al. 2016; Szegedy et al. 2013). For instance, in the image domain, various types of attack algorithms have been shown to fool classifications of well-trained machine learning models by adding perturbation unnoticeable by human eyes to an original image. The existence of adversarial examples is argued to be inevitable in certain classes of problems (Shafahi et al. 2018). There has been an extensive study of adversarial examples in the unconstrained domains, specifically in the image domain, where an adversary can modify any features of the inputs by any amount. However, in constrained domains such as intrusion detection, the features of the network traffic must obey the domain constraints of the network

Game Theory and Machine Learning for Cyber Security, First Edition.
Edited by Charles A. Kamhoua, Christopher D. Kiekintveld, Fei Fang, and Quanyan Zhu.
© 2021 The Institute of Electrical and Electronics Engineers, Inc. Published 2021 by John Wiley & Sons, Inc.

traffic domain. Most of the previous work consider arguable domain constraints or no constraints in their process of crafting adversarial traffic (Hu and Tan 2017; Lin et al. 2018; Piplai et al. 2020; Wang 2018).

In this chapter, we will generate adversarial examples that bypass ML-based NIDS using generative adversarial networks (GANs) (Goodfellow et al. 2014). We evaluated our attack algorithm on the state-of-the-art CICICDS-2017 dataset that contains the most-up-to-date attacks and the benchmark NSL-KDD dataset. In all of the experiments, we assume that the attacker has a full knowledge of the model parameters of the ML-based NIDS or its training data. Our GAN-based attack algorithm crafts adversarial network traffic by adding optimized perturbations to attack traffic while maintaining the domain constraints. The results demonstrate that our GAN-based attack successfully evades the DNN-based IDS models.

17.2 Background

17.2.1 Network Intrusion Detection Systems

A network intrusion detection system (NIDS) is software that detects malicious activities that are intended to steal or corrupt sensitive information in computer networks. The NIDS have two main types: anomaly-based and misuse-based. In this chapter, we focus on the misuse-based detection systems. The misuse-based NIDS are divided into two main categories: knowledge-based and ML-based. In knowledge-based NIDS (also known as a signature-based detection system), attack network traffic flows are directly compared with previously defined rules or attack patterns. One of the drawbacks of signature-based NIDS is that they are not able to detect mutations of known attacks. ML-based misuse detection systems, on the other hand, learn from database signatures and predict the possible mutations of known attacks, therefore, they are widely used for academic and commercial purposes. In this chapter, we will specifically focus on ML-based misuse NIDS.

17.2.2 Adversarial Examples

Deep neural networks (DNNs) can easily be manipulated by adversarial examples (carefully crafted inputs). In 2013, Szegedy et al.(Papernot et al. 2016), and Biggio et al. (2013), first discovered the existence of adversarial examples in the image domain. By applying a small perturbation unnoticeable to human eyes to an original instance of an image, they were able to change the classification of a trained model for the image. By exploiting this vulnerability of neural networks, adversaries can potentially manipulate self-driving cars, smart devices, and intrusion detection systems that rely on DNNs. Given an input $\vec{x}$ and a trained target model $F(\vec{x})$, an adversary tries to find the minimum perturbation $\Delta\vec{x}$ under some norm (l^0, l^2, or l^∞) (Carlini and Wagner 2017) to cause the $F(\vec{x} + \Delta\vec{x})$ to be classified as a specific target label $t \neq F(\vec{x})$ (Szegedy et al. 2013). This can generally be expressed by an optimization problem where c is a constant that is found by binary search and t is the target class (Carlini and Wagner 2017; Szegedy et al. 2013).

$$\min_{\Delta\vec{x}} c|\Delta\vec{x}| + \text{Loss}(F(\vec{x} + \Delta\vec{x}), t)$$
$$\text{such that } (\vec{x} + \Delta\vec{x}) \in [0, 1]^m \tag{17.1}$$

where m is the dimension of the vector $\vec{x}$. As mentioned, most of the work has been done in unconstrained domains, specifically in the image domain. However, most of the domains in the real world are constrained. In this work, we create adversarial examples in a network intrusion detection domain where we determine and enforce the domain constraints. In our problem, F is the target ML-based NIDS we want to fool and the target class t is the benign class.

17.2.3 Generative Adversarial Networks

GANs (Goodfellow et al. 2014), generative models first created in 2014, have achieved unprecedented success in computer vision and natural language processing. GANs consist of two competing neural networks: a generator and a discriminator. The goal of the generator is to create an example that is indistinguishable from examples in the training data. The discriminator is a critic that attempts to output 0 to the generator's fake output and 1 to real example from the training data. The competition between these two neural networks improves each other's performance and eventually, the generator outputs examples that are indistinguishable from the examples of the training data. We use GANs in our algorithm for two reasons: (1) GANs functionality in our objective function will strive to optimize their loss functions such that the generated adversarial examples are from the set of the original attack traffic. (2) GAN-based attack algorithm AdvGAN (Xiao et al. 2018) is known to be very effective at fooling DNN models in the image domain.

17.2.4 Crafting Adversarial Examples Using GANs

There are two main types of GAN-based algorithms to craft adversarial examples: IDSGAN (Lin et al. 2018) and AdvGAN (Xiao et al. 2018). The first algorithm IDSGAN creates adversarial examples in the network intrusion domain. The IDSGAN, depicted in Figure 17.1, first divides features of attack traffic into two groups, important features $\vec{x}_I$ that preserve the semantics of the attack traffic and unimportant features $\vec{x}_N$ that do not have any effect on the semantics of the attack traffic. Then, the IDSGAN's generator takes the unimportant features $\vec{x}_N$ as an input and replaces them with $\vec{x}_C$ without any constraints enforced. This algorithm raises two important questions: (1) Is the perturbed network traffic still valid attack traffic? (2) How do we justify the choice that divides the attack traffic's features into unimportant and important categories? The choice of the important features in this paper is based on prior work, which is insufficient for two reasons. First, the important features of a specific attacks might vary from one environment to a different environment. Second, categorizing attack traffic's features into important and unimportant groups is inherently complex.

The AdvGAN (Figure 17.2) is the first algorithm that crafts adversarial examples by applying perturbations to the original instance. The algorithm outperformed many popular attacks such as PGD, C&W, FGSM, and JSMA in blackbox settings and won first place in Madry's competition held by MIT. However, the algorithm is created for only image domain. We modified this algorithm such that the crafted adversarial example obeys the constraints of the network traffic.

The AdvGAN's goal is to train a generator that crafts a perturbation Δx for an original instance x so that the classification of the $x + \Delta x$ by a target model becomes the class the adversary wants.

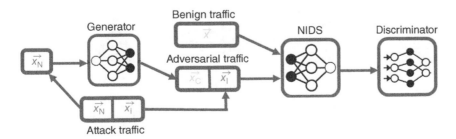

Figure 17.1 IDSGAN.

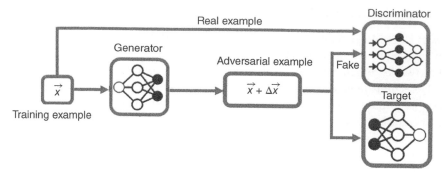

Figure 17.2 AdvGAN.

The optimal such generator can be trained by the competition between the discriminator and the generator networks.

17.3 Methodology

17.3.1 Target NIDS for Detecting Attack Traffic

ML-based NIDS are often trained with a database consisting of both attack and benign traffic. In this work, we evaluated our attack algorithm against two state-of-the-art DNN-based NIDS models as well as our own NIDS model.

To find the optimal DNN model for the NIDS, we vary the number of layers and the number of neurons in the layers depending on the dataset. Theoretically, the optimal number of hidden layers of the DNN is related to the complexity of the datasets. In this work, we choose an optimal architecture of a target NIDS for each dataset based on the evaluations as shown in Tables 17.1 and 17.2. We experimented with five different DNNs differentiated by their number of layers. Each layer of DNNs consists of a number of perceptrons, a ReLU activation function, and a dropout function that prevents overfitting. For each case, we calculate their accuracy, precision, recall, and F-score as shown in Tables 17.3 and 17.4 (where one-layer DNN represents a multiclass logistic regression). For choosing the number of perceptrons in the hidden layers, we used the benchmark rule that the number of perceptrons in the hidden layers should be between the number of neurons in input and output layer (the number of perceptrons are modified for performance improvement). After a number of iterations, we find that the following architectures can provide results as high as the state-of-the-art results shown in the previous work (Abdulraheem and Ibraheem 2019; Gao

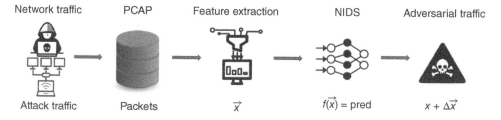

Figure 17.3 Diagram of creating adversarial examples. First, network traffic are simulated on computer networks and stored into PCAP format. Second, traffic features are extracted from the PCAP data using CICFlowMeter. Third, ML-based NIDS models are trained with the network traffic in feature space. At this last step, our attack algorithm crafts adversarial traffic that fools the NIDS models.

Table 17.1   Vinayakumar et al. Abdulraheem and Ibraheem (2019) DNN model.

Type/layer	First	Second	Third	Fourth	Fifth	Output
MLP (ReLU)	1024	768	512	256	128	—
Batch normalization	1024	768	512	256	128	—
Dropout (0.01)	1024	768	512	256	128	—
Fully connected (softmax)	—	—	—	—	—	7 or 5

Table 17.2   Gao et al. (2020) DNN model.

Type/layer	First	Second	Third	Fourth	Output
MLP (ReLU)	256	256	256	256	—
Dropout (0.01)	256	256	256	256	—
Fully connected (softmax)	—	—	—	—	7 or 5

Table 17.3   Target model NIDS for NSL-KDD.

Architecture	Layers	Accuracy	Precision	Recall	F-score
DNN 1 layer	123,5	0.75	0.89	0.62	0.74
DNN 2 layers	123,64,5	0.78	0.89	0.65	0.75
DNN 3 layers	123,64,32,5	0.79	0.89	0.65	0.75
DNN 4 layers	123,80,64,24,5	0.77	0.88	0.63	0.73
DNN 5 layers	123,80,48,32,16,5	0.77	0.88	0.63	0.74

Table 17.4   Target model NIDS for CICIDS-2017.

Architecture	Layers	Accuracy	Precision	Recall	F-score
DNN 1 layer	82,7	0.97	0.88	0.97	0.92
DNN 2 layers	82,42,7	0.980	0.928	0.964	0.946
DNN 3 layers	82,42,21,7	0.985	0.949	0.969	0.959
DNN 4 layers	82,60,42,28,7	0.982	0.942	0.958	0.950
DNN 5 layers	82,60,42,28,14,7	0.96	0.91	0.87	0.89

et al. 2020; Panigrahi and Borah 2018; Rahul Vigneswaran et al. 2018). As shown from the experiments in Tables 17.3 and 17.4, adding more layers to three-layer DNN does not improve the accuracy due to overfitting, but it increases the computational cost.

For these reasons, we choose three-layer DNN as our target NIDS. The prediction accuracy, architecture, and hyperparameters of the target models' are shown in Table 17.5. According to published results of the DNN-models trained on CICIDS and NSL-KDD datasets, the prediction accuracies

Table 17.5 Prediction set up for target models.

Dataset	Layers (MLP)	Batch size	Learning rate	Epochs	Accuracy
CICIDS	82,42,21,7	256	0.001	50	98.5%
NSL-KDD	123,64,32,5	32	0.01	50	79.0%

we found in our models are at the same level as the state-of-the-art predictions (Vinayakumar et al. 2019). In the next section, we introduce domain constraints.

17.3.1.1 Discriminator and Generator Architectures

We choose the number of neurons on each layer to be power of two. We observe that having two hidden layers is sufficient for the training of the algorithm. Having more than two hidden layers does not improve the generator training but significantly increase the training time of the attack algorithm (Table 17.6).

17.3.2 Constraints

In real-world applications, data samples in most domains obey certain rules. For instance, in network datasets, traffic flows will have a protocol type such as TCP. The network traffic of the TCP protocol will have only certain flags and services etc. Here, we determine the two main domain constraints in this work, which are enforced in the process of crafting adversarial examples.

1. Enforcement of constraints on intrinsic features
   (a) Attack type dictates that a set of feature values of traffic that must be kept constant.
   (b) Protocol type dictates that a certain subset of features must be zero. Also, a certain protocol is allowed to have only certain flags and services for feature values.
   (c) One-hot features converted from categorical must not be perturbed.
   To enforce the first constraint, we mask the computed perturbations on the corresponding features of type (a), (b), and (c) at the each iteration of the training as shown in Table 17.7. Here, the perturbation computed by the generator is $\Delta \vec{x} = (\Delta x_1, \ldots, \Delta x_n)^T$. (For the NSL-KDD and the CICIDS-2017 dataset, $n = 123$ and $n = 82$, respectively.) We do this by taking the Hadamard product of the perturbation vector $\Delta \vec{x}$ with a mask vector $\vec{m}$ that enforces these constraints. We implement this functionality within the function of the generator in the code.

$$\vec{m} = (m_1, \ldots, m_n)^T \qquad m_k = \begin{cases} 1, \text{feature included} \\ 0, \text{otherwise} \end{cases} \tag{17.2}$$

2. Enforcement of constraints on valid ranges (constraints of the attacker)
   (a) For each type of attack, we find the minimum and the maximum value of each feature and define the range between the minimum and maximum as the valid range. Then, after

Table 17.6 Target model NIDS for CICIDS-2017.

Network activation function	First layer (leaky ReLU)	Second layer (leaky ReLU)	Output layer (tanh/sigmoid)
Generator	[n,256]	[256,128]	[128,n]
Discriminator	[n,64]	[64,32]	[32,1]

Table 17.7 Depiction of enforcement 1.

Vector type/index	1	2	3	4	5	...	n
Perturbation ($\Delta \vec{x}$)	$\Delta \vec{x}_1$	$\Delta \vec{x}_2$	$\Delta \vec{x}_3$	$\Delta \vec{x}_4$	$\Delta \vec{x}_5$	...	$\Delta \vec{x}_n$
Mask vector ($\vec{m}$)	1	0	1	0	1	...	1
Result ($\Delta \vec{x} \circ \vec{m}$)	$\Delta \vec{x}_1$	0	$\Delta \vec{x}_3$	0	$\Delta \vec{x}_5$	...	$\Delta \vec{x}_n$

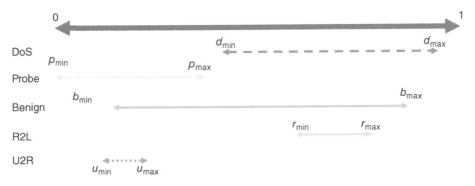

Figure 17.4 Enforcement 2: The valid ranges of dos, probe, benign, r2l, and u2r classes are defined by $[d_{min}, d_{max}], [p_{min}, p_{max}], [b_{min}, b_{max}], [r_{min}, r_{max}], [u_{min}, u_{max}]$, respectively.

applying the perturbation to original attack traffic during the process of crafting adversarial examples, we enforce the valid range for each feature of attack traffic by clipping. Figure 17.4 shows that valid ranges of each class in NSL-KDD dataset for a particular feature x_1. For example, if the feature value of x_1 of a dos attack traffic becomes $x_1 + \Delta x_1 < d_{min}$ after a perturbation is applied, then we clip the value and replace it with d_{min}.

17.3.3 Preprocessing

- *Replacing undefined NaN (not a number) values*: The target DNN model cannot be trained with NaN values in its entries. We replaced NaN the values with the median of the corresponding feature values because the percentage of NaN values in the corresponding feature is almost less than 1%.
- *One hot encoding*: The DNN model cannot be trained with categorical features. We applied one-hot encoding for all categorical features for both datasets.
- *Normalization*: We normalize all the feature values between 0 and 1 because the gradient descent optimization can be much faster with much more straightforward path.

17.3.4 The Attack Algorithm

The main objective of our algorithm (Figure 17.3) is to add a minimum possible perturbation to attack network traffic such that the resulting adversarial traffic successfully bypasses DNN-based NIDS while maintaining its domain constraints. To achieve this, there must be two terms in the optimization (1) L_{adv} is optimized to convert the target model's classification to be benign. (2) L_{pert} is optimized to minimize the perturbation measured with respect to the l^2 norm. We adopt these functionalities from the AdvGAN (Xiao et al. 2018) and add two main modifications on it for crafting adversarial traffic: (1) enforcing constraints and (2) better GAN framework (Wasserstein GAN;

Gulrajani et al. (2017)) with gradient penalty. In each iteration of the algorithm training, we enforce the defined constraints in two steps: (1.a) we clip each feature value to enforce the corresponding valid range of the feature. (1.b) perturbations on the intrinsic features of the traffic are zeroed out. For the original GANs' structure, two loss functions are defined: L_G for the generator and L_D for the discriminator. (2.a) WGAN-GP handles the problems of gradient explosion or vanishing, mode collapse, and nonconvergence issues, therefore, improves the performance of crafting adversarial traffic.

The main loss function (Figure 17.5) is determined by the following equation.

$$L_G = L_{adv} + \alpha L_{GAN} + \beta L_{pert} \tag{17.3}$$

1. The objective of L_{adv} is to modify the generator weights such that the perturbed attack traffic is classified as benign traffic. It measures how far $\vec{x} + \Delta\vec{x}$ is from the boundary of the benign class. If the $\vec{x} + \Delta\vec{x}$ is inside the benign class the loss is certainly 0. Otherwise, it penalizes the loss function based on cross entropy.

$$L_{adv} = \mathop{\mathbb{E}}_{\vec{x} \sim \mathbb{P}_r} [\text{Loss}(F(\vec{x} + \Delta\vec{x}), t)] \tag{17.4}$$

2. The objective of L_{GAN} is to make the perturbed attack traffic $\vec{x} + \Delta\vec{x}$ to be indistinguishable from the original attack traffic. The discriminator D tries to determine the perturbed traffic $\vec{x} + \Delta\vec{x}$ from the real traffic. This helps in two ways: (1) it helps to minimize the perturbation and (2) it helps to maintain the domain constraints. Two types of GAN losses are used in the experiments. L_{GAN} represents the original GAN loss (Goodfellow et al. 2014) and the $L_{WGAN-GP}$ represents the Wasserstain GAN with gradient penalty loss (Gulrajani et al. 2017). WGAN-GP has significantly improved the original GAN's problems such as: gradient vanishing or explosion, mode collapse, and nonconvergence. Therefore, WGAN-GP loss provides better results in our experiments.

$$\begin{aligned}
\vec{x} &\sim \mathbb{P}_r. \\
\Delta\vec{x} &\leftarrow E_1[G(\vec{x})] \\
\vec{x}^* &\leftarrow E_2[\vec{x} + \Delta\vec{x}] \\
\hat{x} &\leftarrow \sigma\vec{x} + (1 - \sigma)\vec{x}^* \\
L_{GAN} &= \min_G \max_D \mathop{\mathbb{E}}_{\vec{x} \sim \mathbb{P}_r} [\log D(\vec{x})] + \mathop{\mathbb{E}}_{\vec{x} \sim \mathbb{P}_r} [\log(1 - D(\vec{x}^*))] \\
&\quad + \mathop{\mathbb{E}}_{\hat{x}}[\lambda(\| \nabla_{\hat{x}}D(\hat{x})\|_2 - 1)^2]
\end{aligned} \tag{17.5}$$

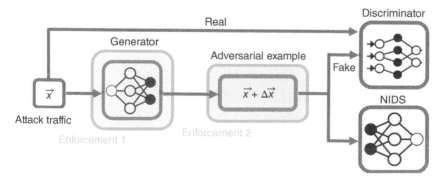

Figure 17.5 Flowchart of creating adversarial examples.

Algorithm 17.1 Crafting adversarial network traffic that bypass DNN-based NIDS.

Require: x-input, F-target NIDS, G-generator, D-discriminator, t-target class, E_1-enforcing type-1 constraints, E_2-enforcing type-2 constraints, n-number of epochs, m-number of batches

Require: initial discriminator parameters w_0, initial generator parameters θ_0

while $\arg\max F(\vec{x}^*) \neq t$ **do**
  for epoch=1,...,n **do**
    for i=1,...,m **do**
      Sample real data $x \sim \mathbb{P}_r$.
      $\Delta \vec{x} \leftarrow E_1[G(\vec{x})]$
      $\vec{x}^* \leftarrow E_2[\vec{x} + \Delta \vec{x}]$
      $L_{GAN} \leftarrow \log D(x) + \log(1 - D(x^*))$
      $w \leftarrow \text{AdamOptimizer}(\nabla_w L_{GAN}^{(i)}, w, \alpha)$
      $\theta \leftarrow \text{AdamOptimizer}(\nabla_\theta L_{GAN}^{(i)} + L_{Pert}^{(i)} + L_{Adv}^{(i)} + L_{Const}^{(i)}, \theta, \alpha)$
    end for
  end for
end while

---

3. The objective of L_{pert} is to minimize the size of the perturbation $\Delta \vec{x}$ so that the perturbed traffic will be as close as to real traffic input. With L_{GAN} optimization, the distribution of $\vec{x} + \Delta \vec{x}$ approaches that of the real traffic distribution (Goodfellow et al. 2014). As specified in the prior work, we bound the l^2 perturbation magnitude of the loss function with following equation where the ϵ- bound is typically chosen to be 0.3. Hinge loss is commonly used for the optimization of the perturbation. Therefore, optimizing for the hinge loss results a better success rate than optimizing for $L_{\text{pert}} = \| \Delta \vec{x} \|_2$.

$$L_{\text{pert}} = \mathop{\mathbb{E}}_{\vec{x} \sim \mathbb{P}_r} [\max(0, \| \Delta \vec{x} \|_2 - \epsilon)] \tag{17.6}$$

The optimal values of α and β coefficients in the generator's loss function are found to be 0.1 and 0.2 with systematic grid search method. The algorithm for crafting adversarial examples that bypass DNN-based NIDS is shown below.

Choice of l^2 norm We pick l^2 norm to bound our perturbations because (1) exceedingly high magnitude of modifications onto a feature would easily break the both the validity of constraints and functionality of attack semantics; (2) large l^2-measured perturbations render higher detectability against statistics-based NIDS in general. For these two reasons, l^2 norm well suits for our attack algorithm's objective as well as help to satisfy the constraints of the attacker.

As shown in Table 17.8, when $\| \Delta x \|_2$ perturbation is 0.3, the maximum number of features we can perturb by 0.3 is only one. When $\| \Delta x \|_\infty$ perturbation is 0.3, there is a possibility that the algorithms change all features by 0.3. Here, we show a rough estimate of how much perturbation can be applied to attack traffic's features by varying the number of features perturbed when $\| \Delta x \|_2$ is 0.3. For instance, 10 features out of 123 features are perturbed, the average perturbation level of 9 features are 0.094, 0.08, 0.07, 0.05, 0.0 when fixing $|\Delta x_1|$ (the one of the 10 perturbed features) values at 0.1, 0.15, 0.2, 0.25, 0.3, respectively. (1) l^∞ vs. l^2: l^2 norm enforces to make small changes to many features while l^∞ norm allows to make large changes to many features. For instance, $l^\infty = 0.3$ allows to perturb all of the features by 0.3 while $l^2 = 0.3$ allows to perturb only one feature by 0.3 or perturb 30 features by 0.055 on average. (2) l^0 vs. l^2: l^0 norm minimizes the number of features to perturb. This can have two downsides: (1) It perturbs the features by any amount, which can make the attack easily detectable by finding an outlier for a particular perturbed feature. (2) When the

Table 17.8 Average perturbation percentage for a single feature based on $\|\Delta x\|_2$ less than 0.3.

Number of features perturbed	Δx_1				
	0.10	0.15	0.20	0.25	0.30
30	0.052	0.04	0.04	0.03	0.0
20	0.064	0.05	0.05	0.03	0.0
10	0.094	0.08	0.07	0.05	0.0
5	0.14	0.12	0.11	0.08	0.0
2	0.28	0.25	0.22	0.16	0.0
1	0.30	0.30	0.30	0.30	0.30

adversary perturbs a small number of features less than five, it could give up the characteristics of the attack traffic that defines its semantics and adopt the behavior that defines the characteristics benign traffic. l^∞ and l^0 norms could result making large changes to features, which is a big risk for the adversary.

17.4 Evaluation

CICIDS-2017. The CICIDS-2017 dataset (Sharafaldin et al. 2018) consists of seven separate files of network traffic simulated within five weekdays. We combined all the seven files and performed preprocessing as follows. To train the target DNN model, the dataset is divided into training and test data containing 75% and 25% of the original dataset, respectively. The dataset is created by simulation based on 25 internet users including email, HTTP, HTTPS, FTP, and SSH protocol traffic. It also includes raw PCAP payload data. The dataset consists of 14 classes: BENIGN, DoS Hulk, PortScan, DdoS, DoS GoldenEye, FTP-Patator, SSH-Patator, DoS slowloris, DoS Slow-httptest, Bot, Web Attack-Brute Force, Web Attack-XSS, Infiltration, Web Attack-Sql Injection, and Heartbleed attacks. Due to high imbalance of the datasets and poor prediction accuracy, same type of attacks are combined into the same class and total of seven classes are generated: benign, bot, pat (Patator), dos (DoS), inf (infiltration), port, and web as recommended (Panigrahi and Borah 2018).

NSL-KDD. The NSL-KDD dataset (Tavallaee et al. 2009) is probably the most popular dataset in the intrusion detection community. It is an improved version of KDD Cup99 which included redundant records in both training and testing sets. NSL-KDD dataset has five classes: dos, probe, r2l, u2r, and benign. The whole dataset contains 125 973 and 22 543 samples for training and testing, respectively. Each sample has 41 features that can be divided into three different categories: connection, traffic, and host features. Connection features include features such as protocol type, traffic features are numeric features that are processed using a 2-second time window.

17.4.0.1 Success Rate
In this chapter, the success rate represents the percentage of adversarial traffic that successfully fooled the ML-based NIDS.

17.4.1 Results

Overall, our experiments performed on the CICIDS-2017 and NSL-KDD datasets demonstrate that our algorithm is an effective method for crafting adversarial examples in the network intrusion

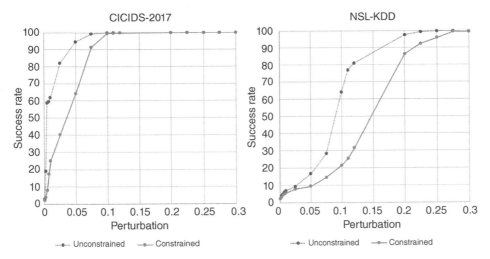

Figure 17.6 Success rate comparison of adversarial network traffic generated on CICIDS-2017 and NSL-KDD dataset in unconstrained and constrained cases.

detection domain, achieving nearly success rates of 100% in a white-box setting with l^2 norm perturbation, $\|\Delta\vec{x}\|_2 = 0.275$ (see Figure 17.6). To show the effect of adding constraints to the process of crafting adversarial examples, graphs of constrained and unconstrained cases are compared. Adding and enforcing the domain constraints in crafting adversarial examples shrinks the space of allowed perturbations. Therefore, crafting adversarial traffic in constrained domains is harder than crafting adversarial traffic in unconstrained domains. In the CICIDS-2017 dataset, the success rate of the constrained case is significantly less than that of unconstrained case when $\|\Delta\vec{x}\|_2$ is less than 0.1. When the $\|\Delta\vec{x}\|_2$ perturbation is 0.01, the difference of success rate is largest at approximately 50%. The difference slowly shrinks as the perturbation increases to 0.1. For any value of the perturbation greater than 0.1, the success rates in both cases are both 100%. In the NSL-KDD dataset, the difference between the success rates of the two cases is largest when perturbation is 0.12. The difference shrinks as the perturbation increases and becomes zero when the perturbation is 0.275. Although the constraints can have a significant effect for crafting adversarial examples with small perturbation, the effect decreases and eventually disappears at a certain perturbation level.

To see the effect of adding constraints, success rate comparisons of unconstrained and constrained cases are plotted in Figures 17.7 and 17.8. Our algorithm achieved 100% success rate with perturbation less than 0.1 for pat, dos, inf, and port attacks. In general, there is not much difference noticed in the success rate between the constrained and unconstrained cases. For web attack, the highest success rate we can achieve is 95%. For bot attack, the success rate reaches only 5% when $\|\Delta\vec{x}\|_2$ perturbation is less than 0.1. The main reason for this phenomenon is that the bot and the web attack traffic are the second and third smallest attacks in the CICIDS dataset behind inf attack traffic in terms of class size. Therefore, the constraint enforced by our algorithm eliminates the huge chunk of perturbation space that can be applied in an unconstrained domain.

With the NSL-KDD dataset, crafting adversarial examples gets harder in constrained domain. Compared to CICIDS-2017, at least perturbation level of 0.2 is needed to achieve high success rate in attack traffic. Enforcing the constraints on the valid range is much harder for the NSL-KDD dataset because the valid range of CICIDS-2017 dataset (Table 17.9) (3 million samples) is much larger than that of NSL-KDD (Table 17.10) (150 thousand samples). Therefore, a NIDS trained with the CICIDS-2017 dataset allows an adversary to have a much larger perturbation space than NIDS

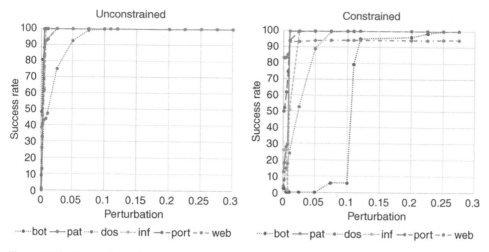

Figure 17.7 Comparison of success rates for each attack on CICIDS-2017 dataset in unconstrained and constrained cases.

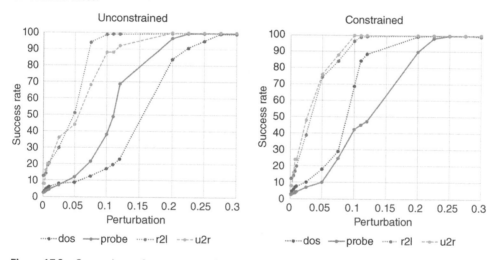

Figure 17.8 Comparison of success rates for each attack on NSL-KDD dataset in unconstrained and constrained cases.

Table 17.9 CICIDS-2017: target model's prediction and size for each class.

Type/class	benign	bot	pat	dos	inf	port	web
Class size	83.3%	0.08%	0.5%	10.4%	0.001%	5.6%	0.07%
Prediction accuracy	99%	38%	98%	100%	45%	100%	93%

Table 17.10 NSL-KDD: target model's prediction and size for each class.

Type/class	benign	dos	probe	r2l	u2r
Class size	52.1%	36.1%	9.4%	0.1%	2.3%
Prediction accuracy	97%	83.8%	66.7%	14.3%	29.9%

Table 17.11 Success rate of the GAN-based attack algorithm in blackbox setting with $\|\Delta x\|_2$ less than 0.3.

NIDS based on ML models	Success Rate		
	VIN et al.	GAO et al.	Our model
Decision tree	62%	70%	97%
Support vector machine	45%	95%	1%
K nearest neighbor	61%	71%	56%
Logistic regression	72%	74.4%	87%
	CICIDS-2017		

trained with the NSL-KDD dataset. In future work, we will modify and evaluate our algorithm for black-box setting where an adversary does not have access to the NIDS model's parameters or its training data.

17.4.2 Transfer-based Attack

Takeaway We evaluate our attack algorithm against NIDS models based on classical ML algorithms: Decision tree, support vector machine, *k*-nearest neighbors, and logistic regression in whitebox setting as shown in Table 17.11. The adversarial traffic crafted by our algorithm based on the three architectures are able to fool the models trained with the classical ML algorithms in most of the cases and outperform prior work for two models (DT and LR).

17.5 Conclusion

Modern NIDS are using ML techniques for the detection attack traffic. However, research focused on the threat of the ML-aided attack algorithm against these systems is scarce. In this work, we introduce GAN-based attack algorithm that crafts adversarial traffic that bypasses DNN-based as well as classical ML models-based NIDS while maintaining the domain constraints. We evaluate our algorithm state-of-the-art DNN-based models as well as our own well-trained DNN-based NIDS and achieve nearly 100% success rates against these models; we achieve 70% success rate on average against the classical ML-based NIDS. Currently, we are looking at how to map the network traffic from feature space to PCAP format space and also the other way around. With these mappings, we will be able to simulate our adversarial traffic in the computer networks and see how much damage it can do to the networks. In the future work, we will investigate a blackbox threat model when the only capability of the adversary is to query network traffic from the NIDS and get a feedback for it. At the final stage, we will study various types of defense mechanisms against both whitebox and blackbox attack scenarios.

References

Mohammed Hamid Abdulraheem and Najla Badie Ibraheem. A detailed analysis of new intrusion detection dataset. *Journal of Theoretical and Applied Information Technology*, 97(17), 2019.

Battista Biggio, Igino Corona, Davide Maiorca, Blaine Nelson, Nedim Šrndić, Pavel Laskov, Giorgio Giacinto, and Fabio Roli. Evasion attacks against machine learning at test time. In *Joint European Conference on Machine Learning and Knowledge Discovery in Databases*, pp. 387–402. Springer, 2013.

Nicholas Carlini and David Wagner. Towards evaluating the robustness of neural networks. In *2017 IEEE Symposium on Security and Privacy (SP)*, pp. 39–57. IEEE, 2017.

Zhihua Cui, Fei Xue, Xingjuan Cai, Yang Cao, Gai-ge Wang, and Jinjun Chen. Detection of malicious code variants based on deep learning. *IEEE Transactions on Industrial Informatics*, 14(7):3187–3196, 2018.

Minghui Gao, Li Ma, Heng Liu, Zhijun Zhang, Zhiyan Ning, and Jian Xu. Malicious network traffic detection based on deep neural networks and association analysis. *Sensors*, 20(5):1452, 2020.

Ian Goodfellow, Jean Pouget-Abadie, Mehdi Mirza, Bing Xu, David Warde-Farley, Sherjil Ozair, Aaron Courville, and Yoshua Bengio. Generative adversarial networks. In *Advances in Neural Information Processing Systems*, pp. 2672–2680, 2014. arXiv preprint, arXiv:1406.2661

Ishaan Gulrajani, Faruk Ahmed, Martin Arjovsky, Vincent Dumoulin, and Aaron C. Courville. Improved training of wasserstein gans. In *Advances in Neural Information Processing Systems*, pages 5767–5777, 2017. arXiv preprint, arXiv:1704.00028.

Weiwei Hu and Ying Tan. Generating adversarial malware examples for black-box attacks based on gan. *arXiv preprint, arXiv:1702.05983*, 2017.

C. Lee, Y. Su, Y. Lin, and S. Lee. Machine learning based network intrusion detection. In *2017 2nd IEEE International Conference on Computational Intelligence and Applications (ICCIA)*, pp. 79–83. IEEE, 2017.

Zilong Lin, Yong Shi, and Zhi Xue. IDSGAN: Generative adversarial networks for attack generation against intrusion detection. *arXiv preprint, arXiv:1809.02077*, 2018.

Varian Luong. Intrusion detection and prevention system: SQL-injection attacks. 2010. scholarworks.sjsu.edu.

Mithun Mukherjee, Rakesh Matam, Lei Shu, Leandros Maglaras, Mohamed Amine Ferrag, Nikumani Choudhury, and Vikas Kumar. Security and privacy in fog computing: Challenges. *IEEE Access*, 5:19293–19304, 2017.

Ranjit Panigrahi and Samarjeet Borah. A detailed analysis of CICIDS2017 dataset for designing intrusion detection systems. *International Journal of Engineering & Technology*, 7(3.24):479–482, 2018.

Nicolas Papernot, Patrick McDaniel, Somesh Jha, Matt Fredrikson, Z. Berkay Celik, and Ananthram Swami. The limitations of deep learning in adversarial settings. In *2016 IEEE European Symposium on Security and Privacy (EuroS&P)*, pp. 372–387. IEEE, 2016.

Aritran Piplai, Sai Sree Laya Chukkapalli, and Anupam Joshi. Nattack! adversarial attacks to bypass a GAN based classifier trained to detect network intrusion. In *2020 IEEE 6th International Conference on Big Data Security on Cloud (BigDataSecurity), IEEE International Conference on High Performance and Smart Computing, (HPSC) and IEEE International Conference on Intelligent Data and Security (IDS)*, 2020.

Frank S. Rietta. Application layer intrusion detection for SQL injection. In *ACM-SE 44: Proceedings of the 44th Annual Southeast Regional Conference, March 2006*, pp. 531–536, 2006. https://doi.org/10.1145/1185448.1185564.

Imtithal Saeed, Ali Selamat, and Ali Abuagoub. A survey on malware and malware detection systems. *International Journal of Computer Applications*, 67:25–31, 2013.

Ali Shafahi, W. Ronny Huang, Christoph Studer, Soheil Feizi, and Tom Goldstein. Are adversarial examples inevitable? *arXiv preprint, arXiv:1809.02104*, 2018.

Iman Sharafaldin, Arash Habibi Lashkari, and Ali A. Ghorbani. Toward generating a new intrusion detection dataset and intrusion traffic characterization. In Proceedings of the 4th International Conference on Information Systems Security and Privacy (ICISSP 2018), pp. 108–116, 2018.

Sami Smadi, Nauman Aslam, and Li Zhang. Detection of online phishing email using dynamic evolving neural network based on reinforcement learning. Decision Support Systems, 107:88–102, 2018.

Christian Szegedy, Wojciech Zaremba, Ilya Sutskever, Joan Bruna, Dumitru Erhan, Ian Goodfellow, and Rob Fergus. Intriguing properties of neural networks. arXiv preprint, arXiv:1312.6199, 2013.

Mahbod Tavallaee, Ebrahim Bagheri, Wei Lu, and Ali A. Ghorbani. A detailed analysis of the KDD CUP 99 data set. In 2009 IEEE Symposium on Computational Intelligence for Security and Defense Applications, pp. 1–6. IEEE, 2009.

K. Rahul Vigneswaran, R. Vinayakumar, K. P. Soman, and Prabaharan Poornachandran. Evaluating shallow and deep neural networks for network intrusion detection systems in cyber security. In 2018 9th International Conference on Computing, Communication and Networking Technologies (ICCCNT), pp. 1–6. IEEE, 2018.

R. Vinayakumar, Mamoun Alazab, K.P. Soman, Prabaharan Poornachandran, Ameer Al-Nemrat, and Sitalakshmi Venkatraman. Deep learning approach for intelligent intrusion detection system. IEEE Access, 7:41525–41550, 2019.

Zheng Wang. Deep learning-based intrusion detection with adversaries. IEEE Access, 6:38367–38384, 2018.

Chaowei Xiao, Bo Li, Jun-Yan Zhu, Warren He, Mingyan Liu, and Dawn Song. Generating adversarial examples with adversarial networks. arXiv preprint, arXiv:1801.02610, 2018.

Mahdi Zamani and Mahnush Movahedi. Machine learning techniques for intrusion detection. arXiv preprint, arXiv:1312.2177, 2013.

Appendix 17.A
Network Traffic Flow Features

Table 17.A.1 Network traffic features of NSL-KDD.

Feature name	Description
Duration	Length (no. of seconds) of the connection
Protocol_type	length (no. of seconds) of the connection
Service	Network service on the destination
Flag	Normal or error status of the connection
Src _bytes	No. of data bytes from source to dest
Dst_bytes	No. of data bytes from dest to source
Land	1 if connection is from/to the same host/port
Wrong_fragment	No. of wrong fragments
Urgent	No. of urgent packets
Hot	No. of hot indicators
Num_failed_logins	No. of failed login attempts

(Continued)

Table 17.A.1 (Continued)

Feature name	Description
Logged_in	1 if successfully logged in
Num_compromised	No. of compromised conditions
Root_shell	1 if root shell is obtained; 0 otherwise
Su_attempted	1 if su root command attempted
Num_root	No. of root accesses
Num_file_creations	No. of file creation operations
Num_shells	No. of shell prompts
Num_access_files	No. of operations on access control files
Num_outbound_cmds	No. of outbound commands in an ftp session
Is_host_login	1 if the login belongs to the hot list
Is_guest_login	1 if the login is a guest login
Count	No. of conn to the same host as the current conn
Srv_count	No. of conn to the same service as the current conn
Serror_rate	No. of conn to the same host as the current conn
Srv_serror_rate	% of conn that have SYN errors
Rerror_rate	% of conn that have REJ errors
Srv_rerror_rate	% of conn that have REJ errors
Same_srv_rate	% of conn to the same service
Diff_srv_rate	% of conn to different services
Srv_diff_host_rate	% of conn to different hosts
Dst_host_count	No. of conn to same host to dest host (current)
Dst_host_srv_count	No. of conn from same service to dest host
Dst_host_same_srv_rate	% of conn from same services to dest host
Dst_host_diff_srv_rate	% of conn from different services to dest host
Dst_host_same_src_port_rate	% of conn from port services to dest host
Dst_host_srv_diff_host_rate	% of conn from different hosts from same service to dest host
Dst_host_serror_rate	% of conn having SYN errors from same host to dest host
Dst_host_srv_serror_rate	% of conn having SYN errors from same service to dest host
Dst_host_rerror_rate	% of conn having REJ errors from same host to dest host
Dst_host_srv_rerror_rate	% of conn having REJ errors from same service to dest host

Table 17.A.2 Network traffic features of CICIDS-2017.

Feature name	Description
Feduration	Duration of the flow in micsec
Flow feduration	Duration of the flow in micsec
Total fwwd pkt	Total pkts
Total bwd pkts	Total pkts
Total len of fwd pkt	Total size of pkt
Total len of bwd pkt	Total size of pkt
Fwd pkt len min	Min Size of pkt
Fwd pkt len max	Max size of pkt
Fwd pkt len mean	Mean size of pkt
Fwd pkt len std	Std dev size of pkt
Bwd pkt len min	Min size of pkt
Bwd pkt len max	Max size of pkt
Bwd pkt len mean	Mean size of pkt
Bwd pkt len std	Std dev size of pkt
Flow byte/s	No. of flow pkts/sec
Flow pkts/s	No. of flow bytes/sec
Flow IAT mean	Mean time btwn 2 pkts sent
Flow IAT std	Std dev time btwn 2 pkts sent
Flow IAT max	Max time btwn 2 pkts sent
Flow IAT min	Min time btwn 2 pkts sent
Fwd IAT min	Min time btwn 2 pkts sent
Fwd IAT max	Max time btwn 2 pkts sent
Fwd IAT mean	mean time btwn 2 pkts sent
Fwd IAT std	Std dev time btwn 2 pkts sent
Fwd IAT total	Total time btwn 2 pkts sent
Bwd IAT min	Min time btwn 2 pkts sent
Bwd IAT max	Max time btwn 2 pkts sent
Bwd IAT mean	Mean time btwn 2 pkts sent
Bwd IAT std	Std dev time btwn 2 pkts sent
Bwd IAT total	Total time btwn 2 pkts sent
Fwd PSH flag	No. of times the PSH flag was set
Bwd PSH flag	No. of times the PSH flag was set
Fwd URG flag	No. of times the URG flag was set
Bwd URG flag	No. of times the URG flag was set
Fwd header len	Total bytes used for headers
Bwd header len	Total bytes used for headers
FWD pkts/s	No. of fwd pkts/sec
Bwd pkts/s	No. of bwd pkts/sec

(Continued)

Table 17.A.2 (Continued)

Feature name	Description
Min pkt len	Min len of a pkt
Max pkt len	Max len of a pkt
Pkt len mean	Mean len of a pkt
Pkt len std	Std dev len of a pkt
Pkt len variance	Variance len of a pkt
FIN flag count	No. of pkts with FIN
SYN flag count	No. of pkts with SYN
RST flag count	No. of pkts with RST
PSH flag count	No. of pkts with PUSH
ACK flag count	No. of pkts with ACK
URG flag count	No. of pkts with URG
CWR flag count	No. of pkts with CWE
ECE flag count	No. of pkts with ECE
Down/up ratio	Download and upload ratio
Avg pkt size	Avg size of pkt
Avg fwd segment size	Avg size observed
Avg bwd segment size	Avg no. of bytes bulk rate
Fwd header len	Len of header for fwd pkt
Fwd avg bytes/bulk	Avg no. of bytes bulk rate
Fwd avg pkt/bulk	Avg no. of pkts bulk rate
Fwd avg bulk rate	Avg no. of bulk rate
Bwd avg bytes/bulk	Avg no. of bytes bulk rate
Bwd avg pkt/bulk	Avg no. of pkts bulk rate
Bwd avg bulk rate	Avg no. of bulk rate
Subflow fwd pkts	The avg no. of pkts in a sub flow
Subflow fwd bytes	The avg no. of bytes in a sub flow
Subflow bwd pkts	The avg no. of pkts in a sub flow
Subflow bwd bytes	The avg no. of bytes in a sub flow
Init_win_bytes_fwd	The total no. of bytes sent
Init_win_bytes_bwd	The total no. of bytes sent
Act_data_pkt_fwd	Count of pkts with at least 1 byte
Min_seg_size_fwd	Min segment size observed
Act min	Min time a flow was act before idle
Act mean	Mean time a flow was act before idle
Act max	Max time a flow was act before idle
Act std	Std dev time a flow was act before idle
Idle min	Min time a flow was idle before act
Idle mean	Mean time a flow was idle before act

Table 17.A.2   (Continued)

Feature name	Description
Idle max	Max time a flow was idle before act
Idle std	Std dev time a flow was idle before act
Total_fpkts	Total pkts
Total_bpkts	Total pkts
Total_fpktl	Total size of pkt
Total_bpktl	Total size of pkt
Min_fpktl	Min size of pkt
Min_bpktl	Min size of pkt
Max_fpktl	Max size of pkt
Max_bpktl	Max size of pkt
Mean_fpktl	Mean size of pkt
Mean_bpktl	Mean size of pkt
Std_fpktl	Std dev size of pkt
Std_bpktl	Std dev size of pkt
Total_fiat	Total time btwn 2 pkts sent
Total_biat	Total time btwn 2 pkts sent
Min_fiat	Min time btwn 2 pkts sent
Min_biat	Min time btwn 2 pkts sent
Max_fiat	Max time btwn 2 pkts sent
Max_biat	Max time btwn 2 pkts sent
Mean_fiat	Mean time btwn 2 pkts sent
Mean_biat	Mean time btwn 2 pkts sent
Std_fiat	Std dev time btwn 2 pkts sent
Std_biat	Std dev time btwn 2 pkts sent
Fpsh_cnt	No. of times the PSH flag was set
Bpsh_cnt	No. of times the PSH flag was set
Furg_cnt	No. of times the URG flag was set
Burg_cnt	No. of times the URG flag was set
Total_fhlen	Total bytes used for headers
Total_bhlen	Total bytes used for headers
Fpkt/sec	No. of fwd pkts/sec
Bpkt/sec	No. of bwd pkts/sec
Flowpkt/sec	No. of flow pkts/sec
Flowbyte/sec	No. of flow bytes/sec
Min_flowpktl	Min len of a flow
Max_flowpktl	Max len of a flow
Mean_flowpktl	Mean len of a flow
Std_flowpktl	Std dev len of a flow

(Continued)

Table 17.A.2 (Continued)

Feature name	Description
Min_flowiat	Min inter-arrival time of pkt
Max_flowiat	Max inter-arrival time of pkt
Mean_flowiat	Mean inter-arrival time of pkt
Std_flowiat	Std dev inter-arrival time of pkt
Flow_fin	No. of pkts with FIN
Flow_syn	No. of pkts with SYN
Flow_rst	No. of pkts with RST
Flow_psh	No. of pkts with PUSH
Flow_ack	No. of pkts with ACK
Flow_urg	No. of pkts with URG
Flow_cwr	No. of pkts with CWE
Flow_ece	No. of pkts with ECE
Downupratio	Download and upload ratio
Avgpktsize	Avg size of pkt
Favgsegmentsize	Avg size observed
Favgbyte/bulk	Avg no. of bytes bulk rate
Favgpkt/bulk	Avg no. of pkts bulk rate
Favgbulkrate	Avg no. of bulk rate
Bavgsegmentsize	Avg size observed
Bavgbytebulk	Avg no. of bytes bulk rate
Bavgpktbulk	Avg no. of pkts bulk rate
Bavgbulkrate	Avg no. of bulk rate
Sflow_fpkt	The avg no. of pkts in a sub
Sflow_fbytes	The avg no. of bytes in a sub
Sflow_bpkt	The avg no. of pkts in a sub
Sflow_bbytes	The avg no. of bytes in a sub
Min_act	Min time a flow was act
Mean_act	Mean time a flow was act
Max_act	Max time a flow was act
Std_act	Std dev time a flow was act
Min_idle	Min time a flow was idle
Mean_idle	Mean time a flow was idle
Max_idle	Max time a flow was idle
Std_idle	Std dev time a flow was idle
Init_win_bytes_fwd	The total no. of bytes sent
Init_win_bytes_bwd	The total no. of bytes sent
Act_data_pkt_fwd	Count of pkts with at least 1 byte
Min_seg_size_fwd	Min segment size observed

18

Concealment Charm (ConcealGAN): Automatic Generation of Steganographic Text Using Generative Models to Bypass Censorship

Nurpeiis Baimukan[1] and Quanyan Zhu[2]

[1] *New York University Abu Dhabi, Abu Dhabi, United Arab Emirates*
[2] *Department of Electrical and Computer Engineering, NYU Tandon School of Engineering, New York University, Brooklyn, NY, USA*

18.1 Censorship

A constant exchange of essential textual information through social media, email services, and other virtual means raises a concern about the extent to which our data is protected, especially under extensive censorship. Kelly et al. (2012) have found that out of 60 researched countries, only 17% has no political censorship, whilst more than one-third of the government oppressed netizens in some way.

Based on the recent observation of M. E. Roberts (2014), there are two types of mechanisms through which censorship can effectively impede the spread of information. On the one hand, the first mechanism is *fear*, where the government explicitly urges the netizens to abstain from spreading information on specific topics by showing possible ramifications. On the other hand, there is *information friction*, which does not require the users to be aware of the censorship. The latter mechanism makes access of information more difficult. As such, even negligible delays in the acquisition of information in an online world can have a significant impact on the traction of the online service. For instance, Google's market research showed that slowing down the search by 400 ms creates a 0.44% decrease in the search queries (Hoelzle 2012).

Comparing the efficacy of the above-mentioned mechanisms in the Chinese context, Roberts (2014) found in that information friction, not the knowledge of which topics are censored, is the main way through which censorship can be effective. The results imply that small costs of access to information have important implications for the spread of information online.

Here, we discuss how censorship works in messenger and emailing services. Most message exchange services, such as email, messengers, and others, are backed by client–server architectures. Thus, the content of the message is available for the service providers, even if there is end-to-end encryption. As a consequence, it is observed in Senftleben et al. (2016) that the confidentiality of the users is vulnerable since the service operators can analyze all of the content sent by users.

One of the most apparent examples of detrimental censorship can be seen in China with its most popular application called WeChat. In February 2018, WeChat reportedly hit one billion monthly active users during the Chinese Lunar New Year. Over the past two years, WeChat has transformed

Game Theory and Machine Learning for Cyber Security, First Edition.
Edited by Charles A. Kamhoua, Christopher D. Kiekintveld, Fei Fang, and Quanyan Zhu.
© 2021 The Institute of Electrical and Electronics Engineers, Inc. Published 2021 by John Wiley & Sons, Inc.

beyond a commercial social media platform and become part of China's e-governance initiatives. A continuous gain in the popularity of Chinese applications has given motivation for their authorities to introduce a tighter content control policy (Knockel et al. 2018).

The Citizen Lab of the University of Toronto has been continuously demonstrating how censorship abridges freedom of speech of WeChat users (Knockel et al. 2018). They have identified some words/phrases that, if written, will not be able to pass to the recipient. The system automatically blocks it. Based on the Citizen Lab, WeChat performs censorship on the server-side, which means that messages sent via this app pass through a remote server that implements censorship rules (Knockel et al. 2018). Recently, WeChat users have been arrested for "insulting police" or "threatening to blow up a government building" on Moments, which indicates that the feature may be subject to monitoring by the authorities or the company. These findings depict how censorship removes privacy from users; hence, the research in the field of enhancing privacy through machine learning is relevant in this century.

End-to-end encryption (E2EE) is a system of communication where only communicating users can read the messages. In principle, it prevents potential eavesdroppers—including telecom providers, Internet providers, and even the provider of the communication service—from accessing the cryptographic keys needed to decrypt the conversation (Senftleben et al. 2016).

In many messaging systems, including email and many chat networks, messages pass through intermediaries and are stored by a third party, from which they are retrieved by the recipient. Even if the messages are encrypted, they are typically only encrypted "in transit," and are stored in decrypted form by the third party. This approach allows the third party to provide search and other features, or to scan for illegal and unacceptable content. It also means they can be read and misused by anyone who has access to the stored messages on the third-party system, whether this is by design or via a backdoor. It is the way, for instance, WeChat can filter certain messages and introduce information friction, which is shown to be an effective way to impede the spread of information.

This research assumes the worst-case scenario in which our messages that are stored on the servers can be used against us or to abridge our freedom of speech. The main purpose is not just to hide the secret message but to ensure that the services are collecting inaccurate data on us, which will ensure safe communication means. Throughout the process, the data must retain usability, ability to be decrypted to the original message by the receiver.

This chapter first explores the linguistic steganography solution on a high level to tackle the problem described above. The latter half of the chapter describes the details of the proposed ConcealGAN system and presents the results of the system. The main contribution of this work is to use the recent advances in machine learning for steganography for social messaging systems. The proposed ConcealGAN integrates LeakGAN and implements multilayered encoding system. ConcealGAN aims to improve the privacy of the users when they use messaging apps such as WeChat and allow users to bypass the censorships to achieve freedom of speech. In the last section, we will conclude and discuss the directions for future works.

18.2 Steganography

In order to exchange secret data in an open medium, such as social media, without detection by anyone, one can use a covert channel. Communication through a covert channel is hidden in a legitimate communication channel. In the monograph on information security (Lampson 1973), Shannon has summarized three basic information security systems: encryption system, privacy system, and concealment system. Encryption system encodes information in a special way so that only authorized parties can decode it while unauthorized ones cannot. It ensures the security of information by making the message indecipherable. The privacy system mainly restricts access to information so that only authorized users have access to important information. Although these two systems ensure information security, they also *expose* the existence and importance of information, making it more vulnerable to attacks, such as interception and cracking (Jacobs and Bean 1963; Senftleben et al. 2016). The concealment system is very much different from these two secrecy systems. Steganography is the key technology in a concealment system.

The main purpose of steganography is to conceal secret data in a legitimate medium. A general model of a steganographic system (i.e. stegosystem) can be explained as follows. The embedded data M is the message that Bob wants to send secretly to Alice. It is hidden in an innocuous message X, usually named cover-object, in the control of a stego-key K, producing the stego-object X. And the receiver, Alice, can extract M from X with the stego-key K.

In steganography, the secret message can be hidden in various communication channels, such as images, audio, and text. Sequential data, such as daily messages that we exchange, the script of the software, or measurement of time series, is one of the most prominent types of data in the open medium. Nevertheless, texts have a higher degree of information coding, which means less redundant information when compared to images and audio files. Therefore, it is more *challenging* to hide information inside text or any other sequential data.

In order to avoid comparison with a known plaintext, linguistic steganography often relies on generating their cover texts. On the one hand, this idea resolves the problem of a known cover attack. On the other hand, the properties of the generated text may give rise to suspicions that the text is not legitimate (Krista 2014). This kind of generation aims to simulate some property of the text, usually by approximating some arbitrary statistical distribution found in real text.

The main purpose of this chapter is to propose the system architecture of ConcealGAN to generate natural cover texts without modifying the known plaintext. It is done through using a state-of-the-art natural language generation models that use deep learning techniques. Firstly, it builds on RNN and LeakGAN to enhance imperceptibility or reduce detectability. Secondly, ConcealGAN proposes a new architecture of a double-layer encoding of the secret message to enhance the difficulty of extraction.

18.3 Methodology

18.3.1 Previous Works Using Machine Learning Techniques

There are works proposed by Tina Fang et al. (2017), and RNN-Stega by Zhong-Liang Yang et al. (2019) that describe various techniques to use RNN to generate cover text that would hide the secret message. Both of the works use standard long short-term memory RNN cells to generate text. Nevertheless, the way they hide the secret bits differs from each other. There are two main contributions of this work to linguistic steganography that make it different from previous works. First, in addition to RNNs, our work employs Generative Adversarial Nets, LeakGAN, whilst the previous works used only RNNs to generate the cover text. Second, we propose a double layer of encoding in order to enhance the difficulty of extraction since the adversary should decode two different layers of encoding to get the original message. As such we propose that by increasing the number of layers, we can increase the difficulty of extracting the original message. This is in contrast to the systems where encoding is done through only one layer.

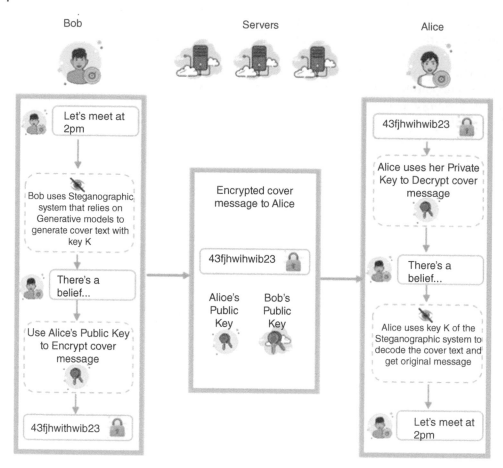

Figure 18.1 Linguistic steganographic system with generative models under messenger services that use E2E encryption.

18.3.2 High Level of Working Mechanism of ConcealGAN

Figure 18.1 depicts a high-level workflow of our proposed system that is being used under messenger services with E2E encryption system. As can be seen from the diagram, the servers collect not the original message but the cover text. Therefore, to get the original message, the attackers have to decode the cover message.

18.3.3 Double Layer of Encoding

Figure 18.2 illustrates the architecture of ConcealGAN. When generating the cover text, Conceal-GAN undergoes two main steps twice: compression and generating texts through embedding algorithms. After initial compression of the data, LSTM is used to generate intermediate text based on the limits set by the bit blocks. Further, the intermediate text is compressed again into binary representations, and this last binary string is used by LeakGAN to generate the final cover text. The words in each bit block are randomly distributed for each iteration of generating sentences. Two-bit blocks constitute the keys used by the receiver to decode the original message of the sender. One advantage of this system is that potential attackers will need to determine two keys for each message

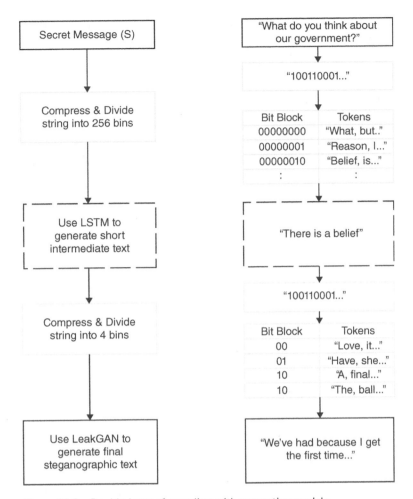

Figure 18.2 Double layer of encoding with generative models.

every time because bit blocks change for every message. Hence, even if the attacker has a lot of data, the keys for every message vary from each other.

18.3.4 Compression of Data

One of the integral parts of our proposed linguistic steganographic system is to compress data in the smallest representation as possible without losing any information when the receiver decodes compressed data. In other words, we have used a lossless compression technique. In the initial step, given textual data, it is divided into 64-bit groups or 8 characters. The compression technique used during this step is converting ASCII character into 8-bit binary representation.

After we have extracted the intermediate text that consists of somewhere between 8 and 5 words, depending on the number of bins used when dividing 64-bit secret data, the text is compressed into binary representation using more space effective technique. As such, the vocabulary space for intermediate text is fixed, which gives an opportunity to assign every single word a unique binary representation instead of having to assign a binary string to every single character. For instance, the EMNLP News dataset's vocabulary size is a 5715 word, which means every single word can be represented using 13 bits.

After the intermediate text has been compressed, it undergoes through the generative model to generate the final cover text. Throughout the process of generating intermediate and final texts, two kinds of embedding algorithms have been used.

18.3.5 Embedding Algorithms

18.3.5.1 RNN

The initial embedding algorithm uses a modified word-level LSTM for generating intermediate text (Mikolov et al. 2010). To encode the secret-containing bit string, S, we consider one-bit block B at a time and have our LSTM select one token from bin WB; hence, the candidate stegotext has as many tokens as the number of bit blocks in S. Even though we restrict the LSTM to select a token from a particular bin, each bin should offer sufficient variety of tokens, allowing the LSTM to generate text that looks natural. This technique is similar to the one proposed by Fang et al. (2017).

18.3.5.2 LeakGAN

Generating coherent sequential data that mimic real data, in particular meaningful text, has been a crucial area of research in Natural Language Processing field since the applications extend to machine translation (Yang et al. 2017), image captioning (Fang et al. 2015), and linguistic stegano-graphic field (Fang et al. 2017). The most common method is to train a recurrent neural network (RNN) with long-shot memory term (LSTM) or gated recurrent units (GRU), where the training is done by maximizing the log predictive likelihood of each true token in the training sequence given the previous token (Salakhutdinov 2009). Nevertheless, as was depicted by Bengio et al. (2015), maximizing the log predictive likelihood suffers from so-called exposure because of the discrepancy between training and inference stage. Furthermore, a scheduled sampling approach by Bengio et al. (2015) was proposed to alleviate this issue but has shown inconsistent results (Huszar 2015). These problems have motivated to use Generative Adversarial Nets (GAN) (Goodfellow et al. 2014), which were initially used for generating continuous data, such as images, but later it has been applied to sequential decision-making processes (Yu et al. 2017). In this system, the state is previously generated words, whilst the action is the next to be generated by the generative net, G, which is a stochastic policy that maps the current state to a distribution over the action space. Only after the whole sequence (the whole text) has been generated discriminator, D, is trained to distinguish between real and generated samples.

Firstly, to make the binary guiding signal more informative, besides providing final reward value, as was proposed by Guo et al. (2018), generator G is trained by learning high-level feature representations of real and generated text by D, i.e. feature map of the CNN. Secondly, to ease the sparsity problem, the hierarchical reinforcement learning idea proposed by Vezhnevets et al. (2017) is employed in LeakGAN. The architecture of Hierarchical Generator, G, consists of a high-level Manager module and a low-level Worker module. The Manager module uses LSTM to serve as a mediator, where it takes D's high-level feature representation and uses it to form the guiding goal for the Worker module in each timestep. On the other hand, the Worker module first encodes the current state (generated sequence) with another LSTM, then combines the output of the LSTM and the goal embedding to take a final action at the current state. This approach implies that the guiding signals are not only available to G at the end as scalar reward signals, but also available in terms of a goal embedding vector during the generation process. As a result of these two major propositions, LeakGAN has shown the state-of-the-art performance when generating long texts (more than 10 tokens) (Guo et al. 2018). Generating long and coherent texts is one of the main objectives of generating the cover text; therefore, at the second step encoding intermediate text LeakGAN was employed. Although still the generated text is not close to the human-readable level.

Table 18.1 Examples of the outputs of ConcealGAN given different numbers of secret bits.

Number of secret bits	Cover text generated by ConcealGAN
16	– "mediterranean easily band reid better aides ranks ufc" – "mean shots understand boat 21 campaigns december reception"
32	– "campaigns december mate tennis longtime appeared super easily treating criminal decline legally attracted decline hundreds short" – "super loss shoot 92 parties fines african roberts different mate traditional report appeared shots bills affects"
64	– "fines easily 60 better health birmingham criminal appeared maintained better already mate eye wasn EOS health african fines criminal registered purchase december better fines ranks entirely line mate african fines workplace maintained" – "report easily fines better short bills 120 executives december bills efforts mate fines bills learned pet speaker criminal december touch legally mate appeared dylan criminal criminal wasn maintained loss december super"

18.4 Results

This work provides qualitative results of the ConcealGAN. The training was done on the EMNLP News Dataset, and thus the generated text reads like a news report. Examples of the output of ConcealGAN are provided in Table 18.1. The training for LSTM and LeakGAN was done separately, and during training, the generative models were not constrained by the bit string vocabulary space. It means that the training and inference from the generative models had different environments.

It can be seen that the cover text sounds like a news report. However, this can be improved if the model gets trained on the user's previous messages. If the user wants to reduce the detectability, they can have the option of exporting their textual data from messengers to train the model. Using this data, the model will generate cover texts that sound like the user's real message. Further, bit blocks set boundaries when generating the text. The token has the highest probability of being used as the next word might not be chosen, because it is not in the vocabulary space of that particular bit block. Therefore, the examples above do not sound sufficiently coherent.

18.5 Conclusion and Future Work

This chapter has shown that linguistic steganography is a field that has a huge potential to bring disruptive technology to enhance the privacy of users when exchanging textual messages. With the ever-growing user base in messaging applications, such as WhatsApp, WeChat, Telegram, and others, online privacy, particularly when exchanging textual messages, will become more relevant. ConcealGAN or any other linguistic steganographic system can be integrated into those applications and provide an additional layer of security.

To conclude the integration of generative models with a linguistic steganographic system will enable a higher level of security when exchanging textual data. Besides enhancing privacy, it also harms the service providers by providing inaccurate information about the user. Nevertheless, as future work, we would justify quantitatively that the steganographic text generated using our system enhances the privacy and difficulty of extraction of the original message. For the quality of text BLEU score can be used, whilst to test the difficulty of extraction, we can train a model to decode steganographic text. We would integrate encoder–decoder architecture to generate context-specific steganographic data, which means given the current message of another user, the system should produce coherent reply that would enhance the quality of steganographic text. Furthermore, the text can be done on textual message data. The proposed system is integrated with messengers so that users can use it to enhance their privacy.

References

S. Bengio, O. Vinyals, N. Jaitly, and N. Shazeer. Scheduled sampling for sequence prediction with recurrent neural networks. In C. Cortes, N. Lawrence, D. Lee, M. Sugiyama, and R. Garnett, editors. *Advances in Neural Information Processing Systems*, vol. 28, pp. 1171–1179. Curran Associates, Inc., 2015. https://proceedings.neurips.cc/paper/2015/file/e995f98d56967d946471af29d7bf99f1-Paper.pdf (accessed 21 March 2021).

H. Fang, S. Gupta, F. Iandola, R. K. Srivastava, L. Deng, P. Dollar, J. Gao, X. He, M. Mitchell, and J. C. Platt. From captions to visual concepts and back. In *Proceedings of the IEEE Conference on Computer Vision and Pattern Recognition*, pp. 1473–1482. IEEE, 2015.

T. Fang, M. Jaggi, and K. Argyraki. Generating steganographic text with LSTMs, 2017, arXiv preprint, arXiv:1705.10742.

I. Goodfellow, J. Pouget-Abadie, M. Mirza, B. Xu, D. Warde-Farley, S. Ozair, A. Courville, and Y. Bengio. Generative adversarial nets. In *Advances in Neural Information Processing Systems (NIPS)*, pp. 2672–2680. 2014.

J. Guo, S. Lu, H. Cai, W. Zhang, Y. Yu, and J. Wang. Long text generation via adversarial training with leaked information. In *Proceedings of the AAAI Conference on Artificial Intelligence*, 32(1), 2018. https://ojs.aaai.org/index.php/AAAI/article/view/11957 (accessed 2 March 2021).

U. Hoelzle. Why speed is essential to search results. Google, 2012. https://www.thinkwithgoogle.com/marketing-resources/the-google-gospel-of-speed-urs-hoelzle/ (accessed 2 March 2021).

F. Huszar. How (not) to train your generative model: Scheduled sampling, likelihood, adversary?, 2015, arXiv preprint, arXiv:1511.05101.

I. S. Jacobs and C.P. Bean. Fine particles, thin films and exchange anisotropy. In *Magnetism*, volume III, pp. 271-350, G. T. Rado and H. Suhl, Eds. Academic, New York, 1963

S. Kelly, S. Cook, and M. Truong. *Freedom on the Net 2012: A Global Assessment of Internet and Digital Media*. Freedom House, 2012.

J. Knockel, L. Ruan, M. Crete-Nishihata, and R. Deibert. *(Can't) Picture This*. University of Toronto, Citizen Lab, 2018. https://citizenlab.ca/2018/08/cant-picture-this-an-analysis-of-image-filtering-on-wechat-moments/

B. Krista. *Linguistic Steganography: Survey, Analysis, and Robustness Concerns for Hiding Information in Text*. CERIAS Tech Report, 2014.

B. W. Lampson. A note on the confinement problem. *Communications of the ACM*, 16(10):613–615, 1973.

T. Mikolov, M. Karafiat, L. Burget, J. Cernocky, and S. Khudanpur. Recurrent neural network based language model. In *11th Annual Conference of the International Speech Communication Association (INTERSPEECH), Makuhari, Chiba, Japan, 26–30 September 2010*, pp. 1045–1048, 2010.

M. E. Roberts. Fear or friction? How censorship slows the spread of information in the digital age. Unpublished manuscript, 2014. http://scholar.harvard.edu/files/mroberts/files/fearfriction_1.pdf

R. Salakhutdinov. Learning deep generative models. PhD dissertation, University of Toronto, 2009.

M. Senftleben, A. Barroso, M. Bucicoiu, M. Hollick, S. Katzenbeisser, and E. Tews. On the privacy and performance of mobile anonymous microblogging. *IEEE Transactions on Information Forensics and Security*, 11(7):1578–91, 2016. doi:https://doi.org/10.1109/TIFS.2016.2541633.

M. Senftleben, A. Barroso, M. Bucicoiu, M. Hollick, S. Katzenbeisser, and E. Tews. On the privacy and performance of mobile anonymous microblogging. *IEEE Transactions on Information Forensics and Security*, 11(7):1578–1591, 2016. doi:https://doi.org/10.1109/TIFS.2016.2541633.

A. S. Vezhnevets, S. Osindero, T. Schaul, N. Heess, M. Jaderberg, D. Silver, and K. Kavukcuoglu. Feudal networks for hierarchical reinforcement learning, 2017, arXiv preprint, arXiv:1703.01161.

Z. Yang, W. Chen, F. Wang, and B. Xu. Improving neural machine translation with conditional sequence generative adversarial nets. 2017, arXiv preprint, arXiv:1703.04887.

Z. Yang, X. Guo, Z. Chen, Y. Huang, and Y. Zhang. RNN-stega: Linguistic steganography based on recurrent neural networks. *IEEE Transactions on Information Forensics and Security*, 14(5):1280–1295, 2019.

L. Yu, W. Zhang, J. Wang, and Y. Yu. SeqGAN: Sequence generative adversarial nets with policy gradient. In *Thirty-First AAAI Conference on Artificial Intelligence, San Francisco, CA, USA, 4–9 February 2017*, 2017.

Part V

Reinforcement Learning for Cyber Security

19

Manipulating Reinforcement Learning: Stealthy Attacks on Cost Signals

Yunhan Huang and Quanyan Zhu

Department of Electrical and Computer Engineering, NYU Tandon School of Engineering, New York University, Brooklyn, NY, USA

19.1 Introduction of Reinforcement Learning

Reinforcement learning (RL) is a powerful paradigm for online decision-making in unknown environment. Recently, many advanced RL algorithms have been developed and applied to various scenarios including video games (e.g. Mnih et al. 2015), transportation (e.g. Arel et al. 2010), network security (e.g. Huang and Zhu, 2019a; Zhu and Başar, 2009), robotics (e.g. Kober et al. 2013), and critical infrastructures (e.g. Ernst et al. 2004). However, the implementation of RL techniques requires accurate and consistent feedback from environment. It is straightforward to fulfill this requirement in simulation while in practice, accurate and consistent feedback from the environment is not guaranteed, especially in the presence of adversarial interventions. For example, adversaries can manipulate cost signals by performing data injection attack and prevent an agent from receiving cost signals by jamming the communication channel. With inconsistent and/or manipulated feedback from environment, the RL algorithm can either fail to learn a policy or misled to a pernicious policy. The failure of RL algorithms under adversarial intervention can lead to a catastrophe to the system where the RL algorithm has been applied. For example, self-driving platooning vehicles can collide with each other when their observation data are manipulated (see Behzadan and Munir, 2019); drones equipped with RL techniques can be weaponized by terrorists to create chaotic and vicious situations where they are commanded to collide to a crowd or a building (see Huang and Zhu, 2019b; Xu and Zhu, 2015).

Hence, it is imperative to study RL with maliciously intermittent or manipulated feedback under adversarial intervention. First, it is important to understand the adversarial behaviors of the attacker. To do so, one has to establish a framework that characterizes the objective of the attacker, the actions available to the attacker, and the information at his disposal. Secondly, it is also crucial to understand the impacts of the attacks on RL algorithms. The problems include how to measure the impacts, how to analyze the behavior of the RL algorithms under different types of attacks, and how to mathematically and/or numerically compute the results of RL algorithms under attack. With the understanding of the adversarial behavior and the impacts of the adversarial behavior on RL algorithms, the third is to design defense mechanisms that can protect RL algorithms from being degenerated. This could be done by designing robust and secure RL algorithms that can automatically detect and discard corrupted feedback, deploying cryptographic techniques to ensure confidentiality, integrity, and building backup communication channels to ensure availability.

Despite the importance of understanding RL in malicious setting, very few works have studied RL with maliciously manipulated feedback or intermittent feedback. One type of related works

Game Theory and Machine Learning for Cyber Security, First Edition.
Edited by Charles A. Kamhoua, Christopher D. Kiekintveld, Fei Fang, and Quanyan Zhu.

studies RL algorithms under corrupted reward signals (see Everitt et al. 2017; Wang et al. 2018). In Everitt et al. (2017), the authors investigate RL problems where agents receive false rewards from environment. Their results show that reward corruption can impede the performance of agents and can result in disastrous consequences for highly intelligent agents. Another type of work studies delayed evaluative feedback signals without the presence of malicious adversaries (see Tan et al. 2008; Watkins, 1989). The study of RL under malicious attacks from a security point of view appeared in the recent past (Behzadan and Munir, 2018; Huang and Zhu, 2019b; Lin et al. 2017; Ma et al. 2019). In Huang and Zhu (2019b), the authors study RL under malicious falsification on cost signals and introduces a quantitative framework of attack models to understand the vulnerabilities of RL. Ma et al. (2019) focuses on security threats on batch RL and control where attacker aims to poison the learned policy. Lin et al. (2017) and Behzadan and Munir (2018) focus on deep RL which involves deep natural networks (DNNs) for function approximation.

In this chapter, we first introduce RL techniques built on a Markov decision process framework and provide self-contained background on RL before we discuss security problems of RL. Among RL techniques, of particular interest to us are two frequently used learning algorithms: TD learning and Q-learning. We then introduce general security concerns and problems in the field of RL. Security concerns arise from the fact that RL technique requires accurate and consistent feedback from environment, timely deployed controls, and reliable agents, which are hard to guarantee under the presence of adversarial attacks. The discussion of general security concerns in this chapter invokes a large number of interesting problems yet to be done. In this chapter, we focus on one particular type of problems where the cost signals that the RL agent receives are falsified or manipulated as a result of adversarial attacks. In this particular type of problems, a general formulation of attack models is discussed by defining the objectives, information structure, and the capability of the adversary. We analyze the attacks on both TD learning and Q-learning algorithms. We develop important results that tell the fundamental limits of the adversarial attacks. For TD learning, we characterize the bound on the approximation error that can be induced by the adversarial attacks on the cost signals. The choice of λ does not impact the bound of the induced approximation error. In the Q-learning scenario, we aim to address two fundamental questions. The first is to understand the impact of the falsification of cost signals on the convergence of Q-learning algorithm. The second is to understand how the RL algorithm can be misled under the malicious falsifications. This chapter ends with an educational example that explains how the adversarial attacks on cost signals can affect the learning results in TD learning.

The rest of this chapter is organized as follows. Section 19.1.1 gives a basic introduction of Markov decision process and RL techniques with a focus on TD(λ) learning and Q-learning. In Section 19.2, we discuss general security concerns and problems in the field of RL. A particular type of attacks on the cost signals is studied on both TD(λ) learning and Q-learning in Section 19.3. Section 19.4 comes an educational example that illustrates the adversarial attacks on TD(λ) learning. Conclusions and future works are included in Section 19.5.

19.1.1 Setting of RL

Consider an RL agent interacts with an unknown environment and attempts to find the optimal policy minimizing the received cumulative costs. The environment is formalized as a Markov decision process (MDP) denoted by $\langle S, A, g, P, \alpha \rangle$. The MDP has a finite state space denoted by S. Without loss of generality, we assume that there are n states and $S = \{1, 2, \dots, n\}$. The state transition depends on a control. The control space is also finite and denoted by A. When at state i, the control must be chosen from a given finite subset of A denoted by $U(i)$. At state i, the choice of a control $u \in U(i)$ determines the transition probability $p_{ij}(u)$ to the next state j. The state transition information is encoded in P. The agent receives a running cost that accumulates additively over time and depends on the states visited and the controls chosen. At the kth transition, the agent incurs a cost $\alpha^k g(i, u, j)$, where $g : S \times A \times S \to \mathbb{R}$ is a given real-valued function that describes the cost associated with the states visited and the control chosen, and $\alpha \in (0, 1]$ is a scalar called the discount factor.

The agent is interested in policies, i.e. sequences $\pi = \{\mu_0, \mu_1, \dots\}$ where $\mu_k : S \to A, k = 0, 1, \dots$, is a function mapping states to controls with $\mu_k(i) \in U(i)$ for all states i. Denote i_k the state at time k. Once a policy π is fixed, the sequence of states i_k becomes a Markov chain with transition probabilities $P(i_{k+1} = j | i_k = i) = p_{ij}(\mu_k(i))$. In this chapter, we consider infinite horizon problems, where the cost accumulates indefinitely. In the infinite horizon problem, the total expected cost starting from an initial state i and using a policy $\pi = \{\mu_0, \mu_1, \dots\}$ is

$$J^\pi(i) = \lim_{N \to \infty} E\left[\sum_{k=0}^{N} \alpha_k g\left(i_k, \mu_k(i_k), i_{k+1}\right) \mid i_0 = i\right],$$

the expected value is taken with respect to the probability distribution of the Markov chain $\{i_0, i_1, i_2, \dots\}$. This distribution depends on the initial state i_0 and the policy π. The optimal cost-to-go starting from state i is denoted by $J^*(i)$, that is,

$$J^*(i) = \min_\pi J^\pi(i).$$

We can view the costs $J^*(i), i = 1, 2, \dots$, as the components of a vector $J^* \in \mathbb{R}^n$. Of particular interest in the infinite-horizon problem are stationary policies, which are policies of the form $\pi = \{\mu, \mu, \dots\}$. The corresponding cost-to-go vector is denoted by $J^\mu \in \mathbb{R}^n$.

The optimal infinite-horizon cost-to-go functions $J^*(i), i = 1, 2, 3, \dots$, also known as value functions, arise as a central component of algorithms as well as performance metrics in many statistics and engineering applications. Computation of the value functions relies on solving a system of equations:

$$J^*(i) = \min_{u \in U(i)} \sum_{j=1}^{n} p_{ij}(u)(g(i, u, j) + \alpha J^*(j)), \quad i = 1, 2, \dots, n, \tag{19.1}$$

referred to as Bellman's equation (Bertsekas and Tsitsiklis, 1996; Sutton et al. 2018), which will be at the center of analysis and algorithms in RL. If $\mu(i)$ attains the minimum in the right-hand side of Bellman's equation (19.1) for each i, then the stationary policy μ should be optimal (Bertsekas and Tsitsiklis, 1996). That is, for each $i \in S$,

$$\mu^*(i) = \arg \min_{\mu(i) \in U(i)} \sum_{j=1}^{n} p_{ij}\left(\mu(i)\right)\left(g(i, \mu(i), j) + \alpha J^*(j)\right).$$

The efficient computation or approximation of J^* and an optimal policy μ^* is the major concern of RL. In MDP problems where the system model is known and the state space is reasonably large, value iteration, policy iteration, and linear programming are the general approaches to find the value function and the optimal policy. Readers unfamiliar with these approaches can refer to Chapter 2 of Bertsekas and Tsitsiklis (1996).

It is well known that RL refers to a collection of techniques for solving MDP under two practical issues. One is the overwhelming computational requirements of solving Bellman's equations because of a colossal amount of states and controls, which is often referred to as Bellman's "curse of dimensionality." In such situations, an approximation method is necessary to obtain suboptimal solutions. In approximation methods, we replace the value function J^* with a suitable approximation $\tilde{J}(r)$, where r is a vector of parameters which has much lower dimension than J^*. There are two main function approximation architectures: linear and nonlinear approximations. The approximation architecture is linear if $\tilde{J}(r)$ is linear in r. Otherwise, the approximation architecture is nonlinear. Frequently used nonlinear approximation methods include polynomial-based approximation, wavelet-based approximation, and approximation using neural network. The topic of deep reinforcement learning studies the cases where approximation $\tilde{J}(r)$ is represented by a deep neural network (Mnih et al. 2015).

Another issue comes from the unavailability of the environment dynamics; i.e. the transition probability is either unknown or too complex to be kept in memory. In this circumstance, one alternative is to simulate the system and the cost structure. With given state space S and the control space A, a simulator or a computer that generates a state trajectory using the probabilistic transition from any given state i to a generated successor state j for a given control u. This transition accords with the transition probabilities $p_{ij}(u)$, which is not necessarily known to the simulator or the computer. Another alternative is to attain state trajectories and corresponding costs through experiments. Both methods allow the learning agent to observe their own behavior to learn how to make good decisions. It is clearly feasible to use repeated simulations to find the approximate of the transition model of the system P and the cost functions g by averaging the observed costs. This approach is usually referred to as model-based RL. As an alternative, in model-free RL, transition probabilities are not explicitly estimated, but instead the value function or the approximated value function of a given policy is progressively calculated by generating several sample system trajectories and associated costs. Of particular interest to us in this chapter is the security of model-free RL. This is because firstly, model-free RL approaches are the most widely applicable and practical approaches that have been extensively investigated and implemented; secondly, model-free RL approaches receive observations or data from environment successively and consistently. This makes model-free RL approaches vulnerable to attacks. An attack can induce an accumulative impact on succeeding learning process. The most frequently used algorithms in RL are TD learning algorithms and Q-learning algorithms. Hence, in this chapter, we will focus on the security problems of these two learning algorithms as well as their approximate counterparts.

19.1.2 TD Learning

Temporal difference (TD) learning is an implementation of the Monte Carlo policy evaluation algorithm that incrementally updates the cost-to-go estimates of a given policy μ, which is an important subclass of general RL methods. TD learning algorithms, introduced in many references, including Bertsekas and Tsitsiklis (1996), Sutton et al. (2018) and Tsitsiklis and Van Roy (1997), generates an infinitely long trajectory of the Markov chain $\{i_0, i_1, i_2, \dots\}$ from simulator or experiments by fixing a policy μ, and at time t iteratively updates the current estimate J_t^μ of J^μ using an iteration that depends on a fixed scalar $\lambda \in [0, 1]$, and on the temporal difference

$$d_t(i_k, i_{k+1}) = g(i_k, i_{k+1}) + \alpha J_t^\mu(i_{k+1}) - J_t^\mu(i_k), \forall t = 0, 1, \dots, \forall k \le t.$$

The incremental updates of TD(λ) have many variants. In the most straightforward implementation of TD(λ), all of the updates are carried out simultaneously after the entire trajectory has been simulated. This is called the off-line version of the algorithm. On the contrary, in the online implementation of the algorithm, the estimates update once at each transition. Under our discount MDP, a trajectory may never end. If we use an off-line variant of TD(λ), we may have to wait infinitely long before a complete trajectory is obtained. Hence, in this chapter, we focus on an online variant. The update equation for this case becomes

$$J_{t+1}^\mu(i) = J_t^\mu(i) + \gamma_t(i)z_t(i)d_t(i), \quad \forall i, \tag{19.2}$$

where the $\gamma_t(i)$ are non-negative stepsize coefficients and $z_t(i)$ is the eligibility coefficients defined as:

$$z_t(i) = \begin{cases} \alpha\lambda z_{t-1}(i), & \text{if } i_t \neq i, \\ \alpha\lambda z_{t-1}(i) + 1, & \text{if } i_t = i. \end{cases}$$

This definition of eligibility coefficients gives the every-visit TD(λ) method. In every-visit TD(λ) method, if a state is visited more than once by the same trajectory, the update should also be carried out more than once.

Under a very large number of states or controls, we have to resort to approximation methods. Here, we introduce TD(λ) with linear approximation architectures. We consider a linear parametrization of the form:

$$\tilde{J}(i, r) = \sum_{k=1}^{K} r(k)\phi_k(i).$$

Here, $r = (r(1), \dots, r(K))$ is a vector of tunable parameters and $\phi_k(\cdot)$ are fixed scalar functions defined on the state space. The form can be written in a compact form:

$$\tilde{J}(r) = (\tilde{J}(1, r), \dots, \tilde{J}(n, r)) = \Phi r,$$

where

$$\Phi = \begin{bmatrix} | & & | \\ \phi_1 & \cdots & \phi_K \\ | & & | \end{bmatrix} = \begin{bmatrix} - & \phi'(1) & - \\ \cdots & \cdots & \cdots \\ - & \phi'(n) & - \end{bmatrix},$$

with $\phi_k = (\phi_k(1), \dots, \phi_k(n))$ and $\phi(i) = (\phi_1(i), \dots, \phi_K(i))$. We assume that Φ has linearly independent columns. Otherwise, some components of r would be redundant.

Let η_t be the eligibility vector for the approximated TD(λ) problem which is of dimension K. With this notation, the approximated TD(λ) updates are given by:

$$r_{t+1} = r_t + \gamma_t d_t \eta_t, \tag{19.3}$$

where

$$\eta_{t+1} = \alpha\lambda\eta_t + \phi(i_{t+1}). \tag{19.4}$$

Here, $d_t = g(i_t, i_{t+1}) + \alpha r'_t \phi(i_{t+1}) - r'_t \phi(i_t)$.

The almost-sure convergence of r_t generated by (19.3) and (19.4) is guaranteed if the conditions in assumption 6.1 in Bertsekas and Tsitsiklis (1996) hold. It will converge to the solution of

$$Ar + b = 0,$$

where $A = \Phi'D(M - I)\Phi$ and $b = \Phi'Dq$. Here, D is a diagonal matrix with diagonal entries $d(1), d(2), \dots, d(n)$, and $d(i)$ is the steady-state probability of state i; the matrix M is given by $M = (1 - \lambda)\sum_{m=0}^{\infty} \lambda^m(\alpha P_\mu)^{m+1}$ and the vector b is given by $b = \Phi'Dq$ with $q = \sum_{m=0}^{\infty} (\alpha\lambda P_\mu)^m \bar{g}$, where $\bar{g}$ is a vector in $\mathbb{R}^n$ whose ith component is given by $\bar{g}(i) = \sum_{j=1}^{n} P_{ij}(\mu(i))g(i, \mu(i), j)$. The matrix P_μ is the transition matrix defined by $P_\mu := [P_\mu]_{ij} = p_{ij}(\mu(i))$. A detailed proof of the convergence is provided in Tsitsiklis and Van Roy (1997) and Bertsekas and Tsitsiklis (1996).

Indeed, TD(λ) with linear approximations is a much more general framework. The convergence of J_t^μ in TD(λ) without approximation follows immediately if we let $K = n$ and $\Phi = I_n$ where I_n is $n \times n$ identity matrix.

19.1.3 Q-Learning

Q-learning method is initially proposed in Watkins and Dayan (1992) which updates estimates of the Q-factors associated with an optimal policy. Q-learning is proven to be an efficient computational method that can be used whenever there is no explicit model of the system and the cost structure. First, we introduce the first notion of the Q-factor of a state-control pair (i, u), defined as:

$$Q(i, u) = \sum_{j=0}^{n} p_{ij}(u)(g(i, u, j) + \alpha J(j)). \tag{19.5}$$

The optimal Q-factor $Q^*(i, u)$ corresponding to a pair (i, u) is defined by (19.5) with $J(j)$ replaced by $J^*(j)$. It follows immediately from Bellman's equation that:

$$Q^*(i, u) = \sum_{j=0}^{n} p_{ij}(u) \left(g(i, u, j) + \alpha \min_{v \in U(j)} Q^*(j, v) \right). \tag{19.6}$$

Indeed, the optimal Q-factors $Q^*(i, u)$ are the unique solution of the above system by Banach fixed-point theorem (see Kreyszig, 1978).

Basically, Q-learning computes the optimal Q-factors based on samples in the absence of system model and cost structure. It updates the Q-factors as follows:

$$Q_{t+1}(i, u) = (1 - \gamma_t)Q_t(i, u) + \gamma_t \left(g(i, u, \bar{\zeta}) + \alpha \min_{v \in U(\bar{\zeta})} Q_t(\bar{\zeta}, v) \right), \tag{19.7}$$

where the successor state $\bar{\zeta}$ and $g(i, u, \bar{\zeta})$ is generated from the pair (i, u) by simulation or experiments, i.e. according to the transition probabilities $p_{i\bar{\zeta}}(u)$. For more intuition and interpretation about Q-learning algorithm, one can refer to chapter 5 of Bertsekas and Tsitsiklis (1996), Sutton et al. (2018), and Watkins and Dayan (1992).

If we assume that stepsize γ_t satisfies $\sum_{t=0}^{\infty} \gamma_t = \infty$ and $\sum_{t=0}^{\infty} \gamma_t^2 < \infty$, we obtain the convergence of $Q_t(i, u)$ generated by (19.7) to the optimal Q-factors $Q^*(i, u)$. A detailed proof of convergence in provided in Borkar and Meyn (2000) and chapter 5 of Bertsekas and Tsitsiklis (1996).

19.2 Security Problems of Reinforcement Learning

Understanding adversarial attacks on RL systems is essential to develop effective defense mechanisms and an important step toward trustworthy and safe RL. The reliable implementation of RL techniques usually requires accurate and consistent feedback from the environment, precisely and timely deployed controls to the environment and reliable agents (in multi-agent RL cases). Lacking any one of the three factors will render failure to learn optimal decisions. These factors can be used by adversaries as gateways to penetrate RL systems. It is hence of paramount importance to understand and predict general adversarial attacks on RL systems targeted at the three doorways. In these attacks, the miscreants know that they are targeting RL systems, and therefore, they tailor their attack strategy to mislead the learning agent. Hence, it is natural to start with understanding the parts of RL that adversaries can target at.

Figure 19.1 illustrates different types of attacks on the RL system. One type of attack aims at the state which is referred to as state attacks. Attacks on state signals can happen if the remote sensors in the environment are compromised or the communication channel between the agent and the sensors is jammed or corrupted. In such circumstances, the learning agent may receive a false state observation $\bar{\zeta}_k$ of the actual state i_k at time k and/or may receive a delayed observation of the actual state or even never receive any information regarding the state at time k. An example of effortless

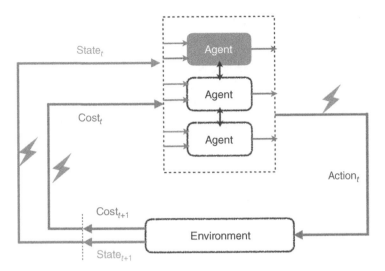

Figure 19.1 Attacks on RL systems: adversaries can manipulate actions, costs, and state signals in the feedback system. Agents can also be compromised in the learning process.

state attacks is sequential blinding/blurring of the cameras in a deep RL-based autonomous vehicle via lasers/dirts on lens, which can lead to learning false policies and hence lead to catastrophic consequences. Based on its impact on the RL systems, state attacks can be classified into two groups: (i) denial of service (DoS) and (ii) integrity attacks. The main purpose of DoS attacks is to deny access to sensor information. Integrity attacks are characterized by the modification of sensor information, compromising their integrity.

Another type of attacks targets at cost signals. In this type of attacks, the adversary aims to corrupt the cost signals that the learning agent has received with a malicious purpose of misleading the agent. Instead of receiving the actual cost signal $g_k = g(i_k, u_k, j_{k+1})$ at time k, the learning agent receives a manipulated or falsified cost signal $\tilde{g}_k$. The corruption of cost signals comes from the false observation of the state in cases where the cost is predetermined by the learning agent. In other cases where cost signals are provided directly by the environment or a remote supervisor, the cost signal can be corrupted independently from the observation of the state. The learning agent receives falsified cost signals even when the observation of the state is accurate. In the example of autonomous vehicle, if the cost depends on the distance of the deep RL agent to a destination as measure by GPS coordinates, spoofing of GPS signals by the adversary may result in incorrect reward signals, which can translate to incorrect navigation policies (see Behzadan and Munir, 2018).

The corruption of cost signals may lead to a doomed policies. For example, Clark and Amodei (2016) has trained an RL agent on a boat racing game. High observed reward misleads the agent to go repeatedly in a circle in a small lagoon and hit the same targets, while losing every race. Or even worse, the learning agent may be misled to lethal policies that would result in self-destruction.

The learning agent influences the environment by performing control u via actuators. There is a type of attacks targeting the actuators. If the actuator is exposed to the adversary, the control performed will be different than the one determined by the agent. Hence, during the learning process, the agent will learn a deteriorated transition kernel and a matched reward function. An example of this type of attack is RL-based unmanned aerial vehicles may receive corrupted control signals which would cause malfunctioning of its servo actuators.

To understand an attack model, one needs to specify three ingredients: objective of an adversary, actions available to the adversary, and information at his disposal. The objective of an adversary describes the goal the adversary aims to achieve out of the attacks. Objectives of an adversary include but not limit to maximizing the agent's accumulative costs, misleading the agent to learn certain policies, or diverging the agent's learning algorithm. To achieve the objective, adversaries need a well-crafted strategy. An attack strategy is developed based on the information and the actions at his disposal. An attack strategy is a map from the information space to the action space. The information structure of the adversary describes the knowledge the adversaries have during the learning and the attack process. Whether the adversaries know which learning algorithms the agent implements, whether the adversaries know the actual state trajectory, controls and costs are decisive for the adversaries to choose their strategies. The scope of the attack strategy also depends on the actions at the adversary's disposal. Due to adversaries' capabilities, adversaries can only launch certain attacks. For example, some adversaries can only manipulate the costs induced at certain states or add white noise to the observations.

In next section, we focus on cases where adversaries target at the cost signals received by the agent. We analyze the performance of TD learning and Q-learning under manipulated costs, respectively, and provide fundamental limits under the attacks on cost signals.

19.3 Reinforcement Learning with Manipulated Cost Signals

Under malicious attacks on cost signals as we have discussed in Section 19.2, the RL agent will fail to observe the actual cost feedback from the environment. Instead, the agent receives a cost signal that might be falsified by the adversary. Consider the following MDP with falsified cost denoted by $\langle S, A, g, \tilde{g}, P, \alpha \rangle$. Under the falsification, the agent, instead of receiving the actual cost signal $g_t \in \mathbb{R}$ at the tth update, observes a falsified cost signal denoted by $\tilde{g}_t \in \mathbb{R}$. The remaining aspects of the MDP framework stay the same. The adversary's task here is to design falsified cost signals $\tilde{g}$ based on his information structure and the actions available to him so that he can achieve certain objectives.

Suppose the adversary is an insider and he knows what the agent knows at time t. In other words, at time t, the adversary knows the state trajectory, the control trajectory, and the cost signals up to time t. The adversary may or may not know the system model. Suppose that the adversary falsifies the cost signals in a stealthy way. In this case, the cost associated with each state-control-state triple is consistently falsified in the same way. The following definition gives a formal definition of the attack.

Definition 19.1 (Stealthy Attacks) If $\tilde{g}_t$ takes the same value for the same state-control-state triple for all t; i.e. for $t \neq \tau$, we have $\tilde{g}_t = \tilde{g}_\tau$ if $(i_t, u_t, j_{t+1}) = (i_\tau, u_\tau, j_{\tau+1})$, then we say the attacks on the cost signals are stealthy.

Under stealthy attacks, the falsified cost signals can be given by a function $\tilde{g} : S \times A \times S \to \mathbb{R}$. Then, at time t, the falsified cost the agent receives is $\tilde{g}_t = \tilde{g}(i_t, u_t, i_{t+1})$. Since the transition from i_t to i_{t+1} depends on the transition probability, without loss of generality, we consider only the cost structure defined on the state-control pair, i.e. $g(i, u)$ for $i \in S, u \in A$.

19.3.1 TD Learning with Manipulated Cost Signals

If the RL agent updates the estimates of the cost-to-go function of a given policy μ according to (19.3) and (19.4), then under the stealthy attacks, the algorithm can be written as:

$$\tilde{r}_{t+1} = \tilde{r}_t + \gamma_t \tilde{d}_t \eta_t,$$
$$\eta_{t+1} = \alpha\lambda\eta_t + \phi(i_{t+1}), \tag{19.8}$$

where $\tilde{d}_t = \tilde{g}(i_t, u_t) + \alpha\tilde{r}'_t \phi(i_{t+1}) - \tilde{r}'_t \phi(i_t)$.

Suppose the sequence of parameters $\{\tilde{r}_t\}$ generated by (19.8) under the falsified cost signals is convergent and converges to $\tilde{r}^*$. (We will show the conditions under which the convergence of $\{\tilde{r}_t\}$ is guaranteed.) Let r^* be the solution of $Ar + b$. In TD(λ), the agent aims to estimate the cost-to-go function J^μ of a given policy μ. In approximated TD(λ) with linear approximation architecture, $\tilde{J}(i, r) = \phi'(i)r$ serves as an approximation of $J^\mu(i)$ for $i \in S$. One objective of the adversary can be to deteriorate the approximation and estimation of J^μ by manipulating the costs. One way to achieve the objective is to let $\tilde{r}_t$ generated by (19.8) converge to $\tilde{r}^*$ such that $\Phi'\tilde{r}^*$ is as a worse approximate of J^μ as possible.

Lemma 19.1 If the sequence of parameters $\{\tilde{r}_t\}$ is generated by the TD(λ) learning algorithm (19.8) with stealthy and bounded attacks on the cost signals, then the sequence $\{\tilde{r}_t\}$ converges to $\tilde{r}^*$ and $\tilde{r}^*$ is a unique solution of $Ar + \tilde{b} = 0$, where $\tilde{b} = \Phi'D \sum_{m=0}^{\infty} (\alpha\lambda P_\mu)^m \tilde{g}$ and $\tilde{g}$ is vector whose ith component is $\tilde{g}(i, \mu(i))$.

The proof of Lemma 19.1 follows the proof of proposition 6.4 in Bertsekas and Tsitsiklis (1996) with $g(i, \mu(i), j)$ replaced by $\tilde{g}(i, \mu(i))$. If the adversary performs stealthy and bounded attacks, he can mislead the agent to learn an approximation $\Phi'\tilde{r}^*$ of J^μ. The distance between $\Phi'\tilde{r}^*$ and J^μ with respect a norm $\|\cdot\|$ is what the adversary aims to maximize. The following lemma provides an upper bound of the distance between $\Phi'\tilde{r}^*$ and J^μ.

Lemma 19.2 Suppose that $\tilde{r}^*$ is the parameters learned from the manipulated TD(λ) (19.8). Then, the approximation error under the manipulated satisfies

$$\|\Phi'\tilde{r}^* - J^\mu\|_D \leq \|\Phi'\tilde{r}^* - \Phi'r^*\|_D + \frac{1 - \lambda\alpha}{\sqrt{(1-\alpha)(1+\alpha-2\lambda\alpha)}}\|(\Pi - I)J^\mu\|_D, \tag{19.9}$$

where $\|\cdot\|_D$ is the weighted quadratic norm defined by $\|J\|_D^2 = J'DJ = \sum_{i=1}^{n} d(i)J(i)^2$ and $\Pi = \Phi(\Phi'D\Phi)^{-1}\Phi'D$.

Proof: A direct application of triangle inequality gives us

$$\|\Phi'\tilde{r}^* - J^\mu\| = \|\Phi'\tilde{r}^* - \Phi'r^* + \Phi'r^* - J^\mu\| \leq \|\Phi'\tilde{r}^* - \Phi'r^*\| + \|\Phi'r^* - J^\mu\|.$$

This indicates that the distance between $\Phi'\tilde{r}^*$ and J^μ is bounded by the distance between the falsified approximation $\Phi'\tilde{r}^*$ of J^μ and the true approximation $\Phi'r^*$ of J^μ plus the approximation error $\|\Phi'r^* - J^\mu\|$. Moreover, we know from theorem 1 in Tsitsiklis and Van Roy (1997) that

$$\|\Phi'r^* - J^\mu\|_D \leq \frac{1 - \lambda\alpha}{\sqrt{(1-\alpha)(1+\alpha-2\lambda\alpha)}} \min_r \|\Phi'r - J^\mu\|_D. \tag{19.10}$$

From Lemma 6.8, we know there exists $\Pi = \Phi(\Phi'D\Phi)^{-1}\Phi'D$ such that for every vector J, we have $\|\Pi J - J\|_D = \min_r \|\Phi r - J\|_D$. Hence, we arrive at

$$\|\Phi'\tilde{r}^* - J^\mu\|_D \leq \|\Phi'\tilde{r}^* - \Phi'r^*\|_D + \frac{1 - \lambda\alpha}{\sqrt{(1-\alpha)(1+\alpha-2\lambda\alpha)}}\|(\Pi - I)J^\mu\|_D. \qquad \square$$

Note that $\|(\Pi - I)J^\mu\|$ can be further bounded by $\|(\Pi - I)J^\mu\| = \|J^\mu\| - \|\Pi J^\mu\| \leq \|J^\mu\|$. But (19.9) provides a tighter bound. From (19.10), we know that for the case $\lambda = 1$, the TD(λ) algorithm under the true costs (19.3) and (19.4) gives the best approximation of J^μ, i.e. $\Phi'r^* = \Pi J^\mu$ and $\|\Phi'r^* - J^\mu\|_D = \min_r \|\Phi'r - J^\mu\|_D$. As λ decreases, $(1 - \alpha\lambda)/\sqrt{(1-\alpha)(1+\alpha-2\lambda\alpha)}$ increases and the bound deteriorates. The worst bound, namely, $\|(\Pi - I)J^\mu\|/\sqrt{1-\alpha^2}$ is obtained when $\lambda = 0$. Although the result provides a bound, it suggests that in the worst-case scenario, as λ decreases, the TD(λ) under costs manipulation can suffer higher approximation error. From Lemma 19.10, we know that the manipulation of the costs has no impact on the second part of the bound which is $(1 - \lambda\alpha)\|(\Pi - I)J^\mu\|_D/\sqrt{(1-\alpha)(1+\alpha-2\lambda\alpha)}$. In the following theorem, we analyze how much $\|\Phi'(\tilde{r}^* - r^*)\|_D$ will change under a bounded manipulation of the costs.

Theorem 19.1 Suppose the manipulation of the costs is given by $\eta \in \mathbb{R}^n$ where $\eta(i) = \tilde{g}(i, \mu(i)) - g(i, \mu(i))$ for $i \in S$. Then, the distance between the manipulated TD(λ) estimate and the true TD(λ) estimate of J^μ satisfies

$$\|\Phi'\tilde{r}^* - \Phi'r^*\|_D \le \frac{1}{1-\alpha}\|\eta\|_D,$$

and the approximation error of $\Phi'\tilde{r}^*$ satisfies

$$\|\Phi'\tilde{r}^* - J^\mu\|_D \le \|\frac{1}{1-\alpha}\|\eta\|_D + \frac{1-\lambda\alpha}{\sqrt{(1-\alpha)(1+\alpha-2\lambda\alpha)}}\|(\Pi - I)J^\mu\|_D. \tag{19.11}$$

Proof: Note that $\tilde{r}^*$ and r^* satisfies $A\tilde{r}^* + \tilde{b} = 0$ and $Ar^* + b = 0$ where $b = \Phi'D \sum_{m=0}^{\infty}(\alpha\lambda P_\mu)^m \bar{g}$. Here, $\bar{g}$ is vector whose ith component is $g(i, \mu(i))$.

Then, we have $A(\tilde{r}^* - r^*) + \tilde{b} - b = 0$ which implies (because $\Phi'D$ is of full rank)

$$M\Phi(\tilde{r}^* - r^*) + \sum_{m=0}^{\infty}(\alpha\lambda P_\mu)^m(\tilde{g} - \bar{g}) = \Phi(\tilde{r}^* - r^*).$$

Applying $\|\cdot\|_D$ on both side of the equation, we obtain

$$\|M\Phi(\tilde{r}^* - r^*)\|_D + \|\sum_{m=0}^{\infty}(\alpha\lambda P_\mu)^m \eta\|_D = \|\Phi(\tilde{r}^* - r^*)\|.$$

From lemma 6.4 in Bertsekas and Tsitsiklis (1996), we know that for any $J \in \mathbb{R}^n$, we have $\|P_\mu J\|_D \le \|J\|$. From this, it easily follows that $\|P_\mu^m J\|_D \le \|J\|_D$. Note that $M = (1 - \lambda)\sum_{m=0}^{\infty}\lambda^m(\alpha P_\mu)^{m+1}$. Using the triangle inequality, we obtain

$$\|MJ\|_D \le (1 - \lambda)\sum_{m=0}^{\infty}\lambda^m \alpha^{m+1}\|J\|_D = \frac{\alpha(1-\lambda)}{1-\alpha\lambda}\|J\|_D.$$

Hence, we have

$$\|\sum_{m=0}^{\infty}(\alpha\lambda P_\mu)^m \eta\|_D \ge \frac{1-\alpha}{1-\alpha\lambda}\|\Phi(\tilde{r}^* - r^*)\|.$$

Moreover, we can see that

$$\|\sum_{m=0}^{\infty}(\alpha\lambda P_\mu)^m \eta\|_D \le \|\sum_{m=0}^{\infty}(\alpha\lambda P_\mu)^m\|_D\|\eta\|_D \le \sum_{m=0}^{\infty}(\alpha\lambda)^m\|\eta\|_D = \frac{1}{1-\alpha\lambda}\|\eta\|_D,$$

which indicates that

$$\frac{1}{1-\alpha\lambda}\|\eta\|_D \ge \frac{1-\alpha}{1-\alpha\lambda}\|\Phi(\tilde{r}^* - r^*)\|_D.$$

Thus, we have

$$\|\Phi(\tilde{r}^* - r^*)\|_D \le \frac{1}{1-\alpha}\|\eta\|_D.$$

Together with Lemma 19.2, we have

$$\|\Phi'\tilde{r}^* - J^\mu\|_D \leq \frac{1}{1-\alpha}\|\eta\|_D + \frac{1-\lambda\alpha}{\sqrt{(1-\alpha)(1+\alpha-2\lambda\alpha)}}\|(\Pi-I)J^\mu\|_D.$$

$\square$

The distance between the true TD(λ) approximator $\Phi'r^*$ and the manipulated TD(λ) approximator $\Phi'\tilde{r}^*$ is actually bounded by a constant term times the distance of the true cost and the manipulated cost. And the constant term $1/(1-\alpha)$ does not depend on λ. This means the choice of λ by the agent does not affect the robustness of the TD(λ) algorithm. This means in the worst-case scenario, TD(λ) algorithms with different values of λ will suffer the same loss of approximation accuracy. From (19.11), we can conclude that the approximation error of the manipulated TD(λ) algorithm is bounded by the magnitude of the costs manipulation and a fixed value decided by the value of λ, the choice of basis Φ, and the properties of the MDP.

19.3.2 *Q*-Learning with Manipulated Cost Signals

If the RL agent learns an optimal policy by Q-learning algorithm given in (19.7), then under stealthy attacks on cost, the algorithm can be written as:

$$\tilde{Q}_{t+1}(i,u) = (1-\gamma_t)\tilde{Q}_t(i,u) + \gamma_t\left(\tilde{g}(i,u) + \alpha\min_{v\in U(\bar{\zeta})}\tilde{Q}_t(\bar{\zeta},v)\right). \tag{19.12}$$

Note that if the attacks are not stealthy, we need to write $\tilde{g}_t$ in lieu of $\tilde{g}(i_t, a_t)$. There are two important questions regarding the Q-learning algorithm with falsified cost (19.12): (1) Will the sequence of Q_t-factors converge? (2) Where will the sequence of Q_t converge to?

Suppose that the sequence $\tilde{Q}_t$ generated by the Q-learning algorithm (19.12) converges. Let $\tilde{Q}^*$ be the limit, i.e. $\tilde{Q}^* = \lim_{n\to\infty}\tilde{Q}_t$. Suppose the objective of an adversary is to induce the RL agent to learn a particular policy $\mu^\dagger$. The adversary's problem then is to design $\tilde{g}$ by applying the actions available to him/her based on the information he/she has so that the limit Q-factors learned from the Q-learning algorithm produce the policy favored by the adversary $\mu^\dagger$, i.e. $\tilde{Q}^* \in \mathcal{V}_{\mu^\dagger}$, where

$$\mathcal{V}_\mu := \{Q \in \mathbb{R}^{n\times|A|} : \mu(i) = \arg\min_u Q(i,u), \forall i \in S\}.$$

In Q-learning algorithm (19.12), to guarantee almost sure convergence, the agent usually takes tapering stepsize ((Borkar, 2009)) $\{\gamma_t\}$ which satisfies $0 < \gamma_t \leq 1, t \geq 0$, and $\sum_t\gamma_t = \infty, \sum_t\gamma_t^2 < \infty$. Suppose in our problem, the agent takes tapering stepsize. To address the convergence issues, we have the following result.

Lemma 19.3 If an adversary performs stealthy attacks with bounded $\tilde{g}(i,a,j)$ for all $i,j \in S, a \in A$, then the Q-learning algorithm with falsified costs converges to the fixed point of $\tilde{F}(Q)$ almost surely where the mapping $\tilde{F} : \mathbb{R}^{n\times|A|} \to \mathbb{R}^{n\times|A|}$ is defined as $\tilde{F}(Q) = [\tilde{F}_{ii}(Q)]_{i,i}$ with

$$\tilde{F}_{iu}(Q) = \alpha\sum_j p_{ij}(u)\min_v Q(j,v) + \tilde{g}(i,u,j),$$

and the fixed point is unique and denoted by $\tilde{Q}^*$.

The proof of Lemma 1.1 is included in Huang and Zhu (2019b). It is not surprising that one of the conditions given in Lemma 19.3 is that an attacker performs stealthy attacks. The convergence can be guaranteed because the falsified cost signals are consistent over time for each state action pair. The uniqueness of $\tilde{Q}^*$ comes from the fact that if $\tilde{g}(i,u)$ is bounded for every $(i,u) \in S \times A, \tilde{F}$ is a contraction mapping. By Banach's fixed point theorem Kreyszig (1978), $\tilde{F}$ admits a unique fixed point. With this lemma, we conclude that an adversary can make the algorithm converge to a limit point by stealthily falsifying the cost signals.

Remark 19.1 Whether an adversary aims for the convergence of the Q-learning algorithm (19.12) or not depends on his objective. In our setting, the adversary intends to mislead the RL agent to learn policy $\mu^\dagger$, indicating that the adversary promotes convergence and aim to have the limit point $\tilde{Q}^*$ lie in $\mathcal{V}_{\mu^\dagger}$.

It is interesting to analyze, from the adversary's perspective, how to falsify the cost signals so that the limit point that algorithm (19.12) converges to is favored by the adversary. In later discussions, we consider stealthy attacks where the falsified costs are consistent for the same state action pairs. Denote the true cost by matrix $g \in \mathbb{R}^{n \times |A|}$ with $[g]_{i,u} = g(i, u)$ and the falsified cost is described by a matrix $\tilde{g} \in \mathbb{R}^{n \times |A|}$ with $[\tilde{g}]_{i,u} = \tilde{g}(i, u)$. Given $\tilde{g}$, the fixed point of $\tilde{F}$ is uniquely decided; i.e. the point that the algorithm (19.12) converges to is uniquely determined. Thus, there is a mapping $\tilde{g} \mapsto \tilde{Q}^*$ implicitly described by the relation $\tilde{F}(Q) = Q$. For convenience, this mapping is denoted by $f : \mathbb{R}^{n \times |A|} \to \mathbb{R}^{n \times |A|}$.

Theorem 19.2 *(No Butterfly Effect)* Let $\tilde{Q}^*$ denote the Q-factor learned from algorithm (19.7) with falsified cost signals and Q^* be the Q-factor learned from (19.12) with true cost signals. There exists a constant $L < 1$ such that

$$\|\tilde{Q}^* - Q^*\| \leq \frac{1}{1-L}\|\tilde{g} - g\|, \tag{19.13}$$

and $L = \alpha$.

The proof of Theorem 19.2 can be found in Huang and Zhu (2019b). In fact, taking this argument just slightly further, one can conclude that falsification on cost g by a tiny perturbation does not cause significant changes in the limit point of algorithm (19.12), $\tilde{Q}^*$. This feature indicates that an adversary cannot cause a significant change in the limit Q-factors by just applying a small perturbation in the cost signals. This is a feature known as stability, which is observed in problems that possess contraction mapping properties. Also, Theorem 19.2 indicates that the mapping $\tilde{g} \mapsto \tilde{Q}^*$ is continuous, and to be more specific, it is uniformly Lipschitz continuous with Lipschitz constant $1/(1 - \alpha)$.

With Theorem 19.2, we can characterize the minimum level of falsification required to change the policy from the true optimal policy μ^* to the policy $\mu^\dagger$. First, note that $\mathcal{V}_\mu \subset \mathbb{R}^{n \times |A|}$ and it can also be written as:

$$\mathcal{V}_\mu = \{Q \in \mathbb{R}^{n \times |A|} : Q(i, \mu(i)) < Q(i, u), \forall i \in S, \forall u \neq \mu(i)\}. \tag{19.14}$$

Second, for any two different policies μ_1 and μ_2, $\mathcal{V}_{\mu_1} \cap \mathcal{V}_{\mu_2} = \emptyset$. Lemma 19.4 presents several important properties regarding the set $\mathcal{V}_\mu$.

Lemma 19.4

(a) For any given policy μ, $\mathcal{V}_\mu$ is a convex set.
(b) For any two different policies μ_1 and μ_2, $\mathcal{V}_{\mu_1} \cap \mathcal{V}_{\mu_2} = \emptyset$.
(c) The distance between any two different policies μ_1 and μ_2 defined as $D(\mu_1, \mu_2) := \inf_{Q_1 \in \mathcal{V}_{\mu_1}, Q_2 \in \mathcal{V}_{\mu_2}} \|Q_1 - Q_2\|$ is zero.

Proof: (a) Suppose $Q_1, Q_2 \in \mathcal{V}_\mu$. We show for every $\lambda \in [0, 1]$, $\lambda Q_1 + (1 - \lambda) Q_2 \in \mathcal{V}_\mu$. This is true because $Q_1(i, \mu(i)) < Q_1(i, u)$ and $Q_2(i, \mu(i)) < Q_2(i, u)$ imply

$$\lambda Q_1(i, \mu(i)) + (1 - \lambda) Q_2(i, \mu(i)) < \lambda Q_1(i, u) + (1 - \lambda) Q_2(i, u),$$

for all $i \in S, u \neq \mu(i)$.

(b) Suppose $\mu_1 \neq \mu_2$ and $\mathcal{V}_{\mu_1} \cap \mathcal{V}_{\mu_2}$ is not empty. Then there exists $Q \in \mathcal{V}_{\mu_1} \cap \mathcal{V}_{\mu_2}$. Since $\mu_1 = \mu_2$, there exists i such that $\mu_1(i) \neq \mu_2(i)$. Let $u = \mu_2$. Since $Q \in \mathcal{V}_{\mu_1}$, we have $Q(i, \mu_1(i)) < Q(i, \mu_2(i))$. Hence, $Q \notin \mathcal{V}_{\mu_2}$, which is a contradiction. Thus, $\mathcal{V}_{\mu_1} \cap \mathcal{V}_{\mu_2} = \emptyset$.

(c) Suppose $\mu_1 \neq \mu_2$. Construct Q_1 as a matrix whose entries are all one except $Q(i, \mu_1(i)) = 1 - \epsilon/2$ for every $i \in S$ where $\epsilon > 0$. Similarly, construct Q_2 as a matrix whose entries are all one except $Q(i, \mu_2(i)) = 1 - \epsilon/2$ for every $i \in S$. It is easy to see that $Q_1 \in \mathcal{V}_{\mu_1}$ and $Q_2 \in \mathcal{V}_{\mu_2}$. Then $\inf_{Q_1 \in \mathcal{V}_{\mu_1}, Q_2 \in \mathcal{V}_{\mu_2}} \|Q_1 - Q_2\|_\infty \leq \|Q_1 - Q_2\|_\infty = \epsilon$. Since ϵ can be arbitrarily small, $\inf_{Q_1 \in \mathcal{V}_{\mu_1}, Q_2 \in \mathcal{V}_{\mu_2}} \|Q_1 - Q_2\|_\infty = 0$. Since norms are equivalent in finite-dimensional space (see section 2.4 in Kreyszig, 1978), we have $D(\mu_1, \mu_2) = \inf_{Q_1 \in \mathcal{V}_{\mu_1}, Q_2 \in \mathcal{V}_{\mu_2}} \|Q_1 - Q_2\| = 0$. □

Suppose the true optimal policy μ^* and the adversary desired policy $\mu^\dagger$ are different; otherwise, the optimal policy μ^* is what the adversary desires, there is no incentive for the adversary to attack. According to Lemma 19.4, $D(\mu^*, \mu^\dagger)$ is always zero. This counterintuitive result states that a small change in the Q-value may result in any possible change of policy learned by the agent from the Q-learning algorithm (19.12). Compared with Theorem 19.2 which is a negative result to the adversary, this result is in favor of the adversary.

Define the point Q^* to set $\mathcal{V}_{\mu^\dagger}$ distance by $D_{Q^*}(\mu^\dagger) := \inf_{Q \in \mathcal{V}_{\mu^\dagger}} \|Q - Q^*\|$. Thus, if $\tilde{Q}^* \in \mathcal{V}_{\mu^\dagger}$, we have

$$0 = D(\mu^*, \mu^\dagger) \leq D_{Q^*}(\mu^\dagger) \leq \|\tilde{Q}^* - Q^*\| \leq \frac{1}{1 - \alpha} \|\tilde{g} - g\|, \tag{19.15}$$

where the first inequality comes from the fact that $Q^* \in \mathcal{V}_{\mu^*}$ and the second inequality is due to $\tilde{Q}^* \in V_{\mu^\dagger}$. The inequalities give us the following theorem.

Theorem 19.3 *(Robust Region)* To make the agent learn the policy $\mu^\dagger$, the adversary has to manipulate the cost such that $\tilde{g}$ lies outside the ball $B(g; (1 - \alpha) D_{Q^*}(\mu^\dagger))$.

The *robust region* for the true cost g to the adversary's targeted policy $\mu^\dagger$ is given by $B(g; (1 - \alpha) D_{Q^*}(\mu^\dagger))$ which is an open ball with center c and radius $(1 - \alpha) D_{Q^*}(\mu^\dagger)$. That means the attacks on the cost needs to be "powerful" enough to drive the falsified cost $\tilde{g}$ outside the ball $B(g; (1 - \alpha) D_{Q^*}(\mu^\dagger))$ to make the RL agent learn the policy $\mu^\dagger$. If the falsified cost $\tilde{g}$ is within the ball, the RL agent can never learn the adversary's targeted policy $\mu^\dagger$. The ball $B(g; (1 - \alpha) D_{Q^*}(\mu^\dagger))$ depends only on the true cost g and the adversary desired policy $\mu^\dagger$ (Once the MDP is given, Q^* is uniquely determined by g). Thus, we refer this ball as the robust region of the true cost g to the adversarial policy $\mu^\dagger$. As we have mentioned, if the actions available to the adversary only allows him to perform bounded falsification on cost signals and the bound is smaller than the radius of the robust region, then the adversary can never mislead the agent to learn policy $\mu^\dagger$.

Remark 19.2 First, in discussions above, the adversary policy $\mu^\dagger$ can be any possible polices and the discussion remains valid for any possible policies. Second, set $\mathcal{V}_\mu$ of Q-values is not just a convex set but also an open set. We thus can see that $D_{Q^*}(\mu^\dagger) > 0$ for any $\mu^\dagger \neq \mu^*$ and the second inequality in (19.15) can be replaced by a strict inequality. Third, the agent can estimate his own robustness to falsification if he can know the adversary desired policy $\mu^\dagger$. For attackers who have access to true cost signals and the system model, the attacker can compute the robust region of the true cost to his desired policy $\mu^\dagger$ to evaluate whether the objective is feasible or not. When it is not feasible, the attacker can consider changing his objectives, e.g. selecting other favored policies that have a smaller robust region.

We have discussed how falsification affects the change of Q-factors learned by the agent in a distance sense. The problem now is to study how to falsify the true cost in a right direction so that the resulted Q-factors fall into the favored region of an adversary. One difficulty of analyzing this problem comes from the fact that the mapping $\tilde{g} \mapsto \tilde{Q}^*$ is not explicit known. The relation between $\tilde{g}$ and $\tilde{g}^*$ is governed by the Q-learning algorithm (19.12). Another difficulty is that due to the fact that both $\tilde{g}$ and $\tilde{Q}^*$ lies in the space of $\mathbb{R}^{n \times |A|}$, we need to resort to Fréchet derivative or Gâteaux derivative (Cheney, 2013) (if they exist) to characterize how a small change of $\tilde{g}$ results in a change in $\tilde{Q}^*$.

From Lemma 19.3 and Theorem 19.2, we know that Q-learning algorithm converges to the unique fixed point of $\tilde{F}$ and that $f : \tilde{g} \mapsto \tilde{Q}^*$ is uniformly Lipschitz continuous. Also, it is easy to see that the inverse of f, denoted by f^{-1}, exists since given $\tilde{Q}^*$, $\tilde{g}$ is uniquely decided by the relation $\tilde{F}(Q) = Q$. Furthermore, by the relation $\tilde{F}(Q) = Q$, we know f is both injective and surjective and hence a bijection which can be simply shown by arguing that given different $\tilde{g}$, the solution of $\tilde{F}(Q) = Q$ must be different. This fact informs that there is a one-to-one, onto correspondence between $\tilde{g}$ and $\tilde{Q}^*$. One should note that the mapping $f : \mathbb{R}^{n \times |A|} \to \mathbb{R}^{n \times |A|}$ is not uniformly Fréchet differentiable on $\mathbb{R}^{n \times |A|}$ due to the min operator inside the relation $\tilde{F}(Q) = Q$. However, for any policy μ, f is Fréchet differentiable on $f^{-1}(\mathcal{V}_\mu)$ which is an open set and connected due to the fact that $\mathcal{V}_\mu$ is open and connected (every convex set is connected) and f is continuous. In the next lemma, we show that f is Fréchet differentiable on $f^{-1}(\mathcal{V}_w)$ and the derivative is constant.

Lemma 19.5 The map $f : \mathbb{R}^{n \times |A|} \to \mathbb{R}^{n \times |A|}$ is Fréchet differentiable on $f^{-1}(\mathcal{V}_\mu)$ for any policy μ and the Fréchet derivative of f at any point $\tilde{g} \in \mathcal{V}_\mu$, denoted by $f'(\tilde{g})$, is a linear bounded map $G : \mathbb{R}^{n \times |A|} \to \mathbb{R}^{n \times |A|}$ that does not depend on $\tilde{g}$, and Gh is given as

$$[Gh]_{i,u} = \alpha P_{iu}^T (I - \alpha P_\mu)^{-1} h_\mu + h(i, u) \tag{19.16}$$

for every $i \in S, u \in A$, where $P_{iu} = (p_{i1}(u), \ldots, p_{in}(u)), P_\mu := [P_\mu]_{i,j} = p_{ij}(\mu(i))$.

The proof of Lemma 19.5 is provided in Huang and Zhu (2019b). We can see that f is Fréchet differentiable on $f^{-1}(\mathcal{V}_\mu)$ and the derivative is constant, i.e. $f'(\tilde{g}) = G$ for any $\tilde{g} \in f^{-1}(\mathcal{V}_\mu)$. Note that G lies in the space of all linear mappings that maps $\mathbb{R}^{n \times |A|}$ to itself and G is determined only by the discount factor α and the transition kernel P of the MDP problem. The region where the differentiability may fail is $f^{-1}(\mathbb{R}^{n \times |A|} \setminus (\cup_\mu \mathcal{V}_\mu))$, where $\mathbb{R}^{n \times |A|} \setminus (\cup_\mu \mathcal{V}_\mu)$ is the set $\{Q : \exists i, \exists u \neq u', Q(i, u) = Q(i, u') = \min_v Q(i, v)\}$. This set contains the places where a change of policy happens, i.e. $Q(i, u)$ and $Q(i, u')$ are both the lowest value among the ith row of Q. Also, due to the fact that f is Lipschitz, by Rademacher's theorem, f is differentiable almost everywhere (w.r.t. the Lebesgue measure).

Remark 19.3 One can view f as a "piece-wise linear function" in the norm vector space $\mathbb{R}^{n \times |A|}$. Actually, if the adversary can only falsify the cost at one state-control pair, say (i, u), while costs at other pairs are fixed, then for every $j \in S, v \in A$, the function $\tilde{g}(i, u) \mapsto [\tilde{Q}^*]_{j,v}$ is a piece-wise linear function.

Given any $g \in f^{-1}(\mathcal{V}_\mu)$, if an adversary falsifies the cost g by injecting value h, i.e. $\tilde{g} = g + h$, the adversary can see how the falsification causes a change in Q-values. To be more specific, if Q^* is the Q-values learned from cost g by Q-learning algorithm (19.7), after the falsification $\tilde{g}$, the

Q-value learned from Q-learning algorithm (19.12) becomes $\tilde{Q}^* = Q^* + Gh$ if $\tilde{g} \in f^{-1}(\mathcal{V}_\mu)$. Then, an adversary who knows the system model can utilize (19.16) to find a way of falsification h such that $\tilde{Q}^*$ can be driven to approach a desired set $\mathcal{V}_{\mu^\dagger}$. One difficulty is to see whether $\tilde{g} \in f^{-1}(\mathcal{V}_\mu)$ because the set $f^{-1}(\mathcal{V}_\mu)$ is now implicit expressed. Thus, we resort to the following theorem.

Theorem 19.4 Let $\tilde{Q}^* \in \mathbb{R}^{n \times |A|}$ be the Q-values learned from the Q-learning algorithm (19.12) with the falsified cost $\tilde{g} \in \mathbb{R}^{n \times |A|}$. Then $\tilde{Q}^* \in \mathcal{V}_{\mu^\dagger}$ if and only if the falsified cost signals $\tilde{g}$ designed by the adversary satisfy the following conditions

$$\tilde{g}(i, u) > (\mathbf{1}_i - \alpha P_{iu})^T (I - \alpha P_{\mu^\dagger})^{-1} \tilde{g}_{\mu^\dagger}. \tag{19.17}$$

for all $i \in S, u \in A \setminus \{\mu^\dagger(i)\}$, where $\mathbf{1}_i \in \mathbb{R}^n$ a vector with n components whose ith component is 1 and other components are 0.

With the results in Theorem 19.4, we can characterize the set $f^{-1}(\mathcal{V}_\mu)$. Elements in $f^{-1}(\mathcal{V}_\mu)$ have to satisfy the conditions given in (19.17). Also, Theorem 19.4 indicates that if an adversary intends to mislead the agent to learn policy $\mu^\dagger$, the falsified cost $\tilde{g}$ has to satisfy the conditions specified in (19.17).

If an adversary can only falsify at certain states $\tilde{S} \subset S$, the adversary may not be able to manipulate the agent to learn $\mu^\dagger$. Next, we study under what conditions the adversary can make the agent learn $\mu^\dagger$ by only manipulating the costs on $\tilde{S}$. Without loss of generality, suppose that the adversary can only falsify the cost at a subset of states $\tilde{S} = \{1, 2, \ldots, |\tilde{S}|\}$. We rewrite the conditions given in (19.17) into a more compact form:

$$\tilde{g}_u \geq (I - \alpha P_u)(I - \alpha P_{\mu^\dagger})^{-1} \tilde{g}_{\mu^\dagger}, \forall u \in A, \tag{19.18}$$

where $\tilde{g}_u = (\tilde{g}(1, u), \ldots, \tilde{g}(n, u))$, $\tilde{g}_{\mu^\dagger} = (\tilde{g}(1, \mu^\dagger(1)), \tilde{g}(2, \mu^\dagger(2)), \ldots, \tilde{g}(n, \mu^\dagger(n)))$ and $P_u = [P_u]_{i,j} = p_{ij}(u)$. The equality only holds for one component of the vector, i.e. the ith component satisfying $\mu(i) = u$. Partition the vector $\tilde{g}_u$ and $\tilde{g}_{\mu^\dagger}$ in (19.18) into two parts, the part where the adversary can falsify the cost denoted by $\tilde{g}_u^{\text{fal}}, \tilde{g}_{\mu^\dagger}^{\text{fal}} \in \mathbb{R}^{|\tilde{S}|}$ and the part where the adversary cannot falsify $g_u^{\text{true}}, g_{\mu^\dagger}^{\text{true}} \in \mathbb{R}^{n - |\tilde{S}|}$. Then (19.18) can be written as

$$\begin{bmatrix} \tilde{g}_u^{\text{fal}} \\ \hline g_u^{\text{true}} \end{bmatrix} \geq \begin{bmatrix} R_u & Y_u \\ \hline M_u & N_u \end{bmatrix} \begin{bmatrix} \tilde{g}_{\mu^\dagger}^{\text{fal}} \\ \hline g_{\mu^\dagger}^{\text{true}} \end{bmatrix}, \forall u \in A, \tag{19.19}$$

where

$$\begin{bmatrix} R_u & Y_u \\ \hline M_u & N_u \end{bmatrix} := (I - \alpha P_u)(I - \alpha P_{\mu^\dagger})^{-1}, \forall u \in A$$

and $R_u \in \mathbb{R}^{\tilde{S} \times \tilde{S}}, Y_u \in \mathbb{R}^{|\tilde{S}| \times (n - |\tilde{S}|)}, M_u \in \mathbb{R}^{(n - |\tilde{S}|) \times |\tilde{S}|}, N_u \in \mathbb{R}^{(n - |\tilde{S}|) \times (n - |\tilde{S}|)}$.

If the adversary aims to mislead the agent to learn $\mu^\dagger$, the adversary needs to design $\tilde{g}_u^{\text{fal}}, u \in A$ such that the conditions in (19.19) hold. The following results state that under some conditions on the transition probability, no matter what the true costs are, the adversary can find proper $\tilde{g}_u^{\text{fal}}, u \in A$ such that conditions (19.19) are satisfied. For $i \in S \setminus \tilde{S}$, if $\mu(i) = u$, we remove the rows of M_u that correspond to the state $i \in S \setminus \tilde{S}$. Denote the new matrix after the row removals by $\bar{M}_u$.

Proposition 19.1 Define $H := [\bar{M}'_{u_1} \ \bar{M}'_{u_2} \ \cdots \ \bar{M}'_{u_{|A|}}]' \in \mathbb{R}^{(|A|(n - |\tilde{S}|) - |\tilde{S}|) \times |\tilde{S}|}$. If there exists $x \in \mathbb{R}^{|\tilde{S}|}$ such that $Hx < 0$, i.e. the column space of H intersects the negative orthant of $\mathbb{R}^{|A|(n - |\tilde{S}|) - |\tilde{S}|}$, then for any true cost, the adversary can find $\tilde{g}_u^{\text{fal}}, u \in A$ such that conditions (19.19) hold.

The proof can be found in Huang and Zhu (2019b). Note that H only depends on the transition probability and the discount factor, if an adversary who knows the system model can only falsify cost signals at states denoted by $\tilde{S}$, an adversary can check if the range space of H intersects with the negative orthant of $\mathbb{R}^{|A|(n-|\tilde{S}|)}$ or not. If it does, the adversary can mislead the agent to learn $\mu^\dagger$ by falsifying costs at a subset of state space no matter what the true cost is.

Remark 19.4 To check whether the condition on H is true or not, one has to resort to Gordan's theorem (Broyden, 2001): Either $Hx < 0$ has a solution x, or $H^T y = 0$ has a nonzero solution y with $y \geq 0$. The adversary can use linear/convex programming software to check if this is the case. For example, by solving

$$\min_{y \in \mathbb{R}^{|A|(n-|\tilde{S}|)}} \|H^T y\| \text{ s.t. } \|y\| = 1, \ y \geq 0, \tag{19.20}$$

the adversary can know whether the condition about H given in Proposition 19.1 is true or not. If the minimum of (19.20) is positive, there exists x such that $Hx < 0$. The adversary can select $\tilde{g}_{\mu^\dagger}^{\text{fal}} = \lambda x$ and choose a sufficiently large λ to make sure that conditions (19.19) hold, which means an adversary can make the agent learn the policy $\mu^\dagger$ by falsifying costs at a subset of state space no matter what the true costs are.

19.4 Case Study

Here, we consider TD learning on random walk. Given a policy μ, an MDP can be considered as a Markov cost process, or MCP. In this MCP, we have $n = 20$ states. The transition digram of the MCP is given in Figure 19.2. At state $i = i_k, k = 2, 3, \ldots, n - 1$, the process proceed either left to i_{k-1} or right to i_{k+1}, with equal probability. The transition at states from i_2 to i_{n-1} is similar to symmetric one-dimensional random walk. At state i_1, the process proceed to state i_2 with probability $1/2$ or stays at the same state with equal probability. At state i_n, the probabilities of transition to i_{n-1} and staying at i_n are both $\frac{1}{2}$. That is, we have $p_{i_k i_{k+1}}(\mu(i_k)) = p_{i_k i_{k-1}}(\mu(i_k)) = \frac{1}{2}$ for $k = 2, 3, \ldots, n - 1, p_{i_1 i_1} = p_{i_1 i_2} = \frac{1}{2}$ and $p_{i_n i_n} = p_{i_n i_{n-1}} = \frac{1}{2}$. The cost at state i_k is set to be k if $k \leq 10$ and $21 - k$ if $k > 10$. That is,

$$g(i_k, \mu(i_k)) = \begin{cases} k & \text{if } k \leq 10 \\ 21 - k & \text{else} \end{cases}.$$

We consider the discount factor $\alpha = 0.9$. The task here is to use approximate TD(λ) learning algorithm to esitimate and approximate the cost-to-go function J^μ of this MCP. We consider a linear parametrization of the form

$$\tilde{J}(i, r) = r(3)i^2 + r(2)i + r(1), \tag{19.21}$$

and $r = (r(1), r(2), r(3)) \in \mathbb{R}^3$. Suppose the learning agent updates r_t based on TD(λ) learning algorithm (19.3) and (19.4) and tries to find an estimate of J^μ. We simulate the MCP and obtain a trajectory that long enough and its associated cost signals. We need an infinite long trajectory ideally. But here, we set the length of the trajectory to be 10^5. We run, respectively, TD(1) and TD(0) on

Figure 19.2 The diagram of the MCP task with 20 states denoted by $i_1, \ldots, i_{20}$.

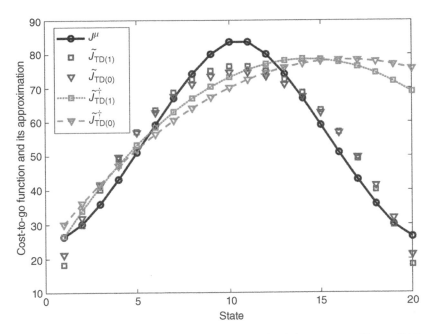

Figure 19.3 The cost-to-go function of a given policy J^μ is represented by a solid line and circle markers; $\tilde{J}_{TD}$ is the approximations of the cost-to-go function under true cost signals represented by square and triangle markers (no line); $\tilde{J}^\dagger$ is the approximations of the cost-to-go function under manipulated cost signals marked by dash lines and markers; the subscript TD(1) and TD(0) denote the TD parameter λ is set to be 1 and 0, respectively.

the same simulated trajectory based on rules given in (19.3) and (19.4). The black line indicates the cost-to-go function of the MCP. The blue markers are the approximations of the cost-to-go function obtained by following the TD(λ) algorithm (19.3) and (19.4) with $\lambda = 1$ and $\lambda = 0$. We can see that $\tilde{J}_{TD(1)}$ and $\tilde{J}_{TD(0)}$ is a quadratic function of i as we set in (19.21). Both $\tilde{J}_{TD(1)}$ and $\tilde{J}_{TD(0)}$ can serve a fairly good approximation of J^μ as we can see. The dimension of the parameters we need to update goes from $n = 20$ in the TD(λ) algorithm (19.2) to $K = 3$ in the approximation counterpart (19.3) which is more efficient computationally.

Suppose the adversary aims to deteriorate the TD(λ) algorithm by stealthily manipulating the cost signals. Suppose the adversary can only manipulate the cost signals at state i_{20} and the manipulated cost is $\tilde{g}(i_{20}, \mu(i_{20})) = 20$. We can see from Figure 19.3 that the TD(λ) learning under manipulated cost signals fails to provide accurate approximation of J^μ. Although only the cost signal at one state is manipulated, the approximation of cost-to-go function at other states can also be largely deviated from the accurate value.

We generate 100 random falsifications denoted by η. Note that $\eta = \tilde{g} - g$. For each falsification, we plot the norm $\|\eta\|_D$ of η and its associated approximation error $\|\Phi' \tilde{r}^* - J^\mu\|_D$ and the pair is marked by black circle in Figure 19.4. Note that $\tilde{r}^*$ is the parameter learned by the agent using TD(1) learning algorithm (19.8) under the falsified costs. The blue line describes the map $\|\eta\|_D \mapsto \frac{1}{1-\alpha}\|\eta\|_D + \|\Pi J^\mu - J^\mu\|_D$. The results in Figure 19.4 collaborate the results we proved in Theorem 19.1. This demonstrate how the falsification on cost signals in only one state can affect the approximated value function on every single state and how the resulted learning error is bounded.

For case study of Q-learning with falsified costs, one can refer to Huang and Zhu (2019b).

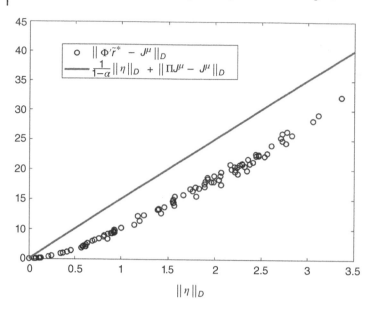

Figure 19.4 The approximated TD(1) learned from 100 random falsifications of the costs. For each falsification, we plot the distance of the falsification and the approximation error of its associated approximate of J^{μ}. The solid line is a demonstration of the bound developed in (19.11).

19.5 Conclusion

In this chapter, we have discussed the potential threats in RL and a general framework has been introduced to study RL under deceptive falsifications of cost signals where a number of attack models have been presented. We have provided theoretical underpinnings for understanding the fundamental limits and performance bounds on the attack and the defense in RL systems. We have shown that in TD(λ) the approximation learned from the manipulated costs has an approximation error bounded by a constant times the magnitude of the attack. The effect of the adversarial attacks does not depend on the choice of λ. In Q-learning, we have characterized a robust region within which the adversarial attacks cannot achieve its objective. The robust region of the cost can be utilized by both offensive and defensive sides. An RL agent can leverage the robust region to evaluate the robustness to malicious falsifications. An adversary can leverage it to assess the attainability of his attack objectives. Conditions given in Theorem 19.4 provide a fundamental understanding of the possible strategic adversarial behavior of the adversary. Theorem 19.1 helps understand the attainability of an adversary's objective. Future work would focus on investigating a particular attack model and develop countermeasures to the attacks on cost signals.

References

Itamar Arel, Cong Liu, Tom Urbanik, and Airton G Kohls. Reinforcement learning-based multi-agent system for network traffic signal control. *IET Intelligent Transport Systems*, 4(2):128–135, 2010.

Vahid Behzadan and Arslan Munir. The faults in our pi stars: Security issues and open challenges in deep reinforcement learning. *arXiv preprint, arXiv:1810.10369*, 2018.

Vahid Behzadan and Arslan Munir. Adversarial reinforcement learning framework for benchmarking collision avoidance mechanisms in autonomous vehicles. *IEEE Intelligent Transportation Systems Magazine*, 2019. https://ieeexplore.ieee.org/document/8686215 (accessed 21 March 2021).

Dimitri P. Bertsekas and John N. Tsitsiklis. Neuro-dynamic Programming, volume 5. Athena Scientific, Belmont, MA, 1996.

Vivek S. Borkar. *Stochastic Approximation: A Dynamical Systems Viewpoint*, volume 48. Springer, 2009.

Vivek S. Borkar and Sean P. Meyn. The ode method for convergence of stochastic approximation and reinforcement learning. *SIAM Journal on Control and Optimization*, 38(2):447–469, 2000.

C.G. Broyden. On theorems of the alternative. *Optimization Methods and Software*, 16(1–4):101–111, 2001.

Ward Cheney. *Analysis for Applied Mathematics*, volume 208. Springer Science & Business Media, 2013.

Jack Clark and Dario Amodei. Faulty reward functions in the wild. URLhttps://blog.openai.com/faulty-reward-functions, 2016.

Damien Ernst, Mevludin Glavic, and Louis Wehenkel. Power systems stability control: Reinforcement learning framework. *IEEE Transactions on Power Systems*, 19(1):427–435, 2004.

Tom Everitt, Victoria Krakovna, Laurent Orseau, Marcus Hutter, and Shane Legg. Reinforcement learning with a corrupted reward channel. *arXiv preprint, arXiv:1705.08417*, 2017.

Linan Huang and Quanyan Zhu. Adaptive honeypot engagement through reinforcement learning of semi-markov decision processes. In *International Conference on Decision and Game Theory for Security*, pp. 196–216. Springer, 2019a.

Yunhan Huang and Quanyan Zhu. Deceptive reinforcement learning under adversarial manipulations on cost signals. In *International Conference on Decision and Game Theory for Security*, pp. 217–237. Springer, 2019b.

Jens Kober, J. Andrew Bagnell, and Jan Peters. Reinforcement learning in robotics: A survey. *The International Journal of Robotics Research*, 32(11): 1238–1274, 2013.

Erwin Kreyszig. *Introductory Functional Analysis with Applications*, volume 1. Wiley, New York, 1978.

Yen-Chen Lin, Zhang-Wei Hong, Yuan-Hong Liao, Meng-Li Shih, Ming-Yu Liu, and Min Sun. Tactics of adversarial attack on deep reinforcement learning agents. *arXiv preprint, arXiv:1703.06748*, 2017.

Yuzhe Ma, Xuezhou Zhang, Wen Sun, and Jerry Zhu. Policy poisoning in batch reinforcement learning and control. In *Advances in Neural Information Processing Systems*, pp. 14543–14553. Curran Associates, Inc., 2019.

Volodymyr Mnih, Koray Kavukcuoglu, David Silver, Andrei A. Rusu, Joel Veness, Marc G. Bellemare, Alex Graves, Martin Riedmiller, Andreas K. Fidjeland, Georg Ostrovski, et al. Human-level control through deep reinforcement learning. *Nature*, 518(7540):529, 2015.

Richard S. Sutton and Andrew G. Barto. *Reinforcement Learning: An Introduction*, volume 2. MIT Press, Cambridge, 2018.

Ah-Hwee Tan, Ning Lu, and Dan Xiao. Integrating temporal difference methods and self-organizing neural networks for reinforcement learning with delayed evaluative feedback. *IEEE Transactions on Neural Networks*, 19(2):230–244, 2008.

John N. Tsitsiklis and Benjamin Van Roy. An analysis of temporal-difference learning with function approximation. *IEEE Transactions on Automatic Control*, 42(5):674–690, 1997.

Jingkang Wang, Yang Liu, and Bo Li. Reinforcement learning with perturbed rewards. *arXiv preprint, arXiv:1810.01032*, 2018.

Christopher J.C.H. Watkins and Peter Dayan. Q-learning. *Machine Learning*, 8(3–4):279–292, 1992.

Christopher John Cornish Hellaby Watkins. Learning from delayed rewards. PhD thesis, King's College, University of Cambridge, 1989.

Zhiheng Xu and Quanyan Zhu. A cyber-physical game framework for secure and resilient multi-agent autonomous systems. In 2015 54th IEEE Conference on Decision and Control (CDC), pp. 5156–5161. IEEE, 2015.

Quanyan Zhu and Tamer Başar. Dynamic policy-based ids configuration. In Proceedings of the 48h IEEE Conference on Decision and Control (CDC) held jointly with 2009 28th Chinese Control Conference, pp. 8600–8605. IEEE, 2009.

Nomenclature

α	Discounted factor.
η_t	Eligibility vector at time t.
$\gamma_t(i)$	State-dependent non-negative stepsize at time t.
$\mathcal{V}_\mu$	A set of Q-factors that produces policy μ.
μ^*	Optimal policy.
$\mu^\dagger$	A policy desired by the adversary.
μ_k	Control policy at time k.
Φ	Basis functions in compact form.
ϕ	Basis functions of the approximation.
Π	Projection matrix.
$\tilde{\eta}$	Difference between cost signals and compromised cost signals.
$\tilde{F}(Q)$	Right-hand side of Bellman's equation under compromised cost signals.
$\tilde{g}$	Compromised cost function (matrix).
$\tilde{J}$	Approximated value function of the MDP.
$\tilde{Q}^*$	Optimal Q-factors under compromised cost signals.
$\tilde{Q}_t$	Q-factors at time t under compromised cost signals.
$\tilde{r}$	Approximation parameters under compromised cost signals.
$\tilde{r}^*$	Optimal approximation parameters under compromised cost signals.
A	Control space.
D	Diagonal entries of steady-state probabilities.
$d(i)$	Steady-state probability at state i.
d_t	Temporal difference at time t.
f	A mapping from cost function to Q-factors.
f^{-1}	The inverse of mapping f.
g	Cost function (matrix).
g_μ	A vector of costs under policy μ.
i_k	State at time k.
J^*	Value function of the MDP.
L	A mapping from cost function to Q-factors.
P	Transition kernel.
P_μ	Transition matrix given policy μ.
$p_{ij}(u)$	Transition probability from state i to state j under control u.
P_{iu}	Transition probabilities at state i under control u.
$Q(i, u)$	Q-factor of a state-control pair (i, u).
Q^*	Optimal Q-factors.
r	Approximation parameters.
S	State space.
$U(i)$	Control set of state i.
$z_t(i)$	State-dependent eligibility coefficient at state i.

20

Resource-Aware Intrusion Response Based on Deep Reinforcement Learning for Software-Defined Internet-of-Battle-Things

Seunghyun Yoon[1,2], Jin-Hee Cho[1], Gaurav Dixit[1], and Ing-Ray Chen[1]

[1] *Department of Computer Science, Virginia Tech, Falls Church, VA, USA*
[2] *School of Electrical Engineering and Computer Science, Gwangju Institute of Science and Technology, Gwangju, Republic of Korea*

20.1 Introduction

20.1.1 Motivation and Challenges

Military tactical networks often suffer from severe resource constraints (e.g. battery, computing power, bandwidth, and/or storage). This puts a high challenge in designing a network (or system) which should be robust against adversaries under high dynamics in tactical environments. When a network consists of highly heterogeneous entities, such as Internet-of-Things (IoT) devices in a large-scale network, deploying multiple defense mechanisms to protect a system requires high intelligence to meet conflicting system goals of security and performance. When a node fails due to being compromised or functional fault, multiple strategies can be considered to deal with this node failure, such as destruction, repair, or replacement. What strategy to take to deal with this node is vital for the system to complete a given mission as well as to defend against adversaries because nodes themselves are network resources for providing destined services and defense/security. If a node detected as compromised is not useful in a given system, it can be discarded (i.e. disconnected or destroyed) based on the concept of "disposable security" (Kott et al. 2016) or "self-destruction" (Brueckner et al. 2014; Curiac et al. 2009; Zeng et al. 2010). However, if the node is regarded as a highly critical asset which keeps highly confidential information or provides critical services (e.g. web servers or databases), its removal will cause a critical damage to service provision and/or may introduce security breach. To obtain optimal intrusion response strategies to deal with nodes detected as compromised/failed, we propose a bio-inspired multilayer network structure that consists of three layers including the core layer, the middle layer, and the outer layer where the nodes are placed in each layer according to their importance (i.e. place the most important nodes in the core layer, the medium important nodes in the middle layer, and the least important nodes in the outer layer). This network design is bio-inspired by mimicking a defense system of the human body upon pathogen attacks and how the body deals with an infection aiming to reach inner organs (Janeway et al. 1996).

20.1.2 Research Goal

In this work, we aim to solve the following problems:

- Build a multilayered defensive network structure (i.e. a three-layer structure with core, middle, and outer layers) that can deploy nodes based on their importance levels in a network. This

Game Theory and Machine Learning for Cyber Security, First Edition.
Edited by Charles A. Kamhoua, Christopher D. Kiekintveld, Fei Fang, and Quanyan Zhu.
© 2021 The Institute of Electrical and Electronics Engineers, Inc. Published 2021 by John Wiley & Sons, Inc.

multilayered network structure will be designed to protect the given network based on a different level of priorities under resource constraints in terms of network availability and vulnerability caused by interconnectivity.

- Identify an optimal response action (i.e. destruction, repair, or replacement) to nodes detected as compromised/failed (NDCF) in the network based on deep reinforcement learning (DRL). The DRL agent will receive system state information from the core, middle, and outer layer SDN controllers and an intrusion detection system (IDS). Accordingly, the DRL agent will identify an optimal action to maximize the accumulated reward that can achieve the required levels of system security (i.e. the number of the importance-weighted legitimate nodes, which are not compromised in a network) and performance (i.e. the average criticality levels of nodes considering how they are connected with other nodes, such as centrality weighted by their importance based on their roles).

20.1.3 Research Questions

We have the following **key research questions** in this study:

1. What are the key node features (e.g. importance, centrality, vulnerability) that significantly affect system performance and security introduced by the proposed multilayered defense network structure?
2. What are the key design factors that affect an optimal policy (i.e. a probability distribution of the set of actions available) for a defense response action (i.e. destruction, repair, or replacement) which can lead to maximum system security (i.e. low security vulnerability) and performance (i.e. high service availability or mission completion)?

20.1.4 Key Contributions

In this work, we made the following **key contributions**:

1. The multilayered defensive network structure is inspired by a biological immune system where a human body has multiple layers of protection from skin to inner organs, where the outer layer (e.g. skin) can be easily damaged but quickly recoverable while the inner organs are not easily accessible and are well protected with the internal immune system (Janeway et al. 1996). This biology-inspired, multilayer network structure has not been proposed as a defensive network structure aiming to be resilient against epidemic attacks (e.g. malware spreading) as well as targeted attacks (i.e. attacks targeting for a specific node to introduce maximum damage to a system, such as nodes in the core layer) in the literature.
2. Deep reinforcement learning (DRL) has not been applied to provide optimal resource-aware defensive response strategies in IoBT environments in the literature. Our proposed DRL-based resource-aware intrusion response decision framework is the first that leverages DRL for centralized learning based on input from multiple agents (i.e. agents of each layer and an IDS) in resource-constrained tactical IoBT environments. In this work, we compare the performance of multiple DRL algorithms under a different level of attack severity in the network. We consider both value-based (i.e. Deep-Q-learning network, DQN) and policy-based (i.e. Proximal Policy Optimization, PPO, and Actor-Critic) methods along with a baseline random decision by the DRL agent and compare their performance.
3. We leverage software-defined networking (SDN) technology to consider IoBT where an SDN controller can be employed to deploy the proposed defense framework. This is the first work that adopts SDN technology to realize multilayer defensive, resource-aware intrusion response strategies using DRL.

20.1.5 Structure of This Chapter

The rest of this chapter is organized as follows:

- Section 20.2 discusses related work in terms of software-defined IoBT (SD-IoBT), DRL, and resource-aware defense systems.
- Section 20.3 describes our node model, network model, attack model, system failure conditions, and assumptions made in this work.
- Section 20.4 provides the detailed description of the proposed DRL-based resource-aware active defense framework that provides optimal actions as intrusion response strategies upon nodes being detected as compromised/failed (NDCF).
- Section 20.5 describes our experimental setting, including tools used, network setting, comparing schemes, and metrics.
- Section 20.6 demonstrates the experimental results obtained from our simulation model, discusses their overall trends, and analyzes the performance and security of the proposed framework.
- Section 20.7 concludes the paper and suggests future research directions.

20.2 Related Work

20.2.1 Software-Defined Internet-of-Battle-Things

Software-defined networking (SDN) is an emerging technology that manages a network by decoupling a network control plane from a data-forwarding plane. This capability allows a networked system highly flexible, robust, and programmable easily (Kreutz et al. 2015). In traditional network environments, decisions on routing have been made at each switch. This introduced a lack of controls over switches and resulted in making the system less optimized. By leveraging the SDN technology, an SDN controller can be used to centrally control all packet forwarding operations, leading to running the network in a more optimized manner. Due to this merit, SDN-based security and defense mechanisms have been deployed easily with the help of the SDN controllers (Yoon et al. 2017, 2020).

Internet-of-Battle-Things (IoBT) consists of all cyber connected devices and things in a battlefield. For example, various sensors, computers, mobile devices, and so on. IoBT shares similarities with IoT technology although the cyber and physical hostile environment reflects the key characteristic of IoBT (Kott et al. 2016). Further, Azmoodeh et al. (2018) discussed that the attackers of IoBT are more likely state sponsored and resourceful. The authors used deep learning for malware classification in IoT and IoBT environments. Farooq and Zhu (2018) also mentioned the lack of communication infrastructure and a wide variety of devices as a differentiating aspect from IoT networks. Further, Suri et al. (2016) discussed existing works in applying IoT technology for IoBT. They discussed how various emerging capabilities of IoT can be leveraged in IoBT scenarios.

SDN has started to find its use in IoBT networks. Singh et al. (2017) and Nobre et al. (2016) have suggested using SDN in IoBT due to its agility and efficient/effective network management capability. They also discussed how effectively SDN can be deployed in IoBT. Nobre et al. (2016) proposed the so-called *software-defined battlefield networking* (SDBN), which is SDN modified to suit battlefield needs. None of these works have used machine learning or any other learning mechanisms. Our work takes a step further by using DRL for SDN-based IoBT environments to solve more complex problems in IoBT settings.

20.2.2 Deep Reinforcement Learning

Reinforcement learning (RL) is a goal-oriented algorithm to autonomously learn a policy and then tell an agent to take a particular action to achieve a final goal in the future with the maximum accumulated reward. Although RL has achieved high success in some existing studies (Kohl and Stone 2004; Singh et al. 2002), it is limited for solving low-dimensional problems and hard to fix complexity issues. Deep reinforcement learning (DRL) has been proposed to address the disadvantages of RL in solving a problem with huge state spaces or action spaces (Arulkumaran et al. 2017b). To apply deep learning (DL) algorithms to an ordinary RL domain, some improvements have been implemented. One of the important aspects of DRL is the experience replay mechanism, which stores interaction experience information into a replay buffer and randomly samples batches from the experience pool to train the DL's parameters rather than directly using the experience sample in the learning phase. This allows the DL networks to stably learn from the set of past experiences and to reduce the correlation between experience samples. Learning algorithms have a significant impact on bringing huge success for solving sequential decision making and control problems. Well-known examples include deep model-free Q-learning for general Atari game playing (Mnih et al. 2013), a DRL for a driver's decision making in traffic situations (Bai et al. 2019), and continuous control decision making in 3D humanoid locomotion (Duan et al. 2016).

Policy-based DRL is also proposed (e.g. policy gradient) in order to learn a policy network which prevents an agent to be at intermediate states due to the stochastic nature in the policy network (Sutton et al. 1999). DRL also enhances learning decision policies directly from high-dimensional inputs using end-to-end RL. The example applications using DRL include designing a decision making algorithm for dynamic sensor networks using RL (Zhang and Ma 2007), solving a path discovery problem using DRL to learn dynamics of environments for network reconfigurations (Xiong et al. 2017), and development of a DRL-based recommendation system for recommending news articles (Zheng et al. 2018). The actor-critic has been widely considered as an architecture for the policy gradient approaches (Bhatnagar et al. 2007; Degris et al. 2012; Peters and Schaal 2008; Sutton et al. 1999). The basic idea is to adjust the parameters of a policy for the policy improvement in an actor while evaluating the policy in the critic. In this work, we also leverage the natural actor-critic method (Peters and Schaal 2008) to be compared against a value-based DRL using DQN.

Recently, to reduce nonstationarity due to high variances of policy gradient methods and to increase the training stability, Trust Region Policy Optimization (TRPO) has been proposed (Schulman et al. 2015). When the DRL agent updates their policy, if the policy differs too much from the existing policy, calculating policy gradient becomes more complex because the state distribution of the MDP changes significantly. In order to deal with this problem, TRPO restricts the size of policy updates at every iteration by enforcing a Kullback–Leibler (KL) Divergence constraint to prevent the policy from being updated to the worse direction. However, one of the major drawbacks of this algorithm (Schulman et al. 2015) is the lack of scalability because of the second-order optimization. In Schulman et al. (2017), the authors proposed the Proximal Policy Optimization (PPO) algorithm which is more lightweight and reliable algorithm than TRPO while taking some benefits of TRPO. We will also leverage the PPO to solve our problems.

20.2.3 Resource-Aware Defense Systems

Sherwood et al. (2009) developed a new approach called "Flowvisor" based on a network slicing technique to deploy hardware's forwarding for serving multiple logical networks and building the virtual slices of a network. Most existing works have not specifically focused on slicing IoT networks; rather they focused on slicing SDN to support IoT. Kalør et al. (2018) focused on slicing SDN supporting IoT devices to reduce the complexity and improve Quality-of-Service (QoS) parameters.

They limited their research to industrial networks where a variety of devices on the network have mandated such a solution. Nassar and Yilmaz (2019) proposed a method to slice IoT networks using reinforcement learning to manage QoS requirements. Various learning mechanisms were discussed to obtain an optimal decision-making policy that is best suited for IoT environments. Recently, network slicing technologies have been popularly applied in IoT with 5G. Ni et al. (2018) proposed a method for IoT network slicing which also preserves privacy while guaranteeing secure access to IoT services. Unlike the existing network slicing approaches described above (Kalør et al. 2018; Nassar and Yilmaz 2019; Ni et al. 2018; Sherwood et al. 2009), our proposed approach does not limit slicing only to 5G networks. Additionally, to maximize the utilization of defender resources, the proposed multilayered network structure leverages DRL methods to develop resource-aware intrusion response strategies.

Recently, Farooq and Zhu (2018) developed a framework for IoBT networks by using epidemiology as inspiration. Their work focused on developing a multilayer network and analyzing its security effectiveness. However, their work did not consider SDN networks based on DRL decision making for effective defense resource allocation aiming to achieve priority-based, resource-aware defense. Both of these features are considered as novelty in our work.

20.3 System Model

This section will discuss the node model, network model, attack model, system failure conditions, and assumptions made in this work.

20.3.1 Node Model

This work concerns an IoBT network environment consisting of vastly different heterogeneous nodes where the network is assigned with a tactical mission. We propose a multilayered, resource-aware active defense framework by first grouping IoT nodes into multiple layers based on the node type in terms of a required level of security (i.e. how important a node is to ensure system security) and performance (i.e. how critical the node is to ensure a required level of QoS). To this end, we characterize a node (e.g. a web server, a database, or an IoT device) as follows:

- **Importance** (importance$_i$) indicates how important node i is for security and performance in a given network. importance$_i$ characterizes node i's importance in its given role in terms of the consequence of node i being compromised/failed. A system administrator can assign this importance value to each node i ranged in $[1, 10]$ as a real number during the network deployment phrase and it is a static value during the entire mission period.
- **Reachability** (reachability$_i$) refers to how reachable node i is from other nodes in a network. We use the faster betweenness centrality metric (Brandes 2001) estimated in the SDN controller. reachability$_i$ is estimated in the range of $[0, 1]$ as a real number, which may change during the mission period upon the changes introduced in the network topology due to node failure (i.e. detected as compromised/failed) and/or mobility.
- **Vulnerability** (vulnerability$_i$) indicates node i's vulnerability, which is estimated based on the mean value of vulnerabilities associated with multiple software installed in each node. We employ the Common Vulnerability Scoring System (CVSS) severity score (CVE; CVS) to assign the degree of a Common Vulnerabilities and Exposures (CVE) to each node. A node's mean vulnerability is simply obtained by the scaled mean vulnerability across multiple vulnerabilities in $[0, 10]$ where the maximum vulnerability score is 10 in the CVSS. To use this vulnerability score as an infection rate in modeling epidemic attacks (e.g. malware spread), we simply normalize it by dividing the value with 10 so vulnerability$_i$ can be ranged in $[0, 1]$.

Since a legitimate user can be compromised by the attacker, each node's status is indicated by compromise$_i$ with 1 for compromised and 0 for noncompromised. active$_i$ returns 1 when node i is not compromised and active (i.e. not failed); 0 otherwise. In summary, we formulate node i's attributes as:

$$n_i = [\text{importance}_i, \text{reachability}_i, \text{vulnerability}_i, \text{compromise}_i, \text{active}_i]. \tag{20.1}$$

To build a multilayered defense network structure that provides more security protection for more important assets, we group the existing nodes in a given SDN-based IoBT network into three types in terms of its asset importance: high importance nodes (HINs), medium importance nodes (MINs), and low importance nodes (LINs). We leverage a node's importance (i.e. importance$_i$), representing how important the node is in terms of security (e.g. the amount of credential information and its consequence upon being the information leaked out, such as databases) and performance (e.g. the importance of its role in providing services such as web servers). These three different types of node importance are considered for designing the proposed multilayered network structure detailed in Section 20.4.

20.3.2 Network Model

We concern a software-defined IoBT network (SD-IoBT) with an SDN controller available to centrally and flexibly manage security and performance based on the separation of the data plane from the control plane. Specifically, the SDN controller decides data forwarding decisions and can play a key role in effectively deploying various network defense mechanisms, e.g. intrusion detection, firewall, moving target defense, etc.

SDN-enabled switches handle forwarding packets, where they encapsulate packets without exact matching flow rules in flow tables in which the encapsulated packets, "OFPT_PACKET_IN" messages in OpenFlow (OF) protocol (i.e. standard communication protocol between SDN-enabled switches and the SDN controller), are provided to the SDN controller handling the flow. The IoBT devices, which are resource constrained, depend on servers to collect and communicate data from the battlefield. The data is collected and processed at remote cloud servers, which in turn provide data-query services using multihop communications.

We assume that the SD-IoBT environment needs to execute a mission requiring communications to execute multiple mission tasks. In addition, we consider the presence of attackers that can hinder mission execution and/or normal operations/functionalities of the given SD-IoBT. To this end, the SD-IoBT environment consists of a multilayer network structure in which each layer is controlled by its own SDN controller and the central security coordinator (i.e. DRL agent) manages the entire system security. The security coordinator collects network and node information from SDN controllers and IDSs. Based on the result of DRL algorithms in the security coordinator, it sends security-related commands (i.e. destruction, repair, replacement) to SDN controllers. The overall description of the considered SD-IoBT network is described in Figure 20.1 along with the detailed attack behaviors described in Section 20.3.3.

20.3.3 Attack Model

In this work, we consider the following two types of attacks:

Outside attackers: A node in the outer layer (see Figure 20.1) can be attacked by outside attackers based on the degree of a node' vulnerabilities, represented by the average probability that the node is vulnerable (v_i) due to the vulnerabilities of multiple software installed in it. We consider the initial number of nodes to be attacked by the outside attackers as $P_a \times n$ where P_a is the probability of a node being initially attacked and n is the total number of outer layer nodes in the network. Then, the node being compromised becomes an inside attacker, as explained below.

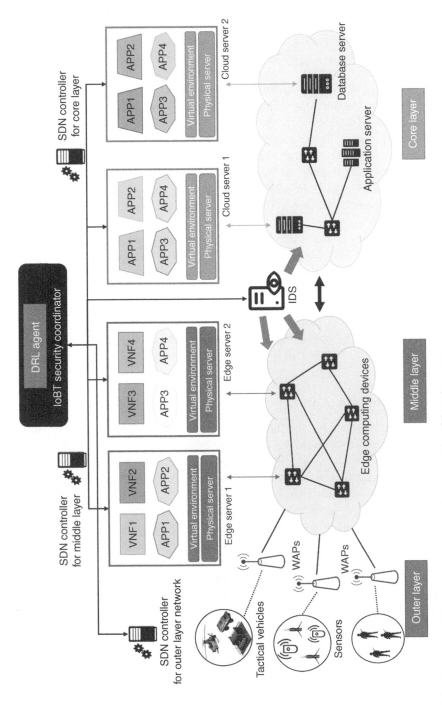

Figure 20.1 Example multilayered network structure of an SD-IoBT environment.

Inside attackers: When a node in the given SD-IoBT network is being compromised, we call it an inside attacker. An inside attacker can perform epidemic attacks to its neighboring nodes (i.e. 1-hop neighbors) based on an epidemic model called the SIR (Susceptible-Infected-Removed/Recovered) model (Newman 2010). That is, an inside attacker can compromise the nodes directly connected to itself, called its direct neighbors, without access rights to their settings or files. Typical example scenarios include the spread of malware or viruses. Botnets can spread malware or viruses via mobile devices. A mobile device can use a mobile malware, such as a Trojan horse, playing a role of a botclient to receive a command and control from a remote server (Mavoungou et al. 2016). Further, worm-like attacks are popular in wireless sensor networks where the sensor worm attacker sends a message to exploit the software vulnerability to cause crash or take a control of sensor nodes (Yang et al. 2008, 2016).

We consider that an inside attacker can compromise other neighboring nodes based on their vulnerability probabilities. To save attack cost, we assume that an inside attacker chooses a neighboring node with the highest vulnerability. We use the node's vulnerability probability for the inside attacker to successfully compromise the neighboring node. The ultimate goal of this inside attacker is to compromise a target node located in the core layer and exfiltrate the information from the target node to the outside.

We assume that compromised nodes that have not been detected by the IDS due to false-negative errors can perform message dropping attacks, such as gray whole attacks with probability P_{pd} (i.e. the probability that the inside attacker drops a packet) or message modification attacks. Hence, any messages passing through the undetected compromised nodes can be either dropped or modified by inside attackers. If a legitimate node is misidentified as compromised due to false-positive errors, it can either increase unnecessary cost to take a response action toward the legitimate node or may reduce the number of nodes for mission execution which may cause a system failure (as triggering Byzantine Failure condition, discussed in Section 20.3.4) if the legitimate node is not recovered in a timely manner.

20.3.4 System Failure Condition

We define a system failure state based on the below condition.

Byzantine Failure (BF): We adopt the concept of Byzantine Failure (BF) (Gärtner 2003) to define the failure of a system in that at least one-third of legitimate nodes are compromised. The occurrence of this system failure is because there are too many compromised nodes, ultimately resulting in the lack of control over compromised nodes. This BF condition is commonly used to define a general system failure in the literature (Doudou et al. 2002; Driscoll et al. 2003). We make this BF condition more generic by considering the impact introduced by nodes detected as compromised/failed with different importance levels in a network. Specifically, we count the number of compromised or failed nodes weighted by their importance and determine whether a system fails or not based on a threshold ρ as a fraction of compromised nodes in the system. That is, when at least ρ fraction of the importance-weighted nodes in the system is compromised, a Byzantine failure has occurred. BF is defined by:

$$
\mathrm{BF} = \begin{cases} 1 & \text{if } \rho \leq \frac{\sum_{i \in V} \text{importance}_i \times \text{compromised}_i}{\sum_{i \in V} \text{importance}_i} \\ 0 & \text{otherwise.} \end{cases} \tag{20.2}
$$

Here importance$_i$ is node i's importance level, compromise$_i$ is node i's status being compromised or not (i.e. 1 for being compromised; 0 otherwise), and V is a set of nodes in the given network. This mainly leads to the loss of system integrity and availability. This system failure condition will be utilized in order to measure the system-level metric, such as "mean time to security failure (MTTSF)," as detailed in Section 20.5.2.

20.3.5 Assumptions

In this section, we discuss the assumptions made in terms of key management and trusted entities in a given network.

Key Management: We assume that legitimate nodes use a group key (i.e. a secret cryptography) to ensure secure communications between users while prohibiting outsiders from accessing the network resources. In order for an outsider to be a legitimate node, the outsider needs to go through the authentication process and obtain the group key. In addition, each node's access to network resources should follow the procedures to obtain an access right based on a privilege to be granted. When an inside attacker attempts to compromise a legitimate node, it must first obtain a privilege to access the legitimate node.

In the proposed multilayered network structure (see Section 20.4.1), HINs can be directly connected only with either MINs or HINs, not LINs; MINs can be directly connected with all node types; and LINs can be directly connected with LINs or MINs, but not HINs. The characteristics of each layer are detailed in Section 20.4.1. We assume that nodes with the same importance type (i.e. HINs, MINs, or LINs) will share a secret key (a.k.a. a group key) to communicate with each other in the same importance type. Hence, for a node to communicate with a different type node, it should be given another secret key for the nodes communicating in the different layers. In our proposed network structure, HINs are not given a secret key for LINs and vice-versa. Therefore, the network maintains five different secret keys to secure communications (i.e. three keys for three different node types; one key between HINs and MINs; and one key between MINs and LINs). In addition, we assume that each secret key will be rekeyed upon membership changes due to nodes being destructed in the corresponding layer. Examining the effect of key management (i.e. issuance, distribution, update, and revocation) is beyond the scope of this work.

Trusted SDN Controllers and Switches: In the IoBT network we concerned in this work, we allow three SDN controllers to govern the corresponding three layers. The SDN controllers and control switches are assumed trusted. Building defense against SDN controllers/switches is beyond the scope of this work. Since the SDN controller should be well informed of basic network information under its control, the SDN controller periodically updates the network topology and software vulnerabilities of nodes under its control. Via this process, the SDN controller can periodically check the status of overall system security and take actions accordingly.

Intrusion Detection System (IDS): We assume that an IDS exists and interacts with the SDN controller in order to inform nodes detected as compromised/failed (NDCF) so that the DRL framework in the SDN controller can take an appropriate response action toward the NDCF. The development of an efficient/effective IDS is beyond the scope of this work. We simply characterize the IDS by false negative and false-positive probabilities, denoted by P_{fn} and P_{fp}, respectively, so that we can investigate how effectively and efficiently the DRL-based intrusion response decision framework can learn an optimal strategy to deal with the NDCF.

20.4 The Proposed DRL-Based Resource-Aware Active Defense Framework

In this section, we describe our proposed resource-aware active defense framework based on the following two components: (1) a multilayered defense network structure; and (2) a DRL-based intrusion response framework leveraging the proximal policy optimization (PPO) method (a class of DRL algorithms) which allows the DRL agent to identify the best policy to maximize the accumulated reward.

20.4.1 MultiLayered Defense Network Structure

We simply deploy HINs, MINs, and LINs to core, middle, and outer layers, correspondingly. To be specific, each layer consists of the corresponding types of nodes as follows:

1. **The core layer** consists of nodes with high importance (i.e. HINs). Since nodes in this layer have high importance levels given, if they are compromised and/or failed, it may easily lead to a system failure, as high importance nodes are weighted more in estimating the BF condition, as shown in Eq. (20.2). The example nodes may include central database servers or authentication servers in the cloud. The importance range of HINs is in [8–10] as an integer and each HIN's importance value is assigned based on this range. Due to the high importance of nodes in this layer, all nodes in the core layer will be targeted by attackers. Recall that HINs can be directly connected only with either MINs or HINs, not LINs. This network topology is designed to add more security protection toward HINs in the core layer.
2. **The middle layer** consists of nodes with medium importance (i.e. MINs). The example nodes may include edge servers which primarily connect the server in the core layer to sensors and IoT devices in the outer layer, such as routers and switches. The importance range of MINs is set in [4–7] as an integer and each MIN's importance is assigned based on this range. MINs can be directly connected with either HINs or LINs as they mainly serve as a coordinator between the core and outer layers. The failure or compromise of MINs may hinder service availability upon a lack of MINs, in addition to possibly causing a system failure due to triggering the BF condition in Eq. (20.2).
3. **The outer layer** consists of nodes with low importance (i.e. LINs). The example nodes may include low-cost IoT devices and/or sensors. Nodes in this layer (i.e. LINs) will have their importance ranged in [1, 3] as an integer. Similar to MINs, the compromise or failure of LINs may trigger the BF condition. LINs can be directly connected with LINs or MINs, but not HINs, as the LINs in the outer layer is the first layer attackers can access to penetrate into a given network. A lack of LINs may lead to a shallow protection layer, which may lead to easy access for the attackers to get into the deeper layers.

To maintain this priority-based, multilayered network structure, even if nodes move around, they need to check which nodes to communicate with while not violating the multilayered network structure. Figure 20.1 shows an example SD-IoBT network structure described in this section.

20.4.2 DRL-Based Intrusion Response Strategies

In this section, we will consider three types of response strategies: destruction, replacement, or repair under a given resource-constrained, dynamic, and hostile environment. A DRL agent will aim to identify an optimal intrusion response action which can maximize the levels of both security (e.g. security vulnerability) and performance (e.g. service availability). We describe how these are modeled in this work.

20.4.2.1 Intrusion Response Strategies
The three response strategies are considered to ensure system security and performance while minimizing defense cost incurred. When a node is detected as compromised/failed, a DRL agent can take one of the following strategies:

- **Destruction**: The node detected as compromised/failed (NDCF) will be simply destroyed. In this case, any security or performance benefit to be introduced by the node when it is successfully repaired or replaced will be zero due to its immediate destruction. However, this action immediately removes the node from the network, reducing the likelihood that an attack compromising other neighboring nodes in the network. In addition, there will be a minimum cost in the destruction process.

- **Repair**: NDCF i can be repaired by taking delay d_i^r where d_i^r is a function of node i's importance, which is given by:

$$d_i^r = \zeta \times \text{importance}_i, \tag{20.3}$$

where importance_i is given as one of node attributes (see Eq. (20.3.1)) and ζ is a normalized maximum repair delay. This implies that higher node importance introduces longer delay. This d_i^r parameter is considered when estimating the gain by repairing node i, which is detailed in Section (20.8).

- **Replacement**: NDCF i can be replaced by taking delay d_i^p, assuming that there is a constant delay ξ, which refers to the normalized maximum delay introduced to replace node i. We assume that the replacement delay is longer than the repair delay.

20.4.2.2 Selection of an Intrusion Response Policy

Formulation of the MDP: In Markov Decision Process (MDP), we assume complete state information available for the environment and the next state of the environment only depends on the current state and the chosen action.

This work considers an MDP as a mathematical framework for DRL, by a tuple $\langle \mathcal{S}, \mathcal{A}, \mathcal{T}, \mathcal{R}, \gamma \rangle$ for a DRL agent, consisting of (1) a set of states $\mathcal{S}$ with a distribution of starting states $p(\mathbf{s}_0)$ where $\mathbf{S}_t$ refers to a set of states at time t; (2) a set of actions, $\mathcal{A} = \{\mathbf{a}_1, \dots, \mathbf{a}_t, \dots \mathbf{a}_T\}$ where $\mathbf{a}_t$ is a set of actions available at time t; (3) a transition probability, $\mathcal{T}_i(\mathbf{s}_t, \mathbf{a}_t, \mathbf{s}_{t+1})$, from state $\mathbf{s}_t$ to state $\mathbf{s}_{t+1}$ via $\mathbf{a}_t$; (4) an immediate reward function, $\mathcal{R}(\mathbf{s}_t, \mathbf{a}_t, \mathbf{s}_{t+1})$; (5) a reward discount factor (e.g. γ); and (6) policy π mapping from states to a probability distribution of actions, $\pi : \mathcal{S} \rightarrow (\mathcal{A} = \mathbf{a} \mid \mathcal{S})$. An RL agent aims to maximize its own total expected return $\mathcal{R} = \sum_{t=0}^{T} \gamma_t R_t$ where γ is a discounting factor and T is the time horizon. The MDP can generate an optimal policy π^* that maximizes the expected reward from states (Arulkumaran et al. 2017a).

To select an optimal intrusion response action that can maximize both security and performance of a given network, we formulate the MDP by:

- **States** ($\mathcal{S}$): A state at time t by taking an action, denoted by S_t, refers to a system state in terms of system security and performance (i.e. SS_t and PS_t), which are detailed later in this section.
- **Actions** ($\mathcal{A}$): When a legitimate, internal node is detected as compromised/failed, one of the following three actions can be taken: destruction, repair, or replacement. Each action is detailed in Section 20.4.2. The reward associated with each action taken is formulated later in this section.
- **Transitions** ($\mathcal{T}$): For simplicity, no preference is considered for the DRL agent to choose a particular action. The DRL agent will select one of three actions.
- **Discount factor** (γ): We will consider a discount factor, γ, ranged in (0, 1] as a real number.

Objectives of the Multi-Layered Network Structure: At each layer, an agent sitting on a corresponding SDN controller will keep track of the number of nodes detected as compromised/failed (NDCFs) and the number of messages correctly delivered in the layer it governs. Based on the information informed by each agent to the DRL agent, the DRL agent will estimate the system state. For simplicity, we consider system security state and performance states to represent a system state, $S_t = (SS_t, PS_t)$.

Each state is detailed as follows:

- **System security state** (SS_t): This state (per step t) measures the fraction of the number of noncompromised, legitimate nodes weighted by their importance ($\sum_{i \in V}$importance$_i \times (1 - \max[\text{compromised}_i, \text{active}_i])$) over the total number of importance-weighted nodes initially deployed ($\sum_{i \in V}$importance$_i$) in the network can be only measured as long as the system is still alive (until the system does not reach the BF condition). SS_t is formulated by:

$$SS_t = \begin{cases} \frac{\sum_{i \in V}\text{importance}_i \times (1-\text{compromised}_i)}{\sum_{i \in V}\text{importance}_i} & \text{if BF is not met;} \\ 0 & \text{otherwise.} \end{cases} \quad (20.4)$$

- **System performance state** (PS_t): This state (per step t) captures the level of a system's service availability per step t. PS_t is captured by:

$$PS_t = \begin{cases} \frac{N_{MD}^t}{N_M^t} & \text{if BF is not met;} \\ 0 & \text{otherwise.} \end{cases} \quad (20.5)$$

Here N_{MD}^t refers to the number of messages correctly delivered and N_M^t is the number of messages transmitted per step t.

The overall procedures of the proposed policy are described later in this section.

Immediate Reward Formulation (R_i): We estimate agent i's immediate reward associated with a given action based on the extent of the net gain consisting of the gain (G_i) minus loss (L_i), i.e. $R_i = G_i - L_i$. The gain and loss are obtained as follows:

- **Gain** (G_i): This gain includes the benefits introduced in terms of system security and performance. The security gain (SG_i) refers to the difference between the current and expected system security states when taking an action toward node i (i.e. NDCF), which is given by:

$$SG_i = \tilde{SS}_i - SS_i, \quad (20.6)$$

where $\tilde{SS}_i$ refers to the expected system security state and SS_i is the current system security state. The time unit is omitted for simplicity.

The performance gain (PG_i) is the difference between the average current and expected node criticality where node i's criticality (criticality$_i$) is computed by importance$_i \times$ reachability$_i$. PG_i is estimated by:

$$PG_i = \tilde{c}_i - c_i, \quad (20.7)$$

where $\tilde{c}_i$ and c_i refer to node i's expected and current average criticality in the network, respectively. The gain, G_i, is estimated by:

$$G_i = \frac{(SG_i + PG_i)}{d_i}, \quad (20.8)$$

where $1/d_i$ is the penalty introduced by delay d_i associated with an action taken (i.e. either repair delay d_i^r or replacement delay d_i^p). Note that we set $G_i = 0$ for destruction.

- **Loss** (L_i): This refers to cost associated with a given action. For simplicity, we consider economic cost (i.e. economic loss due to node destruction or repairing fees or machine cost for replacement). In this sense, destruction cost (D_C) is considered to be the most expensive due to the loss of the resource in a given network; repair cost (R_C) is related to fixing fees; and replacement cost (P_C) is the associated expense to purchase a new machine or replace parts in a current machine. We model each cost based on Gaussian distribution considering a variety of degrees in possible cost. We assume $D_C \geq \text{mean}(R_C) \geq \text{mean}(P_C)$ for our experiment.

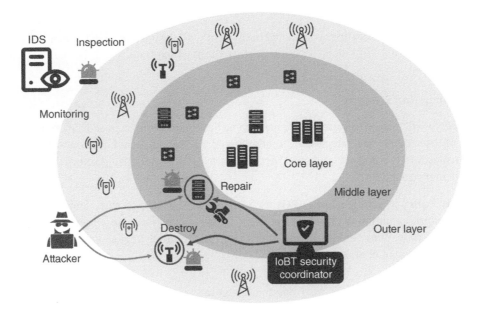

Figure 20.2 Multilayered intrusion response strategies.

The Overall Procedures of the DRL-Based Intrusion Response Policy: Figure 20.2 describes the overall idea of the proposed multilayered intrusion response strategies in this work. For the proposed DRL-based intrusion response framework to take an action, a single episode is composed of the following events:

1. When node i is detected as compromised/failed by the IDS (i.e. NDCF), the IDS will inform the DRL agent. Then, the DRL agent will collect the current network/system information (i.e. a number of nodes detected as compromised and the correct message delivery ratio occurred in each layer to estimate the security and performance states in Eqs. (20.4) and (20.5)) from the multiple SDN controllers under the multiple layers of the proposed network structure where each SDN controller monitors and collects security and performance information based on the status of nodes it governs in its corresponding layer.
2. The DRL agent identifies an optimal action to take per NDCF (where the IDS may detect more than one NDCF) by considering: (i) a system state consisting of both security and performance states, as described in Section 20.4.2.2; (ii) node i's profile which is determined by its connectivity, status in being compromised/failed, and the importance of the node for security and performance in a given network; (iii) system security and performance level; and (iv) estimated gain and loss to calculate reward (R_i) based on the net gain (i.e., $G_i - L_i$). For simplicity, we consider the reward by taking an action toward each node i.
3. The action toward the given NDCF node is taken and the corresponding reward is fed back to the DRL agent.

When the IDS detects more than one NDCF node (i.e. k nodes) at the same time, the DRL agent needs to take k actions (i.e. steps) within the episode. If the system fails by meeting the BF condition, the episode will end immediately.

20.5 Experimental Setting

For the network and node settings considered in this work, all design parameters and their default values are listed in Table 20.1 along with their corresponding meanings.

20.5.1 Comparing Schemes

The DRL agent will take an action toward a node being detected as compromised/failed (i.e. NDCF) from one of three response strategies, destruction, repair, or replacement based on each strategy's expected reward. To this end, we compare the performance of various DRL algorithms to investigate their performance as follows:

- **Proximal Policy Optimization (PPO)** (Schulman et al. 2017): In this algorithm, only the first-order approximation of the objective function is used while maintaining data efficiency and running robustness, which are the merits of TRPO (Trust Region Policy Optimization) algorithm (Schulman et al. 2015).

Table 20.1 Design parameters, their meanings, and default values

Param.	Meaning	Default values
N	Total number of nodes in a given network	50 nodes
P_{core}	Ratio of nodes in the core layer	0.2 (10 nodes)
P_{middle}	Ratio of nodes in the middle layer	0.3 (15 nodes)
P_{outer}	Ratio of nodes in the outer layer	0.5 (25 nodes)
P_a	Fraction of the initial attackers in the outer layer	0.2 (10 nodes)
$v_{i,k}$	Vulnerability of software k installed in node i. Core nodes in range[1, 3], [3, 7] mid layer; [6, 10] outer layer nodes	Selected at random
importance$_i$	Node i's importance level based on its role; [1, 3] for low; [4, 7] for medium; [8, 10] for high	Selected at random
P_{pd}	Probability that an inside attacker drops a packet	0.5
P_{fp}, P_{fn}	False-positive and -negative probabilities, respectively	0.2
ρ	Threshold of the fraction of compromised to indicate the BF condition in Eq. (20.2)	1/3
ζ	Normalized maximum repair delay in Eq. (20.3)	1
ξ	Normalized maximum replacement delay	10
mean(D_C)	Mean destruction cost	8
mean(R_C)	Mean repair cost	2
mean(P_C)	Mean replacement cost	1
γ	A discount factor in RL	0.0005
$N_{epi-max}$	Maximum number of episodes considered (same as the mission duration)	30
$N_{step-max}$	Maximum number of steps considered (same as the actions taken by the DRL agent per episode)	300
N_r	Total number of simulation runs	50
ϵ	The probability of exploration	0.0005–0.01

- **Actor-Critic** (Peters and Schaal 2008): This algorithm is one of well-known policy gradient methods. In the Actor-Critic, two neural networks are separately maintained to learn the policy (i.e. an actor network) and value function (i.e. a critic network) by an agent.
- **Deep Q-Learning (DQN)** (Mnih et al. 2015): DQN utilizes neural networks parameterized by θ to represent the action-value function (i.e. Q function). DQN assumes that the agent can fully observe the environment.
- **Random**: This method simply allows the agent to take a random action (i.e. destruction, repair, and replacement) to each compromised node being detected (i.e. NDCF).

20.5.2 Metrics

We use the following metrics to examine the performance of the proposed resource-aware active defense mechanism:

- **Instantaneous Accumulated Reward** (R_t): This metric refers to an accumulated reward at time t based on the rewards obtained by the DRL agent.
- **Final Accumulated Reward** (R): This metric is the estimate of the final accumulated reward at the end of the simulation (i.e. either when the system fails or when the mission finishes as the maximum number of episodes are performed).
- **Mean Time to Security Failure (MTTSF)**: This metric measures the average time that a given network fails by meeting BF in Section 20.3.4. MTTSF is computed by:

$$\text{MTTSF} = \int_{t=0}^{\infty} P_t \, dt, \quad P_t = \begin{cases} 1 & \text{if BF is not met;} \\ 0 & \text{otherwise.} \end{cases} \tag{20.9}$$

Here P_t represents whether a system is alive or not.

- **Average Fraction of Correct Messages Delivered** (P_{MD}): This metric indicates the average fraction of the number of messages correctly delivered (N_{MD}) over the number of messages transmitted (N_M) during the entire mission duration or until the system fails. P_{MD} is obtained by:

$$P_{MD} = \frac{N_{MD}}{N_M}. \tag{20.10}$$

20.6 Simulation Results & Analysis

In this section, we demonstrate and discuss the experimental results that compare the performance of the four different DRL algorithms addressed in Section 20.5.1 in order to study the following: (1) Effect of the proposed DRL-based intrusion response strategies (DRL-IRS) on the accumulated rewards over episodes; (2) Effect of the DRL-IRS on MTTSF with respect to the ratio of the initial infection (P_a); (3) Effect of the DRL-IRS on the ratio of messages correctly delivered with respect to P_a; and (4) Effect of the DRL-IRS on the converged accumulated rewards with respect to P_a.

20.6.1 Effect of DRL-Based Intrusion Response Strategies on Accumulated Rewards

Figure 20.3 shows how different DRL algorithms can reach an accumulated reward as the DRL agent learns from more episodes. In Figure 20.3, we clearly observe that Proximal Policy Optimization (PPO) outperforms the vanilla Actor-Critic model, the value-based model, DQN, and Random as a baseline model in terms of the instantaneous accumulated reward with respect to episodes. Because PPO is based on the Trust Region Policy Optimization (TRPO) algorithm, it prevents the action selection from going toward the worst direction in a lightweight manner. Recall that taking an action in our work is based on each node being detected as compromised/failed (NDCF). When each action is taken for an NDCF, the DRL agent assumes that other NDCF nodes are not being compromised/failed and estimates a system state (i.e. security and performance states) based on the assumption. This may hinder a lot in accurately estimating Q function values, which can prevent the DRL agent from correctly learning an optimal action in DQN. On the other hand, a policy gradient method, such as AC, aims to identify an optimal action directly from a policy and the corresponding reward estimated based on a taken action. Hence, a policy gradient method allows the convergence to the optimal policy to go relatively faster than value-based methods. Accordingly, we observed the performance order as PPO $\geq$ Actor-Critic $\geq$ DQN $\geq$ Random.

20.6.2 Effect of DRL-Based Intrusion Response Strategies Under Varying Attack Severity (P_a)

Figure 20.4 shows how the DRL agent learns under different levels of hostility by varying P_a in terms of MTTSF when it uses different DRL algorithms. When the hostility is very low (i.e. $P_a = 0.2$), MTTSF is almost the same under all DRL algorithms. However, as the environment becomes more hostile with higher P_a, MTTSF is significantly affected. Similar to the performance in Figure 20.3, PPO outperforms all due to its high convergence under dynamic environments as considered in this work. However, unlike the performance of DQN in Figure 20.3, DQN performs fairly well even if the Actor-Critic still outperforms it. This is partly because our MTTSF is estimated based on the mission duration (i.e. the number of episodes), which is relatively short to reflect a given short-term mission duration. Therefore, even before the system fails, most of episodes are finished without experiencing the system failure even under DQN.

In Figure 20.5, we demonstrate how the DRL agent's decisions achieve the service availability goal of this system in terms of the ratio of messages correctly delivered (P_{MD}) when the ratio of initial infection (P_a) varies. The performance order of the comparing algorithms in P_{MD} is also the same as other results shown in Figures 20.3 and 20.4 with PPO $\geq$ Actor-Critic $\geq$ DQN $\geq$ Random. As expected, we observe the overall trends under all algorithms are lower P_{MD} observed under higher P_a. However, interestingly, unlike Figure 20.4, the outperformance of PPO and Actor-Critic is clear, compared to the performance of DQN and Random. This would be because our formulated reward

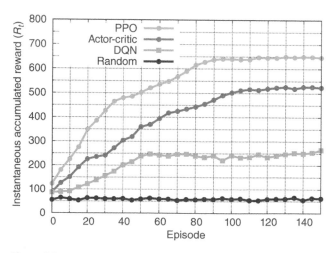

Figure 20.3 Accumulated reward with respect to the number of episodes.

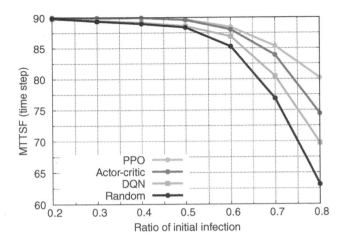

Figure 20.4 MTTSF under varying the ratio of initial infection (P_a).

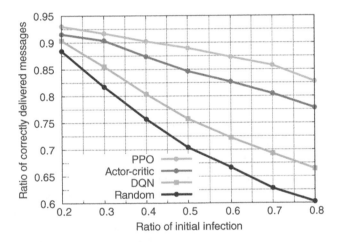

Figure 20.5 Ratio of messages delivered under varying the ratio of initial infection (P_a).

function is more tuned to optimize the performance in the presence of attackers while MTTSF is more affected by the mission duration, which is given based on the number of episodes. In addition, recall that a system failure is defined based on the BF condition which is controlled by the threshold ρ (i.e. the threshold of the fraction of compromised nodes above which the system is considered failed). Hence, a subtle security difference caused by a different number of compromised nodes in the system may not be captured because the system failure is a binary decision based on the threshold ρ.

Figure 20.6 shows the resilience of each DRL algorithm when P_a varies in terms of the final accumulated reward (R). The observed trends of the results are consistent with those observed in Figure 20.3. That is, the final accumulated reward maintains at about the same performance level at the end of the mission duration (or until the system fails) even under different settings. The reasons of the outperformance in the PPO can be explained based on the same reasons discussed in Figures 20.3 and 20.5. It is also noticeable that R in PPO and Actor-Critic slightly decreases as P_a increases. This is because as the DRL agent takes actions toward many NDCF nodes, the reward tends to be slightly lower as the number of destroyed nodes increases.

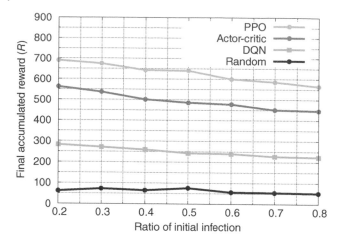

Figure 20.6 Final rewards under varying the ratio of initial infection (P_a).

20.7 Conclusion & Future Work

In this work, we proposed a resource-aware active defense framework that can provide a set of deep reinforcement learning-based intrusion response strategies for a software-defined networking (SDN)-based Internet-of-Battle-Things (IoBT) environment. The proposed defense framework achieved the following goals: (1) The proposed defense framework is built based on a highly attack-resistant network based on multilayered structuring in order to provide higher security protection toward more critical system assets while maintaining acceptable service availability during mission execution; (2) The proposed defense framework is resource-aware in taking intrusion response strategies against nodes being detected as compromised/failed based on decisions made by the agent using a set of deep reinforcement learning (DRL) algorithms; and (3) We conducted extensive simulation experiments to identify the best DRL algorithm that can perform the best in terms of accumulated rewards among a class of DRL algorithms against a baseline random selection model, and provided physical interpretations to the experimental results. Through these key contributions, we hope this work can provide insightful and helpful angles in simulation and modeling of a cyber-hardened system which leverages machine/deep learning-based defense strategies.

The **key findings** from our study are summarized as: (1) The DRL agent learns better when using a policy gradient method (i.e. PPO and Actor-Critic) than when using a value-based method (i.e. DQN), showing higher accumulated rewards; (2) Although DQN does not learn well in dealing with NDCF nodes, due to the definition of a system failure defined based on the total number of compromised/inactive nodes, called Byzantine failure condition, MTTSF is much less affected with DQN due to poor learning; (3) PPO particularly performs well in maintaining high service availability in terms of the ratio of messages correctly delivered. This is because PPO is also capable of capturing security vulnerability by keeping track of compromised/failed nodes in the network and discouraging message passing between these problem nodes; and (4) We also observed that the PPO even performs better in more hostile environments with a higher fraction of initial infection (P_a). This is because the DRL agent can have more chances to learn and make proper decisions as more nodes are being detected as compromised/failed.

We suggest the following **future work directions** in terms of extending our current work: (1) Consider a DRL agent's imperfect observability toward the environment by considering Partially Observable Markov Decision Process (POMDP) (Zhu et al. 2017) in order to reflect high dynamics and potential attacks that could cause packet dropping or uncertainty in the IoBT environment. A possible direction is to investigate the applicability of Recurrent Neural Networks (RNN) to enhance the agent's learning ability; (2) Refine the attack model that can consider more sophisticated attack behaviors such as advanced persistent threat (APT) attacks; and (3) Develop a multiagent framework so that multiple DRL agents can cooperate or compete to each other in a large-scale network where a single centralized entity is not feasible for scalability.

References

K. Arulkumaran, M. P. Deisenroth, M. Brundage, and A. A. Bharath. Deep reinforcement learning: A brief survey. *IEEE Signal Processing Magazine*, 34 (6):26–38, 2017a.

Kai Arulkumaran, Marc Peter Deisenroth, Miles Brundage, and Anil Anthony Bharath. A brief survey of deep reinforcement learning. *arXiv preprint, arXiv:1708.05866*, 2017b.

Amin Azmoodeh, Ali Dehghantanha, and Kim-Kwang Raymond Choo. Robust malware detection for internet of (battlefield) things devices using deep eigenspace learning. *IEEE Transactions on Sustainable Computing*, 4(1): 88–95, 2018.

Zhengwei Bai, Baigen Cai, Wei Shangguan, and Linguo Chai. Deep reinforcement learning based high-level driving behavior decision-making model in heterogeneous traffic. *arXiv preprint, arXiv:1902.05772*, 2019.

Shalabh Bhatnagar, Richard S. Sutton, Mohammad Ghavamzadeh, and Mark Lee. Incremental natural actor-critic algorithms. In Proceedings of the 20th International Conference on Neural Information Processing Systems, NIPS'07, pp. 105–112. Curran Associates Inc., Red Hook, NY, USA, 2007.

Ulrik Brandes. A faster algorithm for betweenness centrality. *Journal of Mathematical Sociology*, 25(2):163–177, 2001.

Stephen K. Brueckner, Robert A. Joyce, Carl Manson, Hajime Inoue, and Kenneth J. Thurber. Fight-through nodes with disposable virtual machines and rollback of persistent state. US Patent 8,839,426, September 2014.

Common Vulnerabilities and Exposures (CVE). n.d. https://cve.mitre.org/.

Common Vulnerability Scoring System (CVSS). n.d. https://www.first.org/cvss/.

Daniel-Ioan Curiac, Madalin Plastoi, Ovidiu Banias, Constantin Volosencu, Roxana Tudoroiu, and Alexa Doboli. Combined malicious node discovery and self-destruction technique for wireless sensor networks. In *2009 Third International Conference on Sensor Technologies and Applications*, pp. 436–441. IEEE, 2009.

T. Degris, P. M. Pilarski, and R. S. Sutton. Model-free reinforcement learning with continuous action in practice. In *2012 American Control Conference (ACC)*, pp. 2177–2182, June 2012.

Assia Doudou, Benoît Garbinato, and Rachid Guerraoui. Encapsulating failure detection: From crash to byzantine failures. In Johann Blieberger and Alfred Strohmeier, editors, *Reliable Software Technologies—Ada-Europe 2002*, pp. 24–50. Springer, Berlin, Heidelberg, 2002.

Kevin Driscoll, Brendan Hall, Håkan Sivencrona, and Phil Zumsteg. Byzantine fault tolerance, from theory to reality. In Stuart Anderson, Massimo Felici, and Bev Littlewood, editors, *Computer Safety, Reliability, and Security*, pp. 235–248. Springer, Berlin, Heidelberg, 2003.

Yan Duan, Xi Chen, Rein Houthooft, John Schulman, and Pieter Abbeel. Benchmarking deep reinforcement learning for continuous control. In *ICML*, pp. 1329–1338, 2016.

Muhammad Junaid Farooq and Quanyan Zhu. On the secure and reconfigurable multi-layer network design for critical information dissemination in the internet of battlefield things (iobt). *IEEE Transactions on Wireless Communications*, 17(4):2618–2632, 2018.

Felix C Gärtner. Byzantine failures and security: Arbitrary is not (always) random. Technical report. Mit Sicherheit Informatik, 2003.

Charles A. Janeway, Paul Travers, Mark Walport, and Mark Shlomchik. *Immunobiology: The Immune System in Health and Disease*, volume 7. Current Biology London. Garland Science, 1996.

Anders Ellersgaard Kalør, Rene Guillaume, Jimmy Jessen Nielsen, Andreas Mueller, and Petar Popovski. Network slicing in industry 4.0 applications: Abstraction methods and end-to-end analysis. *IEEE Transactions on Industrial Informatics*, 14(12):5419–5427, 2018.

Nate Kohl and Peter Stone. Policy gradient reinforcement learning for fast quadrupedal locomotion. In Proceedings. ICRA'04. 2004, volume 3, pp. 2619–2624. IEEE, 2004.

A. Kott, A. Swami, and B. J. West. The internet of battle things. *Computer*, 49 (12):70–75, 2016.

D. Kreutz, F. M. V. Ramos, P. E. Veríssimo, C. E. Rothenberg, S. Azodolmolky, and S. Uhlig. Software-defined networking: A comprehensive survey. *Proceedings of the IEEE*, 103(1), 2015. doi: 10.1109/JPROC.2014.2371999.

S. Mavoungou, G. Kaddoum, M. Taha, and G. Matar. Survey on threats and attacks on mobile networks. *IEEE Access*, 4:4543–4572, 2016.

Volodymyr Mnih, Koray Kavukcuoglu, David Silver, Alex Graves, Ioannis Antonoglou, Daan Wierstra, and Martin Riedmiller. Playing Atari with deep reinforcement learning. *arXiv preprint, arXiv:1312.5602*, 2013.

Volodymyr Mnih, Koray Kavukcuoglu, David Silver, Andrei A. Rusu, Joel Veness, Marc G. Bellemare, Alex Graves, Martin Riedmiller, Andreas K. Fidjeland, Georg Ostrovski, et al. Human-level control through deep reinforcement learning. *Nature*, 518(7540):529–533, 2015.

Almuthanna Nassar and Yasin Yilmaz. Reinforcement learning for adaptive resource allocation in fog ran for IoT with heterogeneous latency requirements. *IEEE Access*, 7:128014–128025, 2019.

M. E. J. Newman. *Networks: An Introduction*, 1st edition. Oxford University Press, 2010.

Jianbing Ni, Xiaodong Lin, and Xuemin Sherman Shen. Efficient and secure service-oriented authentication supporting network slicing for 5G-enabled iot. *IEEE Journal on Selected Areas in Communications*, 36(3):644–657, 2018.

Jeferson Nobre, Denis Rosario, Cristiano Both, Eduardo Cerqueira, and Mario Gerla. Toward software-defined battlefield networking. *IEEE Communications Magazine*, 54(10):152–157, 2016.

Jan Peters and Stefan Schaal. Natural actor-critic. *Neurocomputing*, 71(7–9): 1180–1190, 2008.

John Schulman, Sergey Levine, Pieter Abbeel, Michael Jordan, and Philipp Moritz. Trust region policy optimization. In *International Conference on Machine Learning*, pp. 1889–1897, 2015.

John Schulman, Filip Wolski, Prafulla Dhariwal, Alec Radford, and Oleg Klimov. Proximal policy optimization algorithms. *arXiv preprint, arXiv:1707.06347*, 2017.

Rob Sherwood, Glen Gibb, Kok-Kiong Yap, Guido Appenzeller, Martin Casado, Nick McKeown, and Guru Parulkar. Flowvisor: A network virtualization layer. *OpenFlow Switch Consortium, Technical Report*, 1:132, 2009.

Dhananjay Singh, Gaurav Tripathi, Antonio M. Alberti, and Antonio Jara. Semantic edge computing and iot architecture for military health services in battlefield. In *2017 14th IEEE Annual Consumer Communications & Networking Conference (CCNC)*. IEEE, 2017.

Satinder Singh, Diane Litman, Michael Kearns, and Marilyn Walker. Optimizing dialogue management with reinforcement learning: Experiments with the njfun system. *Journal of Artificial Intelligence Research*, 16:105–133, 2002.

N. Suri, M. Tortonesi, J. Michaelis, P. Budulas, G. Benincasa, S. Russell, C. Stefanelli, and R. Winkler. Analyzing the applicability of internet of things to the battlefield environment. In *2016 International Conference on Military Communications and Information Systems (ICMCIS)*. IEEE, 2016.

Richard S. Sutton, David McAllester, Satinder Singh, and Yishay Mansour. Policy gradient methods for reinforcement learning with function approximation. In *Proceedings of the 12th International*

Conference on Neural Information Processing Systems (NIPS'99), pp. 1057–1063. MIT Press, Cambridge, MA, USA, 1999.

Wenhan Xiong, Thien Hoang, and William Yang Wang. Deeppath: A reinforcement learning method for knowledge graph reasoning. *arXiv preprint, arXiv:1707.06690*, 2017.

Y. Yang, S. Zhu, and G. Cao. Improving sensor network immunity under worm attacks: A software diversity approach. In *Proceedings of the 9th ACM International Symposium on Mobile Ad Hoc Networking and Computing*, MobiHoc '08, pp. 149–158. ACM, 2008.

Y. Yang, S. Zhu, and G. Cao. Improving sensor network immunity under worm attacks: A software diversity approach. *Ad Hoc Networks*, 47(Supplement C): 26–40, 2016.

Seunghyun Yoon, Taejin Ha, Sunghwan Kim, and Hyuk Lim. Scalable traffic sampling using centrality measure on software-defined networks. *IEEE Communications Magazine*, 55 (7):43–49, 2017. doi: 10.1109/MCOM.2017.1600990.

Seunghyun Yoon, Jin-Hee Cho, Dong Seong Kim, Terrence J. Moore, Frederica Free-Nelson, and Hyuk Lim. Attack graph-based moving target defense in software-defined networks. *IEEE Transactions on Network and Service Management*, 2020.

Lingfang Zeng, Zhan Shi, Shengjie Xu, and Dan Feng. Safevanish: An improved data self-destruction for protecting data privacy. In *2010 IEEE Second International Conference on Cloud Computing Technology and Science*, pp. 521–528. IEEE, 2010.

Dongmei Zhang and Huadong Ma. A q-learning-based decision making scheme for application reconfiguration in sensor networks. In *2007 11th International Conference on Computer Supported Cooperative Work in Design*, pp. 1122–1127. IEEE, 2007.

Guanjie Zheng, Fuzheng Zhang, Zihan Zheng, Yang Xiang, Nicholas Jing Yuan, Xing Xie, and Zhenhui Li. Drn: A deep reinforcement learning framework for news recommendation. In *International World Wide Web Conference WWW)*, pp. 167–176. ACM, 2018.

Pengfei Zhu, Xin Li, Pascal Poupart, and Guanghui Miao. On improving deep reinforcement learning for pomdps, 2017. arXiv preprint, arXiv:1704.07978.

Part VI

Other Machine Learning Approach to Cyber Security

21

Smart Internet Probing: Scanning Using Adaptive Machine Learning

Armin Sarabi, Kun Jin, and Mingyan Liu

Department of Electrical Engineering and Computer Science, University of Michigan, Ann Arbor, MI, USA

21.1 Introduction

Network scanning is a widely studied topic, ranging from partial scans of the Internet (Fan and Heidemann 2010; Murdock et al. 2017), to global scans of the IPv4 address space (Leonard and Loguinov 2010; Durumeric et al. 2013, 2015). This has led to the development of network scanning tools such as ZMap (Durumeric et al. 2013) and NMap (Lyon 2009), which have provided researchers with large amounts of information on arbitrary Internet hosts. Data resulting from network scans have been used in a wide range of security studies, e.g., to probe and characterize machines utilized in the Mirai botnet (Antonakakis et al. 2017), to gauge the security posture of networks for cyber-risk forecasting (Liu et al. 2015), and to study hosts susceptible to the Heartbleed vulnerability (Durumeric et al. 2014). Internet scanning is a crucial tool for giving visibility into the security of Internet-connected entities, as it can measure the attack surface of networks by revealing (potentially misconfigured/vulnerable) networked devices accessible on the public Internet. Additionally, network scanning has been used in many Internet measurement studies, including studies for examining trends and adoption rates of different technologies (Felt et al. 2017; Kotzias et al. 2018; Kumar et al. 2018), to detect discoverable hosts and to categorize them (e.g., IoT devices) (Bano et al. 2018; DeMarinis et al. 2018; Feng et al. 2018; Scheitle et al. 2018), and to map network topologies (Shavitt and Shir 2005; Claffy et al. 2009; Beverly et al. 2018).

However, the current approach to Internet scanning involves exhaustively sending probes to every scanned IP address (possibly the entire IPv4 address space), regardless of whether the target host is reachable on the public Internet. Therefore, network scans strain the targeted networks/hosts, as they can produce large amounts of traffic, especially when multiple ports of a host are being probed. In addition, global scanning of the IPv6 address space is not feasible using such exhaustive methods, forcing researchers to come up with techniques for producing scan targets, in order to obtain a representative subset of publicly discoverable hosts for characterizing and studying IPv6 networks (Murdock et al. 2017; Gasser et al. 2018).

Note that a large majority of probes sent during a scan will go unanswered, since most IP addresses are *inactive*, meaning that they are not running any Internet-facing service, or do not respond to outside probes. This gets more pronounced as multiple ports are scanned, since a single *active* IP address may only have a few number of active/open ports, i.e., ports that respond to probes. In fact, the Censys database (Durumeric et al. 2015) which contains global scans of

Game Theory and Machine Learning for Cyber Security, First Edition.
Edited by Charles A. Kamhoua, Christopher D. Kiekintveld, Fei Fang, and Quanyan Zhu.
© 2021 The Institute of Electrical and Electronics Engineers, Inc. Published 2021 by John Wiley & Sons, Inc.

IPv4 address space across 37 different port contains roughly 161 million records in its snapshots on 1/1/2019, meaning that ~94.3% of the announced Border Gateway Protocol (BGP) prefixes (~170/8 blocks, or ~2.8 billion addresses) are inactive, or do not respond to requests on any of the scanned ports. For active IP addresses, the corresponding hosts are only responding to requests for 1.8 ports on average.

Given the above context, a question arises as to whether probes can be sent in a more judicious manner, i.e., if an Internet scanner can preemptively predict whether a probe will be left unanswered, and thus refrain from sending it in the first place. This would then lead to an overall reduction in bandwidth for an Internet scan, reducing its intrusive nature. Moreover, reducing the bandwidth of Internet scans allows one to probe IP addresses at a faster rate, which in turn increases the hit rate for discovering active hosts. This is important as more hosts are migrating to the IPv6 address space (Akamai 2021; Internet Society 2018), where global scanning is not feasible by existing methods, and increasing the rate at which hosts can be probed (e.g., using a hitlist; Murdock et al. 2017; Gasser et al. 2018) will lead to more discovered active devices in this space.

Motivated by the above, we develop a framework that leverages machine learning to predict whether a given host will respond to probes on different ports. In a machine learning setting, port responses can be seen as a set of binary labels and we can use classification models to perform these multi-label predictions. In situations where we can predict port responses accurately, we are able to save unnecessary probes and improve the efficiency of the scanner. Our first set of features for prediction includes geolocation and autonomous system (AS) information of IP addresses. These features can provide a machine learning model with information about the underlying network, e.g., to distinguish between residential, educational, and hosting networks, which can help the scanner adjust its probes accordingly. As an example, we observe that hosts belonging to residential networks are more likely to respond to requests on port 7547 (CWMP), a protocol commonly used by routers/modems.

Moreover, we observe dependencies between the responses of different ports, i.e., whether a host is responding to probes on the scanned ports. For instance, if we receive a response from a host on port 443 (HTTPS), then it is likely that it will also respond to probes on port 80 (HTTP), since both ports are used to serve web content to clients. We also measure and observe high correlation between ports associated with mail servers. We can then leverage these correlations to improve classification accuracy by scanning different ports of an IP address in sequence, and appending the obtained labels/features resulting from each probe to the features vectors for predicting the remaining port responses. We observe different levels of correlation between different ports, in other words, some port responses are highly dependent on one another, while others act more independently. Therefore the efficacy of the scanner is highly dependent on the order in which ports are scanned. For this, we develop an adaptive technique for generating an optimal order for probing different ports of a host.

We evaluate our technique over scans collected over 20 different ports between January and May of 2019 by the Censys database and show that using only geolocation and autonomous system information, we can achieve bandwidth savings (the reduction in number of probes) of 26.7–72.0%, while still achieving 90–99% true positive rates for detecting active ports. We further train and evaluate a sequence of classifiers, gradually adding the information obtained from previous scans for predicting the responses of remaining ports and show that we can increase bandwidth savings to 47.4–83.5% at the same true positive rates. We show that using only a single feature from probed ports, i.e., whether the host has responded on said port, is sufficient for achieving the aforementioned boost in performance, while minimizing the computational overhead for performing predictions. Adding more information, e.g., features resulting from a stateful, or application layer probe (The ZMap Project 2021), only results in marginal benefits, while significantly increasing the computation required for performing and processing the information from such scans. Additionally, capturing the demographics of networks through geolocation and AS information is crucial for sequential scans, as conducting predictions based on only port responses results in poor performance.

Our main contribution is summarized as follows:

- We develop and evaluate two methods for conducting *smart* network scans by predicting whether a host will respond to probes on a certain port: (1) parallel scans using a priori (i.e., location and AS) attributes of hosts as features for prediction, and (2) sequential scans that take advantage of cross-protocol dependencies, using the result of one scan as features, in order to adaptively adjust the probes for consequent scans.
- For sequential scans, we develop a novel technique to find an optimal order for scanning ports. We achieve this by first training a set of classifiers, measuring the contribution of one port for predicting the responses of remaining ports, and probing ports by decreasing order of their importance.
- We evaluate this framework by simulating it over global scans of the public Internet conducted between January and May of 2019 over 20 ports, and show that we can reduce the number of probes by 26.7–72.0% (47.4–83.5%) using parallel (sequential) scans with negligible computational overhead, while maintaining a 90–99% true positive rate for discovering active devices across all ports. We also examine the coverage of scans over vulnerable/misconfigured IP addresses, and observer high true positive rates (>98.5%) along these subpopulations of active IPs, suggesting that our method can be reliably used for discovering vulnerable devices and assessing networks' security posture.

The remainder of this chapter is organized as follows. In Section 21.2, we go over the data sets used in our study, and how we preprocess the data to prepare it for our analysis. In Section 21.3, we go over the models used in our study and define metrics for evaluating the performance of our framework. Section 21.4 details our methodology for combining machine learning models with network scanners under different scenarios. In Section 21.5, we evaluate the performance of our technique for reducing the number of probes sent by a network scanner. We discuss our results in Section 21.6, go over related works in Section 21.7, and conclude in Section 21.8.

21.2 Data Sets

In this section, we go over the database used for obtaining global scans of the Internet, and explain in detail how we curate and process measurements for evaluating the performance of our technique in the real-world.

21.2.1 Global Internet Scans

For obtaining scans of the public Internet, we use Censys (Durumeric et al. 2015), a database containing results from global scans of the IPv4 address space across 37 different ports for our observation window between January and May of 2019. Each snapshot in the Censys database contains records (stored using JSON documents) on discoverable hosts, i.e., hosts that respond to at least one of the sent probes. For this study, we use snapshots corresponding to the following five dates from 2019: 1/1, 2/1, 3/1, 4/1, and 5/1. We mainly use the snapshot from 1/1/2019 to evaluate our framework, but use the more recent snapshots to measure the performance degradation of our models over time in Section 21.6.3.

Each record in a Censys snapshot contains attributes that have been extracted from port responses, such as headers and banners, SSL/TLS certificates, and even highly granular attributes such as the choice of ciphers for encryption and specific vulnerabilities probed for by Censys (e.g., Heartbleed). In addition, Censys also reports geolocation and ownership (AS) information about hosts in their database provided by the MaxMind GeoLite2 database (2021), Merit Network (2021), and Team Cymru (2021). We use these records for extracting features of Internet hosts, and training/evaluating our machine learning models for predicting port responses.

21.2.2 Data Curation

To evaluate our framework, we generate information for randomly drawn IP addresses in the following manner. We first select 17.5 million random IP address from announced Border Gateway Protocol (BGP) prefixes corresponding to each snapshot date, captured by CAIDA from Routeviews data (CAIDA 2021), about 170/8 blocks or ~2.8 billion addresses. This is done to remove reserved and private IP addresses, as well as address spaces not announced on BGP. For each selected IP address, we then check whether it has a corresponding record in a Censys snapshot. For IP addresses that do have a Censys record (i.e., an *active* IP), we append the Censys record to our curated data set. For addresses that do not have a corresponding Censys record (i.e., an *inactive* IP), we query its geolocation and autonomous system information from Censys using the following technique. We first find the two closest active IPs in Censys to the inactive IP, i.e., one with a smaller IP address, and one with a larger IP address. We then find the smallest Classless Inter-Domain Routing (CIDR) blocks that contain the inactive IP address and each of its active neighbors. If the corresponding CIDR block for one neighbor is smaller than the other, we then decide that the inactive IP belongs to the same network as that neighbor, and use the AS and geolocation properties of the corresponding neighbor for the inactive IP. If all three addresses are contained within the same CIDR block, then we copy AS and geolocation information from the closest neighbor, or the one with a larger IP address if both neighbors have the same distance to the inactive IP address.

The above procedure yields about one (16.5) million randomly drawn active (inactive) IP addresses from each snapshot (note that only ~5.7% of all IP addresses are active according to Censys probes). We further sub-select one million addresses from the inactive IPs to obtain a more balanced data set, resulting in a curated data set containing roughly one million active and one million inactive IPs for each snapshot. We use these data sets for training and evaluating the performance of our scanning techniques.

21.2.3 Data Processing

Records from the Censys database are stored using JSON documents with deeply nested fields, containing location and ownership (AS) properties, as well as attributes extracted from parsed responses, including headers, banners, certificate chains, and so on. However, while these documents contain a wide range of characteristics about Internet hosts, the information cannot be fed into a classification model out of the box, and we need to convert these documents to numerical feature vectors for analysis by a machine learning model.

JSON documents follow a tree-like structure, allowing different fields to be nested inside one another, e.g., properties regarding the location of a host, including country, city, latitude, and longitude. Therefore, simply extracting tokens from the string corresponding to a JSON document

```
"AS" : {                                    AS.Organization has "akamai": True
  "Organization" : "Akamai Technologies",   AS.Organization has "vodafone": False
},                                          Location.Country = "China": False
"Location" : {                              Location.Country = "United States": True
  "Country" : "United States",              has HTTP: True
},                                        → has HTTP.Headers: True
  "HTTP": {                                 HTTP.Headers has "apache": True
    "Headers" : {                           HTTP.Headers has "microsoft": False
    "Server" : "Apache/2.4.18 (Ubuntu)",    has property HTTPS: False
  }                                         has property SSH: False
}                                           ...
```

Figure 21.1 An example JSON document with nested fields (left), and binary features extracted for analysis in a machine learning model (right).

fails to recognize its structure, and does not provide any information about the field from which the token was extracted.

To address the above problem, we use the approach developed by Sarabi and Liu (2018) to extract high-dimensional binary features vectors from these documents. This feature extraction algorithm first learns the schema (JSON Schema 2021) of JSON documents in the Censys database by inspecting a number of sample documents and then extracts binary features from each field according to the learned schema. This then produces features that can be attributed to fields of the original JSON documents and are extracted according to the data type of those fields (i.e., string, categorical, Boolean). Furthermore, for optional fields we can also generate features that reflect their existence in a document, e.g., open ports, or if a host is returning headers/banners for different protocols.

Figure 21.1 shows an example of how a JSON document can be transformed into a binary vector representation using this approach. Note that each generated feature is assigned to a certain field of the original JSON document, allowing us to separate features extracted from location and ownership (AS) information, as well as features extracted from different port responses. This allows us to gradually add the information of scanned ports to our models for performing predictions of remaining ports.

We train the feature extraction model from Sarabi and Liu (2018) on one million randomly drawn records from the 1/1/2019 Censys snapshot (chosen independently from the dataset detailed in Section 21.2.2), producing 14 443 binary features extracted from 37 different ports. To control the number of generated features, we impose a limit of 0.05% on the sparsity of extracted features. These features are in the form of tags assigned to a host, e.g., if a host responds to probes on a certain port, if it belongs to a particular country, or if we observe certain tokens in fields inside the document, e.g., AS names, headers/banners, etc.

We exclude features that are extracted from Censys documents' metadata, which are added by Censys by processing the information gathered from all scanned ports and cannot be assigned to a certain port. We further remove 11 ports that have been observed on less than 0.3% of active IP addresses, since we cannot collect enough samples on these ports for training robust models. We also remove port 3389 (RDP protocol), observed on 1.9% of active IPs, due to poor prediction performance, indicating that our feature set is not effective in predicting responses for this port. After pruning the feature set we obtain 13 679 features from 20 ports, as well as location and AS properties, for training/evaluating our framework. Table 21.1 contains the fields/ports used for our analysis, as well as the number of features extracted from each field, and frequencies of active/open ports among active IP addresses.

Table 21.1 Sections from Censys documents, number of features generated from each section, and frequencies of active (open) ports among active IP addresses; note that 5.7% of all IPv4 addresses are active according to Censys measurements.

Section	# of features	Frequency
Geolocation	1513	N/A
Ownership (AS)	425	N/A
21/ftp	141	6.5%
22/ssh	622	11.6%
23/telnet	125	2.4%
25/smtp	875	4.1%
53/dns	41	5.1%
80/http	810	33.3%
110/pop3	790	3.0%
143/imap	827	2.8%
443/https	3263	35.3%
445/smb	1	1.8%
465/smtp	588	2.5%
587/smtp	993	3.4%
993/imaps	808	2.7%
995/pop3s	792	2.6%
2323/telnet	20	0.4%
3306/mysql	201	3.0%
5432/psql	68	0.4%
7547/cwmp	202	12.5%
8080/http	417	12.7%
8888/http	157	4.2%

21.3 Model and Metrics

In this section, we go over the classification algorithm used for our proposed framework and the features used for training models and define the metrics used for evaluating the performance of our classifiers.

21.3.1 Classification Algorithm

Classification is one of the most well-studied machine learning tasks. A wide range of algorithms with various levels of complexity has been designed to tackle this problem, including logistic regression, support vector machines (SVM), neural networks (Goodfellow et al. 2016), and decision tree-based models such as random forests (Scikit-learn 2021) and gradient-boosted trees (Chen and Guestrin 2016). Simple linear algorithms such as logistic regression are fast to train, but often achieve less accuracy than more complex and recent models, especially over large data sets. Neural networks and deep learning models achieve state-of-the-art performance for many tasks such as image/text classification, but are often slow to train. With tabular data (or when not dealing with image, video, or natural language processing tasks) tree-based algorithms are often preferred to deep learning models, due to their superior speed/performance. For this study, we use a gradient-boosted trees model, more specifically XGBoost (Chen and Guestrin 2016) for predicting whether a host will respond to requests on a certain port. We train one classifier for predicting the label assigned to each port, resulting in a collection/sequence of classifiers that can be used to predict the responses of all ports for a specific IP address.

21.3.2   Features for Model Training

For training a model, we first produce labels for each port of an IP address by observing whether Censys has reported a response under said port for its record of that IP address. Note that for an inactive IP, all the produced labels are zero, meaning that no port is responding to requests. We then use different subsets of the binary features discussed in Section 21.2.3 for training binary classifiers, as detailed below.

Prescan features   These include features extracted from location and AS properties, which are available before performing any scans. These features provide a priori information about each host, which can be used as initial attributes for predicting port responses. Location information can help detect patterns in the behavior of IPs in different regions, while AS properties can help predict labels based on the type/owner of the IP address. For instance, observing the word "university" in an AS name can indicate an educational network, while "cable" can help recognize residential/ISP networks.

Postscan features   Assuming that probes are performed sequentially, classifiers can also leverage features extracted from previous probes of an IP address for predicting the responses of the remaining ports. These then provide a posteriori features for classification. Note that using a stateless scanner such as ZMap (Durumeric et al. 2013), we only record whether a host has responded on a certain port, resulting in a single binary feature. However, with a stateful scan such as ZGrab (The ZMap Project 2021), a full handshake is completed with the server, and subsequent classifiers can also make use of parsed responses, resulting in a richer feature set. We evaluate both of these cases to determine the improvement provided by machine learning for stateless and stateful scans.

21.3.3   Metrics

In Section 21.1, we pointed out that our learning task can be transformed into a binary classification problem. Traditional metrics to measure the performance of classifiers include accuracy and the AUC score. Note that the latter can only be applied when the classifier is able to produce probabilities or scores, and not just a zero or one prediction, which is the case for gradient-boosted trees. The AUC score is a good metric to measure a classifier's ability in rank ordering samples when labels are highly unbalanced. Indeed, our task is a highly unbalanced classification problem, since most IP addresses are inactive or do not respond to probes. However, for our study we are focusing on reducing the number of probes for relatively high true positive rates, while the AUC score factors in the accuracy of the model over all operating points or true positive rates. Therefore, to evaluate the performance of trained classifiers, we instead estimate the probing rate at different true positive rates for discovering active ports. Take $y_i^k \in \{0, 1\}$ to denote the label for IP $i \in \{1, \dots, N\}$ and port $k \in \{1, \dots, M\}$ (i.e., whether IP i responds to probes on port k), $0 \le \hat{y}_i^k \le 1$ to be the prediction of the true label generated by a trained classifier, and $0 \leq t_r^k \leq 1$ to be the threshold corresponding to the true positive rate $0\% \leq r \leq 100\%$. Further assume that $\mathbb{S}_a$ denotes the set of active IP addresses in our data set, and p_a to be the percentage of active IPs in-the-wild (for Censys p_a is approximately 5.7%). We can then define the probing rate as follows:

$$
\begin{aligned}
PR_r^k = p_a \frac{\sum_{i \in \mathbb{S}_a} \mathbb{1}\{\hat{y}_i^k >= t_r^k\}}{|i \in \mathbb{S}_a|} \\
+ (1 - p_a) \frac{\sum_{i \notin \mathbb{S}_a} \mathbb{1}\{\hat{y}_i^k >= t_r^k\}}{|i \notin \mathbb{S}_a|}
\end{aligned}
\tag{21.1}
$$

Note that the probing rate is defined for a certain port (k), and a target true positive rate (r). In Eq. (21.1), we are computing the weighted average of probing rates over active and inactive IPs, since in our curated data sets, detailed in Section 21.2.2, the proportions of active/inactive IPs do not reflect their proportions in-the-wild, i.e., we oversampled active IP addresses to obtained a more balanced data set for training. $1 - PR_r^k$ then denotes the bandwidth savings for port k at the true positive rate r, which we will use to report the performance of our models in Section 21.5.

21.4 Methodology

In this section, we propose two methods for combining machine learning with networks scans, namely parallel and sequential scanning. Parallel scans can be done independently on multiple ports, but use minimal information for predicting port responses. On the other hand, sequential scans use a richer feature set which in turn leads to more bandwidth savings, but require scanning multiple ports in a predetermined order.

21.4.1 Parallel Scanning

Currently, most Internet scans (e.g., scans in the Censys database) are performed separately and independently across different ports. In other words, the entire IPv4 address is sweeped multiple times, each time sending probes to all IP addresses on a certain port. This allows different ports to be scanned independently, possibly at different times, thereby reducing the amount of traffic sent to networks/hosts. In this scenario, our method can only use the location and AS properties of the targeted IP addresses for predicting the responses of hosts, as depicted in Figure 21.2a. In this diagram, the geolocation (GL) and AS features are fed to each trained model in order to produce the prediction $\hat{y}_i^k$ of the true label y_i^k for sample i and port k, i.e., the estimated likelihood that IP address i will respond to probes on port k. These predictions are then fed to the scanner, which will decide whether to scan different IP/port pairs depending on the prediction of the model. In this study, we make decisions by thresholding $\hat{y}_i^k$; if $\hat{y}_i^k < t_r^k$ the scanner refrains from sending the probe. Note that t_r^k (specific to port k) is the threshold for reaching a target true positive rate r.

While this approach uses a minimal amount of information for prediction, applying machine learning to parallel scans is fairly straightforward, since the predictions of trained models can simply be translated into blacklists that can be fed to network scanners for refraining from sending probes to certain IP/port pairs. Moreover, due to the crude granularity of geolocation and AS features, we do not need to perform predictions for every IP address, but only for IP blocks in which all IP addresses share the same features, therefore reducing the computational overhead of our approach. We will discuss this point in more detail in Section 21.6.4.

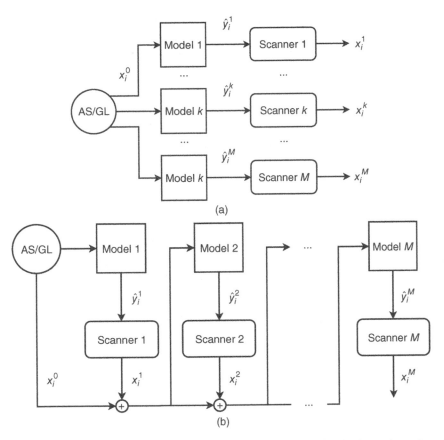

Figure 21.2 Diagram of scanning when all port labels are predicted using only geolocation and ownership (AS) properties (left), and when information obtained from port responses are used as features for subsequent probes (right). $\hat{y}_i^k$ denotes the predicted label for IP address i and port k from a trained classification model (model k), and x_i^k is the feature vector resulting from the probe of IP i on port k; x_i^0 is the feature vector for AS/GL features. (a) Parallel scans. (b) Sequential scans.

21.4.2 Sequential Scanning

In contrast to parallel scanning, one can also design a scanner to scan different ports in a sequential manner. In this setting, we can take advantage of the responses of previously scanned ports for predicting the remaining labels. Cross-protocol dependencies have been observed by Bano et al. (2018), but were not directly used for bootstrapping network scans. This is due to the fact that cross-protocol correlations by themselves are not sufficient for predicting other port labels, as we will further discuss in Section 21.6.1. However, we show that when combined with prescan features, i.e., location and AS properties, cross-protocol information can help improve the efficiency of sequential scans, as compared to parallel scans.

Assume $x_i^k, k \in \{1, \dots, M\}$ to denote the feature vector resulting from probing IP i on port k, and x_i^0 to denote prescan features. Then for sequential scanning, the classifier for port k is trained using $\{x_i^l, l < k\}$ as features, i.e., GL/AS features, as well as ports scanned earlier in the sequence. Note that for parallel scans in Section 21.4.1 we are only performing predictions using x_i^0 as features. We evaluate and compare this approach to parallel scanning in Section 21.5, resulting in more bandwidth savings. Figure 21.2b depicts the process used for sequential scans. Similar to parallel scanning, each model in this figure is generating a prediction $\hat{y}_i^k$ for an IP/port pair, which is then fed to the scanner for thresholding. The features resulting from each scan (i.e., the postscan features in Section 21.3.2) are then appended to the model's input features and used for all subsequent models. This allows models toward the end of the sequence to make predictions based on a richer feature set, which can result in more bandwidth savings for their corresponding scans.

21.4.2.1 Finding an Optimal Scan Order

Note that to achieve the most bandwidth savings, we first need to obtain an optimal order for scanning different ports of an IP address, since the dependency between port responses can vary between different pairs of ports. Moreover, the relationship is not necessarily symmetric, since a prominently used port such as port 80 (HTTP) can provide a lot of information regarding the responses of more uncommon ports, while the reverse might not be true.

To this end, we need to find an optimal scanning order to achieve the lowest possible average probing rate across all ports. Note, however, that exhaustively evaluating all permutations of ports is not feasible; instead, we use the following heuristic approach for quantifying port dependencies and finding a semi-optimal order. We first train a set of classifiers that use the responses of all remaining ports for prediction. For instance, if we want to predict the labels assigned to port 21 (FTP), we use features obtained from all the remaining 19 ports, as well as location/AS features, for training/evaluating our classifiers. We then compute the importance of each port for predicting the labels of all other ports. In ensembles of decision trees, such as gradient-boosted trees, this can be achieved by summing the improvement in accuracy brought by a feature in all the splits that it is used for, yielding its contribution to the predictions of the model. For each trained classifier, we then compute and normalize the contribution from all features used in the model (so that all feature importances sum up to one), and then compute the contribution of each port for the model's predictions.

Let $A_{M \times M}$, where M is the number of ports ($M = 20$ for this study), be a square matrix, where a_{ij} represent the importance of port i for predicting the label of port j. Our goal is to find an ordering of ports, so that ports that pose high importance for other ports are scanned first. In other words, if a_{ij} is high, then we prefer to scan port i prior to port j. We can reformulate this problem to finding a permutation A_p of both rows and columns in A, such that the sum of elements in the upper triangle of A is maximized. We compare two techniques for finding an optimal ordering, namely sorting ports by decreasing order of their total contribution to other classifiers ($a_i = \sum_{j \neq i} a_{ij}$), and sorting by increasing order of total contribution received from other ports ($a_i = \sum_{i \neq j} a_{ji}$). The former prioritizes scanning ports that pose high importance for predicting other port response first, while the latter prioritizes scanning ports that are highly dependent on other ports last. In our experiments, we found that the first approach resulted in higher overall bandwidth savings, and therefore we report our results using this approach. While the proposed heuristic approach is not guaranteed to find the best possible order (which would require exhaustive evaluation of all port permutations), we found it to perform well in practice; we will further elaborate on this in Section 21.5.3.

21.4.2.2 Training the Sequence of Classifiers

Note that the aforementioned models are only used for finding the order in which ports are scanned and are not used for running the actual probes. Once the order had been determined, we retrain our classifiers using the obtained sequence. Each classifier is then trained using geolocation and AS features, as well as features resulting from scans that are placed earlier in the sequence. It is worth mentioning that for false negatives (i.e., when an active port is not scanned due to an incorrect prediction) the true features of the corresponding port are not revealed, and subsequent models are making predictions based on partially masked features. Therefore when evaluating our classifiers, we also mask the features corresponding to false negatives,[1] to get an accurate estimate of our technique's performance; for training, we use the true features of each port without any masking.

21.5 Evaluation

In this section, we evaluate the bandwidth savings of our framework for both parallel and sequential scans and compare the performance of stateless sequential scans with stateful scans and scanning without using a priori (i.e., location and AS) information.

21.5.1 Setup

We use cross-validation to train, tune, and evaluate the performance of our framework. We split the curated data sets from each snapshot (detailed in Section 21.2) into a training set containing 60% of samples, used for training classifiers, and a test set containing 40% of samples for evaluating performance according to the metrics defined in Section 21.3.3.

For XGBoost models, we use 100 boosting rounds (i.e., the number of trees in the ensemble), a learning rate of 0.2, and a maximum depth of 20 for each tree in the ensemble; these parameters have been chosen by cross-validation. We use the logistic objective to produce probabilities between zero and one for each sample.

Note that for curating data sets, we undersampled inactive IP addresses to prevent our data sets from being dominated by inactive samples. However, simply training a classifier on this dataset means that the training algorithm gives equal weight to predictions on active/inactive IPs, while in reality a false prediction on an inactive sample should be given higher weight than an active sample. Therefore, we adjust the weights for active and inactive samples to reflect their true population in the real world.

Finally, we weight positive labels for each port by computing the ratio between the sum of weights for IPs with an inactive port, and the sum of weights for IPs with an active port, i.e., $\sum_{y_i^k=0} w_i^k / \sum_{y_i^k=1} w_i^k$, where w_i^k is the weight assigned to sample i for port k. This is done to handle the imbalanced nature of our data set, where positive labels are scarce. Without this scaling, the model will be heavily biased toward predicting a label of zeros for all samples, resulting in poor performance over samples with an active port (XGBoost 2021).

21.5.2 Parallel Scanning

We first evaluate the performance of our technique for improving parallel scans, depicted in Figure 21.2a. Table 21.2 summarizes our results. Each row in Table 21.2 corresponds to a certain target true positive rate, reporting the bandwidth savings $(1 - PR_r^k)$ for different ports, that is the percentage of probes that were not sent while achieving the target true positive rate. We have also reported the overall bandwidth savings by averaging bandwidth savings across all ports in the last row in Table 21.2.

Our evaluation results suggest that we can achieve an overall bandwidth savings of 26.7% while detecting 99% of IP/port pairs that respond to requests; this can be further increased to 72.0% when using a lower true positive rate of 90%. Note that this true positive rate is consistent across all

---

1 Note that false negatives are dependent on the choice of the classifier's operating point, or the target true positive rate. Therefore, we generate different masks and rerun the sequence for evaluating each true positive rate.

Table 21.2 Bandwidth savings for parallel scans using AS and geolocation features. Each cell reports the percentage of IPs for which we can refrain from sending a probe, while still achieving the corresponding true positive rate for detecting active ports. Overall bandwidth savings are computed by averaging over all ports. Ports 445 (SMB) and 7547 (CWMP) receive the most bandwidth savings.

Port/protocol	Target TPR				
	90%	95%	98%	99%	99.9%
21/ftp	60.8%	43.9%	27.7%	17.6%	2.9%
22/ssh	59.8%	44.5%	28.7%	20.2%	6.9%
23/telnet	51.7%	36.1%	21.5%	15.3%	4.3%
25/smtp	69.1%	50.3%	30.8%	21.4%	4.0%
53/dns	59.7%	43.7%	27.2%	16.9%	3.3%
80/http	56.6%	42.3%	27.0%	19.5%	7.7%
110/pop3	81.8%	64.2%	39.4%	23.1%	3.0%
143/imap	82.9%	66.0%	43.7%	27.3%	3.6%
443/https	53.9%	39.7%	26.0%	19.2%	7.5%
445/smb	92.3%	78.0%	56.9%	46.8%	24.4%
465/smtp	82.1%	67.3%	42.8%	29.5%	3.4%
587/smtp	77.2%	61.1%	37.6%	23.8%	5.0%
993/imaps	83.9%	69.1%	46.3%	31.6%	3.1%
995/pop3s	86.1%	71.4%	48.0%	31.5%	4.1%
2323/telnet	62.7%	46.0%	31.1%	24.1%	1.0%
3306/mysql	81.8%	63.3%	43.8%	28.1%	4.6%
5432/psql	66.4%	45.2%	20.8%	14.3%	1.0%
7547/cwmp	92.4%	86.7%	78.6%	70.0%	39.2%
8080/http	61.0%	48.5%	33.1%	23.9%	8.0%
8888/http	78.8%	63.1%	41.9%	30.5%	8.5%
Overall	72.0%	56.5%	37.6%	26.7%	7.3%

ports; for each port, we are computing the threshold for achieving the target true positive rate by computing the ROC curve of samples in our test set, and choosing the corresponding operating point for the target true positive rate.

Our results suggest that location and AS properties of IP addresses can be effectively used to predict whether a host will respond to requests on a certain port. Interestingly, we observe that responses of some ports can be predicted more accurately, resulting in higher overall bandwidth savings. For instance, while the savings at 99% true positive rate in Table 21.2 are ~20–30% for most ports, we can obtain higher savings of 70.0% for port 7547 (corresponding to the CWMP protocol). Note that the CWMP is protocol used by modems and router, which are more common on ISP networks. Indeed, we observe that IP addresses that have the token "charter" (i.e., Charter Communications) in their AS description field are 15 times more likely to respond to probes on port 7547, while observing the token "amazon" makes a host more than 400 times less likely to have an open CWMP port.

Table 21.3 Bandwidth savings when using all port responses (except the one that is being predicted) as features. Note that these results provide a lower bound for the bandwidth savings for sequential scanning (Table 21.4).

Port/protocol	Target TPR				
	90%	95%	98%	99%	99.9%
21/ftp	95.2%	82.0%	58.5%	42.8%	13.8%
22/ssh	85.7%	66.5%	43.3%	31.4%	10.5%
23/telnet	73.4%	56.4%	37.1%	22.5%	5.0%
25/smtp	98.7%	94.7%	69.2%	45.2%	8.7%
53/dns	79.2%	63.3%	39.6%	25.9%	4.5%
80/http	87.6%	69.6%	48.8%	34.9%	11.6%
110/pop3	99.8%	99.8%	99.8%	99.5%	60.9%
143/imap	99.9%	99.8%	99.8%	99.7%	82.8%
443/https	78.7%	59.8%	39.1%	28.8%	9.8%
445/smb	94.0%	81.9%	63.6%	53.4%	30.1%
465/smtp	99.9%	99.8%	99.7%	96.7%	50.3%
587/smtp	99.2%	94.0%	72.7%	52.8%	12.9%
993/imaps	99.9%	99.9%	99.8%	99.7%	48.9%
995/pop3s	99.9%	99.9%	99.8%	99.8%	95.2%
2323/telnet	69.4%	52.3%	39.3%	28.9%	15.2%
3306/mysql	98.8%	96.8%	84.4%	67.3%	21.9%
5432/psql	90.1%	73.5%	34.4%	19.9%	5.3%
7547/cwmp	92.9%	87.2%	80.2%	72.6%	41.1%
8080/http	70.8%	58.5%	42.5%	31.3%	10.6%
8888/http	85.2%	72.4%	53.5%	41.2%	10.8%
Overall	89.9%	80.4%	65.3%	54.7%	27.5%

21.5.3 Sequential Scanning

We then evaluate the performance for sequential scans, depicted in Figure 21.2b. For this section, we only use a single feature from each port, i.e., whether it has responded to the scanner. This makes our technique compatible with stateless (IP layer) scanners such as ZMap, while at the same time minimizing the processing required for constructing features for subsequent probes.

As discussed in Section 21.4.2, we first need to measure the importance of each port for predicting other port labels. Therefore, we train a set of classifiers by using all port responses (except the one for which we are predicting labels) as features for prediction. Our results are reported in Table 21.3. We observe that the bandwidth savings are significantly higher than those in Table 21.2; note that the bandwidth savings in Table 21.3 are an upper bound for the savings we can achieve by predicting port responses, since each classifier is using the most amount of information that can be available to them for prediction. However in reality, some ports are inevitably scanned before others,

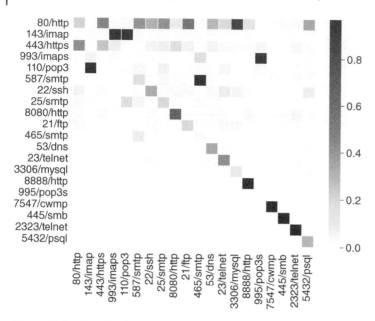

Figure 21.3 Contribution of port responses for predicting other port labels. Ports are scanned from left to right (top to bottom). Each column displays the importances of other ports for predicting the label of the corresponding port. The importances on the diagonal correspond to the contribution of location and AS features. We observe that high importance cells are placed in the upper triangle, enabling classifiers to use them for prediction.

meaning that they need to generate predictions based on a subset of the information that is used in Table 21.3.

We then compute the importance of each port to find an optimal ordering according to the process detailed in 21.4.2, so that ports that pose high importance for predicting the remaining labels are scanned first. Figure 21.3 displays the pairwise importance of ports. Ports that are scanned first are placed to the left (top) of the figure. We observe that after ordering, most of the cells with a large contribution are placed in the upper triangle. Note, however, that in some instances (e.g., ports 80 and 143, corresponding to the HTTP and IMAP protocols) we inevitably have to forgo some high importance cells, e.g., when two ports are mutually important for one another. Interestingly, we observe that ports corresponding to mail servers (IMAP/IMAPS, POP3/POP3S, and SMTP) are highly correlated with one another, which explains the high bandwidth savings for these ports (often more than 90%) in Table 21.3. The HTTP protocol also exhibits high importance for a wide range of ports, which is the main reason why it is scanned first, followed by the IMAP protocol, which provides major benefits for predicting the responses of other mail protocols. Ports such as 7547 (CWMP) and 445 (SMB) are isolated, meaning that they do not provide nor receive any benefits for/from the labels of other ports and are placed toward the end of the sequence.

We then train a sequence of classifiers using the obtained order from Figure 21.3, the bandwidth savings of our trained models are included in Table 21.4. Compared to Table 21.2 we observe between 10 and 20% more bandwidth savings for different true positive rates; this is largely due to the significant performance benefits over mail server ports. However, we also obtain more bandwidth savings across almost all ports, with notable savings on ports 21 (FTP), 3306 (MySQL), and 5432 (PSQL). Our results justify the use of the sequential structure in Figure 21.2b for reducing the probing rate of scans. For instance, at 99% true positive we can refrain from sending approximately 1/4 probes for parallel scans, while using sequential scans this increases to dropping slightly less

Table 21.4 Bandwidth savings when ports are sequentially scanned according to the diagram in Figure 21.2b, yielding ~10–20% more bandwidth savings compared to Table 21.2. Mail server ports (IMAP/IMAPS, POP3/POP3S, and SMTP) receive the most benefits, reducing the number of probes by more than 90% in some instances. FTP, MySQL, and PSQL protocols also receive considerable boost compared to Table 21.2.

Port/protocol	Target TPR				
	90%	95%	98%	99%	99.9%
21/ftp	82.1%	68.7%	47.7%	36.3%	11.2%
22/ssh	69.3%	52.7%	35.3%	25.6%	9.7%
23/telnet	62.9%	49.0%	32.3%	22.3%	7.9%
25/smtp	85.8%	69.9%	48.8%	37.1%	10.1%
53/dns	72.1%	57.0%	36.5%	25.4%	5.9%
80/http	55.8%	41.4%	26.1%	18.8%	7.0%
110/pop3	99.1%	97.6%	87.3%	72.8%	27.8%
143/imap	98.0%	91.0%	68.0%	51.7%	6.3%
443/https	61.7%	45.3%	30.8%	22.5%	7.3%
445/smb	94.1%	82.9%	65.1%	53.1%	22.6%
465/smtp	99.8%	99.7%	97.0%	90.6%	50.1%
587/smtp	90.7%	75.6%	53.8%	39.9%	12.6%
993/imaps	99.5%	99.2%	97.8%	88.4%	37.1%
995/pop3s	99.9%	99.8%	99.8%	99.7%	94.0%
2323/telnet	68.1%	50.7%	38.6%	35.3%	12.0%
3306/mysql	98.2%	95.9%	80.6%	67.2%	23.4%
5432/psql	90.1%	70.2%	34.0%	21.6%	6.4%
7547/cwmp	92.6%	86.8%	79.3%	70.8%	37.9%
8080/http	65.6%	53.7%	37.7%	28.4%	10.8%
8888/http	84.3%	72.7%	53.2%	40.5%	14.9%
Overall	83.5%	73.0%	57.5%	47.4%	20.8%

than 1/2 probes. At 95% true positive rate, Table 21.4 suggests that we can drop roughly 3/4 probes, with a probing rate of 27.0%, while using parallel scans we achieve a probing rate of 43.5%.

We also observe that the obtained scan order puts our bandwidth savings close to the upper bounds in Table 21.3. Note that as mentioned before, we inevitably have to forgo the dependency of some pairs of ports (e.g., the dependency of port 80 on other ports, and port 143 on other mail server ports), in order to utilize their high predictive power for ports placed further in the classifier sequence. In fact, the difference between bandwidth savings for ports 80 and 143 in Tables 21.3 and 21.4 accounts for 44% of the difference in overall bandwidth saving at 99% true positive rate. This suggests that we are indeed taking full advantage of cross-protocol information for reducing probing rates, and further justifies our proposed technique for finding an optimal order for scanning ports.

21.6 Discussion

In this section, we provide further motivation for our proposed method for conducting smart scans by comparing our methodology to other approaches with different levels of information, inspecting coverage of scans over vulnerable and misconfigured IP addresses, discuss how often models need to be retrained to keep them up-to-date, as well as the practical utility of our framework and its computational performance.

21.6.1 Comparison with Other Approaches

Our results on parallel and sequential scans demonstrate the ability of using location, AS, and cross-protocol information for predicting active IP addresses. In previous work, Klick et al. (2016) have shown that it is possible to reduce scan traffic by conducting full scans and identifying low density prefixes. However, this method relies on conducting full scans to be able to identify low density prefixes, especially when inspecting small prefixes or rarely active ports, where random sampling of a small subset would not yield an accurate estimation. We address this by conducting partial scans of 17.5 million IP addresses (~0.6% of the address space announced on BGP) to train classification models based on location and ownership properties, combining them with cross-protocol information to further reduce the traffic generated by scans. Moreover, Klick et al. (2016) report that their accuracy (true positive rate) drops at a rate of 0.3–0.7% per month, while our method can guarantee coverage by retraining models, or readjusting thresholds as discussed in Section 21.6.3.

While the dependency between active ports has been observed in previous work (Bano et al. 2018), our method attempts to utilize this property for improving the efficiency of scans by combining them with a priori attributes. To examine the boost achieved by appending geolocation and AS properties to port responses, in this section we compute bandwidth savings when using only cross-protocol information for sequential scan.

Additionally, our results so far on sequential scans have been obtained by adding only a single binary feature for each scanned port: whether the probed host has responded to the request. While this assumption makes our technique compatible with stateless scanners such as ZMap, it leads to the following question, can we achieve a better performance by completing a full handshake with the probed host, and record and append the resulting features for prediction? Note that for active IP addresses, Censys also records details of stateful (application layer) scans conducted using ZGrab (The ZMap Project 2021), which in turn is converted to a rich feature set as detailed in Table 21.1. These features include tokens extracted from headers/banners of different protocols, parsed SSL/TLS certificate chains served by secure protocols such as HTTPS, and even granular attributes such as the choice of encryption ciphers, misconfiguration such as open DNS resolvers, etc.

To answer the previous questions, we also train a sequence of classifiers without using geolocation and AS features, and another sequence utilizing the full set of features extracted from stateful scans. Our results are included in Table 21.5, where we have included average bandwidth savings across all ports. Note that for inactive IPs/ports, stateful and stateless scans provide the same level of information, and since the majority (94.3%) of IP addresses are inactive, the overall performance of both methods will be similar. Therefore for a more thorough comparison, we are also reporting the bandwidth savings of all three methods over only active IP addresses in Table 21.5. We have plotted the overall bandwidth savings of the examined scanning strategies in Figure 21.4.

Comparing stateless scans with and without a priori features, we observe that the latter achieves significantly less bandwidth savings, even lower than parallel scans. This demonstrates the

Table 21.5 Overall bandwidth savings across all ports for parallel, stateless sequential, and stateful sequential scans. Savings are reported over all IP addresses, as well as active IPs only. Conducting stateless scans without the use of location and AS feature results in poor performance. Additionally, we do not observe a significant improvement by using stateful scans.

Scan type		Target TPR				
		90%	95%	98%	99%	99.9%
Parallel	All	72.0%	56.5%	37.6%	26.7%	7.3%
	Active	57.6%	43.5%	27.9%	19.4%	5.6%
Stateless sequential	All	83.5%	73.0%	57.5%	47.4%	20.8%
	Active	60.9%	50.1%	37.2%	29.2%	12.0%
Stateless sequential	All	49.7%	37.1%	25.0%	19.6%	3.5%
(w/o prescan features)	Active	32.9%	25.0%	20.5%	14.2%	3.6%
Stateful sequential	All	83.9%	73.1%	58.2%	47.9%	20.2%
	Active	67.9%	58.0%	45.3%	36.7%	15.0%

Figure 21.4 Overall bandwidth savings of different feature sets for bootstrapping scans. Prescan (location and AS) and cross-protocol information complement each other well, producing the largest savings in the sequential case.

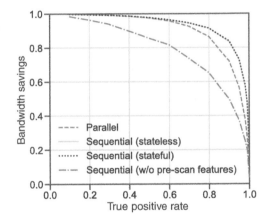

importance of using geolocation and AS features for conducting machine learning enabled scans, i.e., characterizing the network that an IP address belongs to plays an important role for predicting active hosts. This also suggests that that prescan and postscan features complement each other well, boosting the performance of our framework when using both feature sets. Cross-protocol information by itself does not provide good predictive power for most port responses, with the exception of mail protocols which are strongly correlated. In fact, mail protocols account for almost all of the bandwidth savings when not using any prescan features. On the other hand, location and AS properties offer bandwidth savings across all ports, even for isolated ports such as 7547 (CWMP) and 445 (SMB). Consequently, using both prescan and cross-protocol information, we can further reduce probes by leveraging all available information, including biases in different regions, and strong correlations between certain ports. Note that while cross-protocol dependencies have been observed in previous work (Bano et al. 2018), our framework leverages these correlations for bootstrapping scans by combining them with a priori attributes.

Comparing stateless and stateful scans in Table 21.5, we observe similar overall performance,[2] and slightly higher bandwidth savings for active IPs in the stateful scenario. Note that applying our technique for stateful scans also leads to a significant computational overhead, since the results of each scan must be parsed, followed by feature extraction, and the increase in the number of features also leads to complex, and therefore slower models. Due to the small performance benefit offered by stateful scans, we conclude that using stateless scans is sufficient for our sequential approach in Figure 21.2b.

Note that another approach for reducing probes of an Internet scan is to refrain from sending probes to hosts that have not been responsive in previous scans. However, this method relies on periodically performing global scans, while our proposed method only needs a partial exhaustive scan for training the underlying classifiers. Moreover, using historical data fails to recognize new active IP addresses. On the five snapshot used for this study, we observe that an average of 19.7% of active IPs in each snapshot are not present in the previous month's snapshot. Previous work has also reported that the accuracy of this approach drops to 80% within one month (Klick et al. 2016). We have also included a breakdown of new active ports in Table 21.6, where the percentage of new active ports can be as high as 50.6% for port 445 (SMB). This suggests that naively using historical data for producing scan target can lead to low true positive rates, while our proposed approach is also effective at detecting new active IPs. Nevertheless, it is interesting to examine whether adding historical data to our feature set can further boost the performance of classifiers, a problem that we leave for future work.

21.6.2 Coverage on Vulnerable IP Addresses

Network scanning is often used to give visibility into the security posture of networks by revealing vulnerable Internet-facing machines. While simply being accessible on the public Internet poses risk of being targeted by attackers, and some ports (e.g., port 7547 for routers) generally should not be left open, other protocols such as HTTP(S) and mail protocols are used to offer services to clients, and simply responding to probes is not necessarily an indication of a vulnerability. In this scenario, the scanner should be able to discover vulnerable subpopulations with good coverage (i.e., true positive rate) to be able to accurately assess the attack surface of networks. Therefore in this section, we examine the true positive rate of our scanning technique for three types of vulnerable and misconfigured machines, namely open DNS resolvers, HTTPS servers susceptible to the Heartbleed vulnerability, and Exim mail servers vulnerable to CVE-2018-6789 disclosed on 2/8/2018.

Open DNS resolvers are utilized in DNS amplification attacks, where an attacker turns small queries into large payloads, resulting in a reflection Distributed Denial of Service (DDoS) attack by eliciting responses from open DNS resolvers to spoofed IP addresses (Cloudflare 2021). While an open DNS resolver is not a direct vulnerability, it endangers the security of the Internet, and previous work has shown that it is correlated with high cyber-risk due to being an indication of mismanagement and poor security policies (Liu et al. 2015). We extract 3 520 442 (43.4% of all DNS servers) open DNS resolvers for 1/1/2019 as indicated by Censys. The Heartbleed vulnerability (CVE-2014-0160) (Durumeric et al. 2014) is a security bug in the OpenSSL library that was discovered in April 2014, allowing attackers to remotely read protected memory from HTTPS servers. While this vulnerability was promptly patched following its disclosure, Censys reports 101 083 (0.18% of HTTPS servers) vulnerable HTTPS sites in its 1/1/2019 snapshot. CVE-2018-6789 affecting Exim SMTP mail server versions $\leq$4.90, allows attackers to remotely execute arbitrary code.

---

2 The small decrease in overall performance for the 99.9% true positive rates for stateful scans is possibly due to overfitting because of the large number of features, and variations due to randomness in the training process.

Table 21.6 Average percentage of active ports that are inactive in the previous month's snapshot, averaged over all five snapshots in this study. Note that the average percentage of new active IPs over all snapshots is 19.7%.

Port	Protocol	Percentage
21	ftp	17.6%
22	ssh	20.4%
23	telnet	31.6%
25	smtp	12.2%
53	dns	28.8%
80	http	13.5%
110	pop3	6.9%
143	imap	5.2%
443	https	14.6%
445	smb	50.6%
465	smtp	7.2%
587	smtp	5.0%
993	imaps	7.8%
995	pop3s	5.2%
2323	telnet	28.5%
3306	mysql	12.2%
5432	psql	10.8%
7547	cwmp	33.3%
8080	http	17.8%
8888	http	29.7%

By parsing the banner of the SMTP protocol on ports 25, 465, and 587, we extract 400 199 (6.0% of all servers), 523 134 (13.0%), and 685 191 (12.4%) susceptible servers, respectively.

Table 21.7 displays the observed true positive rates for the aforementioned vulnerabilities at different operating points for parallel and sequential scans. We observe similar coverage to the overall target true positive rates for open resolvers and higher coverage for Exim servers. For servers vulnerable to Heartbleed, coverage is lower (~75%) at 90% target TPR, however this improves for higher operating points. Overall, we do not observe a large bias that would impede our technique from discovering vulnerable subpopulations, even for rare vulnerabilities such as Heartbleed affecting 0.18% of HTTPS sites; this further justifies the efficacy of our technique for security applications. Note, however, that while we observe consistent discovery rates for the examined vulnerabilities, there may exist other subpopulations of interest with low coverage. Nevertheless, one can guarantee discovery rates by adjusting the threshold for sending out probes, or training classifiers specifically targeting said subpopulations by applying the same methodology.

21.6.3 Keeping Models Up-to-Date

The Internet is an ever-changing ecosystem, where structures and patterns can change over time. For instance, ownerships of different networks, or the behavior of system administrators

Table 21.7 Coverage (true positive rate) over vulnerable and misconfigured IP addresses at different target true positive rates.

Scan type		Open resolver 53/dns	Heartbleed 443/https	Exim ≤ 4.90 25/smtp	465/smtp	587/smtp
Parallel	90%	87.9%	74.8%	96.1%	92.6%	95.0%
	95%	93.6%	91.3%	98.2%	97.7%	98.4%
	98%	97.4%	95.7%	99.4%	99.5%	99.6%
	99%	98.6%	98.5%	99.8%	99.8%	99.8%
	99.9%	99.9%	100.0%	99.9%	100.0%	99.9%
Sequential	90%	87.7%	75.2%	98.1%	93.6%	94.1%
	95%	94.0%	92.9%	99.2%	97.7%	97.8%
	98%	97.6%	98.4%	99.8%	99.3%	99.8%
	99%	98.7%	99.9%	100.0%	99.9%	99.9%
	99.9%	99.9%	100.0%	100.0%	100.0%	99.9%

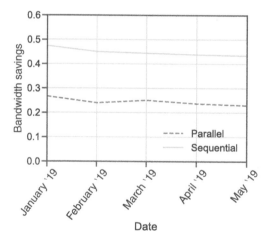

Figure 21.5 Performance of parallel and sequential models trained at 1/1/2019, over data from different dates. Bandwidth savings are reported for a target true positive rate of 99%. We observe gracefully degrading performance, suggesting that models only need to be retrained once every few months.

can change over time, varying the patterns which are utilized by our models for predicting port responses; this in turn warrants retraining models in order to keep them up-to-date. To determine how often models should be retrained, and to evaluate the performance of trained models for predicting port responses in future scans, we evaluate the overall bandwidth savings over our data sets for 2/1, 3/1, 4/1, and 5/1 of 2019, when models are trained on data from 1/1/2019.

We have included our results in Figure 21.5; we are reporting bandwidth savings corresponding to a 99% true positive rate. To achieve the target true positive rate for each scan, we use partial exhaustive scans on 17.5 million addresses for each date (the same amount used for training the machine learning models), and readjust thresholds t_r^k detailed in Section 21.3.3; this allows us to guarantee overall coverage while removing the need to retrain models. We observe a graceful degradation of performance, suggesting that models only need to be retrained once every few months to keep them up-to-date.

21.6.4 Practical Utility

We now discuss how our framework can be combined with existing scanners. The first step is to collect exhaustive measurements on a random subset of IP addresses for training classification models. Note that tools such as ZMap already traverse the IPv4 address space in a randomized order, in order to spread their traffic and reduce the strain on different networks; therefore we can simply perform a partial scan first, and then pause for training models. For this study, we used data corresponding to 17.5 million IP addresses, which only account for ~0.6% of the utilized IPv4 space announced on BGP. For parallel scans, the predictions of trained models can simply be provided to the scanner as a blacklist, telling the scanner to avoid sending probes to IPs where it is fairly certain that the probe is not going to be answered. For sequential scans, the same process can be achieved in sequence, simply taking the results of one full scan, performing predictions, and providing the results as a blacklist for the subsequent scan.

One important aspect of our framework is its computational overhead for performing predictions. For our experiments, we used the GPU-accelerated algorithm for XGBoost for both training and evaluating our models, using a NVIDIA GeForce GTX 1080 Ti GPU. It takes 20–40 seconds to train a model for a single port; we observe similar times for both parallel and sequential models. Note, however, that according to our results in Section 21.6.3, we do not need to retrain models after each scan.

For a trained model it takes approximately 10 μs to perform a prediction on a single sample. While this would add up to a significant time if predictions are performed for every single IP address (i.e., 2.8 billion addresses for the entire IPv4 address space), note that AS/GL information are typically constant across large networks, and there are a limited number of port configurations, leading to duplicate feature vectors. In fact, we observe that for an entire Censys snapshot, there are only ~130 000 unique GL/AS feature vectors; adding the labels for all 20 ports we observe roughly 3 million unique configurations of both GL/AS features and port labels. This suggests that the entire computation for a parallel scan can be done in ~1.3 seconds, while for a sequential scan we can perform the required predictions in approximately 30 seconds.[3] Therefore, the computational overhead of applying our proposed techniques is negligible; for comparison, ZMap scans the entire IPv4 address space for a single port in 45 (5) minutes with a (10) gigabit connection.

21.7 Related Work

There are numerous measurement studies that analyze Internet scan data. Security studies focus on measuring vulnerabilities, misconfiguration, and analyzing malicious hosts; these include, e.g., errors made by Certificate Authorities (CAs) when issuing certificates (Kumar et al. 2018), a study on the Mirai botnet (Antonakakis et al. 2017), and a measurement-based analysis of the Heartbleed vulnerability (Durumeric et al. 2014). Studies on trends and adoption rates include measuring HTTPS adoption (Felt et al. 2017), TLS deployment (Kotzias et al. 2018), and a study of Certification Authority Authorization (CAA) (Scheitle et al. 2018). Studies on discovering different types of Internet-facing devices include scanning for instances of the Robot Operating System (ROS) (DeMarinis et al. 2018), and developing an engine for discovering IoT devices (Feng

---

3 Note that this is a worst-case estimate. For ports toward the start of the sequence, the number of unique feature vectors is closer to the unique vectors for GL/AS features, since only a fraction of port responses have been revealed and are being used for prediction.

et al. 2018). Beverly et al. (2018) and Shavitt and Shir (2005) use networks scans for mapping the topology of the Internet.

While there exist a wide range of measurement studies utilizing Internet scans, there are fewer works focusing on applications of machine learning for processing network scan data. Liu et al. (2015) use symptoms of mismanagement (e.g., open resolvers and untrusted certificates) discovered on an organization network, to predict their risk of suffering from a data breach. Sarabi and Liu (2018) develop a framework for processing Internet scan data in machine learning models. In this chapter, we use the feature extraction techniques from Sarabi and Liu (2018) to generate features for training classifiers.

Generating scanning targets and analyzing active (or live) IPs has also received attention in the past. Bano et al. (2018) conduct an analysis of live (active) IPs, and (similar to our results) find cross-protocol correlations between the liveness of different ports. Klick et al. (2016) periodically scan the entire IPv4 address space, omitting low density prefixes from future scans. In comparison with Klick et al. (2016), our method does not require full scans by using machine learning to identify low density networks and leverages cross-protocol information for further improving the efficiency of scans. Fan and Heidemann (2010) develop an automated technique for generating a representative Internet hitlist for scanning. Murdock et al. (2017) identify dense address space regions and generate candidates for IPv6 scanning. Our proposed framework can complement target generation techniques, using machine learning to probe hosts more prudently, which in turn results in a higher hit rate, allowing one to scan hitlists at a faster rate.

Note that for our problem we are trying to predict multiple (possibly correlated) labels, making our learning task a multi-label classification problem. Read et al. (2011) suggest training a chain of classifiers in sequence, appending the output of each model in the chain for subsequent predictions. We use a similar approach to predict port labels; however, instead of using the predictions of the model, we append the true label of each sample (only if the corresponding classifier tells us to perform a probe) to subsequent predictions, since probing then reveals the true label of the IP address. Another method for dealing with correlated labels for multi-label classification includes applying a label transformation, e.g., to produce uncorrelated labels (Chen and Lin 2012; Tai and Lin 2012; Bi and Kwok 2014). Dealing with imbalanced classes has been studied by (Chawla et al. 2002), where the authors suggest a combination of undersampling the majority class with oversampling the minority class (by creating synthetic samples). In this study, we undersample the majority class (inactive IPs) and also weight the minority class (active IPs/ports) to deal with imbalanced classes.

21.8 Conclusions and Future Work

In this chapter, we developed and evaluated a framework for reducing the bandwidth of network scans by predicting whether a host will respond to requests on a certain port, using location and ownership (AS) properties, as well as cross-protocol information. We demonstrated that using only location and AS features we can achieve overall bandwidth savings of 26.7–72.0% at 90–99% true positive rates for detecting active/open ports, averaged over 20 port scans from the Censys database. Moreover, we developed a novel technique for finding an optimal order for scanning ports and training a sequence of classifiers, appending the responses of scanned ports for predicting active IPs over subsequent scans. We show that using this technique we can increase the bandwidth savings of Internet scans to 47.4–83.5% at 90–99% coverage levels. This reduction in bandwidth is due to the high dependency between the responses of certain ports, for instance ports corresponding to mail servers. We further show that our technique can be applied on top of current scanning tools with little computational overhead, providing blacklists in order to refrain from sending probes to certain IP/port pairs.

We compared our methodology to other strategies for conducting machine learning enabled scans, concluding that ignoring location and AS properties results in poor performance, while using the full set of features from stateful scans only provides marginal benefits, while significantly increasing computational requirements. We also showed that scans have consistent coverage along vulnerable and misconfigured subpopulations and are therefore appropriate for efficient and accurate assessment of the attack surface of networks.

We intend to apply our developed techniques to develop smart scanners that can efficiently scan IPv4 networks and IPv6 hitlists, increasing the hit rate for discovering active IPs. This allows scanners to scan IPv4 faster and less intrusively compared to exhaustive scans, while covering larger histlists for discovering more devices on IPv6. Additionally, using other sources of information, e.g., historical data, and local patterns in how devices are placed on the Internet (for example some networks might tend to put active devices at the start of their allocated IP blocks, while others might use random placement) can also help improve the efficiency of scans. Note that for sequential scans we are using a static order for probing different ports. However, it might be more efficient to change the order of scans for different networks, for instance scanning modem/router protocols first for consumer networks, while prioritizing web protocols for hosting networks. Using a dynamic order for scans is another direction for future work.

Acknowledgments

This work is supported by the NSF under grants CNS-1939006, CNS-2012001, and by the ARO under contract W911NF1810208.

References

Akamai. State of the Internet IPv6 adoption visualization, https://www.akamai.com/uk/en/resources/our-thinking/state-of-the-internet-report/state-of-the-internet-ipv6-adoption-visualization.jsp (accessed 26 February 2021).

M. Antonakakis, T. April, M. Bailey, M. Bernhard, E. Bursztein, J. Cochran, Z. Durumeric, J. A. Halderman, L. Invernizzi, M. Kallitsis *et al.* Understanding the Mirai botnet. In *USENIX Security Symposium, Vancouver, Canada, 16–18 August 2017*, pp. 1092–1110, 2017.

S. Bano, P. Richter, M. Javed, S. Sundaresan, Z. Durumeric, S. J. Murdoch, R. Mortier, and V. Paxson. Scanning the Internet for liveness. *ACM SIGCOMM Computer Communication Review*, 48(2):2–9, 2018.

R. Beverly, R. Durairajan, D. Plonka, and J. P. Rohrer. In the IP of the beholder: Strategies for active IPv6 topology discovery. In *Internet Measurement Conference 2018*, pp. 308–321. ACM, 2018.

W. Bi and J. T. Kwok. Multilabel classification with label correlations and missing labels. In *Twenty-Eighth AAAI Conference on Artificial Intelligence, Quebec City, Canada, 27–31 July 2014*, 2014.

CAIDA. Routeviews prefix to AS mappings dataset (pfx2as) for IPv4 and IPv6. https://www.caida.org/data/routing/routeviews-prefix2as.xml (accessed 26 February 2021).

N. V. Chawla, K. W. Bowyer, L. O. Hall, and W. P. Kegelmeyer. SMOTE: Synthetic minority over-sampling technique. *Journal of Artificial Intelligence Research*, 16:321–357, 2002.

T. Chen and C. Guestrin. XGBoost: A scalable tree boosting system. In *ACM SIGKDD International Conference on Knowledge Discovery and Data Mining*, pp. 785–794. ACM, 2016.

Y.-N. Chen and H.-T. Lin. Feature-aware label space dimension reduction for multi-label classification. In *Advances in Neural Information Processing Systems, Lake Tahoe, NV, 3–8 December 2012*, pp. 1529–1537, 2012.

K. Claffy, Y. Hyun, K. Keys, M. Fomenkov, and D. Krioukov. Internet mapping: From art to science. In *Cybersecurity Applications & Technology Conference for Homeland Security*, pp. 205–211. IEEE, 2009.

Cloudflare. DNS amplification (DDoS) attack. https://www.cloudflare.com/learning/ddos/dns-amplification-ddos-attack (accessed 26 February 2021).

N. DeMarinis, S. Tellex, V. Kemerlis, G. Konidaris, and R. Fonseca. Scanning the Internet for ROS: A view of security in robotics research. *arXiv preprint, arXiv:1808.03322*, 2018.

Z. Durumeric, D. Adrian, A. Mirian, M. Bailey, and J. A. Halderman. A search engine backed by Internet-wide scanning. In *ACM Conference on Computer and Communications Security*, pp. 542–553. ACM, 2015.

Z. Durumeric, F. Li, J. Kasten, J. Amann, J. Beekman, M. Payer, N. Weaver, D. Adrian, V. Paxson, M. Bailey *et al.* The matter of heartbleed. In *Internet Measurement Conference*, pp. 475–488. ACM, 2014.

Z. Durumeric, E. Wustrow, and J. A. Halderman. ZMap: Fast Internet-wide scanning and its security applications. *USENIX Security Symposium*, 8:47–53, 2013.

X. Fan and J. Heidemann. Selecting representative IP addresses for Internet topology studies. In *ACM SIGCOMM Conference on Internet Measurement*, pp. 411–423. ACM, 2010.

A. P. Felt, R. Barnes, A. King, C. Palmer, C. Bentzel, and P. Tabriz. Measuring HTTPS adoption on the web. In *USENIX Security Symposium, Vancouver, Canada, 16–18 August 2017*, pp. 1323–1338, 2017.

X. Feng, Q. Li, H. Wang, and L. Sun. Acquisitional rule-based engine for discovering Internet-of-things devices. In *USENIX Security Symposium, Baltimore, MD, 15–17 August 2018*, pp. 327–341, 2018.

O. Gasser, Q. Scheitle, P. Foremski, Q. Lone, M. Korczyński, S. D. Strowes, L. Hendriks, and G. Carle. Clusters in the expanse: Understanding and unbiasing IPv6 hitlists. In *Internet Measurement Conference*, pp. 364–378. ACM, 2018.

I. Goodfellow, Y. Bengio, and A. Courville. *Deep Learning*. MIT Press, 2016. http://www.deeplearningbook.org.

Internet Society. State of IPv6 deployment, 2018. https://www.internetsociety.org/resources/2018/state-of-ipv6-deployment-2018.

JSON Schema. http://json-schema.org (accessed 26 February 2021).

J. Klick, S. Lau, M. Wählisch, and V. Roth. Towards better Internet citizenship: Reducing the footprint of Internet-wide scans by topology aware prefix selection. In *Internet Measurement Conference*, pp. 421–427. ACM, 2016.

P. Kotzias, A. Razaghpanah, J. Amann, K. G. Paterson, N. Vallina-Rodriguez, and J. Caballero. Coming of age: A longitudinal study of TLS deployment. In *Internet Measurement Conference*, pp. 415–428. ACM, 2018.

D. Kumar, Z. Wang, M. Hyder, J. Dickinson, G. Beck, D. Adrian, J. Mason, Z. Durumeric, J. A. Halderman, and M. Bailey. Tracking certificate misissuance in the wild. In *IEEE Symposium on Security and Privacy*, pp. 785–798. IEEE, 2018.

D. Leonard and D. Loguinov. Demystifying service discovery: Implementing an Internet-wide scanner. In *ACM SIGCOMM Conference on Internet Measurement*, pp. 109–122. ACM, 2010.

Y. Liu, A. Sarabi, J. Zhang, P. Naghizadeh, M. Karir, M. Bailey, and M. Liu. Cloudy with a chance of breach: Forecasting cyber security incidents. In *USENIX Security Symposium, Washington, DC, 12–14 August 2015*, pp. 1009–1024, 2015.

G. F. Lyon. *Nmap Network Scanning: The Official Nmap Project Guide to Network Discovery and Security Scanning*. Insecure, 2009. https://dl.acm.org/doi/10.5555/1538595.

Maxmind. GeoLite2 database. https://dev.maxmind.com/geoip/geoip2/geolite2 (accessed 26 February 2021).

Merit Network. https://www.merit.edu (accessed 26 February 2021).

A. Murdock, F. Li, P. Bramsen, Z. Durumeric, and V. Paxson. Target generation for Internet-wide IPv6 scanning. In *Internet Measurement Conference*, pp. 242–253. ACM, 2017.

J. Read, B. Pfahringer, G. Holmes, and E. Frank. Classifier chains for multi-label classification. *Machine Learning*, 85(3):333, 2011.

A. Sarabi and M. Liu. Characterizing the Internet host population using deep learning: A universal and lightweight numerical embedding. In *Internet Measurement Conference*, pp. 133–146. ACM, 2018.

Q. Scheitle, T. Chung, J. Hiller, O. Gasser, J. Naab, R. van Rijswijk-Deij, O. Hohlfeld, R. Holz, D. Choffnes, A. Mislove *et al.* A first look at certification authority authorization (CAA). *ACM SIGCOMM Computer Communication Review*, 48(2):10–23, 2018.

Scikit-learn. Ensemble methods: Random forests. http://scikit-learn.org/stable/modules/ensemble .html\LY1\textbackslash#forest (accessed 26 Februay 2021).

Y. Shavitt and E. Shir. DIMES: Let the Internet measure itself. *ACM SIGCOMM Computer Communication Review*, 35(5):71–74, 2005.

F. Tai and H.-T. Lin. Multilabel classification with principal label space transformation. *Neural Computation*, 24(9):2508–2542, 2012.

Team Cymru. http://www.team-cymru.org (accessed 26 Februay 2021).

The ZMap Project. ZGrab 2.0. https://github.com/zmap/zgrab2 (accessed 26 Februay 2021).

XGBoost. Parameter tuning. https://xgboost.readthedocs.io/en/latest/tutorials/param_tuning.html (accessed 26 Februay 2021).

22

Semi-automated Parameterization of a Probabilistic Model Using Logistic Regression—A Tutorial

Stefan Rass[1], Sandra König[2], and Stefan Schauer[2]

[1] *Institute for Artificial Intelligence and Cybersecurity, Universitaet Klagenfurt, Klagenfurt, Austria*
[2] *Center for Digital Safety & Security, Austrian Institute of Technology, Vienna, Austria*

22.1 Introduction

This chapter is about the parameterization of stochastic simulation models. Generally, such models involve a (sometimes large) number of probability parameters. While their semantic and usefulness in the model are often made clear, many model descriptions leave the reader without advice on how to choose a particular value for a parameter. For example, probabilities of certain player types in Bayesian games (Chapter 2) or conditional distributions in stochastic games (Chapter 10). Further examples include (but are not limited to), the efficacy of honeypots (Chapters 3 and 4) or for intrusion detection systems (Chapters 11 and 21), sometimes measured as a probability for the intruder to (not) recognize the deception or detection system as such. Likewise, human error (Chapter 5), sensor and monitoring imperfectness (Chapters 8 and 23), or information uncertainty may be quantified in probabilities, for implementing deception mechanisms (Chapter 6).

Learning algorithms, say to train a Bayesian network, are usually designed very specifically for a particular stochastic model, but there is no general method useable to estimate parameters for an arbitrary stochastic model.

This chapter will take a first step toward such a method, irrespectively of actual use for the parameter underestimation, e.g. needed as a direct value of interest in a model, or to determine another probability distribution (similar to what is done in Chapter 7 or Chapter 9), or to describe imperfectness of code analysis for software security (see Chapter 12). The method is based on the idea of "parameterization by example," i.e. instead of letting the user "guess" the value of a variable p (which can be a joint or marginal probability, conditional probability, or other), we pursue the idea of letting the user specify cases in which an event occurred and cases in which it did not occur. Under the hypothesis that the semantic of a probability parameter p is often well-defined, meaning that the variable often refers to a very concrete event that may or may not occur, the user may be capable of giving example configurations under which the respective event occurs or not. In the best case, this information is directly available from data, and if so, we can use logistic regression (LR) techniques to systematically estimate a value for such a parameter.

To ease and at the same time showcase our exposition, we will develop and exemplify our method using the parameterization problem for a model that uses stochastic Mealy automata to simulate cascading effects of security incidents in networks of critical infrastructures. Pursuing a tutorial style, we will use the model laid out in König et al. (2019), SAURON Consortium (2019), and Schauer et al. (2020) as a *running example throughout the chapter*. For brevity, we refrain from reproducing a complete description of assets here and refer the reader to the cited references.

Game Theory and Machine Learning for Cyber Security, First Edition.
Edited by Charles A. Kamhoua, Christopher D. Kiekintveld, Fei Fang, and Quanyan Zhu.

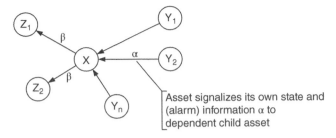

Figure 22.1 Dependency graph (example structure).

Figure 22.2 Inner asset dynamics model as a
Mealy automaton.

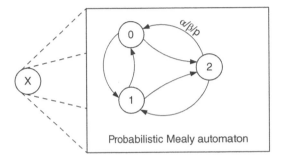

In a nutshell, the model considered is a network of coupled probabilistic Mealy automata, each representing an asset, who mutually signalize incidents to each other, and reactively undergo changes in their individual health status. Such an automaton has lots of probability parameters to set, and our interest in the following is how to set these values. Note that this is explicitly different from defining the automaton itself! Our interest here is merely on *parameterizing a given model*, and *not* finding or defining the automaton model itself! The Mealy automaton model is nothing else than a replaceable example; here composed from two nested models:

1. A directed graph $G = (V, E)$, with V being the set of all assets, and E describing dependencies among them. This is the *outer model*; an example of which is shown as Figure 22.1.
2. Each asset, i.e. node in V, has an *inner model* to describe its dynamics in terms of state changes that are probabilistic and depend on signals that an asset receives from its related assets. That is, if an asset Y_1, on which asset X depends on, experiences a state change, it signalizes this change as an information α to the dependent (subordinate) assets. In turn, the dependent asset X may change its internal state accordingly. It does so …
   a. … only with a certain probability p,
   b. … and by itself emitting a further signal β (possibly being an empty information) to its own subordinate assets.

An inner model is exemplified in Figure 22.2.

The problem for a practitioner seeking to use this model is to compute the probabilities of such changes, based on the current health status and signals from other automata. One, among many parameters in the so-created stochastic model, could be a value p describing the "probability for some infrastructure to run into trouble, if its supplier structures experience problems at the moment." Such qualitative descriptions hardly lend themselves to a numeric estimation or learning technique to estimate a value for p, and the problem is often amplified by there being many such parameters calling for values. However, using the same qualitative and vague description of what the parameter means, we can estimate a value for it using examples (as many as we can or are available from data).

The main object of this chapter is thus a statistical estimation of the parameter p here used in a probabilistic automaton model, but generally allowed to play any role in a stochastic model. The general shape of the stochastic model is herein unrestricted and can have as many parameters as desired. Automaton models are just an example, among Markov chains, and many others. To assign a proper value, however, it is necessary to understand what the parameter does. Hence, we will dive into more details about the model itself in the following.

22.1.1 Context, Scope, and Notation

We assume a set V of cyber and physical assets, letting $X \in V$ denote a general asset. Furthermore, we let X depend on a set pa(X) of parent assets, with inline notation $X \leftarrow$ pa(X). A *dependency graph* is the directed graph $G = (V, E)$ with the edge set $E \subseteq V \times V$ having its elements denoted as $X \leftarrow Y$ to mean that X depends on Y. The parent set of a node X is pa(X) = $\{Y \in V : X \leftarrow Y \in E\}$.

A few more assumptions may be useful to bear in mind for the data collection and specification by example, since the method proposed is only concerned with labeling edges in a directed graph, but agnostic of the graph's topological properties. Thus, the assumptions made next are not technical requirements but shall detail the context of the modeling:

- We allow circular, and in particular mutual, dependencies in the graph. Note that this may indeed be problematic for a simulation model, since it may induce oscillating or other non-stationary behavior.

- We allow for dependencies among the parents of X that are relevant only for X, but not for its parents. For example, let X have two parents Y_1, Y_2. The functionality of X hinges on both of them to work, whereas an outage of Y_1 leaves Y_2 completely unaffected and vice versa. More complex dependencies of X on the parents are allowed and the main subject of interest to model throughout the chapter.

The point in these two hints is that during the modeling, one needs not to worry about circular dependencies, or, when modeling the behavior of X depending on several nodes $Y_1, ..., Y_n$, mutual dependencies between the Y's themselves are *not* bothered with when describing how X depends on these.

22.1.2 Assumptions on Data Availability

The dependency structure is herein considered as an input artifact that, for example, can be based on standard risk management methods. Taking IT Grundschutz (BSI 2016) as an example, Table 22.1 shows an example overview of how assets may initially be listed with relevant properties for a risk assessment. From this data,

Description of IT application		IT systems						
Application (information)	Has sensitive data?	S1	S2	...	S6	C1	C2	...
Human resources department	X	X						
Accounting	X	X	X					
SCADA control	X	X			X		X	
Active directory and user management	X		X		X	X		
System-management			X		X	X	X	

Table 22.1 Example of structure analysis.

No.	Description	Platform	Count	Physical location	Status	User/admin
S1	Server for human resources department	Windows NT-Server	1	City X, R 1.01	Running	Human resources
S2	Primary domain controller	Windows NT-Server	1	City X, R 3.10	Running	All IT applications
...	...	...	...	...	...	...
S6	Backup domain controller	Windows NT-Server	1	City Y, R 8.01	Running	All IT users
C1	Group of office clients in HR deparment	Windows NT-Workstation	5	City X, R 1.02–R 1.06	Running	Human resources
C2	Office network group for SCADA control department	Windows NT-Workstation	10	City Y, R 1.07–R 1.16	Running	Office management

Source: According to IT Grundschutz (BSI 2016).

Table 22.2 Example of dependency specification based on a prior structure analysis.

Description of IT application		IT-systems						
Application (information)	Have sensitive data?	S1	S2	...	S6	C1	C2	...
Human resources department	X	X						
Accounting	X	X	X					
SCADA control	X	X			X		X	
Active directory and user management	X		X		X	X		
System-management			X		X	X	X	

Source: Adapted from examples given by the BSI on how to apply IT Grundschutz.

Table 22.2 shows an example of how the dependencies of assets on one another may be specified. We emphasize that there is no obligation to follow any of practice like this in the collection of the data, and the two example tables serve only for illustration of how the data *could* look like.

Regarding the assets themselves, data about these are also assumed available, as part of the enterprise's domain knowledge. We will have to say more about this in Section 22.4.2, as the discussion is, for the moment, focused on the relations *between* the assets.

Collecting data like in Table 22.1 and

Description of IT application		IT-systems						
Application (information)	Have sensitive data?	S1	S2	...	S6	C1	C2	...
Human resources department	X	X						
Accounting	X	X	X					
SCADA control	X	X			X		X	
Active directory and user management	X		X		X	X		
System-management			X		X	X	X	

Table 22.2 is part of the manual labor for the *model construction* (not parameterization). It may include rating the criticality of a dependency directly based on personal expertise, experience, and available background domain knowledge. This method is partly facilitated, and should be complemented by, risk management procedures and artifacts therefrom. Early stages of risk management, here referring to the "Structure Analysis" of the Grundschutzkompendium (BSI 2016) usually entail an identification of assets, say, IT systems, together with a collection of related and relevant *attributes*. These attributes will hereafter be unconstrained in their form and summarize (include) any physical or logical (cyber) property that has an impact on health (changes) of an asset. We will revisit example such attributes in Section 22.4.2, when we construct example data to illustrate our methods.

If a risk management process follows these or similar lines, the necessary data for model parameterization can partly be obtained from an augmented version of these output artifacts: simply think of the "X" indicator in the above tables to be replaced or annotated by information regarding the criticality of the dependency. That is, instead of plainly indicating that there is a dependency, the risk manager may in addition tell how strong that dependency may be. For example, if the human resources department's data are primarily hosted on server S1, then this dependency may be critical; but less so if the data from server S1 is known to backup (mirror) its content to another server S6, so that upon an outage of S1 (e.g. for patching, replacement, …), the HR department could temporarily switch access to S6 until S1 comes up running again.

22.2 Method Derivation

We shall develop the final proposal incrementally, by starting from an attempt to describe the transition probability as an "exact" object, so as to reveal the ingredients of a parameterization model in Section 22.2.1.

Section 22.2.2 will pick up on this by discussing how the ingredients could be specified, along which example heuristics and exact techniques will appear. This section, as the previous one, is devoted to highlighting the difficulties on this route, so as to finally motivate the chapter's recommendation as doing the specification "by example," or in other words, parameterize the system based on expert's example input by virtue of logistic regression. Section 22.2.3 describes this general idea, and the appeal that goes with it in the sense of logistic regression offering rich theory useful for model diagnostics and plausibility checks.

22.2.1 Exact Transition Models

Let us assume that causal relationships between the attributes of an asset and the asset's state to be judged (namely the conditional failure probability, or general state change probability, of a node) are unknown a priori. The description of dependencies by edges, especially their annotation with conditional probabilities, is available either in form of personal expertise and domain knowledge, or "by example" in form of a number of example records. The recommended method will be based on the latter setting, whereas we start with an attempt toward a formally exact treatment, to the end of justifying the data-driven specification later.

More specifically, let an asset X depend on one or more assets $Y_1, ..., Y_n$ in a way that is specific for the application. Continuing the example in Tables 22.1 and 22.2, we have $X, Y_1, ..., Y_n \in \{S1, ..., S6, C1, C2, ...\}$. The quantities that we are interested in is the likelihood

$$p = \Pr(\text{transition } X_{\text{current}} \to X_{\text{next}} | \text{status of the parents is } \{Y_1, ..., Y_n\})$$

$$p = \Pr(X_{\text{next}} | X_{\text{current}}, \text{status of the parents is } \{Y_1, ..., Y_n\}), \tag{22.1}$$

i.e. the conditional probability for the asset X to change its status, given that its supplier assets (parents) $Y_1, ..., Y_n$ are jointly in some given configuration. This is part of the annotation of an edge in the dependency graph (cf. Figure 22.1) that we need to set automatically, to the accuracy and extent possible.

22.2.2 Meaning of Dependencies

A dependency like (22.1) can mean diverse things, ranging from slight problems of just one among the Y_i's, up to a joint failure involving interdependencies among the Y_i's, but may also relate to any of the Y_i's being just in some abnormal state without this causing service interruptions (e.g. alerted, under inspection, etc.).

Suppose that any parent asset Y will (on its own) signalize any state change to subordinate (child) assets, such as X (among perhaps others). Then, the child asset X can directly react by its own state changing based on the signal it got from Y. Such a state change, however, is not necessarily occurring, say, if from X's perspective, the signal from Y is only relevant under additional circumstances that did not kick in yet. A simulation model, however, would need exactly the information of how X reacts on signals from its parents, and to the end of a most flexible, abstract, and hence useful model, we may ask for the probability of any state change based on the currently given signals, as (22.1) expresses implicitly.

In the most general yet sloppy form as in (22.1), the value p is thus not directly relatable to the dependency of X on any specific Y_i; it refers to the "joint dependency" of X on the entirety of its predecessors. A simplified approach of just asking for an outage of X if a specific and only one of the Y_{i_0} fails, may in turn be an oversimplification, since the dependency may be more complex than just involving X and the single Y_{i_0}. Still, the numeric examples to follow will adopt this restriction, since the more general case of several parents involved is conceptually identical and thus an obvious extension. In the data/example-driven parameterization proposed here, this amounts to including more variables (for the entire set of parents) in the training data. Using examples saves us from an analytic specification of the dependency, and since there are no hard quantitative requirements on how much training data we require (besides from "the more the better"), we can at least partly escape the issue of exponentially many cases to be specified (such as would be necessary if we go for a full specification of a conditional probability, say, in a Bayesian network). Nonetheless, more variables in the training data will naturally demand more training records for a good coverage of the (exponentially large) lot of possible instances, so the complexity issue is not going away entirely, but at least relieved.

22.2.3 Dependency Descriptions and Information Exchange

Hereafter, in a slight abuse of notation, let the symbol X, Y_i (for $i = 1, 2, \ldots, n$) denote the event of the respective asset to be in any of its possible states, whenever the symbol appears inside a probability. That is, we will write $\Pr(X = s)$ to mean the probability of asset X to be in a certain status s (likewise for conditional probabilities), which includes the normal working condition and outage, but in most cases will also cover different degrees of restricted functionality as intermediates between "normal working condition" and "complete outage." A common choice may be 3–5 such categories in total, e.g. using the 3-category range "working/problems/failure" or the 5-category range "working/small problems/medium problems/severe problems/failure." Naturally, the categories of functionality come as an ordered set, so they can be mapped to an integer range. In addition, let us suppose that each state is further refined by adding information to it, such as, e.g. an asset becoming "alerted" under normal working conditions. This could mean that information has come in putting the asset under explicit supervision or inspection, while this did not yet entail any restriction of its service or functionality (it could just be current audit). This, however, may be relevant to other assets, say, if the audit on Y locks certain databases for writing, so that a subordinate asset X can continue to read information from Y, but cannot update the database as long as Y is under inspection. This in turn may cause a state change for X, even though Y has not failed at this time. Depending on the granularity of the scale used, this can be represented by a state change for Y since not all its services are fully available (on a 3-tier scale, such small disturbance is probably not enough to switch to state "problems" while on a 5-tier scale a change to "small problems" might be reasonable).

Implementing the idea behind this example in the state space of an asset, let us think of the states of an asset to carry *information messages*, which can be interpreted by other assets (this data would be a part or combination of asset data and event data (SAURON Consortium 2018, 2019)). Both, the state and message to it are then a *signal* from one asset to another. To keep matters feasible for practice, let us assume that there is only a finite (and small) set M of such messages defined, i.e. only those that need to be communicated to subordinate assets, and the signals themselves can be ordered in terms of their own criticality. If we let the states be collected as a finite set $S = \{s_1 < s_2 < \ldots\}$ ($|S| < \infty$), with a finite number of message items per state being from $M = \{m_1 < m_2 < \ldots\}$ ($|M| < \infty$).

On the ordering of messages: it may occur that an event, manifested as a signal, affects more than one asset, so assets may change their state concurrently and thus signalize each other. In turn, one asset may receive multiple alarms to react on, and the ordering here serves as a tie-breaking rule. Suppose that a signal carries, among others, the following information items (cf. section 2 in SAURON Consortium (2018)), the compound of which is one of the aforementioned message items $m \in M$:

- Timestamp
- Alarm criticality
- Alarm impact
- …

Then an asset may process alarms in *lexicographic* order of these items, starting with the time (cf. SAURON Consortium 2019): alarms coming in earlier are processed first, and only signals that come in "roughly" at the same time are prioritized based on their criticality, and upon equal criticality, ordered according to their impact, and so on.

Based on such an ordering, we can extend the state space to the, again lexicographically, ordered and still finite set (s_i, m_j) for all i, j. This amounts to the support of the random variable X to be the set of "augmented" states, i.e. *signal space* $\Omega_X = \{(s_i, m_j) : i = 0, \ldots, |S| - 1; j = 0, \ldots, |M| - 1\}$. The general simulation model then works by updating the state of an asset X depending on what signal (s_i, m_j) it receives from a parent asset Y, where X can trigger its own state change upon the current state s_i of the parent asset but also taking the information item m_j into account.

The sets Ω_X, Ω_Y may thus look different for distinct assets $X \neq Y$, though the possible states are similar, the message items may be individual. Consequently, our discussion will distinguish two such sets, being:

- The ordered *state space* $(S, <)$: this is a *common set for all assets*, and describes the functionality of an asset, and is the same range for all assets of interest, including full functionality and complete outage as extremes, with optional and perhaps many intermediate states.
- The ordered *signal space* $(\Omega_X, <)$: this set is *individual for each asset* X, and is the lexicographically ordered Cartesian product $\Omega_X := S \times M_X$, when $(M_X, <)$ denotes the set of all messages that asset X can emit to its subordinates (in the dependency graph).

Both, the ordering and the finiteness of the signal space, are crucial, since it allow to identify a set Ω_X with a subset of the natural numbers, so that we can define a categorical distribution F_X to describe the probabilistic state of an asset X.

Remark on the dual use of the term "state": it will be important to distinguish the term "state space," which is a global object relevant for all assets, from the term "state" that describes the current element from S that describes the current working condition of X. Since we are not going further into discussions about the global set S itself, we will hereafter unambiguously speak about the *state of an asset* X, meaning its *current degree of functionality* (working condition).

The object F_X is then a mapping $F_X : \Omega_X \to [0, 1]$, with value $F_X(t) = \Pr(X \leq t)$, meaning that X is currently in state $(s_X, m_X) \leq (t_1, t_2) = t$ with $\leq$ being the lexicographic order. Observe that the $\leq$-relation entailed by the definition of F_X includes the information item, which is not obviously necessary at first glance. However, we need to bear in mind that the subordinate asset X may further send signals to its own child assets, for which an information item may again be important to know. As such, it will be useful to think of a state of the asset X to be primarily described by the first entry $s_X \in S$, with the second entry $m_X \in M_X$ only being an auxiliary information (a refinement) that is not necessarily relevant for X itself, but perhaps has individually different importance for distinct children that X notifies about its current state. Thus, each node X may define its own (total) ordering on the set of all signals that it can receive from all its parents.

22.2.4 Local and Global Models

This setup allows splitting the specification of (22.1) into two separate tasks:

1. *Local* modeling: Specifying the individual (marginal) distributions for the states in Ω_X and $\Omega_{Y_1}, \ldots, \Omega_{Y_n}$, i.e. the specification of distributions F_X, and $F_{Y_1}, \ldots, F_{Y_n}$ *per asset*, and *independently of other assets*. This task can be based on information that is available on the assets themselves, such as collected along risk management reporting or part of domain knowledge and expertise.
   Regression techniques discussed in Section 22.5 of this chapter can be used for this purpose.

2. *Global* modeling: The specification of the dependency as such, which works separate (and independent) of the "inner dynamics" of an asset. That is, we specify a copula function C that describes the joint state of $(X, Y_1, \ldots, Y_n)$ by a distribution function $F_{X,Y_1,\ldots,Y_n} = C\left(F_X, F_{Y_1}, \ldots, F_{Y_n}\right)$, in which we see that the individual behavior of each parent asset Y_i and the subordinate asset X goes into the "black box" function C, which exclusively describes how the assets depend on each other. The structural such dependency is exactly the dependency graph (the same structure, irrespectively of how the particular copulas look like), which—without edge annotations—lacks the details on how the dependency would manifest in practice. A copula function can fill this gap by specifying the stochastic dependency in functional form, but constructing the copula is a challenge on its own, with at least two possible systematic methods:

   a. Estimate the copula from data: this process is similar to the estimation of empirical distributions from data, and as such relies on massive amounts of training information, which may not be available to the extent required.

   b. Direct construction of the function C. Except for special circumstances, a direct specification entails considerable degrees of freedom: generally, a copula C describes the joint probability distribution of $(X, Y_1, \ldots, Y_n)$ as a function of the marginal $F_X, F_{Y_1}, \ldots, F_{Y_n}$, from which the conditional distribution $F_{(X|Y_1,\ldots,Y_n)}$ is obtained in a second step. Within the conditional probability, the term $\Pr(Y_1, \ldots, Y_n)$ may thus be described with help of a(nother) copula, whose choice can be arbitrary, including among others the product copula $C(y_1, \ldots, y_n) = y_1 y_2 \cdots y_n$ to express stochastic independence or $C(y_1, \ldots, y_n) = \min\{y_1, \ldots, y_n\}$ for an all pessimistic view where a single failure may cause a full breakdown (*maximum principle of system security*).

   In the most general setting, the expert may be able to specify a Boolean condition of AND and OR connectives between the parent assets, so that the asset X will become affected. For example, if X has three independent suppliers Y_1, Y_2, Y_3, among which Y_3 can "substitute" Y_2 (e.g. if Y_3 is a backup server for Y_2), then an outage of X may occur if $Y_1 \wedge (Y_2 \vee Y_3)$ fail. That is, X fails if Y_1 goes down and any (or both) of Y_2, Y_3 have problems. Boolean formulae like this example, however, not obviously translate into a copula function. Although fuzzy logic has lots of methods to offer for converting a Boolean formula into a continuous function, such conversions still call for a - not naturally available or obvious - statistically solid fundament.

22.3 Parameterization by Example

Suppose that instead of asking for a functional relationship to define X's state based on the signals from its parents in the dependency graph, we let the expert give only examples on how X would change its status based on input from the parents. Clearly, the examples need to be exhaustive in the sense that all relevant attributes and values thereof should appear in the data. On a high level, we seek a (black box) function f that we can train on examples provided by experts to provide us with an estimate of p as in (22.1), based on X's current state and the states of the parents $Y_1, \ldots, Y_n$ at time t. Formally, we thus want to have a function f so that

$$p = \Pr\left(X_{\text{next}} = x_{\text{next}} \mid X_{\text{current}} = x_{\text{current}}, Y_1 = y_1, \ldots, Y_n = y_n\right)$$

$$\approx f\left(x_{\text{next}} = i, x_{\text{current}} = j, y_1, \ldots, y_n\right) = f_{ij}\left(y_1, \ldots, y_n\right)$$

on input of concrete states (signals) from each of X's parents, where each transition within X from state $x_{\text{current}} = i \in \Omega_X$ to $x_{\text{next}} = j \in \Omega_X$ would carry its own individual function f_{ij}.

Now, the estimation of p is doable by various techniques only on grounds of example records describing a binary decision on whether or not the transition would occur under the particular configuration of the parent nodes. We discuss the applicability of alternative methods in the appendix to the chapter, and begin with a how the examples should look like and the extent to which state transitions can be specified directly.

22.3.1 Structure of Examples

The data we need to construct the aforementioned function f is a table containing user-specified (say, using data) examples of configurations where the transition of X into some fixed state s_k would happen, or not. This indication is either a "yes/no," in which case we put a 1 or a 0 in the column, or can also be a probability for the transition under the given circumstances, if such a value is available (however, since this is what we seek to estimate, any such value available could just directly go into the model without any need for logistic regression for this given transition). An exception may be cases where there is inconsistent information on the likelihoods, say, deviating opinions of experts, which could be settled by using this data for a regression analysis. In absence of any experience or reported quality of such a method, we here neither recommend nor discourage this treatment of that special case.

Remark: note that to simplify matters of the practical examples later, we shall omit the alarm messages from the data used for our numeric examples (starting from Section 22.4.2), although a real-life instance of the method may indeed take this information into account. Hence, the data specification generally will need to contain both, the current state and alarm (jointly encoded in the variable $\omega = (s, m) \in \Omega$ for an asset), but our illustrations will be kept slightly simpler by omitting the alarm message m in the worked examples.

Since, for a full specification of a state transition automaton, this would be required for all possible target states, the table of examples will need to follow a generic structure, as exemplified in Table 22.3.

Let us here briefly recap the risk management data that is assumed available as an aid for the model parameterization: Suppose that for asset X, the criticality level C_{X,Y_i} of the dependency on its parent Y_i has been specified, say, on a scale from 1 (only mildly critical) to 5 (very critical). The training data as shown in Table 22.3 can then be set with help of such criticality levels, perhaps

Table 22.3 Training data structure (generic).

Transition indicator (yes/no) $p \in \{0, 1\}$	Asset X state		Parent state (current "health status" + alarms raised at the moment), each value $Y_i \in \Omega_i = S_i \times M_i$			
	X_{current}	$X_{\text{next}} \omega = (s, m) \in \Omega_X$	Y_1	Y_2	...	Y_n
0	ω_0	(s_1, m_1)				
1	ω_0	(s_1, m_2)				
⋮	⋮	⋮				
1	ω_0	ω_j				
1	ω_i	ω_j				
⋮	⋮	⋮				
0	ω_k	ω_ℓ				
	⋮	⋮				

using heuristic rules. Note that the criticality is herein *not* considered as a necessary part of the (training) data, but only as an informal aid for an expert during the (manual) labeling, insofar this information is available.

For example, suppose that asset X is highly dependent on parent Y_1 but only mildly dependent on parent Y_2. Then an expert may say that:

- an outage of Y_1 (state s_5) would put X into a medium critical state s_3, as long as Y_2 is intact (s_0)
- an outage of both, Y_1 and Y_2 would put X into a highly critical state

This information would translate into the following two lines in the training data table:

Transition indicator (yes/no) $p \in \{0, 1\}$	Asset X state		Parent state (current "health status" + alarms raised at the moment)			
	$X_{current}$	X_{next} (incl. alarms)	Y_1	Y_2	...	Y_n
1	ω_0	ω_3	ω_5	ω_0		
1	ω_0	ω_5	ω_5	ω_5		

Note that we may leave the other fields empty here in two cases:

1. There is indeed no dependency of X on the parent
2. The transition of X into its next state (as specified in the column "X_{next}") is indeed unaffected by what the other parents do; the dependency may be such that only the two parents determine it completely.

Further rows in the training data table may be set even automatically, say, if an asset is already in a bad state, then a change of the parents may indeed not cause any further trouble (besides, perhaps, forwarding the information in form of an alarm).

The training data table may thus be augmented by records of the generic (automatically producible) records, e.g.:

Transition indicator (yes/no) $p \in \{0, 1\}$	Asset X state		Parent state (current "health status" + alarms raised at the moment)			
	$X_{current}$	X_{next} (incl. alarms)	Y_1	Y_2	...	Y_n
...	ω_5	ω_5	Every $\omega \leq \omega_5$	Every $\omega \leq \omega_5$	...	Every $\omega \leq \omega_5$

Conversely, it may also be possible for an asset to recover back to a "more working" state, based on the parents having recovered, in which case $X_{next} = \omega_5$ in the above table may be replaced by some other state under the proper configuration of the parents.

22.3.2 Modeling Recovery Events

A perhaps interesting aspect is the modeling of recovery processes for an asset. This is a case where state transitions can be specified directly (even if not, the methods developed remain applicable *without change*). Let X be the asset for which we seek to model a recovery. To X, we can associate a "recovery process asset" R_X, with the dependencies $X \leftarrow R_X$ and $R_X \leftarrow X$ (a circular dependency), and with the following state transitions specified:

- A transition of X into a state that demands recovery signalizes to R_X to start its own internal state changes.
- Internally, R_X undergoes a circular sequence of state transitions[1] as follows:
  - It normally stays in "idle" state 0, with the transition $0 \rightarrow 0$ upon no particular (i.e. an empty) signal from its parent asset X.
  - If X signalizes that a repair is needed, R_X switches to state 1, and from there onwards, goes into states $2 \rightarrow 3 \rightarrow 4 \rightarrow \cdots \rightarrow k$, each transition happening with a constant probability p_{R_X}. The expected time until the kth transition is thus geometrically distributed (as we require $k - 1$ transitions corresponding to Bernoulli-distributed events to reach the final state k).
  - From state k, R_X can transit back to 0, but does so with signal to X that it should recover into a healthier state.

Transition indicator (yes/no) $p \in \{0, 1\}$	Asset X state			Parent state (current "health status" + alarms raised at the moment)
	$X_{current}$	X_{next} (incl. alarms)	$\cdots$	R_X
$\cdots$	ω_5	ω_0		0 with signal "repair done"

The training data for X would thus have an example transition occurring with (fixed) probability 1:

We remark that the circular inner structure of the "repair asset R_X" is extensible by inner branches into longer or shorter cycles, each having its final "finishing" transition with a signal for X to fall back into a different state. For example, if the repair turns out to be simpler than expected, the cycle may take a shortcut into the state 0 and signalize to X that everything is back to normal. Likewise, if the repair is more complex (for any reason), then a longer cycle may be taken, with the final transition maybe putting X back into a less healthy state than s_0. Figure 22.3 displays an example of this idea, where the asset X may "inform" the virtual repair asset to switch from state 0 (idle) to either a short (quick) repair cycle with probability p_1, or maybe also to a longer repair phase with probability p_2. The length of the repair is in any case a geometric distribution, induced by random switches between the states (without external signaling, but generalizable to

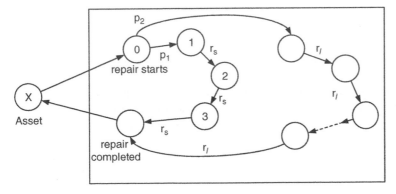

Figure 22.3 Conceptual idea of modeling a repair event.

---

1 Here denoted differently as 0, 1, 2, ..., to distinguish these "virtual" states from those reflecting the health status of an asset; we assumed these to be the same for all assets, and hence chose a different naming.

take shortcuts upon external signals as well, yet this possibility goes unexpanded here), happening in Figure 22.3 either with probability r_S (short) or r_ℓ (long repair).

Finally, even if the asset X itself has been repaired, it may be the case that its final state is still not "fully functional," since the parents may still not be working correctly, in which case the above example specification would need to be adapted with this information accordingly.

Note that the hereby proposed method induces no change to the state transition model based on Mealy automata, since the virtual repair asset is itself nothing else than another Mealy automata, only with a special structure to describe the reparation process.

22.3.3 Constructing the Model Parameter Estimation Function

All methods considered to construct f belong to the realm of machine learning. We discuss a selection of candidate methods in the appendix, where we briefly go into the respective pros and cons. Here, our focus will be on logistic regression.

Logistic regression models fit a function f taking values in the interval $[0, 1]$, so as to approximate a relation between parent states and the likelihood p, based on examples, but unlike the other two methods, provide a structure for f that is sufficiently simple to tune the model and allow for a direct interpretation of the model components.

The major advantages are:

1. The model is entirely transparent in the sense of assigning an explicit statistical meaning to all its variables, which can be (statistically) tested.
2. As such, there is rich theory available to check the model plausibility and to "validate" its parameter setting. Moreover, unlike membership functions in fuzzy logic, there is lots of theory available on estimating statistical parameters in a consistent way. Most importantly, indications of insufficient or inconsistent training data can be obtained. We will look into such model diagnostics in Section 22.6.1.1.

We therefore recommend logistic regression as the method of choice to accomplish semi-automatic parameterization of the state transition probabilities in a simulation model of a critical infrastructure asset.

22.4 Data Gathering and Preparation

In the preliminary stages of statistical data evaluation, it is generally necessary to clarify whether sufficient data/observations are available for all possible attributes or their combinations in order to be able to make a statistically substantiated statement.

The discussion applies likewise to any form of data analytics and is thus not bound to the proposed method. We illustrate the general method of logistic regression by constructing an artificial data set for our running example model sketched in the introductory sections of this chapter. We emphasize, however, that the general idea of parameterization by logistic regression enjoys much wider applicability, and the stochastic Mealy automata simulation framework (Schauer et al. 2020) is only our chosen showcase.

22.4.1 Public Sources of Data

Due to the model being rather recent (König et al. 2019) (developed along the lines of the project SAURON; see the Acknowledge at the end of the chapter), we cannot expect public data sources to provide us with data that we can straightforwardly use. Instead, we must use public data on threats and vulnerabilities to define our models with attributes of sufficient "genericity" to (i) apply to our assets as they are, and (ii) to be with values that we can find.

For that sake, we may investigate several different sources of data, such as (but not limited to):

- Configuration Management Databases (CMDB) (Müller and Buchsein 2016; SourceForge 2019): this may be a particularly rich source of information about assets, since the main purpose of CMDBs is to store all sorts of useful information about system components. The maintenance of such a database is indeed among the best practices of security management.
- The National Vulnerability Database (NVD) (Information Technology Laboratory 2019)
- The Common Vulnerability and Exposures (CVE) (MITRE Corporation 2019)
- Common Vulnerability Scoring System Data (Houmb and Franqueira 2009), obtained from (topological) vulnerability scanners (OpenVAS 2019),
- Data about the performance of surveillance equipment (e.g. Rass et al. 2017)

None of these sources will provide data directly as we need it, but for example, the use of CVE could be the following: let a cyber-attack be an exploit of some vulnerability. Then, knowing the vendor and version of a particular computer or other IT component, we can look into the CVE database to give us a list of vulnerabilities and CVSS scores on how "severe" the vulnerability is. The data about the asset can be augmented by such information to ease the assessment for the expert.

For example, if an asset has very many vulnerabilities, a transition into a malfunctioning state may be more likely; moreover, the alarm communicated to dependent assets may relate to the security breach caused by the exploit, which can be a violation of confidentiality (say, an unauthorized data access found in the logs), temporary unavailability, or data integrity violations, e.g. if checksums turn out to be incorrect.

The usefulness of such public data is thus indirect, by augmenting the asset data with it in order to make the expert's life easier when preparing the training data. We show this approach below, when we construct example training data for illustrating the method throughout the rest of this chapter. Nonetheless, we emphasize that the data constructed for our exposition here is in any case (and necessarily for the reasons just stated) purely artificial.

22.4.2 Getting Training Data

The constructed data consists of attributes chosen arbitrarily for demonstration, so they need tailoring to the specific application. Furthermore, the subjective expert assessment carried out on the data will be "simulated" here by implementing a heuristic rule on how an expert could judge the transitions to occur or not. This heuristic is intentionally oversimplified and designed to not take into consideration all the attributes available (unlike a real expert would probably do); rather the example will show how logistic regression would indeed recognize and point out this "ignorance" of certain attributes in the assessment. Besides, the data will be designed with *perfect separation*, which in a practical application may likely to occur if the expert (perhaps implicitly) follows certain heuristics that make the assessment explainable in a deterministic fashion. In such cases, logistic regression (in a standard instance) will numerically fail, as we shall demonstrate later in Section 22.6, to explain how we can overcome this issue.

First, we construct the data set used throughout, to illustrate the way of automated model parameterization.

Let us consider an asset with a single parent. The case of multiple parents proceeds alike, only with several copies and perhaps other additional relevant attributes, leaving the method itself exactly the same. For our illustration, let us think of an asset state transition to depend on the following attributes, referring to (and extending thereby) the structure shown in Table 22.3 (Section 22.3.1). More concrete instances of potentially relevant attributes depend on the application, and in real life, can (in first instance) be obtained from CMDBs (Müller and Buchsein 2016). The ones chosen here are for illustration, but in any case should include the current and next states, since those characterize the state transition arc in our probabilistic automaton that we seek to label with a probability (22.1).

- `CurrentState`: the current state $\in \{s_1, s_2, \ldots, s_5\}$ of the asset
- `NextState`: the state into which the asset may move, also over the same range of values as `CurrentState`. The transition whose likelihood we want to compute based on the data is thus the arc from `CurrentState` $\rightarrow$ `NextState`; the probability annotation to which the logistic regression shall compute for us.

Observe that we herein simplified the general structure laid out by Table 22.3, in the sense that we exclude the output alarm message of an asset in the specification of `NextState`. It may indeed be the case that practically, the output alarm can (deterministically) depend on the current health status, as well as the alarms from the parent. For example, if an asset gets notified, it may process the alarm (i.e. undergoes the proper state transition) and then forward the received information to its subordinate assets (after perhaps augmenting it with some own notifications). It appears thus as a viable simplification of practicalities, to just use the state without alarms here, and let the output signalization be up to a separate mechanism, not discussed here.[2]

- `Parent1State`: this is part of the variable $Y_1 \in S_1 \times M_1$ in Table 22.3, actually the left component in S_1. It ranges over the same values as the states above.
- `Parent1Msg`: this is the second half of Y_1, and lives in M_1, where we assume the following possible messages to be communicated (for simplicity): an alarm refers to a breach in terms of confidentiality, availability, or integrity, or can be "empty." The values chosen here relate to the data obtained from the CVE database (MITRE Corporation 2019). Assume a cyber-attack to exploit some vulnerability, then each such vulnerability (in the CVE database) carries attributes telling whether an exploit leads to a breach of confidentiality, availability, or integrity. Thus, if the exploit occurs at the parent node, we here assume that the impact (as told by the CVE database about the exploited vulnerability) is communicated further to dependent assets as an alarm, or specifically as `Parent1Msg`, here.
  To avoid enlarging the message space unnecessarily for the example, our messages in the following will not include the entirety of information mentioned in Section 22.2.1 (such as alarm timestamp, alarm criticality, etc.).
- `criticality`: this is an auxiliary value that we may get with the specification of the asset, telling us whether the dependency on the parent is "critical" or "non-critical." It is as such not part of the modeling but may be included to avoid it having influence on the expert assessment as a hidden variable. Indeed, the regression model can express the outcome only on variables that we supply explicitly, and knowing about the potential relevance of a variable, we can effortlessly include it in the table for the training data (as long as we have the same value available for the prediction later, since this variable will also appear in the final model from which we compute the probabilities, and by then requires a value). This sort of generalizes the modeling outlined before but causes no harm.

---

2 A rigorous formalization of this simplification would be letting the output symbol depend on the state only, and not the transition. This amounts to changing the Mealy- into a Moore automaton, which is a computationally equivalent concept. Still, the Mealy automaton offers greater flexibility for the modelling, and retains consistency (here) with the cited literature.

- dep.type: the type of dependency; for example, the parent may provide data or a service, in which case alarms related to confidentiality and integrity may be handled different to those related to availability. Also, a parent may just provide backup services, in which case it is only required occasionally. It is up to the expert on how this value is interpreted. We include it in the illustration example for two reasons: (i) it is an attribute to be generally considered for relevance and for being generically enough to maybe apply in many contexts, and (ii) it will be an attribute that our hypothetical (simulated) expert does not use, and we want the regression model to point this out (in Section 22.6 we will see how).
- Transition: a 0/1-valued indicator whose value 1 expresses that the asset will move into state NextState under the specified configuration (Parent1State, Parent1Msg, critical-ity, dep.type).

The way to create training data is by constructing a random set of instances for all parameters (the ones just listed for our example). Using random values for the attribute values, we get the chance of a "reasonable cover" of all the (exponentially many) cases of parameter configurations (actually $2^{O(n)}$ many for n attributes), and it is doable by a humble random value generator. If available, the random instances should (but do not need to) resemble the natural distribution of parameter configurations to make the training data more representative to the data used for prediction; we will have more to say about this in Section 22.4.3.

For the example being, the CurrentState column runs from 1 to 4 here,[3] with new states being chosen at random in the interval [CurrentState + 1, 5]. Our example will furthermore proceed by encoding all the (textual) values for the above attributes by numbers, since these values do have a natural ordering on them, which would be unnecessarily cumbersome to tell R when we program the hypothetical expert to do the assessment for us. Thus, to ease matters of programming, we will adopt a numerical encoding for the factor levels (Table 22.4), stressing the fact that these numeric values will, in the regression, remain factors (and not numbers). For the illustration, this has the additional appeal that we can explicitly show how to tell R that what it reads is a factor, rather than a number.[4] The choice to let some numeric scales start from 1 while others start from 0 is arbitrary and has no effect whatsoever on the model or results.

The demonstration data is loaded as follows:

```
fullData <- read.csv(file="training_data.csv", sep=";",
                     header = TRUE)
```

Table 22.4 Example of different representations for attribute values (factors).

Attribute (variable)	Factor levels (strings, ordered)	Numeric representatives (in the same order)
State (Current, Next, Parent1)	$s_1, s_2, ..., s_5$	1, 2, ..., 5
Parent1Msg	conf.impact, avail.impact, integr.impact, empty	0, 1, 2, 3
criticality	low, high	0, 1
dep.type	data, service, backup	0, 1, 2

---

[3] Note that unless we model repair events here too, the last (fifth) state s_5 is "absorbing" and thus excluded here; there is simply no transition back from the failure without a repair.
[4] The system would recognize all string-values as factors, but would otherwise treat numeric factors as numbers, unless told explicitly not to do so.

Now, our (here hypothetically simulated) expert comes into play and is asked to judge whether the proposed transitions will occur in the given (random) example configuration will occur or not. This person is thus supposed to fill the ? values in a table like the following:

Transition	CurrentState	NextState	Parent1State	Parent1Msg	Criticality	Dep.type
?	2	5	3	1	0	1
?	2	3	3	3	0	2

Our demo-expert will follow some simple rules, making a decision upon the first rule that applies in order of appearance:

- *Any message not related to availability will not cause a transition* (because the asset may be able to check data for integrity, and confidentiality is not needed for this asset to function correctly)
- *If the parent is in a state worse than s_3, then a transition will in any case be from s_i into s_{i+2}, but at most into s_5,* since though the dependence is not critical, this is interpreted as a longer outage of the parent, which our asset may not be able to permanently compensate.
- *If the parent is in any state worse than s_1, then the criticality determines whether the transition happens; it does only if the dependency is critical.*

For our example training data, we can implement this heuristic in R by a simple function (expert), and apply it to all rows in the training data to simulate what an expert may say about our data. In reality, such a method may indeed be legitimately followed if the dependence is as clear and deterministic like we just described. We expect, however, that most realistic cases and experts will not merit any such simple rule on when transitions occur or not; nor may there be certainty about it. Thus, we emphasize that the rules above are for illustration only, and do not constitute any practical guidance or recommendation.

Our "automated expert" is thus the following R function, with the three if-clauses corresponding to the above heuristic rules:

Listing 22.1 Example hypothetical expert for illustration purposes.

```
expert <- function(x) {
  if (x$Parent1Msg != 1) return(0);
  if (x$Parent1State >= 3 &&
        x$NextState >= min(5, x$CurrentState + 2)) return(1)
  if (x$Parent1State > 1 && x$criticality > 0) return(1)
  return(0)
}
```

With this function being applied (in a loop) over all records of the fullData, filling the Transition column therein. Still, their form as being deterministic will generally cause troubles in the logistic regression, as we will later illustrate in Section 22.6, where we also show the remedy.

Important: observe that we do not ask the expert to give us any probability! Instead, we just ask for a subjective assessment of whether or not the transition will occur. The answer is really just 0 or 1, and different experts may even disagree here on the same configuration, which can be included in the training data without any problem. That is, we may have a record with equal values for all attributes, but different values set in the column `Transition`, simply because two experts were differently opinionated here. The essence of the expert's assessment, however, is that probabilities are never asked for—these values are exactly computed by the logistic regression.

Alternative to this kind of poll, we may ask the expert for what s/he thinks the next transition would be, and then set the indicator `Transition` to 0 or 1, depending on the expert's confirmation or rejection of the example proposed. If the expert disagrees but has an opinion on what would the correct next state would be instead, we may include the record (`CurrentState`, `NextState` *as told by the Expert*, {…other attributes…}) in the training data. This method has the appeal of occasionally adding two training records (one with a zero for `Transition` if the expert disagreed, and a likewise record with `Transition = 1` but another `NextState` as the expert thinks is right) instead of only one.

Before moving into the actual regression, we finally need to tell R that the numbers are really factors. For our example, this needs to be the last step, since doing it before would have caused trouble with the comparisons within the expert function (the factor would have need to be declared as "ordered," which causes the linear model fitting routines to treat these variables fundamentally different, and in an unwanted manner here).

```
fullData$CurrentState <- as.factor(fullData$CurrentState)
fullData$NextState <- as.factor(fullData$NextState)
fullData$Parent1State <- as.factor(fullData$Parent1State)
fullData$Parent1Msg <- as.factor(fullData$Parent1Msg)
fullData$criticality <- as.factor(fullData$criticality)
fullData$dep.type <- as.factor(fullData$dep.type)
```

Now, we are ready to start with the data analysis and use for model parameterization.

22.4.3 Explorative Data Analysis

A regular application of (logistic) regression would start with an exploratory data analysis as preparatory step. The purpose is to make sure that all realistic combinations of variable values appear approximately with their natural frequency in the training data; that is, the training data is representative for the real data to be expected for prediction. In providing our experts with randomly chosen configurations, we have explicitly worked toward this goal, best so if the randomly generated values do follow the natural distribution (cf. the remarks above).

Otherwise, we may need to reorganize the variables and attributes to have a decent distribution of configurations. This process is called "feature engineering," and there are different methods in the literature proposed. One is discretizing the values to reduce the number of configurations, and others include merging two or more variables into a single one. Explicit such modifications to the data may be intricate in the context of the project, since the variables will most likely be dictated by the asset properties. So, the recommended method is the one described above: pick random values and do so to the best achievable approximation and coverage of the real asset's configuration. A simple and reliable way of achieving this is to collect all asset's attribute configurations in a table (with their values) and select a feasibly small yet reasonably large set of them to have experts tell about whether transitions will occur or not. This is the basic idea of "Parameterization by Example" announced in Section 22.3.

22.5 Logistic Regression (LR)—Basics

Returning to (22.1), the explicit functional form of how p depends on its parent asset is taken as a linear relationship (a weighted sum), expressing the log-odds, hereafter denoted as q, for X to switch into the fixed state $\omega_k \in \Omega_X$ with probability $p_{kj} := \Pr(X_{\text{next}} = \omega_k \mid X_{\text{current}} = s_j, Y_1, \ldots, Y_n)$ as follows:

$$\log\left(\frac{p_{kj}}{1 - p_{kj}}\right) =: q_{kj} = \beta_0 + \beta_1 s_k + \beta_2 s_j + \beta_3 y_1 + \ldots + \beta_{n+2} y_n + \varepsilon, \tag{LRM}$$

in which ε is an error term of zero mean and constant variance, and Ω_{Y_i} for $i = 1, 2, \ldots, n$ is the state space of each parent of asset X.

More precisely, the *logistic regression model* (LRM) has the following components:

1. The variable to be explained is the log-odd q, which corresponds to the indicator "Transition (yes/no)" in the training data. The score $p = \Pr(\text{Transition occurs})$ relevant for the final valuation can be obtained directly from the value $q = \log\left(\frac{p}{1-p}\right)$ by using the inverse function (the so-called link function) $p = \frac{\exp(q)}{1+\exp(q)}$, whose example plot over $q \in [-10, 10]$ represents the S-curve often shown for the logistic regression (see Figure 22.4).

2. The terms $\beta \cdot s$ and $\beta \cdot y$ represent the values of the input variables on the right-hand side of (LRM). These independent variables are called predictors or regressors (in the remainder of this article and in the literature).

3. The error term ε accounts for the fact that no model can (or should) represent the data completely and without error (this would be overfitting, equivalent to the model just reproducing the data "as is"). This value is subject to the assumption that it has a mean value of 0 and constant variance but is otherwise a random value about which no further assumption is made (except that the error for a good model should be "small").

Convention: The training data set is a table of examples for which the relation (LRM) should hold. That is, we are given a table with tuples of the form (p_i, y_i) in which $p_i \in \{0, 1\}$ for all rows i tells whether or not the transition occurred, and the vector $y_i = (y_{i,1}, \ldots, y_{i,m}) = (s_{ij}, s_{ik}, y_{i1}, \ldots, y_{in})$, *to ease notation*, will henceforth include the states of X before and after the transition, as well as the states of all parents of X, $\text{pa}(X) = \{Y_1, \ldots, Y_n\}$.

We assume N such training records to be available, each such set referring to a single state transition between fixed states $X = s_k$ to $X = s_j$. The procedure is thus repeated for all existing transitions $(s_j, s_k) = (X_{\text{current}}, X_{\text{next}})$, conditional on the parents' $\text{pa}(X)$ states.

Fitting the model amounts to finding optimal values $\beta^* = (\beta_{i,k})_{i=1,\ldots,m}$ so that the training data $x_1, \ldots, x_N$ set is best approximated by the values $f(y_1) = q_1, \ldots, f(y_N) = q_n$ being the log-odds, in the sense that it optimizes the log-likelihood function[5]:

$$\beta^* = \underset{\beta}{\text{argmax}} \sum_{i=1}^{N} \log \ell\left(y_i, p_i \mid \beta\right) \tag{LSQ}$$

The fitting is done by least-squares approximations and automated in R via the function `glm` ("generalized linear model"). Hereafter, we will refer to the model's approximation error, i.e the difference between the computed odds and the training data, as the model's *residues*.

Remark: We indeed *do not* convert the binary valued indicators p data to log-odds—that would be nonsense, as the conversion either gives $\log\left(\frac{0}{1-0}\right) = \log(0) = -\infty$ or $\log\left(\frac{1}{1-1}\right) = \log\left(\frac{1}{0}\right) =$

---

5 Crudely speaking, ℓ is the probability of the model to reproduce the given data under the given parameterization.

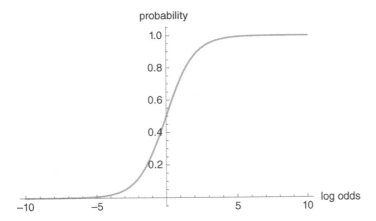

Figure 22.4 S-curve, for backward conversion of odds into probabilities.

undefined—but rather let the linear model express log-odds from which we subsequently compute the sought probabilities (via the S-shaped function in Figure 22.4).

A particularly useful feature of logistic regression is its ability to handle categorical data, i.e. variables y_i that can take on only finitely many values (even not requiring those to be on an ordered scale). This mechanism is useful to treat missing data and hence described in Section 22.5.1.

22.5.1 Handling Categorical and Missing Data

Following our convention from before, the state of an asset is one out of a finite set Ω_{Y_i}, which makes the variable y_i in the model a factor variable, with Ω_{Y_i} being the set of factor levels. Each factor variable goes into the model as a set of individual, so-called "dummy" variables $y_i^{(1)}, \dots, y_i^{(m_i)}$ with $m_i = |\Omega_Y|$.

For illustration, let us call out Ω_Y as $\Omega_Y = \{A, B, C, D, E\}$ (in an intended oversimplification toward a small number of possible states). Then, the variable y_i can take on only those values, and in the model, would appear as a number of $m_i = 5$ "copies" of the variable y_i, having the respective category name encoded into the variable's name; a popular notation is using a dot or plain string concatenation.

For example, suppose that there are records for y_i taking different values. Then

y_iA		y_iA	y_iB	y_iC	y_iD	y_iE	← one dummy variable per factor level
A		1	0	0	0	0	←in each row, there is a "1" entry
A		1	0	0	0	0	if y_i takes on the respective category
B	→	0	1	0	0	0	value, and zero otherwise
C		0	0	1	0	0	
D		0	0	0	1	0	
D		0	0	0	1	0	
E		0	0	0	0	1	

these translate into the respective set of dummy variables as follows: The dummy variables allow for an easy treatment of categorical data in the model. Note that this does not require nor assume any ordering on the factor levels (thus, simplifying the construction and assumptions made in Section 22.1.2).

22.5.1.1 Treatment of Missing Values

Missing values in the data can be dealt with in different ways (Fung and Wrobel 1989; Meeyai 2016); usually, one of the following three methods is used, which can also be applied individually for each data record differently:

1. **Amputation**: deletion of incomplete records from the database without replacement. This only makes sense insofar as it should not restrict the available data too much.
2. **Imputation**: Filling in missing values based on the remaining data. Various methods are available for this purpose, which we will refer to as "horizontal" or "vertical" imputation below[6]:
   a. Horizontal imputation (cf. Table 22.5): For the incomplete dataset $\mathbf{y} = (y_1, y_2, \ldots, y_{i-1}, ?, y_{i+1}, \ldots, y_m)$ the missing value "?" is estimated by another statistical model (e.g. linear/logistic regression or by other means) from the remaining values in the database record $\mathbf{y}$.
   b. Vertical imputation (see Table 22.6): Let y_i be the variable for which no value is available in the relevant record no. j. Then a random value is used for $y_i =$ "missing," the distribution of which matches that of the attribute y_i ("bootstrapping" or "predictive mean matching"; see van Buuren and Groothuis-Oudshoorn (2011) and CRAN (2013)), which is determined from the entire dataset, but without taking into account the other values in record no. j.

Table 22.5 Imputation by regression: Only the (remaining) gray highlighted attributes of the same data set are relevant (horizontal view).

Record no.	y_1	y_2	$\cdots$	y_{i-1}	y_i	y_{i+1}	$\cdots$	y_n
1								
$\vdots$								
$j-1$								
					Missing			
$j+1$								
$\vdots$								
n								

Table 22.6 Imputation by random values (bootstrapping): determination of the missing value based on the statistical distribution of this attribute in the rest of the database (highlighted in gray; vertical view).

Record no.	y_1	y_2	$\cdots$	y_{i-1}	y_i	y_{i+1}	$\cdots$	y_n
1								
$\vdots$								
$j-1$								
					Missing			
$j+1$								
$\vdots$								
n								

---

6 In this context, the terms "horizontal" and "vertical" are not commonly used in the literature, but are used only in the context of this chapter for better illustration.

Mixtures between the two variants are possible; for example, if the remaining (known) attributes allow the record to be mapped to "similar" records, so that the missing value can be replaced by the mean of the "most similar group(s) of records." This procedure is called "predictive mean matching."

In any case, it should be noted that replacing missing values with calculated imputations only makes sense if the missing places do not occur systematically (in the literature this necessary assumption is termed "missing at random"). If, for example, a value is always missing in connection with the occurrence of a certain combination of attributes, imputations of systematically missing values are generally not advisable. Also, substitute values *cannot* add any new information to the database, since the values are either chosen randomly (and hence independent from the data) or are determined from the remaining values, in which case they can only add information that is already there within the values that the estimate was based on.

3. **Treatment** of missing values as a **separate value category**: this method sometimes offers the greatest flexibility. In particular, it can also be used for systemically missing values as a substitute for imputation procedures. Nonetheless, one should use this method with care, since it basically is reasoning in (and from) "absence of information." It is particularly advisable to consider the combination of attributes and to assess the significance of the category "no information" for the evaluation particularly precisely, for example if the lack of information shows a significant influence on the estimation of the model parameter.

The third method requires R to know that there is another category for the missing values, which R internally encodes as NA ("not available"; a special value that any numeric or string variable can take). For our example, the factor level extensions work by updating the levels function accordingly per variable:

Listing 22.2 Adding missing values as a new category.

```
levels(fullData$CurrentState) <- c(levels(fullData$CurrentState),
levels(fullData$NextState) <- c(levels(fullData$NextState),
                                        "NA")
levels(fullData$Parent1State) <- c(levels(fullData$Parent1State),
levels(fullData$Parent1Msg) <- c(levels(fullData$Parent1Msg), "NA")
levels(fullData$criticality) <- c(levels(fullData$criticality),
levels(fullData$dep.type) <- c(levels(fullData$dep.type), "NA")
```

Note that the string "NA" is interpreted different to the token NA in R.

22.5.1.2 Recommended Treatment of Missing Data

Usual statistical methods of estimating missing data would attempt filling gaps based on the remaining available information in the training data. The particular way in which logistic regression treats categorical data (see Section 22.5.1.1), however, opens up an interesting possibility to treat missing data here in several ways:

1. (Recommended method) Explicitly assign no category value, i.e. put a zero for the values of all dummy variables representing factor levels. In the linear model, this will make all related terms vanish, and the model effectively reduces to one that does not even include this parent anymore. This setting has an appealing interpretation given below.
2. Assign a uniform distribution over all the factor dummy variables, expressing the setting that we "do not know" the actual value of the parent's state, but everything is possible. So, the dummy variables would all have the values $y_i^{(j)} = \frac{1}{m_j}$ for $j = 1, \ldots, m_j$.

3. Assign a non-uniform distribution, say, a hypothesis on how likely certain states of the parent are, if there is no other information available.

We recommend the first of these possibilities as a heuristic since it appears as the most viable one for the following reason: let n count the totality of all variables that appear in the linear model, and suppose that the variables would be "divided" into chunks of factor dummy variables, i.e. the variable set is—in a refined naming—the list $y_1^{(1)}, \ldots, y_1^{(n_1)}, y_2^{(1)}, \ldots, y_2^{(n_2)}, \ldots, y_m^{(1)}, \ldots, y_m^{(n_m)}$ in which each $y_i^{(j)}$ is the jth factor level of the ith variable. The total number of variables in the model is thus $n = n_1 + n_2 + \cdots + n_m$. Furthermore, let $\boldsymbol{\beta}_i$ be a vector of weights assigned to all factor dummy variables, that is, a variable x_i appears in the model though its dummies as $\beta_{i,1} y_i^{(1)} + \beta_{i,2} y_i^{(2)} + \cdots + \beta_{i,m_i} y_i^{(m_i)} = \boldsymbol{\beta}_i^T \boldsymbol{y}_i$, where $\boldsymbol{y}_i$ here denotes the vector of dummy variables representing the categorical variable to describe the state of asset Y_i.

The full model (FM), excluding the error term here for simplicity, then takes the form, with $\boldsymbol{y} = (y_1, \ldots, y_m)$

$$f(\boldsymbol{y}) = \boldsymbol{\beta}_1^T \boldsymbol{y}_1 + \cdots + \boldsymbol{\beta}_m^T \boldsymbol{y}_m \tag{FM}$$

This model is then being fitted to the log-odds. Let us denote these by q_i, corresponding to the training data values p_i from the above tables. Under method 1, if we have no training data for variable $\boldsymbol{y}_i$, we would set $\boldsymbol{\beta}_i := 0$, and the full model reduces to the short model (SM)

$$f_{SM}(\boldsymbol{y}) = \boldsymbol{\beta}_1^T \boldsymbol{y}_1 + \cdots + \boldsymbol{\beta}_{i-1}^T \boldsymbol{y}_{i-1} + \boldsymbol{\beta}_{i+1}^T \boldsymbol{y}_{i+1} + \cdots + \boldsymbol{\beta}_m^T \boldsymbol{y}_m. \tag{SM}$$

Likewise, we can "decompose" the overall logarithm of the likelihood function ℓ as

$$\sum_{i=1}^N \log \ell \left(y_i, p_i \mid \boldsymbol{\beta} \right)$$

into two sums, one over all records for which there is data available on y and the other sum for records for which we do not have values. Moreover, let us assume that y_i is the only variable for which data is missing. Say, if $T_{-i} \subseteq T$ is the subset of the full training data T (of cardinality $|T| = N$) in which there is no data on the ith variable, the sum of log likelihoods would decompose into

$$\sum_{i=1}^N \log \ell \left(y_i, p_i \mid \boldsymbol{\beta} \right) = \sum_{(p,y) \in T_{-i}} \log \ell_{SM} \left(y_i, p_i \mid \boldsymbol{\beta} \right) + \sum_{(p,y) \in T \setminus T_{-i}} \log \ell \left(y_i, p_i \mid \boldsymbol{\beta} \right),$$

Where ℓ_{SM} is the likelihood function corresponding to the respectively smaller model (SM).

Now, let us generalize this thinking to whole sets of variables that are missing in different records. Then we would end up with *one model* (i.e. one function f_i) *per pattern of missing variables* (in the extreme case, we thus would have an exponential number of different models to optimize here, i.e. the above sum decomposition would have a substantially larger number of terms).

Record no.	p	y_1	y_2	$\cdots$	y_{i-1}	y_i	y_{i+1}	$\cdots$	$\cdots$
1			Missing			Missing			
$\vdots$									
j					Missing	Missing			
k							Missing		
ℓ							Missing		
$\vdots$									
m			Missing			Missing			

To carry the example toward more generality, consider the following table: Here, we would have the following patterns of missing variables, where i, j, k, ℓ, and m just mark arbitrary records. We partition the data into T_1, T_2, ..., so that each T_i is a set of records that exclude a set J_i of variables that are all missing in that block, so that the remaining variables are all carrying values.

- y_2 and y_i are jointly missing twice (in records 1 and m), so we put $J_1 := \{2, i\}$. The remaining resulting training record set $T_1 = \{1, 2, \ldots\} \setminus \{1, m\}$ contains the all records that have values for y_2 and y_i, so that the resulting training data is complete.
- y_{i-1} and y_i are missing for one record j, so put $J_2 := \{i-1, i\}$, and define $T_2 = \{1, 2, \ldots\} \setminus \{j\}$ with this record excluded, so that this data has no missing values for the two variables.
- y_{i+1} is also missing in two records (with numbers k and ℓ), so put $J_3 = \{i+1\}$, and define T_3 to include all records except record number k and ℓ, analogously to the previous blocks.

With this, we can partition the training data T into sets $T_1 \cup T_2 \cup \cdots$ for which all records in T_k are fully specified for all variables, except those in J_k. According to method 1 above, we would get one model f_k that excludes exactly the variables in J_k from the full model (FM) above.

The log-likelihood, however, is still taken over all records, only rearranged according to the partitioning, i.e. the regression solves for

$$\beta^* = \underset{\beta}{\operatorname{argmax}} \sum_{i=1}^{N} \log \ell \left(y_i, p_i \mid \beta \right) = \underset{\beta}{\operatorname{argmax}} \sum_{j} \sum_{(p,y) \in T_j} \log \ell \left(y_i, p_i \mid \beta \right)$$

We can plainly divide this goal function by the (constant) number M of sets T_j without changing the optimum β^*, i.e.

$$\beta^* = \underset{\beta}{\operatorname{argmax}} \frac{1}{M} \sum_{j} \sum_{(x,y) \in T_j} \log \ell_j \left(y_i, p_i \mid \beta \right) = \underset{\beta}{\operatorname{argmax}} \sum_{j} \underbrace{\left(\frac{1}{M} \sum_{(x,y) \in T_j} \log \ell_j \left(y_i, p_i \mid \beta \right) \right)}_{=:\mathrm{PLL}_j(\beta)}$$

in which case we can just call the right term in brackets the *j*th *partial log-likelihood* (PLL), as a function of β. But since this is just a convex combination of a number of now individual goal functions per model f_j, this is nothing other than a *Pareto-optimization* over the whole set of models. In other words, the fit of the regression model comes up as if we would do the fit over a set of candidate models, each including only a subset of variables and trained over complete data, with the following optimization constraint: the value β^* is chosen such that it optimizes each individual model fit $\mathrm{PLL}_j(\beta^*)$ *simultaneously*, meaning that any deviation from β^* toward further decreasing the error of model f_i comes with the effect of increasing the error of at least one other model f_j.

This treatment of missing data is essentially different from assigning an extra category for the state being "unknown," since this would be reasoning based on the absence of information. As such, however, it goes being unclear about whether the effect of this is like adopting a hidden assumption here (similar to an open-world or closed-world reasoning).

Summarizing the discussion, the recommended method is justified by two properties:

1. It corresponds to a Pareto-optimal fit of a set of models, each of which is trained with a complete subset of records, excluding variables for which no value is available
2. It does not, unlike the three standard methods, hinge on the assumption of the data not being absent systematically (missing at random). The method is thus also applicable when there is systematically missing data in the training set.

Since this method is not standard and thus not available as a pre-implemented routine in the respective functions in R, we will use the `fastDummies` package (Kaplan and Schlegel 2018) to manually create the dummy variables for us, so that we can manipulate them as we desire.

```
# for "flexible" handling of categorical dummy variables
library(fastDummies)
```

22.6   Application of LR for Model Parameterization

Using logistic regression for the parameterization of probabilistic Mealy automata is now straightforward:

- We have training data, being a set of configurations of asset parameters for which an expert (or a group of experts) has told us that certain transitions will happen or not.
- We fit a logistic regression model (LRM) to the training data; this will be demonstrated in Section 22.6.1. This—crudely speaking—equips us with the "black box" function f sought in Section 22.3, and we can evaluate on any asset data (not only the known training configurations), to label the probabilistic transitions of the Mealy automata in question. We show this in Section 22.6.2. The quality of this fit is a matter of intermediate model diagnostics and plausibility checking, explained in Sections 22.6.1.1 and 22.6.1.2.

All of the upcoming computations are done after having loaded the following two packages, for the sake of improved error messaging and more robust methods of model fitting.

```
library(safeBinaryRegression) # for better error diagnostics
library(arm) # for robust model fitting (to handle separation)
```

As a preliminary step, necessary for the model diagnostics later, we will split the full data into a fraction for training and the rest devoted to verification of the model fit. A common division is 75 : 25 at random, which we can readily accomplish as follows (starting from the pre-loaded data `fullData`):

```
# Selecting 75% of data as sample from total 'n'
# rows of the data
sample <- sample.int(n = nrow(fullData),
                     size = floor(0.75*nrow(fullData)),
                     replace = F)
train <- fullData[sample, ]
validation <- fullData[-sample, ]
```

In all of the next section, we use only the data in train to fit the model, whereas the data in validation will be used for the model diagnostics afterwards.

22.6.1   Step 1: Fitting the Regression Model

The actual implementation of logistic regression can be left to standard implementations like found in R (R Core Team 2018) and the relevant packages therein (here arm (Gelman and Su 2018), but see also car (Fox and Weisberg 2018), mice (van Buuren and Groothuis-Oudshoorn 2011), norm (CRAN 2013), and others (Josse et al. 2018)). We thus describe the workflow on the level of commands sent to the R kernel.

Let us directly jump into model fitting. It is merely an invocation of the `glm` function, telling it to use logistic regression by the parameter `family = "binomial."` The right-hand side specifies the model, here being the variable Transition to depend (operator ~) on all other variables in the table (using the dot as a shorthand notation to mean "all variables.")

```
mdl0 <- glm(Transition ~ ., family = "binomial", data = train)
```

Expanding the . in the above call, we would find the full list of variables in the model via the column names of the data frame `fullData`. It is just the set of attributes that we have in the data.

```
colnames(train)
```

This produces the output

```
[1] "Transition"   "CurrentState" "NextState"    "Parent1State"
    "Parent1Msg"   "criticality"
[7] "dep.type"
```

The direct attempt to fit a logistic regression model to the training data is doomed to failure, as the above call to `glm` throws an error message (which the pre-loaded library `safeLogisticRe-gression` refines toward a much more readable and understandable form than what `glm` would report natively; hence we recommend using that package here).

```
mdl0 <- glm(Transition ~ ., family = "binomial", data = train)

Error in glm(Transition ~ ., family = "binomial", data = train):
  The following terms are causing separation among the sample
points: (Intercept), NextState3, NextState4, NextState5,
Parent1State3, Parent1State4, Parent1State5, Parent1Msg1,
Parent1Msg2, Parent1Msg3, criticality1
```

The error reported explains that some (the ones listed along the error above) of the variables actually suffice to perfectly explain the variable `Transition`, which destabilizes the log-likelihood optimization (roughly speaking, the coefficients within β would unboundedly grow toward infinity, manifesting in either an error thrown by `glm` or at least huge variances of the coefficients estimations reported in the results). Although called so, this is not necessarily an error, since it actually tells us that we may not even need a stochastic model, because the underlying mechanism to determine `Transition` "appears" to be deterministic. However, this information is not about how this determinism looks like, and we—for the example used here—know that only because we have constructed our "artificial expert" to follow a set of heuristic rules. Circumstances causing this error are known as *perfect separation* or *quasi-perfect separation* in the literature and require different treatment than the standard fitting of logistic regression models that `glm` implements. Our example data have intentionally been constructed to lead to this effect, so that we can explain how to handle it.

We leave the technicalities of tackling the issue aside here (for being beyond the scope of this chapter; see Gelman et al. (2008) for an exposition), and mention only that the remedy is adopting a Bayesian approach that assigns a prior probability to the coefficients in β in (LRM), which leads to a generalized version of the log-likelihood that remains well-defined even for data that exhibits separation. In R, the `arm` package (Gelman and Su 2018) provides an implementation generalizing that of `glm` toward this method; the change in our call sequence is most trivial: simply replace `glm` by `bayesglm`, and we're rid of the error:

```
mdl <- bayesglm(Transition ~ ., data = train, family = "binomial")
```

At this point, we are technically ready to directly jump to the automated labeling of automata, but it is advisable to dig into more of the practicalities, especially obstacles, before that.

First, let us simulate the absence of certain values to illustrate the treatment of empty table cells as discussed in Sections 22.5.1.1 and 22.5.1.2. We manually insert missing data in the categorical columns at random, for a total of, say 20, incomplete records per attribute. Note that we shall only illustrate methods of leaving out incomplete records (amputation) and treating the missing values as separate categories (third method in Section 22.5.1.1) or set all dummy variable indicators to zero (recommended method of Section 22.5.1.2). All three methods are applicable when data is systematically missing, unlike imputation methods, which hinge on the data to be missing at random. Our manual insertion would be consistent with that assumption.

Note that we intentionally exclude `CurrentState` and `NextState` here; after all, we have a specific transition in mind that we seek to parameterize which requires that we know the current and next state between the arc.

```
N <- nrow(fullData)
idx1 <- sample(N, 20) # choose 20 random indices (without replace-
ment)
fullData[idx1, 'Parent1State'] <- NA  # mark data as "missing"
via NA value
idx2 <- sample(N, 20)
fullData[idx2, 'Parent1Msg'] <- NA;
idx3 <- sample(N, 20)
fullData[idx3, 'criticality'] <- NA
idx4 <- sample(N, 20)
fullData[idx4, 'dep.type'] <- NA
```

Now, we can reapply the `bayesglm` function, after turning the missing value token into its own factor level, with help of `is.na` to locate the missing data (without that step, the default method of `glm` as well as that of `bayesglm` would be amputation, i.e. excluding incomplete records).

Listing 22.3 Fitting the model with missing variables removal.

```
# introduce "missingness" as its own category
# (also called "NA" here)
fullData[is.na(fullData)] <- "NA"
# declare all (other) variables as factors
fullData$CurrentState <- as.factor(fullData$CurrentState)
fullData$NextState <- as.factor(fullData$NextState)
# ...same for the other variables


# fit the model
mdl2 <- bayesglm(Transition ~ ., data = train,
                 family = "binomial")
```

A detailed output, useful for a first diagnostic, is obtained by

```
summary(mdl2)  # detailed information; for brevity, just use
               # "print(mdl2)"
```

We partition the output in blocks to explain them separately. Note that the variable names correspond to those listed by `colnames`, only for the factor variables to appear with multiple copies, each carrying another factor level as suffix in the name (following the naming convention mentioned in Section 22.5.1).

```
Call:
bayesglm(formula = Transition ~ ., family = "binomial",
data = train)
```

```
Deviance Residuals:
Min    1Q     Median      3Q        Max
-1.88212  -0.09606  -0.03715  -0.01142   1.52285
```

```
Coefficients:
              Estimate Std. Error z value Pr(>|z|)
(Intercept)   -10.6315     2.2288  -4.770 1.84e-06 ***
CurrentState2   0.8134     1.0895   0.747 0.455304
CurrentState3  -0.2957     1.1093  -0.267 0.789792
CurrentState4   0.1877     1.1506   0.163 0.870394
NextState3      0.5341     1.4356   0.372 0.709882
NextState4     -0.1761     1.2760  -0.138 0.890252
```

```
NextState5        0.6602      1.2387    0.533 0.594031
Parent1State2     1.3466      1.1816    1.140 0.254459
Parent1State3     3.8453      1.3526    2.843 0.004469 **
Parent1State4     2.3800      1.3008    1.830 0.067314 .
Parent1State5     4.4071      1.3354    3.300 0.000966 ***
Parent1StateNA   -0.4477      1.9184   -0.233 0.815462
Parent1Msg1       7.7650      1.5917    4.878 1.07e-06 ***
Parent1Msg2      -0.5648      1.8153   -0.311 0.755715
Parent1Msg3      -0.5984      1.8015   -0.332 0.739755
Parent1MsgNA      2.6075      1.7104    1.525 0.127380
criticality1      2.0480      1.0219    2.004 0.045060 *
criticalityNA     0.4972      1.7452    0.285 0.775712
dep.type1        -0.2163      0.9967   -0.217 0.828213
dep.type2        -0.7416      1.0829   -0.685 0.493442
dep.typeNA        0.5320      1.5867    0.335 0.737403
—
Signif. codes:  0 '***' 0.001 '**' 0.01 '*' 0.05 '.' 0.1 ' ' 1
```

```
(Dispersion parameter for binomial family taken to be 1)

    Null deviance: 148.472  on 224  degrees of freedom
Residual deviance:  22.175  on 204  degrees of freedom
AIC: 64.175

Number of Fisher Scoring iterations: 29
```

The blocks, explained in order of appearance, contain the following information:

- Block 1: Model information
  This part of the output represents only the details of the function call, and serves the comprehensibility of the results, provided that only the output is communicated. In particular, the training data used for this model is included by its name.
- Block 2: Residuals
  The term residuals refers to all the deviations between the fitted model and the data. The values given in this output block represent some parameters of the distribution of residuals, in particular the value range and some selected quantiles. The analysis of these values can be used to check the assumption via the error term ε (mean value 0) if necessary.
- Block 3: estimated model coefficients
  This table provides information about the (now calculated) coefficients β_i. In detail, the results are to be interpreted as follows (using the above output as an example):

	Estimate	Std. Error	z-value	Pr(>|z|)	
(Intercept)	-106.315	22.288	-4.770	1.84e-06	***
Parent1State3	38.453	13.526	2.843	0.004469	**
Parent1State4	23.800	13.008	1.830	0.067314	
Parent1State5	44.071	13.354	3.300	0.000966	***
Parent1Msg1	77.650	15.917	4.878	1.07e-06	***
criticality1	20.480	10.219	2.004	0.045060	*
dep.type1	-0.2163	0.9967	-0.217	0.828213	
dep.type2	-0.7416	10.829	-0.685	0.493442	
...	...	...	...	...	...

- Estimate is the (estimated) value of the coefficient β_i. Its magnitude and sign reflect the strength and type of influence: a positive coefficient represents positive correlation; likewise expresses a negative coefficient a decrease of the likelihood when this value is larger (negative correlation). The strength of this influence is reflected in the value of the variable.
- The column Std. Error shows the standard deviation of the estimate and is therefore a measure of the accuracy with which the value β_i was determined. The smaller this value, the more accurate the estimate.
- The z-value is the numerical value of the test statistics for the Wald test of the hypothesis (see Davidson and MacKinnon (2009) for details) that the associated value is $\beta_i = 0$. This value is particularly relevant in the context of the following number:
- Pr (>|z|): This is the so-called p-value of the statistical Wald test and provides information on whether the data contradict the assertion that the relevant parameter $\beta_i = 0$. Specifically, in the case of $\beta_i = 0$ the associated attribute would therefore not be relevant for the model, especially since omitting this attribute would not result in any significant (in terms of statistically significant) worsening of the model. If the p-value is small, especially smaller than a given significance level α (typically $\alpha = 0.05$), there is a significant deterioration of the model quality if the attribute β_i would be omitted.
- The star indicators given in the last (unnamed) column thus mark the variables important for the model, while the absence of stars in an attribute indicates that the variables in question could be omitted without significantly (statistically significant) reducing the quality of the model's predictive power.

It should be noted in particular that individual factor levels can have different significance, as can be seen in the example of the variable Parent1State: States s_1 and s_2 appear as insignificant, while state s_3 and higher have some significance (but are not equal in this regard). This shows that the expert's assessment apparently depended on certain states to be visible but is not "functionally" depending on the state. In other words, our "expert" cared for a certain state or higher to play a role for the transition, but lower (i.e. healthier) states apparently do not play a substantial role for a transition. Likewise, the variable dep.type has no significance whatsoever, as indicated by the model analysis. Both observations are indeed consistent with how we constructed the training data: take a look back at the function expert from Listing 22.1, embodying the heuristic that our expert followed: that function implements a rule to react on states $\geq s_3$, which makes these three significant for the model. Likewise, the expert (function) did not use dep.type in any way, which the model analysis correctly tells us here, by saying that this variable is insignificant (and could hence be removed from the model). Finally, the significance of criticality is only minor, consistent with the expert reacting on this value only in one out of its three internal rules.

For the security engineer, this may be an indication that certain health levels or alarm types have apparently no substantial effect to trigger a state transition. Hence, one could consider not signaling this information in the simulation model (since the variable seems to have no substantial influence, the simulation needs not send the message practically).

- Block 4: Meta information about the model
  This information block essentially provides information about the model itself. These include the dispersion parameter (which refers to a canonical representation of the model family used and here contains no information directly relevant to the application), as well as the comparison between the logistic model and the trivial model of the form $\log\left(\frac{s}{1-s}\right) = \beta_0 + \varepsilon$, in which no variables occur at all. For this so-called *null model*, residuals can also be determined, which are referred to here as *deviances*.[7]

Remarks:

- In the following, we will use the terms *residuals* and *deviances* synonymously (although the second term is only a special case of the first term).
- The value `AIC` (Akaike Information Criterion) is a score which is composed of the size of the residuals and the number of model parameters. The evaluation of the AIC score serves primarily to avoid overfitting, i.e. the "too precise" adaptation of the model to the data. Typically, a model with a sufficient set of parameters can reproduce any set of data without error (for example, if the data itself is made into parameters in the model). The AIC score evaluates models as "good" if they have low residuals (deviations) and a low number of parameters.
  In practice, therefore, models with small AIC scores are to be preferred, since these
  ○ On the one hand, explain the data well,
  ○ On the other hand, also have a small number of parameters (compared to alternative models).
  The method of Section 22.5.1.2 is implemented by creating the dummy variables manually from the previous data (using "NA" as a category rather than as a missing data token). This turns the "NA" category into another column in the data frame, having 1 indicator wherever there has previously been a missing data. We can extract and remove these easily in R, as follows:

Listing 22.4 Fitting a model with missing data treatment as its own category.

```
#create dummy variables explicitly (see Section 22.5.1)
fullData <- dummy_columns(fullData) # function from the
                                    # fastDummies package
# find all columns whose name ends with "_NA", according
# to how "dummy columns" would do the naming
# (i.e. attaching the factor level to the name, much as we
# did in Section 22.5.1
cols4deletion <- which(endsWith(colnames(fullData), "_NA"))
# also, remove the "original" columns that dummy columns let in
# there, since we do not need these anymore.
```

---

7 In general, a *deviance* is a(ny) function that quantifies the distinctness between two values by a positive number. A common, but not the only, example are squared differences.

```
# Retaining the "Transition" column no 1,
# we thus delete the columns no. 2 … 7
fullData <- fullData[, -c(2:7,cols4deletion)]

# redo the fitting (…like before…)
mdl3 <- bayesglm(Transition ~ ., family = "binomial",
                 data = train)
```

The result of the new fitting in terms of residual deviation is as follows:

```
    Null deviance: 148.472  on 224  degrees of freedom
Residual deviance:  19.823  on 202  degrees of freedom
AIC: 65.823
```

Consistently with the discussion of Section 22.5.1.2, the model fit is best among the three variants in terms of the residual deviations. Nonetheless, an explicit model diagnostic and plausibility checks as we describe in the next section are always advisable.

22.6.1.1 Model Diagnostics and Plausibility Checks

Since the values being returned are in the continuous interval [0, 1], a conversion to the two values 0 and 1 must be made for comparison with the validation data, by rounding down or rounding up. For the example, we select (arbitrarily) the threshold value 0.5:

Listing 22.5 Converting the probabilities back into indicators for verification.

```
glft <- predict(object = mdl, newdata = validation,
type="response")
glftpred <- rep(NA, length(glft))   # allocate an array of same
length as "glft"
glftpred[glft < 0.5] <- 0           # round < 0.5 towards 0
glftpred[glft >= 0.5] <- 1          # round >= 0.5 towards 1
```

The variable `glftpred` ("generalized linear model fit prediction") can now directly be compared to the corresponding values in the validation data. For this purpose, we determine the so-called confusion matrix by the following commands:

```
confusionMatrix <- table(validation$Transition, glftpred)
error_rate <- 1 - sum(diag(confusionMatrix)) /
sum(confusionMatrix)
```

Explanations:

- `table` determines a contingency table on the basis of the two transferred data records. In the present case, `validation$Transition` is an array with 0/1 values, which stands for "Transition" or "no Transition" per data set. The same column is selected from the validation data by accessing the column `Transition`. The second parameter is `glftpred`, i.e. the transition predicted by the model on the basis of the validation data. The result is a contingency table with the following structure, as Figure 22.5 shows:

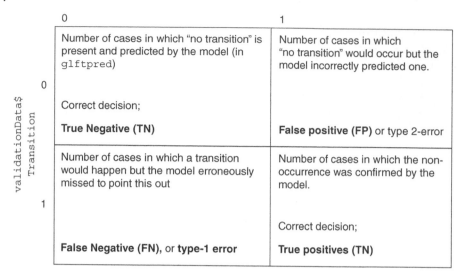

	0	1
0	Number of cases in which "no transition" is present and predicted by the model (in `glftpred`) Correct decision; **True Negative (TN)**	Number of cases in which "no transition" would occur but the model incorrectly predicted one. **False positive (FP)** or type 2-error
1	Number of cases in which a transition would happen but the model erroneously missed to point this out **False Negative (FN)**, or **type-1 error**	Number of cases in which the non-occurrence was confirmed by the model. Correct decision; **True positives (TN)**

(row label: `validationData$` `Transition`)

Figure 22.5 General structure of a confusion matrix.

- The second line of code determines the error rate by summing the main diagonal elements (extraction from the matrix using the `diag` command) $TN + TP$. The division by the total `sum(confusionMatrix)` returns the frequency of correct classifications, or the error rate by taking 1 minus that value.

For the example model `mdl` computed above (see Section 22.6.1), we find the confusion matrix and error rate to be

		Predicted	
		0	1
Expected	0	57	5
	1	0	13

Confusion matrix

Error rate:

$$1 - \frac{57 + 13}{57 + 5 + 0 + 13} \approx 6.6\%$$

In general, models with a too low and a too high error rate should be rejected. A very high error rate indicates *underfitting* and can have several causes:

- The transition may not exclusively depend on the variables considered, so that external factors may be hiddenly present when the data was specified. In that case, a reconsideration of the dependency structure (see Section 22.1.2) may be advisable, or the training data needs to be revised.
- The dependency is nonlinear, which may call for using nonlinear terms like products between the variables. In general, we can add terms like `y1*y2` to, say, model a logical AND between

two indicators (remember that these will most likely be dummy variables that are 0-1-valued, so writing p ~ ... y1.A*y2.B... in the general model supplied to the glm function, would make an influence to the model's outcome only if y1 takes value A and y2 takes value B at the same time. The general existence of such nonlinear relations has been subject of the previous discussion (see Section 22.2 and subsections therein). It may, upon a bad fit, be advisable to consider adding such nonlinear terms to the model, in order to express known interplays between the parent nodes of the asset.

A very low error rate should also be critically questioned, as there is a risk of overfitting. In an extreme case, the model reproduces the test data almost exactly, without necessarily showing internal structures (the model would have "memorized" the data).

In addition to this direct form of quality evaluation of the model, other forms of assessment of goodness exist, such as the *receiver operation characteristic* (ROC) curve. This is a representation of the so-called sensitivity (identical to the true positive rate TP; see the confusion matrix) of the model as a function of the specificity (identical to the true negative rate TN; see the confusion matrix) in the form of a curve. This curve should be above the +45° diagonals and increase differently for individual models (curves near or below the diagonal strongly indicate a bad model). Preference should be given to models with a maximum slope or maximum area under the curve. Both the curve (a graphical representation via plot function in R is shown in Figure 22.6) and the area below it are directly available in R by the command auc ("area under curve").

```
library(pROC)
roccurve <- roc(validation$Transition ~ glft)
q <- auc(roccurve)
```

Explanations:

- The parameter validation$Transition ~ glft describes the comparison of validation to predicted data (as described above)
- The value determined by auc is in the range $0.5 \leq a \leq 1$ and is to be interpreted as follows: a is the probability that a randomly chosen (actually) configuration that causes a transition will achieve a higher scoring than a randomly chosen configuration that does not (actually) lead to a transition. Heuristically, values ≥ 0.8 are typically taken as indications of a decent model quality. Doing the plot for mdl constructed from our full training data (i.e. without the artificially inserted missing data), we find $q \approx 0.9597$, thus confirming the fit to be reasonable.

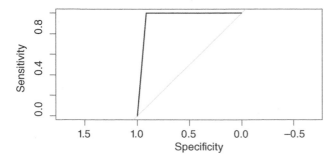

Figure 22.6 ROC curve (example).

22.6.1.2 Choosing/Constructing Alternative Models

If the model turns out as not satisfying or it is a good fit but needs perhaps simplification for better explainability and interpretability, then we can modify it into an alternative to test against our initial try. We hereafter discuss methods to compare different models, where the competitive models will all be simplifications of the original model (thus addressing the second situation just mentioned). This is no necessity, and in real life, we may prefer setting up another model out of educated intuition and background/domain expertise.

If there are two alternative models, `model1` and `model2`, a concrete choice for a model can be made in two ways regarding model quality:

1. Comparison of the respective AIC score. The model with the *lower* AIC score would be preferred. This process can be automated in R through the `step` function, to which at least the following parameters must be passed:
   a. The model to be tested: typically, this includes all available attributes, so that successively smaller models can be generated for an optimization of the AIC score.
   b. `steps`: since the number of possible models is generally exponential in the number of variables, it is recommended to define a limit of maximum models to be tested. If this value is not specified, the system tests 1000 models and then terminates. For example, if there are k parents to X, then the maximum would be testing $O(2^k)$ models. It should be noted, however, that such a model selection will take a long time, since for each model both the adjustment and the statistical quality evaluation are carried out internally by determining the AIC score.

```
step(mdl, trace = 0)  # optimize the AIC score
# supplying "trace = 0" prevents "step"
# from being verbose on every model that it tries;
# it simplifies the output
```

We stress that the application of step is to be treated with care for models coming out of `bayesglm`, since this function (in the version of R that this chapter is based on; see the appendix section after the references), is tailored to models produced by `glm` (and its relatives). Still, we can apply step to the full model `mdl` that we had above to create a smaller model automatically (by having R automatically look for and drop insignificant variables on our behalf): the output of the above call is the following:

```
Call:  bayesglm(formula = Transition ~ Parent1State +
Parent1Msg + criticality,
    family = "binomial", data = train)

Coefficients:
  (Intercept)   Parent1State2   Parent1State3   Parent1State4
Parent1State5       Parent1Msg1       Parent1Msg2
    -11.9135            1.2325            3.8192          2.6314
5.0273          9.1263          -0.1917
  Parent1Msg3    criticality1
    -0.2465          2.2988

Degrees of Freedom: 224 Total (i.e. Null);  216 Residual
Null Deviance:          148.5
Residual Deviance: 19.53     AIC: 37.53
```

Recalling the discussion of the summary output of `mdl`, this minimization of the AIC score consistently with the significance findings from before, drops terms that are insignificant and whose removal doesn't worsen the model too much, relative to the simplification of the model's complexity (as the AIC computes a tradeoff for).

2. Statistical test for significant improvement of the quality of fit. Here we assume that `model2` is the result of deleting variables from `model1`, or vice versa (so-called *nested models*). R_1, R_2 denote the residuals of the two models, then we can (statistically) test the null hypothesis $R_1 = R_2$, i.e. the equality of the residuals. This is equivalent to the assertion (hypothesis) that both models achieve an equally good fit to the data. If the null hypothesis is rejected, the data speak against the equality of the residuals, and the residuals of one of the two models are statistically significantly smaller than those of the alternative model.

Another form of comparison is available in R by the `anova` command. The following example uses only a subset of variables (parents), here being from the example training dataset `full-Data`. The two models, `model1` and `model2`, herein concern the question of whether the smaller `model2` (lacking the variables found insignificant) can explain the data "roughly as good" as the more complex `model1` can. In terms of asset dependencies, this would correspond to the question of whether a smaller dependency graph with less edges (corresponding to asset X having less parents) would do equally well, given the data at hand. In the example following, note that the smaller model is exactly the one provided by step before; practically, this is not necessary, and done here only for the sake of an independent confirmation of how useful step is to simplify the model on demand.

```
model1 <- bayesglm(formula = Transition ~ .,
                   family = "binomial", data = train)
model2 <- bayesglm(formula = Transition ~ Parent1State
                                        + Parent1Msg
                                        + criticality,
                   family = "binomial", data = train)
anova(model1, model2, test="Chisq")
```

The parameter `test="Chisq"` has to be specified in this form due to the model family (parameter `family="binomial"` for the adaptation of the model).

Let us look into the output, separated into logical blocks:

```
Analysis of Deviance Table
Model 1: Transition ~ CurrentState + NextState + Parent1State +
    Parent1Msg + criticality + dep.type
Model 2: Transition ~ Parent1State + Parent1Msg + criticality
```

```
  Resid. Df Resid. Dev Df Deviance Pr(>Chi)
1       208     17.109
2       216     19.532 -8  -2.4231   0.9652
```

The upper block provides details on the models entered (and in particular allows you to check that they are actually models with the same attributes, but in different numbers).

In the lower block, the value `Pr(>Chi)` is decisive: It is the p-value of the χ^2 test for equality of residuals (here referred to as deviances). If this value is:

1. *above* the significance level α, typically $\alpha = 0.05$, the null hypothesis is assumed. Based on the data, the models thus show no significant differences regarding the quality of adjustment. Thus, the "smaller" model can be chosen.
2. *below* the significance level α, the null hypothesis is rejected. In this case, the data contradict the assumption and the test provides the (statistically robust) statement that the "smaller" model represents a significantly less accurate approximation of the training data. On this basis, the larger or more complex model (in terms of the number of attributes) would be preferable.

22.6.2 Step 2: Apply the Regression for Batch Parameterization

Now, let us put the model to work, starting from a humble and possibly very long list of data about Mealy automata that we seek to endow with edge label probabilities in one blow. Suppose that our data looks like shown in Figure 22.7, with the green columns marking the edges `CurrentState` → `NextState`, and the yellow column with the heading "p" is the probability of the edges that we seek to compute; this is the value (*) announced at the very beginning of this chapter. That is, we seek the probability p of asset X to transit from the current state X_{current} to the next state X_{next}, conditional on the states of the parent nodes $Y_1, \ldots, Y_n$ of X, cf. (*).

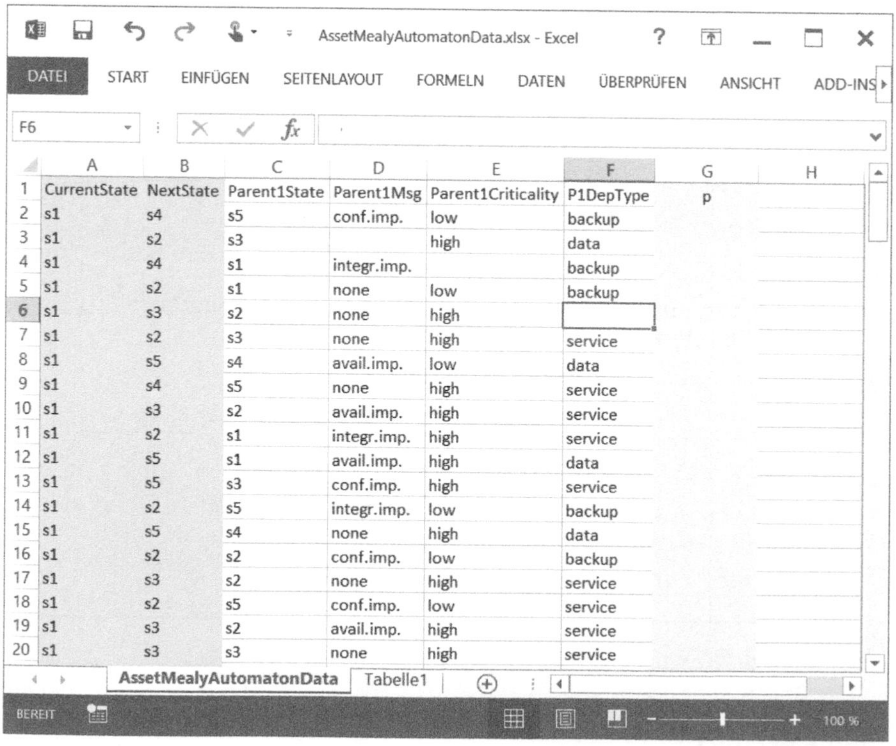

	CurrentState	NextState	Parent1State	Parent1Msg	Parent1Criticality	P1DepType	p
2	s1	s4	s5	conf.imp.	low	backup	
3	s1	s2	s3		high	data	
4	s1	s4	s1	integr.imp.		backup	
5	s1	s2	s1	none	low	backup	
6	s1	s3	s2	none	high		
7	s1	s2	s3	none	high	service	
8	s1	s5	s4	avail.imp.	low	data	
9	s1	s4	s5	none	high	service	
10	s1	s3	s2	avail.imp.	high	service	
11	s1	s2	s1	integr.imp.	high	service	
12	s1	s5	s1	avail.imp.	high	data	
13	s1	s5	s3	conf.imp.	high	service	
14	s1	s2	s5	integr.imp.	low	backup	
15	s1	s5	s4	none	high	data	
16	s1	s2	s2	conf.imp.	low	backup	
17	s1	s3	s2	none	high	service	
18	s1	s2	s5	conf.imp.	low	service	
19	s1	s3	s2	avail.imp.	high	service	
20	s1	s3	s3	none	high	service	

Figure 22.7 Example list of assets for automated parameterization for the respective Mealy automaton.

At this point, let us assume that this list has been exported into R from a comma-separated value (CSV) file.[8] Further, assume that we have already fitted a logistic regression model as above, whose variables are the same as in Figure 22.7 (except for the yellow column of course; this column corresponds to the column Transition in the training data, but unlike this past information having been a 0/1 indicator, we seek a probability (22.1) now; hence the heading "p" instead of `Transition`). To check, one can inspect the model `mdl` by typing `mdl$terms` and verify the model terms to be the same as the columns in the imported data about the assets. For compatibility, we must assure that the data is taken as categorical, since numbers would by default be numeric data for R, but we can tell `read.csv` that they are actually categories by adding the parameter `colClasses = "factor"` when importing the data:

```
assets_to_parameterize <-
read.csv(file="AssetMealyAutomatonData.csv",
        sep = ";", header = TRUE,
        colClasses="factor"))
# display the first four rows, just to see the
# correspondence with Figure 22.7,
# but more importantly, to verify that the loaded data
# has the same columns as the model's variables
assets_to_parameterize[1:4, ]
mdl$terms    # display the model in detail about its formula
```

	CurrentState	NextState	Parent1State	Parent1Msg	criticality	dep.type
1	1	4	5	0	0	2
2	1	2	3	NA	1	0
3	1	4	1	2	NA	2
4	1	2	1	3	0	2

```
Transition ~ CurrentState + NextState + Parent1State + Parent1Msg +
    criticality + dep.type
```

So, knowing that the model can be evaluated with the data now stored in `assets_to_para-meterize`, we can compute the sought probabilities in a batch for as *many assets as we wish* by a *single call* to the function `predict`. Note that the model above was fitted using categorical data, whereas the data loaded from the file, if it is purely numeric, will be taken as numbers, not categories, by default (thus the parameter `colClasses` should not be forgotten).

```
parameters_estimated <- predict(object = mdl,
                                newdata = assets_to_parameterize,
                                type="response")
```

This computes the sought likelihoods p, i.e. expression (22.1) by evaluating the logistic regression model for all rows in the loaded dataset. The parameters are explained as follows:

---

8 We intentionally used a commercial software here to display the results (Figures 22.7 and 22.8), to corroborate the claim that this is doable without requiring a highly specialized software, and the data collection and handling is possible using standard office tools.

- `object` is the model to evaluate; in our previous examples, this was `md1`, `md12`, and `md13`.
- `newdata` are the data upon which we seek to evaluate the model
- `type = "response"` advises the function to internally convert the log-odds back to probabilities (via the S-shaped function; cf. Figure 22.4). That is, the result of predict will be a value within $[0, 1]$ and thus directly useful as $p_k = $ `glft`.

It returns a vector, that we can conveniently display as a column by typing `t(t(parameters_estimated))` on the command prompt.[9] Let us display the first 10 rows (only):

```
       [,1]
1   2.640517e-04
2            NA
3            NA
4   1.636658e-06
5            NA
6   1.425700e-03
7   5.814145e-01
8   3.403247e-03
9   5.408108e-01
10  2.767412e-05
```

This is it already, we have computed the probabilities, wherever R was able to evaluate the model with the data at hand.[10] But observe that there are three rows (no. 2, 3, and 5) for which NA was returned. Upon a quick look back into the data in Figure 22.7, this is easily explained: since the data in the table was incomplete in first place, R simply had no values to plug into the respective variables of the model, so it left these rows unevaluated.

Now, it pays to have taken the dive into the ways of how to treat missing data in Section 22.5.1. These techniques are again applicable to get values even for incomplete asset specifications.

22.6.2.1 Parameterizing the Automaton with Incomplete Data

The available methods were:

1. Amputation: this is not as a viable option, since it means excluding any incomplete records, and thus effectively "giving up" on the parameterization of that asset. This is certainly not what we want.
2. Imputation: this means filling the gaps with values estimated from another regression model; we do not go into details about this method, for the reasons already stated in the previous sections.
3. Treating missing values as their own category: Here, we just fill the empty cells with strings to represent the "NA" category and tell R that it should *not* treat that string as its internal NA (missing data) token. That is, we put the string "NA" into the empty cells in Figure 22.7 and export a CSV file. The `read.csv` function would by default recognize the string "NA" as a missing value and interpret this string alike. However, we want "NA" to be its own category, so we instruct read.csv with the parameter `na.string = " "`, so that it would interpret "NA" really as a string, and automatically make the respective column a factor having that level (among the other values; previously, we had to manually add this factor level; cf. Listing 22.2 or Listing 22.3). The call sequence is the same as above, except for:

---

9 This is admittedly a hack: R displays vectors in horizontal long, line-broken lists. By transposing the vector once (inner call `t(...)`), it internally converts it into a $1 \times n$ matrix, and by transposing it a second time (outer call `t(...)`), it gets displayed as a $n \times 1$ matrix, i.e. a column vector.
10 Note that since we are interested in probabilities, the backward conversion into indicators as done in Listing 22.5 is not necessary here.

∘ The manual filling of the empty cells with the string "NA," so as to make the table in Figure 22.7 all complete.
∘ The loading of the CSV file via `read.csv(..., na.string = " ")`, to avoid R converting the NA strings into missing values internally.

Let us re-load the data with missing values now being their own category:

```
assets_to_parameterize <-
read.csv(file="AssetMealyAutomatonData.csv",
                              sep = ";", header = TRUE,
                              colClasses = "factor",
                              na.strings = "")
```

Now, we are again ready to parameterize our assets, but must do so using the model that we fitted upon treating missing data as its own category. Above, the model has been `mdl2`, which we can directly use now:

```
# this works now, since we have replaced the missing values with
# the string (= factor level)
# "NA", with whom the model was fitted in first place
parameterization <- predict(mdl2,
                              newdata = assets_to_parameterize,
                              type="response")
```

Let us display the first five rows to see that the NAs are gone now:

```
          [,1]
1 7.904953e-04
2 1.061489e-01
3 9.013939e-06
4 6.321525e-06
5 1.147185e-03
```

And that's it again! The only manual labor (besides collecting the asset data) is the—automatable—filling of missing data with the string value "NA" that we used as our category of "missing data." Since the model is essentially different from the previous model mdl, the respective estimates are necessarily also different (but not too different in our example).

Finally, it remains to look into how we would do the parameter estimation based on the technique of Section 22.5.1.2, which maintained a "mix" of models, each one only including the variables for which we do have all values available. The procedure is exactly the same as in that previous section:

1. We create the dummy variables with direct help of the `dummy_variables` function
2. We remove the original columns, and all columns related to the NA category.
3. We redo the prediction exactly as above, but using the proper model fitted under this method. This has been `mdl3` in our exposition.

The call sequence is:

```
assets_to_parameterize <- dummy_columns(assets_to_parameterize)
cols4deletion <- which(endsWith(colnames(assets_to_parameterize),
"_NA"))
# note that we have to remove the "original" variables as before;
# only bearing in mind that the first column "Transition" is not
# existing in this table (hence the columns are numbered
# from 1...6 instead of from 2...7 as in Listing 22.4.
assets_to_parameterize <- assets_to_parameterize[, -
c(1:6,cols4deletion)]


parameterization <- predict(mdl3,
                            newdata = assets_to_parameterize,
                            type="response")
```

For verification, `t(t(parameterization[1:5,]))` now gives

```
           [,1]
1 1.879798e-04
2 2.691824e-02
3 1.408359e-06
4 1.994485e-07
5 9.695761e-04
```

This completes the process, since the call sequences are the same, whether we label only a few or a few thousand automata (assets), a single call to predict will in any case suffice.

It remains to export the results back into a format for passing it onwards; in our case, we will write a CSV file, with a comma as decimal separator (`dec = ","`) and a semicolon to separate values (`sep = ";"`). Also, we do not want quotes around the exported strings everywhere (`quote = FALSE`), and have no use for the record numbers (`row.names = FALSE`). The export is done by `write.table`, since `write.csv` would always use the fixed symbols `"."` and `","` for the decimal point and separation.

```
output <- cbind(assets_to_parameterize, p = parameterization)
write.table(output, file = "output.csv", sep = ";", dec = ",",
row.names = FALSE,
            quote = FALSE)
```

This completes the table that we started from in Figure 22.7 into that of Figure 22.8, after rounding it to two decimals, and after the backward conversion of the numeric categories into the more informative strings (using Table 22.4 again). Note that the "NA" values herein reappear explicitly (as we imported those as categories) but removing them is a trivial matter now. What is important, however, is our ability to extend the automated parameterization to incomplete asset data even, by virtue of the techniques shown.

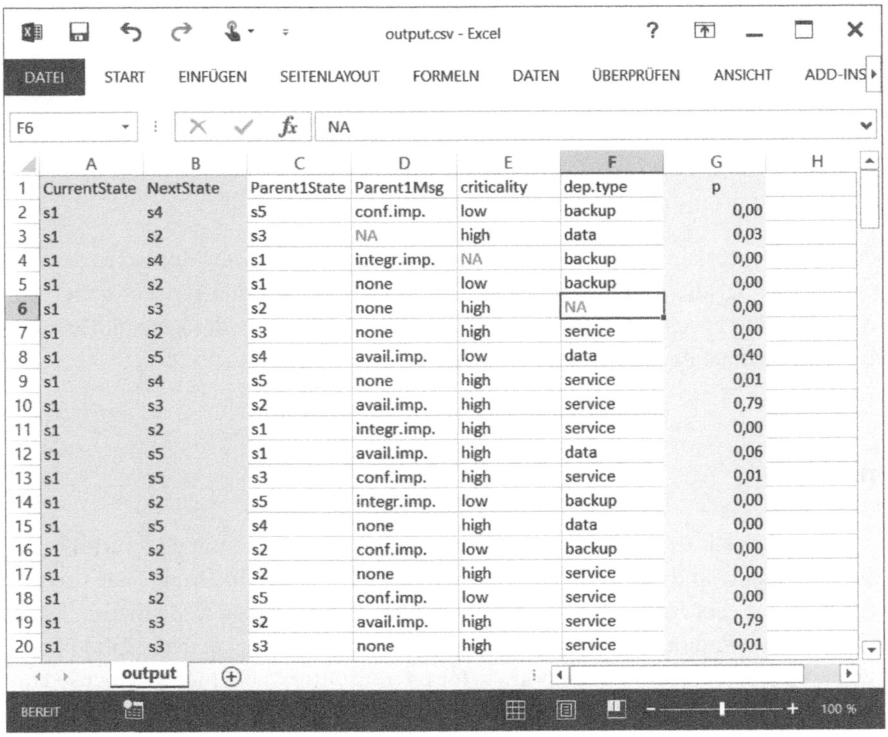

	A	B	C	D	E	F	G	H
1	CurrentState	NextState	Parent1State	Parent1Msg	criticality	dep.type	p	
2	s1	s4	s5	conf.imp.	low	backup	0,00	
3	s1	s2	s3	NA	high	data	0,03	
4	s1	s4	s1	integr.imp.	NA	backup	0,00	
5	s1	s2	s1	none	low	backup	0,00	
6	s1	s3	s2	none	high	NA	0,00	
7	s1	s2	s3	none	high	service	0,00	
8	s1	s5	s4	avail.imp.	low	data	0,40	
9	s1	s4	s5	none	high	service	0,01	
10	s1	s3	s2	avail.imp.	high	service	0,79	
11	s1	s2	s1	integr.imp.	high	service	0,00	
12	s1	s5	s1	avail.imp.	high	data	0,06	
13	s1	s5	s3	conf.imp.	high	service	0,01	
14	s1	s2	s5	integr.imp.	low	backup	0,00	
15	s1	s5	s4	none	high	data	0,00	
16	s1	s2	s2	conf.imp.	low	backup	0,00	
17	s1	s3	s2	none	high	service	0,00	
18	s1	s2	s5	conf.imp.	low	service	0,00	
19	s1	s3	s2	avail.imp.	high	service	0,79	
20	s1	s3	s3	none	high	service	0,01	

Figure 22.8 Resulting asset parameterization exported.

22.6.2.2 Compiling the Results

The models fitted in this section's example have, relative the artificial training and validation data, good predictive accuracy (error rates are all less than 7%, based on the false-positive and false-negative rates as computed in the confusion matrix, see Figure 22.5), so the obtained likelihoods are, based on the available data, plausible.

The values obtained from all three versions of regression slightly differ numerically, though come roughly equal upon rounding to a reasonable number of places behind the comma. To see this, let us round the values to two places after the comma, and display the results of `mdl`, `mdl2`, and `mdl3` next to one another as three columns, for the first 20 records as a snapshot:

```
1  0.00 0.00 0.00
2    NA 0.11 0.03
3    NA 0.00 0.00
4  0.00 0.00 0.00
5    NA 0.00 0.00
6  0.00 0.00 0.00
7  0.58 0.54 0.40
8  0.00 0.01 0.01
9  0.54 0.70 0.79
10 0.00 0.00 0.00
11 0.44 0.46 0.06
12 0.00 0.01 0.01
13 0.00 0.00 0.00
```

```
14  0.00  0.00  0.00
15  0.00  0.00  0.00
16  0.00  0.00  0.00
17  0.00  0.00  0.00
18  0.54  0.70  0.79
19  0.00  0.01  0.01
20  0.00  0.00  0.00
```

The third model is, not surprisingly, somewhat different to the other two, since `mdl3` resembles a mix, as opposed to `mdl` and `mdl2` being "purer" in their goal function for the model fits. The observable numeric differences are attributable to the substantial methodological differences between how the models handle missing data.

22.7 Summary

The bottom line of this chapter is to make readers aware of two aspects: first, many powerful models are, despite their generality and accuracy, difficult to apply in practice due to unclear ways of parameterizing them, and a good model design should always be accompanied by methods and recommendations on how to find the parameters. Second, machine learning can be helpful in this regard, as it can estimate parameters of different kinds for us. This tutorial chapter was focused on the estimation of a general probabilistic parameter, and the methods outlined are sure specific for the modeling of interdependent system components.

The general concept, however, has wider applicability, and other models can be parameterized using similar ideas, and perhaps with machine learning methods of other types than logistic regression. The most general method seems to be using examples to find parameters, and statistics helps to systematize the compilation of a parameter value from (many) examples, i.e. data, even if this data is incomplete or imprecise.

If possible, complete data is generally preferable, and imputations do have received lots of interest and may achieve good results in practice despite the criticism uttered above. The variety of methods to impute data, however, is much beyond the scope of this chapter, and we refer the reader to the packages `mice` (van Buuren and Groothuis-Oudshoorn 2011), `missForest` (Stekhoven and Buehlmann 2012), and related ones, on details about these methods. If data is imputed (for training, validation, or asset parameterization), the remaining procedures run just like we described, except that there is no missing data needing any special care.

In any case (complete or incomplete data), the model diagnostics (Section 22.6.1.1) is a crucial step, since it tells us how well the model approximates the expert's opinion. Note, however, that this cannot be used to confirm the goodness of an imputation, since any such inserted data will, during the fit, be just accepted as "correct," since it *is* training data by then. Nonetheless, whichever model is chosen in the end will need a final expert's judgment, based on domain expertise and experience since statistics cannot replace human intelligence.

Open questions following this outline remain, such as studying other or more advanced machine learning models (such as deep reinforcement learning; cf. Chapter 20) for parameters of various other kinds, e.g. nominal and categorical values, counts (i.e. parameters necessarily being integers), parameters with constraints on their value ranges, and many more. Also open is the issue of vulnerability of the method itself to adversarial machine learning (see Chapters 13–15) or poisoning attacks (see Chapter 19). Statistics does offer lots of estimation methods for parameters of probability distribution families, often also in robust versions, but these are only one class of models among many others. Extending machine learning to applications of model parameterization is an interesting and widely unexplored field at the time of writing this chapter.

Acknowledgment

This work was supported by the European Commission's Project SAURON (Scalable multidimensional situation awareness solution for protecting European ports) under the HORIZON 2020 Framework (Grant No. 740477).

22.A Appendix

22.A.1 On Fuzzy Logic Methods and Copulas

In fuzzy logic, the logical operations $\wedge$, $\vee$, and $\neg$ become replaced by *triangular norms*, specifically *t*-norms for ANDs ($\wedge$), *t*-conorms (a.k.a. *s*-norms) to model OR ($\vee$) and negation functions to model the $\neg$. All these, though having axiomatic characterizations, are usually non-unique and any combination of triangular norms, and negation operators are admissible. Adopting the fuzzy methodology to convert a complex condition on the outage of X into a function will thus not automatically result in a copula, unless the proper *t*- and *s*-norms are chosen, which in particular need satisfy a certain Lipschitz condition (Theorem 1.4.5 in Alsina et al. (2006)): *A t-norm $T(x, y)$ is a copula, if and only if it satisfies $T(x_2, y) - T(x_1, y) \le x_2 - x_1$ whenever $x_1 \le x_2$.* Since *t*-Norms replace the AND-operator and thus, when they coincide with a copula, describe joint probabilities $\Pr(X \cap Y)$, the counterpart being the OR operation would, in probabilistic terms, be $\Pr(X \cup Y) = \Pr(X) + \Pr(Y) - \Pr(X, Y)$. Translating this into the abstract language of *t*- and *s*-norms, we would thus put $\Pr(X \cap Y) = T(x, y)$ and furthermore put $\Pr(X \cup Y) = S(x, y) := x + y - T(x, y)$, whereas we want both functions T and S, to be associative to resemble how joint probabilities $\Pr(Y_1 \cap \cdots \cap Y_n)$ and $\Pr(Y_1 \cup Y_2 \cdots \cup Y_n)$ would behave under stochastic independence. Functions for which both, $T(x, y)$ and $x + y - T(x, y)$ are jointly associative have been studied and classified and are known as *Frank's class* of *t*-Norms. They are given by

$$T(x, y) = \log_b \left[1 + \frac{(b^x - 1)(b^y - 1)}{b - 1} \right]$$

$$\text{and } S(x, y) = 1 - \log_b \left[1 + \frac{(b^{1-x} - 1)(b^{1-y} - 1)}{b - 1} \right]$$

for a parameter $b > 0$. The special cases of $T(x, y) = x \cdot y$ for taking the limit $b \to 1$; moreover, the function T is indeed also a copula (Alsina et al. 2006).

Using Frank's class (the functions T and S above), together with the negation function $N(x) := 1 - x$, it is theoretically possible (details to which are yet unexplored and are outside the scope of this chapter) to find a copula describing the interdependency between the unconditional distributions F_{Y_i} for X's parent nodes $Y_1, \ldots, Y_n$, and the distribution of X over its range of support, under the assumption that the state of X depends only on the states of its parents, and X undergoes no endogenous state changes (i.e. without external triggers). Continuing the example formula $Y_1 \wedge (Y_2 \vee Y_3)$, this would translate into $T(y_1, S(y_2, y_3))$. The practical meaning of such a formula, however, may still be unclear, due to various reasons:

1. The composition of copulas is not automatically a copula again; the translation of a Boolean formula into a copula usually does not work as easy as just described, and can only work

under additional technical conditions (Durante and Sempi 2005); see also (Näther 2010) for the continuation of that discussion.

2. There remain inherent degrees of freedom in the choice of the copula and the negation function. Although the method is much more "concrete" than using fuzzy logic in its full strength, Frank's class offers an entire continuum of possible operators to replace the AND and OR operations. Even more possibilities come in by alternative choices of negation (e.g. Sugeno's function $N(x) := \frac{1-x}{1+\lambda x}$ for $\lambda > -1$ or Yager's negation $N(x) := (1 - x^s)^{\frac{1}{s}}$ for $s > 0$ are only two more candidates).

Every degree of freedom in the modeling can be taken as another parameter for fine-tuning, but against which data should we verify or validate our choice? If there are no data available against which we could test the performance, then each choice of b to select a member from Frank's class would appear equally plausible. Otherwise, if there is data against which we can test the goodness of the model, then there are easier and technically less involved methods to estimate the value p in (22.1).

It is thus here proposed to avoid the technicalities and uncertainty about the model itself in the estimation of expression (22.1), and instead resort to a method that works purely by using examples.

22.A.2   Using Fuzzy Logic to Estimate Parameters

Like logistic regression (see Section 22.3.3) also fuzzy logic can construct a parameter estimation function f from a known (Boolean) functional relationship between $Y_1, \ldots, Y_n$ and p. It is, however, in our case a heuristic, and despite rich theory behind the building blocks (like t-Norms and others), lacks statistical or other systematic methods to check the model plausibility other than by observing that "it just works." Therefore, this method is discouraged here.

22.A.3   Using Neural Networks to Estimate Parameters

Neural networks provide a highly expressive model that can be fitted to approximate the training data (see Chapter 16 for a more comprehensive discussion). Deep learning refers to the fitting of networks with many layers and complex structure, thus achieving high flexibility, at the risk of overfitting the data (if the network is "too complex"). Although deep learning is successful in many applications, prominent ones are generative adversarial networks discussed in Chapters 17 and 18, neural networks come with two major caveats:

1. They are "data hungry," i.e. we need a huge amount of data to carry a training algorithm to convergence.
2. They are "black boxes" in the sense that it is subject of ongoing research to *understand and explain why* (not "how," since this is clear) a network produces a certain outcome based on the inputs. Like for fuzzy logic, the model validation and plausibility checking basically proceed by observing that the network "works as expected," but we have almost no systematic tool to reason about the model structure's validity.

References

C. Alsina, M. J. Frank, and B. Schweizer. *Associative Functions*. World Scientific, 2006.

BSI. *IT-Grundschutz-Kataloge—15. Ergänzungslieferung*. Bundesamt für Sicherheit in der Informationstechnik, 2016.

S. van Buuren and K. Groothuis-Oudshoorn. mice: Multivariate imputation by chained equations in R, *J. Stat. Softw.*, 45(3):1–67, 2011.

CRAN (Ported to R by Alvaro A. Novo Original by Joseph L. Schafe). *norm: Analysis of Multivariate Normal Datasets with Missing Values*, 2013.

R. Davidson and J. G. MacKinnon. *Econometric Theory and Methods*, Internat. ed., Adapted version. Oxford University Press, New York, NY, 2009.

F. Durante and C. Sempi. Copula and semicopula transforms. *Int. J. Math. Math. Sci.*, 2005(4):645–655, 2005. doi: https://doi.org/10.1155/IJMMS.2005.645.

J. Fox and S. Weisberg, *car: Companion to Applied Regression*. Institute for Statistics and Mathematics of WU (Wirtschaftsuniversität Wien), 2018.

K. Y. Fung and B. A. Wrobel. The treatment of missing values in logistic regression. *Biom. J.*, 31(1):35–47, 1989. doi: https://doi.org/10.1002/bimj.4710310106.

A. Gelman and Y.-S. Su, *arm: Data Analysis Using Regression and Multilevel/Hierarchical Models*. Institute for Statistics and Mathematics of WU (Wirtschaftsuniversität Wien), 2018.

A. Gelman, A. Jakulin, M. G. Pittau, and Y.-S Su. A weakly informative default prior distribution for logistic and other regression models. *Ann. Appl. Stat.*, 2(4):1360–1383, 2008. doi: https://doi.org/10.1214/08-AOAS191.

S. H. Houmb and V. N. L. Franqueira. Estimating ToE risk level using CVSS. In Proceedings of the International Conference on Availability Reliability and Security, pp. 718–725. IEEE, 2009.

Information Technology Laboratory. NVD – Home. *National Vulnerability Database* [Online]. https://nvd.nist.gov/ (accessed 28 March 2019).

J. Josse, N. Tierney, and N. Vialaneix, *CRAN Task View: Missing Data*. Institute for Statistics and Mathematics of WU (Wirtschaftsuniversität Wien), 2018.

J. Kaplan and B. Schlegel, fastDummies: Fast creation of dummy (binary) columns and rows from categorical variables, 2018.

S. König, S. Rass, B. Rainer, and S. Schauer. Hybrid dependencies between cyber and physical systems. In K. Arai, R. Bhatia, and S. Kapoor, editors. *Intelligent Computing: Subtitle: Proceedings of the 2019 Computing Conference, volume 2*, pp. 550–565. Advances in Intelligent Systems and Computing, volume 998. Springer International Publishing, 2019.

S. Meeyai. Logistic regression with missing data: A comparison of handling methods, and effects of percent missing values. *Journal of Traffic and Logistics Engineering*, 4(2):128–134, 2016. doi: https://doi.org/10.18178/jtle.4.2.128-134.

MITRE Corporation. CVE—Common Vulnerabilities and Exposures (CVE) [Online]. https://cve.mitre.org/ (accessed 28 March 2019).

T. Müller and R. Buchsein. *A Project Guideline for Implementing a Configuration Management Database*. iET Solutions GmbH Kess DV-Beratung GmbH, 2016.

W. Näther. Copulas and t-norms. *Struct. Saf.*, 32(6):366–371, 2010. doi: https://doi.org/10.1016/j.strusafe.2010.02.001.

OpenVAS. OpenVAS—Open Vulnerability Assessment System [Online]. http://www.openvas.org/ (accessed 28 March 2019).

R Core Team. *R: A Language and Environment for Statistical Computing*. R Foundation for Statistical Computing, 2018.

S. Rass, A. Alshawish, M. A. Abid, S. Schauer, Q. Zhu, and H. de Meer. Physical intrusion games—optimizing surveillance by simulation and game theory. *IEEE Access* 5:8394–8407, 2017. doi: https://doi.org/10.1109/ACCESS.2017.2693425.

SAURON Consortium. Physical and cyber situation awareness fusion models. Project Deliverable D6.2, 2019.

SAURON Consortium. Modelling cyber/physical infrastructures & interdependencies. Project Deliverable D6.1, 2018.

S. Schauer, T. Grafenauer, S. König, M. Warum, and S. Rass. Estimating cascading effects in cyber-physical critical infrastructures. In *Critical Information Infrastructures Security*, volume 11777, pp. 43–56, S. Nadjm-Tehrani, Eds. Springer International Publishing, Cham, 2020.

SourceForge. OneCMDB [Online]. https://sourceforge.net/projects/onecmdb/ (accessed 28 March 2019).

D. J. Stekhoven and P. Buehlmann. MissForest—non-parametric missing value imputation for mixed-type data. *Bioinformatics*, 28(1):112–118, 2012.

23

Resilient Distributed Adaptive Cyber-Defense Using Blockchain

George Cybenko[1] and Roger Hallman[2]

[1] *Thayer School of Engineering, Dartmouth College, Hanover, NH, USA*
[2] *Naval Information Warfare Center Pacific, San Diego, CA, USA*

23.1 Introduction

There is sustained interest in developing cyber and physical systems consisting of multiple coordinated components, each component being simple, inexpensive, and easy to replace. Consider for example the following kind of military vision:

> A military made up of small numbers of large, expensive, heavily manned, and hard-to-replace systems will not survive on future battlefields, where swarms of intelligent machines will deliver violence at a greater volume and higher velocity than ever before. (Brose 2019)

Examples include cyber-bot networks, Internet of Things (IoT), swarms of unmanned autonomous vehicles (airborne, maritime, and land based) as well as cloud computing infrastructures (Campbell 2018; Nguyen et al. 2017; Panli et al. 2018). While many of the arguments for such systems are based on cost-effectiveness and mission performance, we argue in this chapter that there are also compelling analytic arguments based on various advantages that machine learning, game theory, and secure distributed computing offer for such systems. The goal of this chapter is to explain those advantages in analytic terms.

The main question we address is "What provable, analytic advantages do swarm-type adaptive/learning systems have over monolithic systems?" In order to present our arguments, we go over our terminology according to the following:

- A "system" is a collection of components or agents cooperating to achieve a common set of goals, called the "mission" for short; the agents comprising the system are not a priori constrained in terms of their communication or synchronization capabilities;
- An "agent" or component is an atomic component of the system which is capable of independently engaging adversaries in the environment in which the system operates;
- "Online" machine learning is machine learning that strives to improve performance while actively interacting with a real environment, not a simulation; this is in contrast with "offline" machine learning in which the learning is done with a static dataset or simulation and there are no operational penalties for suboptimal performance during training;

Game Theory and Machine Learning for Cyber Security, First Edition.
Edited by Charles A. Kamhoua, Christopher D. Kiekintveld, Fei Fang, and Quanyan Zhu.

- An "engagement" is a sequence of interactions between the system and the environment during which online learning takes place;
- "Adversaries" describe the environment in which a system operates; a stationary adversary does not adapt or perform online learning in the course of an engagement while an adaptive adversary does perform online learning to improve its performance.

Our particular interest is so-called Adaptive Cyber-Defense (ACD) in which cyber-defensive technologies adapt to changes in the attackers' techniques and/or behaviors, as well as to organic changes in the ambient operating environment (Cybenko et al. 2014; Jajodia et al. 2019; Zhu et al. 2014). Organic changes can be due to dynamic re-configurations of the information infrastructure, such as the addition or removal of compute nodes, sensors, applications, and communications links.

The ACD's changes can be made by continuously monitoring the environment, learning its new characteristics, and implementing appropriate new control actions. The basis for making such adaptations in a stationary or slowly changing environment can be based on classical reinforcement learning and adaptive control ideas. However, if the operating environment changes because of adversary adaptations, existing mathematical and algorithmic principles for defensive adaptation do not directly apply with respect to convergence to optimal or near-optimal solutions.

We study the problem of online learning using distributed Upper Confidence Bounds (UCB) algorithms in which cumulative regret (CR) is the performance criterion. CR is an appropriate performance metric because it captures the overall rate at which the systems' performance is improving, not just the asymptotic conditions under which optimality can be reached. The faster a system can approach some notion of optimality, the more it will outperform an adversary that is also changing but at a slower rate.

A key observation and technical contribution is the use of distributed systems to implement UCB-based learning. This entails the deliberate use of "suboptimal" actions in order to fully explore the values of the available actions. In other words, the system must sacrifice some of its capabilities in order to learn faster and perform better. However, because some agents within the distributed system are operating suboptimally, they will be compromised and so some subset of information in the distributed learning system will lack integrity (Figure 23.1).

This requires adding technologies for making such systems robust even under untrusted agent operations. Accordingly, we consider the use of Byzantine Fault tolerance and Blockchain technologies in a distributed adaptive cyber-defensive system to address this challenge.

Section 23.2 reviews basic ideas of reinforcement learning, especially with respect to minimizing cumulative regret during learning. Section 23.3 investigates the manner in which distributed, redundant systems can cooperate to learn faster than a single learner. We refer to this distributed learning as "spatial" in contrast to the "temporal" aspect of classical reinforcement learning over time. Section 23.4 presents some results from simulations of the proposed approaches. Section 23.5 develops approaches for implementing systems in which mission computations can survive even when component failures occur, either through organic failure, adversary compromise or deliberate sacrifice for learning purposes. This section discusses the use of Blockchain techniques and Byzantine algorithms to mitigate such failures. Finally, Section 23.6 summarizes our findings and suggests directions for future work.

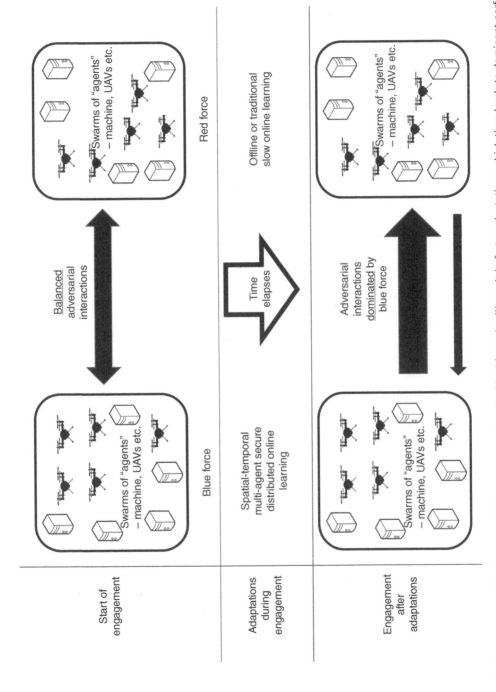

Figure 23.1 The operational implications of the techniques described in this work will result in faster adaptations which then leads to dominant performance during an engagement.

23.2 Temporal Online Reinforcement Learning

In this chapter, we are concerned with operating in adversarial environments. Because operating explicitly involves making sequences of decisions and actions, the problems of *learning* in such situations is typically formulated in terms of reinforcement learning (RL). Readers already familiar with RL concepts can read this section quickly. For readers interested in more background on RL, there are many excellent references to RL and various recent advances (Audibert and Bubeck 2010; Bubeck et al. 2011; Galichet et al. 2013; Menache et al. 2005). In addition, Chapters 19 and 20 discuss RL applications in cyber-defense.

In order to illustrate the concepts we advance here, we will present the very simplest example of reinforcement learning together with several recent enhancements and variants. To that end, consider the so-called two-armed bandit problem. In a casino, there are two slot machines, S_1 and S_2, available for you to play. You win on machine 1 with probability p_1 and on machine 2 with probability p_2 but you do not know whether $p_1 = p_2 \neq 0$, $p_1 > p_2 > 0$ or $0 < p_1 < p_2$. Each play requires you to insert a coin and "winning" means you receive two coins as a reward. If you lose, you receive no coins back so you are penalized by losing the coin you inserted.

You intend to play at the casino in perpetuity and have brought an infinite supply of coins with you. Of course, you would prefer to play the machine which is more likely to win but you don't know which machine that is when you start. A key assumption is that the outcome (win or lose) on each play is statistically independent of previous outcomes. In other words, the win/lose outcome on each machine is a simple Bernoulli random variable with probabilities, p_i respectively. Because you do not know which machine is better to play - that is, which one of the p_i is larger - you must perform experiments to learn some approximate values of the p_i.

At this point, it is important to note that even this simple problem statement can be used to model a variety of cyber-security scenarios. For example, consider two cyber-defenses that are available to a cyber-security administrator who must chose to deploy one or the other because deploying both can be too expensive in terms of performance, inconsistent, and/or technically infeasible. For example, one defense could be to quarantine or sandbox a computer system and the other defense could be to shut down some, but not all, services. The win payoff is a function of mission performance and security status. So while complete quarantine of a machine achieves high confidentiality security, there can be significant mission impact. On the other hand, shutting down some services might sustain mission performance but this still allows for security compromises of the services left operating.

A reasonable, but naive, experiment is to play each machine K times, observe the number of resulting wins on each machine and then continue playing only the machine on which you have won the most. However, it should be evident that with a positive probability, the $2K$ plays results in you playing the machine with lower win probability in perpetuity.

Extrapolating from this simple observation, a decision to play one of the machines in perpetuity after any finite number of plays on either machine can result in suboptimal play forever. As a result, it is necessary to play both machines an unbounded number of times for the probability of selecting the poorer machine to approach zero. Clearly, it makes sense to prefer to play the machine that is empirically outperforming the other - but how should that preference be quantified?

Reinforcement learning as originally developed does not specify how to allocate plays between the two machines so that, in the limit, you are optimizing the total possible winnings. The first rigorous decision rule for allocating play in an optimal way was developed by Roger Gittins in 1974 (Gittins 1974). The resulting so-called Gittins Index approach, while asymptotically optimal, proves to be difficult to implement in practice. A significant breakthrough in online reinforcement learning was the discovery and practical use of the Upper Confidence Bound (UCB) class of reinforcement learning algorithms (Auer 2002; Besbes et al. 2018). The intuition behind UCB-type algorithms is quite easy to understand.

Recall that the independent Bernoulli random variables (the results of plays on the two machines) in our slot machine example have means and variances given by

$$\mu_i = p_i, \ \sigma_i^2 = p_i(1 - p_i)$$

so that the average of K_i plays on machine i would be a random variable with mean and variance given by

$$\mu_{i,K_i} = p_i, \ \sigma_{i,K_i}^2 = p_i(1 - p_i)/K_i.$$

Observe that the means are the expected average rewards and the variances quantify the expected uncertainty in those rewards.

If we pick a machine to play solely based on the largest empirical sample mean, that is *exploiting*, we would be myopic and exclusively using our current estimates of the means. That is, we will play the machine that has the larger $\hat{\mu}_{i,K_i}$. However, we could have had bad luck by having an initial losing streak on the better machine, to the extent that the sample we drew had an empirical win probability that was actually smaller then the worse machine.

On the other hand, if we play the machine that has the larger empirical standard deviation, namely pick machine i for which

$$\hat{\sigma}_{i,K_i} \approx \frac{a_i}{\sqrt{K_i}} \ \text{as} \ K_i \to \infty$$

is largest then we are *exploring* based on the relative uncertainties in the values of p_i. In this case, we are missing the opportunity to exploit the better performing machine.

So neither criterion is suitable as an objective function to maximize and so it would make sense to take a linear combination of the two criteria, specifically with the form

$$\hat{\mu}_{i,K_i} + \frac{a_i}{\sqrt{K_i}}$$

and select the value of i based on which is larger. (Here a_i can be a function of the number of plays K_i and the outcomes of those plays on machine i but the key ingredient is that the second terms decreases as $1/\sqrt{K_i}$. It can also be a constant weighting between the two factors.)

This is precisely the intuition behind UCB-type algorithms-namely, it is a principled balance between exploitation of the best empirical choice given the current information (the $\hat{\mu}_{i,K_i}$ term) and exploration of options according to which option has the most empirical uncertainty (the $a_i/\sqrt{K_i}$ term).

A major difference between "offline" and "online" performance of learning is that in the former, the loss function is based solely on the current performance of the play strategy while in the latter the loss function is cumulative. That is, we cumulatively include the results of every play as we are learning. This is commonly quantified using a cumulative regret function, denoted by CR(T) at time T below, of the form

$$\text{CR}(T) = \sum_{t=1}^{T} E[(\text{Optimal\_Reward}(t) - \text{Realized\_Reward}(t))]$$

$$= T * \text{Optimal\_Reward} - E\left(\sum_{t=1}^{T} \text{Realized\_Reward}(t)\right)$$

where T is the total number of plays, Optimal_Reward is the optimal expected reward per play and E is the expectation operator over the random variables of the underlying machines/systems that comprise the operating environment.

If a reinforcement learning algorithms uses a fixed random selection for decisions (that is, which machine to play) that does not change over time or previous experience, then the realized rewards are time-independent and will grow proportionally with T. That is, $\text{CR}(T) \approx \alpha T$ for some constant α. One of the major advances offered by UCB-type algorithms is that the cumulative regret grows proportionally to $\sqrt{T}$ as opposed to $O(T)$. Moreover, it has been shown using information-theoretic techniques that the growth rate $O(\sqrt{T})$ is the best possible so that UCB-type algorithms are close to optimal within a multiplicative factor.

Finally, recent work (Jin et al. 2018) has shown that UCB-type algorithms are similarly optimal in Q-Learning algorithms. Q-Learning is a variant of reinforcement learning that has been shown to perform better in problems involving large complex state and action spaces (by using deep networks to represent the Q function). As a result, all the asymptotic near-optimal performance results of UCB algorithms carry over to adversarial interactions in which sequential decisions are required, such as in Markov Decision Processes.

In spite of these significant advances in algorithms and performance, Q-Learning (and more generally all current forms of reinforcement learning) using UCB-based action selections explicitly assume that the system is operating in a statistically stationary environment. That is, the probability distributions governing rewards and state transitions are fixed over the learning process.

In the next section, we review some key results concerning reinforcement-type learning in environments that do adapt, most interestingly adversarial environments in which adversaries can learn with unspecified learning algorithms, during online training. In the framework of the two-armed bandit problem introduced at the beginning of this section, this means that the casino operator can change p_1 and p_2 by observing your play. To constrain the casino, we can assume realistically that the casino regulators require that $p_1 + p_2 = 1$ or $p_1 + p_2 \geq p > 0$ to prevent the machines from degenerate situations like $p_1 = p_2 = 0$.

In such a situation, the player improves estimates of the probabilities and incrementally prefers to play one machine over the other, the casino operator observes that play preference and correspondingly decrements the win probability on the machine that you are playing preferentially. While it should be noted that this situation is basically the "Matching Pennies" game for which the Nash Solution is $p_1 = p_2 = 0.5$ for the casino operator and you as the player - both should also play each machine with probability 0.5. However, this Nash equilibrium solution is known only with a bird's eye view of the adversarial situation, that is knowing the game structure and payoffs. The whole point of online learning in adversarial environments is that the structure and payoffs (utilities) of the engagement are not known a priori and must be learned by all players during the course of their interactions.

23.3 Spatial Online Reinforcement Learning

A key aspect of UCB-based sampling on reinforcement learning is that data is processed sequentially and performance is determined by the number of steps (quantified by time for example) taken. However, it is intuitively well understood that a system than can learn faster that an adversary will achieve superior mission outcomes. This has been articulated in many ways but here we quote John Boyd (1987), the inventor of the OODA (Observe-Orient-Decide-Act) Loop concept:

> The key is to obscure your intentions and make them unpredictable to your opponent while you simultaneously clarify his intentions. That is, operate at a faster tempo to generate rapidly changing conditions that inhibit your opponent from adapting or reacting to those changes and that suppress or destroy his awareness. Thus, a hodgepodge of confusion and disorder occur to cause him to over- or under-react to conditions or activities that appear to be uncertain, ambiguous, or incomprehensible.

Section 23.2 summarized the current state-of-the-art in online reinforcement learning (with multiarmed bandits as a special case) using cumulative regret as a loss function. Specifically, UCB-type algorithms are essentially optimal with respect to minimizing the difference between optimal play (which is unknown by the players) and play as directed by the UCB-inspired algorithms. The optimal cumulative regret grows as $O(\sqrt{T})$ where T is the total number of samples processed.

As pointed out above, a critical assumption for those results is that the environment in which an agent operates is stochastically stationary-an assumption violated when the environment is at least partially directed to change by an adversary who typically has different values placed on different outcomes than the player/agent we are considering.

There are several well-known positive examples of reinforcement learning in games in which both players adapt simultaneously but such results dictate the algorithms that all agents must use (Hu and Wellman 2003; Hu et al. 1998). If all agents *cannot* be forced to follow specific algorithms in online learning in games, then convergence problems can arise (Foster and Young 2001; Sato et al. 2002; Young 2002). In fact, there are reasons to believe that this is a generic problem in real world as opposed to artificial recreational games such as chess, Go, and StarCraft. Chapters 13, 14, 15, and 17 discuss some of those limitations of classical game theory models as well.

We now consider a situation in which there are N cooperating agents interacting with the adversarial environment - namely a swarm of virtual/cyber or physical agents. That is we use a multiplicity of agents to accelerate purely temporal learning. We call simultaneous distributed as opposed to sequential centralized exploration of the action space "spatial" in contrast to "temporal."

To this end, consider a system of N agents, distributed logically (as in a networked computer system) or physically (as in an airspace, building, terrain, or body of water) in an environment. The agents communicate and coordinate actions amongst themselves with minimal human supervisory oversight. Most importantly, these agents can simultaneously engage the adversary resulting in N simultaneous samples of the interaction and its results such as state transitions and rewards.

At any given time, before the agents select their actions, assume for the moment that all agents have computed the same values of

$$\hat{\mu}_{i,K_i} + \frac{a_i}{\sqrt{K_i}}$$

from previous interactions. A key question is how do the N agents now use a UCB-type algorithm for selecting which action to try by which agent?

There are several options of which we review a few that are intuitively appealing:

- **Concentrated Maximum Sampling**—all N agents perform action j where

$$j = \text{argmax}_i \ \hat{\mu}_{i,K_i} + \frac{a_i}{\sqrt{K_i}}$$

the intuition being that since this is the "best" myopic action to try with respect to reducing regret, all agents should select that action. The downside of this approach is that too much effort might be expended in reducing uncertainty in one just action, at the expense of learning anything about the other possible action values.

- **Uniform Sampling**—all N agents perform an action drawn uniformly from all actions possible in the system's current state. This provides uniform coverage in terms of exploration but it is not guided by the collected knowledge so far so is not really a UCB-type strategy.

- **Weighted Sampling**—all N agents perform an action drawn randomly from all actions possible in the system's current state but now using a probability distribution reflective of the current UCB-type values. Specifically, if the UCB values for action i are currently

$$v_i = \hat{\mu}_{i,K_i} + \frac{a_i}{\sqrt{K_i}}$$

compute $w_i = e^{\beta v_i}$ where $\beta > 0$ and normalize these values to get a probability function

$$q_i(\beta) = \frac{w_i}{\sum_j w_j}.$$

The N agents independently sample actions from this distribution which can be done in a decentralized asynchronous manner. The parameter $\beta > 0$ can be incremented over time itself to effect a form of annealing.

The role of β here is to vary the sampling bias from uniform to concentrated maximum sampling in the sense that as $\beta \to 0$ the weighted sampling approaches uniform sampling while as $\beta \to \infty$ weighted sampling approaches the concentrated maximum approach. (Note that although optimal actions in an MDP are independent of positive scaling and translation of rewards, the weighted sampling technique described above is not independent of scaling although all of the above sampling techniques are independent of translation.) Accordingly, β is similar to the temperature parameter used in simulated annealing (Van Laarhoven and Aarts 1987).

We note that the order in which Q-values are updated once these simultaneous, possibly randomized, actions occur does not affect convergence because of the robustness of Q-Learning to asynchronous updates (Tsitsiklis 1994). Moreover, because the above-weighted sampling technique with $\beta > 0$ guarantees that all state-action pairs will be sampled infinitely often, it is evident from the general theory of Q-Learning convergence that using weighted sampling Q-values will converge to the optimal as well (Tsitsiklis 1994).

Accordingly, the main remaining issue concerns the rate at which convergence occurs. Using a combination of previous convergence rate results (Auer 2002; Jin et al. 2018), we believe the following is a reasonable conjecture:

> **Conjecture:** The cumulative regret for Weighted Sampling by N agents grows as $CR(T) = O(\sqrt{TN})$ for any $\beta > 0$ where T is the time step. That is, by using N simultaneous samples for Q-Learning, we believe we can effectively accelerate the convergence as if TN as opposed to T time steps occur.

This conjecture is further supported by resent results on distributed agents cooperating to solve a multiarmed bandit problem (Landgren et al. 2016).

23.4 Experimental Results

In this section, we explore the implications of the above observations for multiagent systems that are adapting through a UCB-type algorithm. Of particular interest is the possibility of being compromised (but recoverable) or being impacted to the extent that recovery is impossible in which case the agent component is permanently offline and unavailable for supporting the mission. We illustrate these implications through simulations based on dynamic epidemiological-like models.

We start by noting that if $O(\sqrt{TN})$ is the cumulative regret, the instantaneous regret would be roughly

$$\frac{dCR(T)}{dT} \propto O\left(\frac{1}{\sqrt{TN}}\right).$$

At each iteration or generation of the adversarial interaction, the reward function is the probability of surviving the interaction. If we let the maximal reward per such interaction be 1, denoting the probability of winning/surviving the interaction say, then the probability of losing or being compromised during an interaction would be approximately, according to the conjecture:

$$\frac{\gamma}{\sqrt{TN}}$$

so that roughly

$$1 - \frac{\gamma}{\sqrt{TN}}$$

of the N agents would be compromised on the Tth time step.

This suggests a time-varying dynamical systems model akin to those used in epidemiological studies (Anderson et al. 1992). For the sake of illustration, we propose partitioning the population of agents into three categories: Susceptible (agents that are not compromised and are operating properly), Compromised (agents that have been compromised by an adversary and can be restored to the Susceptible category through regeneration, self-healing or other mechanism) and Inoperable (agents that have failed or compromised so severely that they are incapable of being restored to Susceptible). Denoting the numbers in these populations at time t by $S(t)$, $C(t)$ and $I(t)$ respectively, we can approximate the dynamics of these populations using the following time-varying system of equations:

$$\begin{bmatrix} S(t+1) \\ C(t+1) \\ I(t+1) \end{bmatrix} = \begin{bmatrix} 1 - \alpha_3 - \alpha_1/\sqrt{Nt} & \alpha_2 & 0 \\ \alpha_1/\sqrt{Nt} & 1 - \alpha_2 - \alpha_4 & 0 \\ \alpha_3 & \alpha_4 & 1 \end{bmatrix} \begin{bmatrix} S(t) \\ C(t) \\ I(t) \end{bmatrix}. \tag{23.1}$$

(Note that left multiplication of the above equation by the vector [1 1 1] shows that the total population is fixed over time.) Here the parameter α_1 is a scaling factor for the instantaneous regret, α_2 is the recovery rate from Compromised to Susceptible, α_3 is the failure rate from Susceptible to Inoperable and α_4 is the failure rate from Compromised to Inoperable. Once an agent is inoperable it remains inoperable for the duration of the engagement which is captured by the last column of the matrix. Figures 23.2–23.4 illustrate some of the resulting dynamics for different values of the α_i.

This model and these simulation results demonstrate the variations in performance possible for different parameter values. The goal of course is to design systems for which the Byzantine threshold is crossed as late as possible so that trusted computations and communications can be sustained

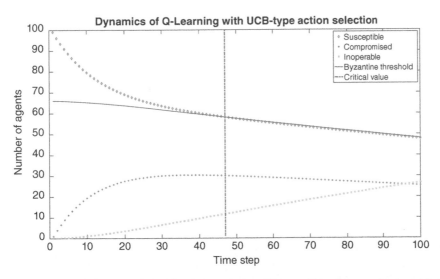

Figure 23.2 This figure shows the time evolution of Susceptible, Compromised, and Inoperable agents by percentages for certain parameter values arising in Eq. 23.1. The figure also shows the "Byzantine threshold" which is two-thirds the number of Susceptible plus Compromised agents. Once the number of Susceptible agents drops below the Byzantine threshold, the system cannot be defended against Byzantine failures. In this example, this event occurs at time step 47 (the "Critical value" depicted by a vertical line in the figure) meaning that after 47 interactions with the environment, the system will not be able to guarantee the integrity of any results it computes collectively, including computations required for learning or system missions.

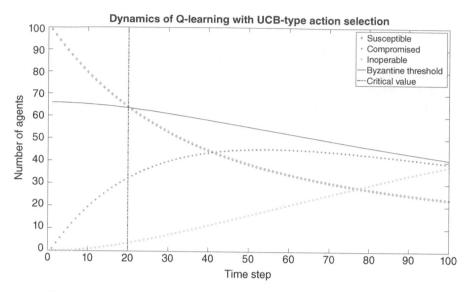

Figure 23.3 This figure shows the time evolution of Susceptible, Compromised and Inoperable agents by percentages for *different* parameter values than used in Figure 23.2 that arise in Eq. 23.1. The figure also shows the "Byzantine threshold" which is two-thirds the number of Susceptible plus Compromised agents. Once the number of Susceptible agents drops below the Byzantine threshold, the system cannot be defended against Byzantine failures. In this example, this event occurs at time step 20 (the "Critical value" depicted by a vertical line in the figure) meaning that after only 20 interactions with the environment, the system will not be able to guarantee the integrity of any results it computes collectively, including computations required for learning or system missions.

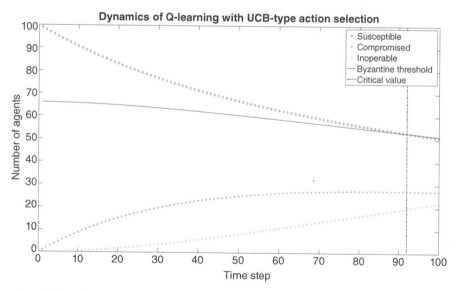

Figure 23.4 This figure shows the time evolution of Susceptible, Compromised and Inoperable agents by percentages for *different* parameter values than used in Figures 23.2 and 23.3 that arise in Eq. 23.1. The figure also shows the "Byzantine threshold" which is two-thirds the number of Susceptible plus Compromised agents. Once the number of Susceptible agents drops below the Byzantine threshold, the system cannot be defended against Byzantine failures. In this example, this event occurs at time step 92 (the "Critical value" depicted by a vertical line in the figure) meaning that after 92 interactions with the environment, the system will not be able to guarantee the integrity of any results it computes collectively, including computations required for learning or system missions.

as long as possible. In particular, one would like to outlast the adversary in an engagement by sustaining a smaller regret which according to our model is a proxy for probability of survival from time step to time step.

One way in which such endurance is enhanced is through the use of larger swarms or more agents comprising the system. The number of agents used appears as a square root in the denominator of the instantaneous regret so, if our conjecture is correct, having four times as many agents or components translates into halving the regret or equivalently, halving the probability of compromise of any single agent.

23.5 Survivable Adaptive Distributed Systems

There are two classes of computations that a system as described above must sustain with high confidence: the underlying missions of the system and the online learning for defensive purposes. We have used a Byzantine failure model to quantify the critical ratio between Susceptible (with assured computing and communications) and Compromised agents for sustaining assured computations. This assumes we do not know which agents are compromised but can discover those that are if we are above the critical threshold and use some kind of Byzantine agreement algorithm (Malkhi et al. 2019).

A particularly simple problem that abstracts aspects of both mission computations and online learning is the "consensus" problem. In particular, consensus about Q-values is required to properly randomize actions during spatial online learning even if some subsequent updates are incorrectly or asynchronously performed by a subset of compromised agents.

On the other hand, one type of system mission could be agents in a distributed intrusion detection system (IDS). Such agents would be distributed throughout an enterprise network, migrating users into high-interaction virtual honeypots for real-time evaluation, which would enable the IDS to make a determination about a user's legitimacy. Agents would learn from previous evaluation decisions and the final determinations following actions so that some sort of consensus, perhaps not 100% agreement, is needed about conditions under which to do the migration and how those decisions are to evolve.

Aspects of the consensus problem, are especial critical to this scenario. Specifically, the consensus problem criteria are (i) *validity*-if a process outputs a decision value v, then v was proposed by some process; (ii) *agreement*-no two correct processes decide differently for the same user; (iii) *termination*-every correct value eventually decides on some output value.

Agents in such an IDS must communicate among themselves to confirm user identities and legitimacy, as well as to decide which users to migrate for evaluation. The decision about which users to migrate and when to migrate them depends on a number of factors, such as the availability of resources, whether a user has a known reputation for legitimacy, and so on. Agents must achieve a consensus on the availability of resources to support evaluation honeypots, including network liveliness (key to ensuring that the migrated users do not detect that they are being evaluated). Furthermore, agents must maintain knowledge of users and their reputation for legitimacy.

The fact that compromised agents might be participating in these decisions as well as the possibility that communications could be stored-and-forwarded through those compromised agents suggests mechanisms for securely agreeing on such decision criteria and decisions must be considered. Blockchains provide an ideal mechanism for achieving consensus on each of these points.

However, a major challenge with using a Blockchain to achieve consensus is choosing the appropriate consensus protocol. Consensus must be reached in a way that wastes neither time nor computing resources. For example, real-time critical decision-making clearly cannot be achieved with

Bitcoin's 10-minute block processing time. Proof-of-Work (PoW) or Proof-of-Stake (PoS) protocols choose a single agent who dictates values adopted by the rest of the Blockchain, which will ensure uniformity but not necessarily the correctness of results. Fortunately, Blockchain protocols exist which enable agents to achieve a timely consensus which is required for critical applications (e.g. geographically dispersed cyber-physical systems).

Recent work has proposed and demonstrated the feasibility of decentralized machine learning schemes over an Ethereum Blockchain (Harris and Waggoner 2019). Other recent work shows how to coordinate multiagent deep reinforcement learning using an Ethereum Blockchain (Liu et al. 2018). Specifically, this work uses an Ethereum Blockchain to create a tamper-proof ledger that is shared among agents. However, a well-known problem when operating on the Ethereum network is that every operation has a cryptocurrency cost. Moreover, Ethereum's PoW protocol - like all PoW and PoS protocols - selects a single node that writes the next block in the chain and distributes that block to the rest of the network, which may lead to the propagation of sub-optimal choices. Implementing a Blockchain in the context of a multiagent autonomous system requires a consensus protocol that efficiently considers results from all agents.

To that end, Federated Byzantine Agreement (FBA) protocols (Mazieres 2015) offer decentralized control and low latency, among other features. At a high level, agents in an FBA system agree to trust a subset of other agents on the network. This subset is known as a *quorum slice* and agents may belong to more than one quorum slice. A *quorum* is a union of quorum slices that includes a slice for every member of the network. Learning, RL in particular, over Blockchain is a relatively new field and we are not currently aware of any learning implementations over FBA systems. However, FBA is the consensus protocol underlying a major cryptocurrency (Mazieres 2015) and has been applied to reputation management for digital manufacturing platforms (Innerbichler and Damjanovic-Behrendt 2018). In future work on this front, we hope to develop an FBA consensus mechanism that is applicable to multiagent autonomous learning systems.

23.6 Summary and Future Work

We have considered the problem of online learning, with cumulative regret minimization, for swarms of agents cooperating to achieve high assurance on missions tasks. To achieve this, we have introduced the concept of "spatial" online learning by which action selection is guided by UCB-type criteria. Because we are dealing with multiple agents, multiple actions are possible by sampling the UCB modified Q-values.

We have identified some regret minimizing challenges in so-called spatial learning and conjectured their solutions but without formal proof. Furthermore, we have performed principled simulations showing how learning and recovery rates can change the overall performance of a multiagent learning system.

While UCB criteria have been used in offline parallel game play learning (Gibney 2015; Mnih et al. 2013; Silver et al. 2016) those uses are for parallel game playing, that is parallel independent engagement, not a single engagement in which adaptation/learning occurs as we have proposed here.

Although we are keenly interested in convergence rates and algorithms for adversarial engagements in which all players are using online learning to improve play, the results here are solely for stationary environments in which adversaries do not adapt or learn during the course of a single engagement. However, if we can adapt faster than our adversary, we can model them as short-term stationary which might be the best possible approach given no information about the adversary of their adaptation mechanisms.

Consequently, a major opportunity for future work is cumulative regret minimizing, online reinforcement learning algorithms that make as few assumptions about the adversary's learning algorithms.

Acknowledgements

George Cybenko's research has been partially supported by ARO MURI Grant W911NF-13-1-042 and the US Air Force Research Laboratory.

Roger A. Hallman is supported by the United States Department of Defense SMART Scholarship for Service Program funded by USD/R&E (The Under Secretary of Defense-Research and Engineering), National Defense Education Program (NDEP)/BA-1, Basic Research.

References

Roy M. Anderson, B. Anderson, and Robert M. May. *Infectious Diseases of Humans: Dynamics and Control.* Oxford University Press, 1992.

Jean-Yves Audibert and Sébastien Bubeck. Best arm identification in multi-armed bandits. 2010. http://sbubeck.com/COLT10_ABM.pdf.

Peter Auer. Using confidence bounds for exploitation-exploration trade-offs. *Journal of Machine Learning Research*, 3(Nov):397–422, 2002.

Omar Besbes, Yonatan Gur, and Assaf Zeevi. Optimal exploration-exploitation in a multi-armed-bandit problem with nonstationary rewards. *Available at SSRN 2436629*, 2018.

John R. Boyd. Organic design for command and control (in "A Discourse on Winning and Losing"). http://www.ausairpower.net/JRB/organic_design.pdf, 1987.

Christian Brose. The new revolution in military affairs: War's sci-fi future. *Foreign Affairs*, 98:122, 2019.

Sébastien Bubeck, Rémi Munos, and Gilles Stoltz. Pure exploration in finitely-armed and continuous-armed bandits. *Theoretical Computer Science*, 412(19):1832–1852, 2011.

Adam Michael Campbell. *Enabling tactical autonomy for unmanned surface vehicles in defensive swarm engagements.* PhD thesis, Massachusetts Institute of Technology, 2018.

George Cybenko, Sushil Jajodia, Michael P. Wellman, and Peng Liu. Adversarial and uncertain reasoning for adaptive cyber defense: Building the scientific foundation. In *International Conference on Information Systems Security*, pp. 1–8. Springer, 2014.

Dean P. Foster and H. Peyton Young. On the impossibility of predicting the behavior of rational agents. *Proceedings of the National Academy of Sciences*, 98(22):12848–12853, 2001.

Nicolas Galichet, Michele Sebag, and Olivier Teytaud. Exploration vs exploitation vs safety: Risk-aware multi-armed bandits. In *Asian Conference on Machine Learning*, pages 245–260. PMLR, 2013.

Elizabeth Gibney. Deepmind algorithm beats people at classic video games. *Nature*, 518(7540):465–466, 2015.

John Gittins and D. Jones. A dynamic allocation index for the sequential design of experiments. In J. Gani,editor, *Progress in Statistics*, pp. 241–266. North-Holland, 1974.

Justin D. Harris and Bo Waggoner. Decentralized and collaborative AI on blockchain. In *2019 IEEE International Conference on Blockchain (Blockchain)*, pp. 368–375. IEEE, 2019.

Junling Hu and Michael P. Wellman. Nash Q-learning for general-sum stochastic games. *Journal of Machine Learning Research*, 4(Nov):1039–1069, 2003.

Junling Hu, Michael P. Wellman, et al. Multiagent reinforcement learning: Theoretical framework and an algorithm. In *Proceedings of the International Conference on Machine Learning (ICML)*, volume 98, pages 242–250, 1998.

Johannes Innerbichler and Violeta Damjanovic-Behrendt. Federated Byzantine agreement to ensure trustworthiness of digital manufacturing platforms. In *Proceedings of the 1st Workshop on Cryptocurrencies and Blockchains for Distributed Systems*, pp. 111–116. Association for Computing Machinery (ACM), 2018.

Sushil Jajodia, George Cybenko, Peng Liu, Cliff Wang, and Michael Wellman. *Adversarial and Uncertain Reasoning for Adaptive Cyber Defense: Control-and Game-Theoretic Approaches to Cyber Security*, volume 11830. Springer Nature, 2019.

Chi Jin, Zeyuan Allen-Zhu, Sebastien Bubeck, and Michael I Jordan. Is Q-Learning provably efficient? In *Advances in Neural Information Processing Systems*, pp. 4863–4873. Curran Associates Inc., 2018.

Peter Landgren, Vaibhav Srivastava, and Naomi Ehrich Leonard. Distributed cooperative decision-making in multiarmed bandits: Frequentist and Bayesian algorithms. In *2016 IEEE 55th Conference on Decision and Control (CDC)*, pp. 167–172. IEEE, 2016.

Chi Harold Liu, Qiuxia Lin, and Shilin Wen. Blockchain-enabled data collection and sharing for industrial IoT with deep reinforcement learning. *IEEE Transactions on Industrial Informatics*, 15(6):3516–3526, 2018.

Dahlia Malkhi, Kartik Nayak, and Ling Ren. Flexible Byzantine fault tolerance. In *Proceedings of the 2019 ACM SIGSAC Conference on Computer and Communications Security*. ACM, 2019.

David Mazieres. The stellar consensus protocol: A federated model for internet-level consensus. *Stellar Development Foundation*, 32:5–39, 2015.

Ishai Menache, Shie Mannor, and Nahum Shimkin. Basis function adaptation in temporal difference reinforcement learning. *Annals of Operations Research*, 134(1):215–238, 2005.

Volodymyr Mnih, Koray Kavukcuoglu, David Silver, Alex Graves, Ioannis Antonoglou, Daan Wierstra, and Martin Riedmiller. Playing Atari with deep reinforcement learning. *arXiv preprint, arXiv:1312.5602*, 2013.

Thanh Nguyen, Michael P. Wellman, and Satinder Singh. A Stackelberg game model for botnet data exfiltration. In *International Conference on Decision and Game Theory for Security*, pp. 151–170. Springer, 2017.

Martina Panfili, Alessandro Giuseppi, Andrea Fiaschetti, Homoud B Al-Jibreen, Antonio Pietrabissa, and Franchisco Delli Priscoli. A game-theoretical approach to cyber-security of critical infrastructures based on multi-agent reinforcement learning. In *2018 26th Mediterranean Conference on Control and Automation (MED)*, pp. 460–465. IEEE, 2018.

Yuzuru Sato, Eizo Akiyama, and J Doyne Farmer. Chaos in learning a simple two-person game. *Proceedings of the National Academy of Sciences*, 99(7): 4748–4751, 2002.

David Silver, Aja Huang, Chris J. Maddison, Arthur Guez, Laurent Sifre, George Van Den Driessche, Julian Schrittwieser, Ioannis Antonoglou, Veda Panneershelvam, Marc Lanctot, et al. Mastering the game of Go with deep neural networks and tree search. *Nature*, 529(7587):484, 2016.

John N. Tsitsiklis. Asynchronous stochastic approximation and Q-Learning. *Machine Learning*, 16(3):185–202, 1994.

Peter J. M. Van Laarhoven and Emile H. L. Aarts. Simulated annealing. In *Simulated Annealing: Theory and Applications*, pp. 7–15. Springer, 1987.

H. Peyton Young. On the limits to rational learning. *European Economic Review*, 46(4–5):791–799, 2002.

Minghui Zhu, Zhisheng Hu, and Peng Liu. Reinforcement learning algorithms for adaptive cyber defense against Heartbleed. In *Proceedings of the First ACM Workshop on Moving Target Defense*, pp. 51–58, 2014.

24

Summary and Future Work

Quanyan Zhu[1] and Fei Fang[2]

[1]*Department of Electrical and Computer Engineering, NYU Tandon School of Engineering, New York University, New York, NY, USA*
[2]*School of Computer Science and Institute for Software Research, Carnegie Mellon University, Pittsburg, PA, USA*

Machine learning has excelled in discovering patterns and knowledge from data. Its revolutionary success has been recently witnessed in various applications, including face recognition, language translation, autonomous driving, and robotics. In the domain of cybersecurity, machine learning has played an important role in detecting anomaly, finding insider threats, and defending against malicious behaviors. Advanced machine learning algorithms, such as deep neural networks, deep reinforcement learning, and generative adversarial networks, have shown to provide security administrators a new set of tools that can process a large volume of real-time security logs and data, assess the risks, and automate detection and response.

Game theory is another active research area that has been used to understand the adversarial behaviors. Starting from the application of game theory to games, such as chess and StarCraft, we have witnessed a surge of interest in its recent application to cybersecurity (Rass et al. 2020; Zhu and Xu 2020; Pawlick and Zhu 2021). Game-theoretic frameworks naturally encapsulate the adversarial models and serve as a suitable design tool for assessing cyber risks and designing defense strategies to mitigate the impact of the attacks (Alpcan and Başar 2010; Manshaei et al. 2013; Zhu and Rass 2018). The first two parts of the book have presented applications of game-theoretic methods for cyber deception and cybersecurity. These chapters have developed a rich set of cybersecurity tools that enable applications in honeypot design, moving target defense, intrusion detection, and authentication.

Game theory and machine learning provides two important sets of AI techniques that are complementary to each other for cybersecurity applications. Machine learning algorithms are conventionally designed without considering an adversary. On the one hand, the integration of game theory into machine learning models will incorporate adversarial behaviors and enrich the learning algorithms. Part 3 of the book has presented recent advances in adversarial machine learning. The authors have addressed the theory and the application of machine learning models in the presence of a malicious adversary. The marriage between machine learning and game theory has spawned generative adversarial networks (GANs) in which two neural networks play a game to generate new data with the same statistics as the training set. Part 4 of this book has presented applications of GANs in cybersecurity. Chapter 17 has assessed the risks of using GANS to evade machine learning-based network intrusion detection systems. Chapter 18 has leveraged GANs to generate steganographic texts to bypass censorships.

On the other hand, the incorporation of machine learning models into game theory will put game-theoretic algorithms into practical use. The payoffs of the players often need to be learned in an adaptive and online fashion. Part 5 of the book has presented reinforcement learning techniques for cybersecurity, and Part 6 of the book has presented recent advances in machine learning to enable adaptive and proactive defenses.

24.1 Summary

24.1.1 Game Theory for Cyber Deception

The goal of cyber deception is to mitigate the information asymmetry between adversaries and defenders and revert the information disadvantage of the defender. To this end, the research of cyber deception addresses the design of decoy assets and camouflage to expend adversaries' resources and time and gather information about the adversaries' strategies, tactics, capabilities, and intent. The five chapters in Part I provide insights into how game-theoretic techniques can be used as a tool to design cyber deception mechanisms. We have kicked off with Chapter 2 presenting an overview of game-theoretic techniques that can be used to capture the interactions between the attackers and the defenders.

In Chapter 3, we have presented a game-theoretic approach for optimizing defensive deception actions with the specific goal of identifying specific attackers as early as possible in an attack. We have shown that strategically using deception can facilitate significantly earlier identification by leading attackers to take different actions early in the attack that can be observed by the defender. Since the optimal algorithm does not scale very well, we have presented a scalable version of the optimal algorithm which reduces the action space by a reasonable margin to handle larger instances of games.

In Chapter 4, we have presented a game-theoretic framework and a scalable allocation algorithm to place honeypots over an attack graph. The game has investigated the trade-off between security cost and deception reward for the defender. The trade-off studied for the attacker is to decide which node to attack to maximize the reward without revealing the identity of the attacker. The chapter has analytically characterized Nash equilibrium defense strategies and discussed the complexity of the general game. A scalable algorithm has been proposed to ensure that the defense approach is applicable to large-scale networks. The chapter has also formulated a dynamic version to study the game dynamics over time and used the Q-minmax algorithm to learn NE strategies.

In Chapter 5, we have conducted human experiments to study the effectiveness of six deceptive algorithms for cyber defense that differ in three dimensions: determinism, adaptivity, and customization to the adversary's actions. We have shown that human participants are capable of learning the patterns of the more static algorithms, and the nondeterministic, adaptive and customizable algorithms have been less effective than the deterministic and noncustomizable algorithms.

Finally, in Chapter 6, we have introduced a game on graph model for capturing the attack-defend interactions in a cyber network for reactive defense, subject to security specifications in temporal logic formula. To understand how the defender can take actions in response to the exploit actions of the attacker, we have introduced a hypergame for games on graphs to capture payoff misperception for the attacker caused by the decoy systems. The solution concept of hypergames has enabled us to synthesize effective defense strategy given the attacker's misperception of the game, without contradicting the belief of the attacker.

24.1.2 Game Theory for Cyber Security

In Part II, we have presented a rich set of game-theoretic techniques to model and design defense mechanisms for a general class of cybersecurity problems. The six chapters have provided insights into game-theoretic techniques to a diverse range of applications, including the security of sensor networks, cyber-physical systems, IoT systems, network authentication, and software security.

In Chapter 7, we have presented a class of minimax detectors that can assess system activities in terms of their riskiness and likeliness across a modern computing system in a cohesive and strategic way. This chapter has designed a detection mechanism that seeks to minimize the detection cost against an adversary maximizing it by intervening into the system activities. The chapter has also introduced a hierarchical scheme where the system level activities are assessed at varying granularity similar to state summarization used to handle big data applications. The detection cost has a nested structure that would enable us to decompose the entire game into nested local sub-games. Hence the equilibrium could be computed efficiently via a dynamic program similar to the backward induction in extensive-form games.

In Chapter 8, we have presented a class of sensor manipulation problems in which an attacker aims to induce a cyber defense system to make the wrong decision. In measurement manipulation games the cyber defense system is forced to make a decision based on a number of sensor recommendations that may have been altered by an attacker, whereas in sensor-reveal games the attacker selects which sensors the defender is allowed to use in making a decision. This chapter has touched upon a rich area of research with numerous open problem that are crucial for the design of cyber defense mechanisms robust to information manipulation.

In Chapter 9, we have studied a setting where a cyber-physical system (CPS) is monitored by a Gaussian process regression-based anomaly detection system. We have presented a stealthy attack on this system that aims to maximize damage while appearing normal to the detector, presenting a novel approach to approximately solve this nontrivial optimization problem. This chapter has presented a model of robust anomaly detection for CPS as a Stackelberg game and developed a novel approach for solving this game, in which the defender can decide which sensors to use as well as how to set anomaly thresholds.

In Chapter 10, we have presented moving target defenses (MTDs) and their applications in cyber security. MTD is a technique that seeks to randomize system components to reduce the likelihood of a successful attack. MTD alters the attack surface (points that could be exploited) in order to harden the attacker's mission. We have studied different theoretical models that were introduced in literature to model MTD systems. In particular, we have focused on the game-theoretic models as they can better model the interactions between attackers and defenders in MTD scenarios. This chapter has focused on a type of games known as single controller stochastic games and their applications to MTD in IoT applications.

In Chapter 11, we have presented a security risk management in a system combining continuous authentication and intrusion detection as a dynamic discrete stochastic leader-follower game with imperfect information between an attacker and a defender. The chapter has derived a backward recursion for the optimal attacker reward to characterize the optimal attack strategy. It has been shown that continuous authentication can be very efficient for security risk reduction, if combined with appropriate incident detection.

In Chapter 12, we have presented a strategy-technique combining approach to reason about software security. The approach has considered both software security techniques and offense and defense strategy. We have introduced the connection between techniques and strategy, and we propose a new methodology to holistically study software security. From the technical perspective, we have presented ByteWeight and ShellSwap, two new techniques for function identification and automatic exploit reuse, respectively. From the strategy perspective, we abstracted software security as autonomous computer security games, and we have shown that the autonomous computer security game model can be used for calculating optimal strategies.

24.1.3 Part 3: Adversarial Machine Learning for Cyber Security

In Part 3, we have presented several adversarial machine learning techniques in which an adversary can manipulate the inputs and the algorithms to attack or cause malfunctions in standard machine learning models. We have kicked off with Chapter 13, presenting an overview of the adversarial models for machine learning and game-theoretic models to understand the interactions between machine learning systems and the adversary. Game theory can provide robustness guarantee for machine learning models that are otherwise vulnerable to application-time data corruption. We have discussed two cases of game theory-based machine learning techniques. In one case, players play a zero-sum game by following a minimax strategy. In the other case, players play a sequential game with one player as the leader and the rest as the followers.

In Chapter 14, we have presented the adversarial machine learning and its use in wireless communications to launch stealthy attacks. We have discussed challenges associated with designing attacks in wireless domains and focused on the vulnerabilities of the 5G communication systems. We investigated a jamming attack and a spoofing attack on 5G communication systems enabled by the adversarial machine learning. This chapter has presented these novel attacks and highlighted the impact of adversarial machine learning on wireless communications in the context of 5G and raise the need for defense mechanisms.

In Chapter 15, we have presented a potential threat that advances through the utilization of machine learning (ML) techniques. We have demonstrated how targeted attacks powered by ML techniques can probe the target system to infer actionable intelligence and customize the attack payload to increase the efficiency of the attack while reducing the time to success. The chapter has discussed the rationale of this attack and the limitations of current mitigation method. This chapter has motivated further research on advanced offensive technologies, not to favor the adversaries, but to know them and be prepared.

In Chapter 16, we have described Trinity, an integrated framework to address the challenges of trust, resilience and interpretability by exploiting the relationship between them. The chapter has shown how the techniques developed for interpretability such as attribution methods can be used to also improve trust and resilience of machine learning models.

24.1.4 Part 4: Generative Models for Cyber Security

In Part 4, we have presented two chapters that have provided insights on recent advances in generative adversarial networks (GANs) and their applications in cyber security. In Chapter 17, we have introduced a GAN-based attack algorithm that crafts adversarial traffic to bypass network intrusion detection systems (NIDS) while maintaining the domain constraints. We have shown that a well-trained NIDS with Deep Neural Networks can achieve nearly 100% success rates against these models, and we can achieve 70% success rate on average against the classical machine learning-based NIDS.

In Chapter 18, we have presented ConcealGAN as a linguistic steganographic tool that can bypass the censorships over messaging applications. Building on Recurrent Neural Networks (RNN) and LeakGAN, ConcealGAN generates neural cover texts without modifying known plaintext. The chapter has shown that the integration of generative models with linguistic steganographic system will enable a higher level of security when exchanging textual data.

24.1.5 Part 5: Reinforcement Learning for Cyber Security

In Part 5, we have presented reinforcement learning techniques and their applications in cyber security. The first chapter in this part has discussed the vulnerabilities of the reinforcement learning algorithms to signal manipulations. The second chapter has focused on the application of deep reinforcement learning algorithms to develop resource-aware active intrusion response strategies for software-defined networks. These two chapters cover two important research directions intersecting security and reinforcement learning.

More specifically, in Chapter 19, we have discussed potential threats in reinforcement learning (RL) algorithms and presented a general framework to study RL under deceptive fascinations of cost signals. The chapter has provided theoretical underpinnings for understanding the fundamental limits and performance bounds on the attack and the defense in RL systems. The chapter has characterized a robust region within which the adversarial attacks cannot achieve its objective. A RL agent can leverage the robust region to evaluate the robustness to malicious falsifications.

In Chapter 20, we have presented a resource-aware active defense framework that can provide a set of deep reinforcement learning-based intrusion response strategies for a software-defined networking (SDN)-based Internet-of-Battle-Things (IoBT) environment. The proposed defense framework is built based on a highly attack-resistant network based on multilayered structuring to provide higher security protection towards more critical system assets while maintaining acceptable service availability during mission execution.

24.1.6 Other Machine Learning approach to Cyber Security

In Part 6, we have presented several recent advances in machine learning and their application to cybersecurity. The three chapters have discussed emerging research areas, including adaptive and cumulative machine learning, semi-automated parametrization, and resilient distributed learning enabled by blockchain. These new techniques have provided insights on network scanning and autonomous cyber defense.

In Chapter 21, we have presented a framework for reducing the bandwidth of network scans by predicting whether a host will respond to requests on a certain port, using location and ownership (AS) properties, as well as cross-protocol information. The chapter has developed a novel technique for finding an optimal order for scanning ports and training a sequence of classifiers, appending the responses of scanned ports for predicting active IPs over subsequent scans. The chapter has further shown that our technique can be applied on top of current scanning tools with little computational overhead, providing blacklists in order to refrain from sending probes to certain IP/port pairs.

In Chapter 22, we have proposed the concept of "parameterization by example" where we have jointly considered a generic parameter of the probability and an application of logistic regression to find a value for the parameter based on available data. We have demonstrated the method using a model of coupled stochastic Mealy automata for which a considerably large set of probabilities needs to be specified. The chapter has shown how large sets of parameters can get values assigned "in batch" by an implementation of our method in the R language for statistical computing.

In Chapter 23, we have tackled the problem of online learning, with cumulative regret minimization, for swarms of agents cooperating to achieve high assurance on missions tasks. The chapter has introduced the concept of "spatial" online learning by which action selection is guided by UCB-type criteria.

24.2 The Future

24.2.1 Game Theory and Cyber Security

In the first two parts of the book, several chapters have focused on game-theoretic approaches to cyber deception and cybersecurity. There are several challenges and possible opportunities for related researches. First, there exist computational challenges with large-scale games and games of incomplete information. Several chapters have discussed the extension of their frameworks to a partially observable defender/attacker. As a result, we will face computational and algorithmic challenges associated with these types of models.

Second, experimental evaluation and validation of the solution concept and the algorithms would play an essential role in the deployment of game-theoretic solution concepts in computer systems. Most of the studies have focused primarily on developing game-theoretic models and algorithms, but there is an urgent need to realize them in practical scenarios. Game theory for cybersecurity is a multidisciplinary topic. Interdisciplinary teamwork between cybersecurity analysts and game theorists would benefit the broader implementation and adoption of the techniques to combat attacks targeted at vulnerable cyber systems. For instance, game-theoretic models often assume a set of parameters as given, for example, the players' utility functions. However, it is sometimes not obvious how to estimate these parameters if the model needs to be deployed. In previous works of applying game-theoretic models to infrastructure security settings such as protecting the airport, the domain experts provided their best estimate of these parameters. In cybersecurity, the cybersecurity analysts can also try their best to provide such input to the game-theoretic model. Also, the models built may ignore or oversimplify some aspects that are critical in practice. Based on the lessons learned from previous research on deploying game-theoretic models for wildlife conservation, the path to real-world deployment is often an iterative process. After the initial solution from the game-theoretic model is tested in a pilot study, the domain experts may find some issues with the solution. These issues will be discussed by both the domain experts and the researchers, which may lead to new research questions, new models, and new solutions. The new solution will be tested again. If the pilot study results are promising, the domain experts and the researchers will discuss what is needed for a large-scale deployment. We expect to see new game-theoretic models inspired by the practical challenges, and eventually large-scale real-world deployment in the future.

24.2.2 Machine Learning and Cyber Security

In Part 3–6 of the book, various machine learning-based techniques for cybersecurity have been introduced. With the development of different general-purpose machine learning models and algorithms, we expect to see more cyber security-specific machine learning tools. For example, in recent years, there has been a rise in the interest of federated learning (Konečný et al. 2016), which focuses on training machine learning models across multiple decentralized computing devices or servers, without explicitly exchanging local data at each server. In the cybersecurity domain, individuals, companies, and institutions are often unwilling to publicly share their own data due to security and privacy concerns. Leveraging federated learning can lead to solutions that enable collaboration of various stakeholders with fewer concerns on data access rights.

As another example, recent advances in multiagent (deep) reinforcement learning (MARL) has shown promising results in virtual environments involving multiagent interaction and sequential decision making (Foerster et al. 2016; Lowe et al. 2017), including in popular real-time strategy games such as StarCraft II (Vinyals et al. 2019). In cybersecurity, we also face a problem with multiple agents and sequential decision making. Thus, it is promising to develop MARL-based techniques for cybersecurity, as illustrated by some recent works (Wright et al. 2019).

There is a lot more to explore in the space of machine learning (and possibly together with game theory) for cybersecurity. There will always be new threat models and new ways of defense, which will lead to new computational challenges that would require novel machine learning techniques. For example, with social media use becoming prevalent in today's society, there will be more cyber attacks through social engineering. On the other hand, attackers may also be social media users. It is thus possible to detect potential attack plans and even attackers by close monitoring of social media posts and accounts. Using machine learning to combat social engineering attacks and identifying potential threats from social media would be exciting problems to investigate. There have been some initial attempts (Khandpur et al. 2017), and we expect more work along this line.

References

Tansu Alpcan and Tamer Başar. *Network Security: A Decision and Game-Theoretic Approach.* Cambridge University Press, 2010.

Jakob Foerster, Ioannis Alexandros Assael, Nando De Freitas, and Shimon Whiteson. Learning to communicate with deep multi-agent reinforcement learning. In *Advances in Neural Information Processing Systems*, pp 2137–2145. Curran Associates, Inc., 2016.

Rupinder Paul Khandpur, Taoran Ji, Steve Jan, Gang Wang, Lu Chang-Tien, and Naren Ramakrishnan. Crowdsourcing cybersecurity: Cyber attack detection using social media. In *Proceedings of the 2017 ACM on Conference on Information and Knowledge Management (CIKM '17), Singapore, 6-10 November 2017*, pp. 1049–1057. Association for Computing Machinery, New York, NY, USA, 2017.

Jakub Konečný, H. Brendan McMahan, Felix X. Yu, Peter Richtárik, Ananda Theertha Suresh, and Dave Bacon. Federated learning: Strategies for improving communication efficiency. *arXiv preprint arXiv:1610.05492*, 2016.

Ryan Lowe, Yi I. Wu, Aviv Tamar, Jean Harb, OpenAI Pieter Abbeel, and Igor Mordatch. Multi-agent actor-critic for mixed cooperative-competitive environments. In *Advances in Neural Information Processing Systems*, pp. 6379–6390. Curran Associates, Inc., 2017.

Mohammad Hossein Manshaei, Quanyan Zhu, Tansu Alpcan, Tamer Bacşar, and Jean-Pierre Hubaux. Game theory meets network security and privacy. *ACM Computing Surveys (CSUR)* 45(3): 1–39, 2013.

Jeffrey Pawlick and Quanyan Zhu. Insights and future directions. In *Game Theory for Cyber Deception: From Theory to Applications*, pp. 171–174. Birkhäuser, Cham, 2021.

Stefan Rass, Stefan Schauer, Sandra König, and Quanyan Zhu. Cyber-Security in Critical Infrastructures. Springer International Publishing, 2020.

Oriol Vinyals, Igor Babuschkin, Wojciech M. Czarnecki, Michaël Mathieu, Andrew Dudzik, Junyoung Chung, David H. Choi et al. Grandmaster level in StarCraft II using multi-agent reinforcement learning. *Nature*, 575(7782):350–354, 2019.

Mason Wright, Yongzhao Wang, and Michael P. Wellman. Iterated deep reinforcement learning in games: History-aware training for improved stability. In *Proceedings of the 2019 ACM Conference on Economics and Computation (EC '19), Phoenix, AZ, USA, 24-28 June 2019*, pp. 617–636. Association for Computing Machinery, New York, NY, USA, 2019. https://doi.org/10.1145/3328526.3329634.

Quanyan Zhu and Stefan Rass. Game theory meets network security: A tutorial. In *Proceedings of the 2018 ACM SIGSAC Conference on Computer and Communications Security*, pp. 2163–2165. ACM, 2018.

Quanyan Zhu and Zhiheng Xu. *Cross-Layer Design for Secure and Resilient Cyber-Physical Systems: A Decision and Game Theoretic Approach*, volume 81. Springer Nature, 2020.

Index

Game Theory and Machine Learning for Cyber Security, First Edition.
Edited by Charles A. Kamhoua, Christopher D. Kiekintveld, Fei Fang, and Quanyan Zhu.
© 2021 The Institute of Electrical and Electronics Engineers, Inc. Published 2021 by John Wiley & Sons, Inc.

Printed and bound by CPI Group (UK) Ltd, Croydon, CR0 4YY